400 Years of Astronomical Telescopes

A Review of History, Science and Technology

Bernhard R. Brandl · Remko Stuik · J.K. (Jet) Katgert-Merkelijn

Editors

400 Years of Astronomical Telescopes

A Review of History, Science and Technology

Previously published in *Experimental Astronomy*,
Volumes 25 and 26, 2009 with additional contributions and poster presentations at extras.springer.com

Springer

Editors
Dr. Bernhard R. Brandl
Leiden Observatory
2300 RA Leiden
Netherlands
brandl@strw.leidenuniv.nl

Dr. Remko Stuik
Leiden Observatory
2300 RA Leiden
Netherlands
stuik@strw.leidenuniv.nl

Dr. J.K. (Jet) Katgert-Merkelijn
Leiden Observatory
2300 RA Leiden
Netherlands
merkelyn@strw.leidenuniv.nl

Additional material to this book can be downloaded from http://extra.springer.com
Corrected 2nd printing 2010, 1st printing 2009 with CD-ROM

ISBN: 978-90-481-2232-5

Library of Congress Control Number: 2009936955

Printed on acid-free paper

springer.com

Section IV – Fundamental Telescope Technologies

Section V – Political and Sociological Aspects

Section VI – Perspectives for Future Telescopes

Section VII – Poster Presentations

Preface

Bernhard Brandl

As with many inventions, the astronomical telescope was probably the result of developments by numerous people. Around the end of the 16th century, the making and polishing of spectacle lenses had been gradually improved, and magnifying optics and combinations of optical components had been developed. Significant progress was made by people like Juan Roget in Spain, Sacharias Janssen and Jacob Metius in the Netherlands, and Raffaelo Gualterotti in Italy. However, it was the spectacle maker Hans Lipperhey from Middelburg (The Netherlands), who came up with the idea to stop down the aperture of the objective lens, which significantly sharpened the magnified image. On 25 September 1608, he presented his instrument to Prince Maurice of Nassau in The Hague and applied for a patent. While the patent was denied (it could be easily copied and used by foreign military) the application placed this important date in the history books.

Almost four centuries later, I overheard a discussion on the first telescope in the coffee room at the Sterrewacht Leiden. Apart from being the oldest still operating university observatory in the world (dating back to 1633, just 25 years past Lipperhey's invention) the Sterrewacht is also geographically the closest Dutch astronomy department to Middelburg. It seemed therefore obvious that the Sterrewacht should take care of the celebration of this milestone in astronomical history. I discussed the idea with Prof. Harm Habing, who proposed a handful of individuals for the scientific organizing committee (SOC). All accepted and were instrumental in making this conference happen.

It soon became clear that historical societies, amateur astronomers, and the world-wide activities for the 'International Year of Astronomy 2009' are likely to generate numerous meetings on telescopes, mainly focused, however, on the historical perspective and on public outreach. Hence, our first challenge was to define an attractive, unique, and focused scope for a meeting on "*400 Years of Astronomical Telescopes*". The SOC agreed to include all types of telescopes that are being used for the detection of astronomical signals (including key technologies), covering the entire electromagnetic spectrum, as well as neutrino detectors[1]. In addition we wanted to include closely related sociological and political topics, which are usually not covered by other conferences.

Bernhard R. Brandl
Conference Chair
Leiden Observatory, Leiden University
e-mail: brandl@strw.leidenuniv.nl

[1]We did not include gravitational wave detectors/telescopes since no detection has yet been made. However, we expect that a future meeting on "500 Years of Astronomical Telescopes" will prominently feature gravitational wave detectors.

As it was our goal that all presentations would provide overviews of the various fields, rather than promoting specific projects, we worked toward a program that consisted exclusively of review talks by speakers who are recognized as leaders of their fields. It was a most pleasant experience that all but two of the speakers we contacted actually accepted our invitation and gave excellent presentations. Among the approximately 130 participants were five (former and current) Directors General of ESO, Nobel Prize winners, institute directors and representatives from industry.

The conference site at the ESA technology center ESTEC in Noordwijk provided an ideal setting, also for stimulating coffee table and dinner discussions. The social program included a reception in the Space Expo at ESTEC (sponsored by ESA), a "final look" at the Herschel telescope, an evening excursion to the excellent collection of historical telescope at the Rijksmuseum Boerhaave in Leiden, and a boat dinner cruise along the Dutch canals. None of these activities, nor the meeting itself, would have been possible without the excellent work of the local organizing committee. With almost 3000 conference-related emails I wonder how people in the past 400 years could have organized such meeting without the internet.

This book contains most of the review talks and a few selected posters (the complete list of posters is included on Springer Extras). I hope that it has captured the essence of this interesting meeting, and that it will serve as a useful reference for many readers, conference participants and beyond. Enjoy the reading!

Opening Address

Karel A. van der Hucht

Dear colleagues, friends, ladies and gentlemen,

Nature has given us eyes. Or more accurately: nature has given itself eyes. In his book Voices from the Sky (New York: Harper & Row, 1965, p. 10), the late Arthur Clarke quotes Derek Lawden from New Zealand: "… I think, man will see himself as one agent by which the whole Universe of matter is slowly becoming conscious of itself …".

It is only a few hundred thousand years ago that our ancestors started to walk on two feet. Presumably, that made it easier than before to look up at the sky and to marvel at the daily appearance of the Sun and the Moon, the planets and the stars. Throughout recorded history, astronomical phenomena have amazed mankind on all continents, and they feature in all cultures throughout the world. Naked-eye observations predating the 17-th century culminated in the scientific work of Ptolemaeus, Aristoteles, Copernicus, Tycho Brahe and Johannes Kepler, to name a few.

The invention and use of the telescope caused a mega jump of progress in observational astronomy. It took a while to get to this point: glass was invented as early as 7000 years ago; clear transparent glass was first produced some 2000 years ago; the magnifying aspects of curved glass were discovered some 900 years ago; and the application of curved glass for spectacle glasses was introduced some 700 years ago (by the Dominican Alessandro di Spino in 1285).

It was the spectacle glass maker Hans Lipperhey who may have produced the first telescope for practical observations in 1608, and in 1609, Galileo Galilei first used a telescope for astronomical observations. The world will celebrate that event in 2009, the International Year of Astronomy.

Since 1608, the construction and use of optical telescopes has blossomed steadily, culminating in the present-day 8-10m-class optical ground-based telescopes and in studies for 30-40m-class telescopes, which you will be discussing during this conference. Moreover, notably in the 2nd half of the previous century, telescopes for the entire electromagnetic spectrum have been built and employed in ground-based and space-based observatories, in near-Earth orbit and in orbits throughout the Solar System. Superb examples are the aging spacecrafts Voyager-1 and Voyager-2, launched 31 years ago to visit and observe the major planets. Having left the Solar System, "to boldly go where no one went before", they are still transmitting signals back to Earth today.

Karel A. van der Hucht
General Secretary of the International Astronomical Union
e-mail: K.A.van.der.Hucht@sron.nl

We are fortunate that the world's tax payers allow us to build and operate great observatories that provide huge amounts of astronomical data. The role of scientific discovery is accelerating continuously. Some of the questions which we could ask ourselves are: Are our telescopes and instrumentation approaching fundamental physical limits? Is astronomical progress limited by the Digital Divide? How can we optimize international cooperation, particularly on major missions? I believe you will address such questions here in Noordwijk this week.

I do wish you all a most constructive and pleasant conference.

Thank you very much.

Participants at the conference, 400 Years of Astronomical Telescopes.
(Credit: Carolina Ödman, Leiden Observatory)

Conference Organizing Committee

Scientific Organizing Committee

Adrian Blaauw (University of Groningen)
Bernhard Brandl (chair) (Leiden University)
Colin Cunningham (UK-ATC)
Dirk van Delft (Leiden University)
Martin Harwit (Cornell University)
Detlef Koschny (ESTEC)
Guy Monnet (ESO)
Andreas Quirrenbach (Landessternwarte Heidelberg)
Woody Sullivan (University of Washington)
Joachim Trümper (Max-Planck-Institut für extraterrestrische Physik)

Local Organizing Committee

Clare Bingham (ESTEC)
Nathalie Boudin (ESTEC)
Bernhard Brandl (Leiden University)
Erik Deul (Leiden University)
Liesbeth van der Veld (Leiden University)
Trami Ho (ESTEC)
Vincent Icke (Leiden University)
Frank Molster (co-chair) (Leiden University/NWO)
Carolina Ödman (Leiden University)
Remko Stuik (chair) (Leiden University)

Sponsors

ASTRON
Dwingeloo, The Netherlands
www.astron.nl

Boerhaave Museum
Leiden, The Netherlands
www.museumboerhaave.nl

Cosine
Leiden, The Netherlands
www.cosine.nl

ESTEC/European Space Agency
Noordwijk, The Netherlands
www.esa.int

International Astronomical Union
Paris, France
www.iau.org

Kayser-Threde
Munich, Germany
www.kayser-threde.de

Nederlandse Onderzoeksschool Voor Astronomie
Leiden, The Netherlands
www.astronomie.nl

Netherlands Organisation for Scientific Research
Den Haag, The Netherlands
www.nwo.nl

Springer
Heidelberg, Germany
www.springer.com

Science and Technology Facilities Council
Swindon, United Kingdom
www.scitech.ac.uk

Leiden Observatory
Leiden, The Netherlands
www.strw.leidenuniv.nl

TNO
Delft, The Netherlands
www.tno.nl

Carl-Zeiss
Oberkochen, Germany
www.zeiss.com

In cooperation with:
SPIE Europe
International Year of Astronomy 2009

Exp Astron (2009) 25:3–16
DOI 10.1007/s10686-009-9160-y

HISTORICAL REVIEW

The beginnings, from Lipperhey to Huygens and Cassini

Albert Van Helden

Received: 9 January 2009 / Accepted: 4 March 2009 / Published online: 8 April 2009

Abstract The first century of telescopic astronomy can be divided into two periods. During the first, from 1609 to ca. 1640, observations were made with a simple "Dutch" or "Galilean" telescope with a concave eyepiece. Galileo made all his discoveries with this instrument. Its limited field of view, however, made magnifications of more than about 20 impractical, and therefore this instrument's limit had been reached within a few years. During the second period, ca. 1640–ca. 1700, the simple astronomical telescope came into use, almost immediately augmented with a field lens and an erector lens (the latter used only for terrestrial purposes). Magnifications were increased by increasing the focal lengths of objectives, and this quickly led to very long telescopes, often used without a tube. The astronomical discoveries made possible by this form of the instrument were, however, made with instruments of relatively modest lengths. By the end of the century, very long telescopes fell out of use, while shorter ones were adapted for measurements. Further discoveries became possible only with the reflecting telescope in the second half of the eighteenth century.

Keywords Telescope · Galileo · Lipperhey · Huygens · Cassini

1 The invention of the telescope

Over the four centuries that separate us from the events in The Hague, in October 1608, a number of men have been proposed as the inventor of the telescope: Hans Lipperhey, Sacharias Janssen, Jacob Metius in the Netherlands,

A. Van Helden (✉)
University of Utrecht, Utrecht, the Netherlands
e-mail: a.vanhelden@uu.nl

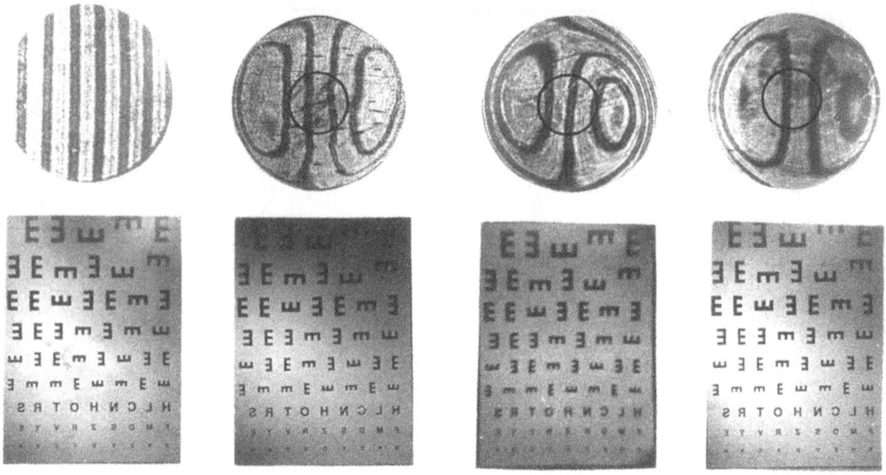

Fig. 1 Ronchi interference tests of sixteenth-century convex spectacle lenses. When the aperture is restricted as shown (ca. 1 cm) these lenses are good enough to produce a reasonably sharp, enlarged image (*bottom row*). Courtesy Rolf Willach

Raffaelo Gualterotti and Giambaptista della Porta in Italy, an unidentified spectacle maker in the Barcelona region, yes, even Galileo himself.[1] Lately, some clarity has come to this confusion.

The story has to begin with spectacle lenses. Devices that magnified, made out of quartz or glass, were present in Europe in the first millennium. Spectacles as we know them came into use around the turn of the thirteenth century. These were produced by cutting round discs out of a blown glass ball before it cooled; the discs were then ground and polished on one side. Although these lenses were adequate for use as spectacles—the focused gaze, after all, used only an area of at most 1 cm^2 of the lens—they were useless for telescopic purposes. Not until the sixteenth century did some lens-makers begin to work both surfaces. This procedure improved the optical quality immeasurably, so much so that modern tests actually produce recognizable interference patterns, see Fig. 1.[2]

It has recently been shown that learned—and not so learned—men in Europe were searching actively for a telescopic device that would magnify objects many times. Especially in Italy, a group of men examined the optics of mirrors and clearly understood that it required a combination of optical devices. Quite naturally, they looked toward mirrors as the primary receptor,

[1]On Lipperhey, Janssen, and Metius, see below; on Gualterotti see [1]. When Porta saw his first spyglass he called it a hoax and that it was taken from one of his works; see Porta to Federico Cesi, 26 August 1610, *Le Opere di Galileo Galilei*, Edizione Nazionale, X (1968), p. 252. On the Spanish claim, see [2].

[2]For a complete account, see Rolf Willach, *The Long Route to the Invention of the Telescope*, American Philosophical Society, *Transactions*, 98, part 5, 2008.

not knowing that mirror surfaces had to be shaped much more precisely than lens surfaces to be useful for this purpose.[3] But at least it was possible to make large concave mirrors. Lenses could be made only in small sizes, 3–5 cm.

The example of the so-called Elizabethan telescope is illustrative of the situation. It concerned a combination of a large convex lens, a foot in diameter, as the primary receptor, and a small concave mirror as the eyepiece—the opposite of what Italians were thinking of. But here, too, it was impossible to make a 30 cm convex lens with reasonable optical properties when William Bourne penned his little treatise ca. 1580.[4] It is somewhat reminiscent of the large optical mirrors pursued in Italy, partly real and partly mythical. Moreover, the small concave mirror in this combination, presumably a few centimeter in diameter, was just as impossible to figure sufficiently accurately. Yet, like the Italian pursuit of large mirrors, it testifies to the fact that theory was not the problem: these men knew perfectly well that a device that produced sharply defined enlarged images had to consist of a combination of (at least) two optical components, where the primary receptor could be either a convex lens or a concave mirror, and the eyepiece could be a lens or mirror.

The solution came from the spectacle makers. Historians of the subject have often had in mind a scenario somewhat like this: someone, a spectacle maker or his children, happens to combine a convex and a concave lens of the appropriate curvature and notices that the weather vane on a nearby church is magnified, and thus the telescope is invented by accident. But this is to suppose that the hundreds of spectacle makers in Europe were not familiar with combinations of spectacles, especially when the lens closest to the eye was concave. There is evidence that suggests that around the turn of the seventeenth century a number of men in Europe were familiar with such combinations of lenses, perhaps even in a tube, but that the enlarged image was fuzzy and blurred. The invention of the telescope was not some lucky accident; it was the result of gradual improvement of the optical quality of spectacle lenses.[5]

In 1634 the spectacle maker Johannes Sachariassen, son of Sacharias Janssen, one of the claimants for the honor of having invented the telescope, stated that his father had made the first spyglass in the Netherlands in 1604, after one in the possession of an Italian dated 1[5]90.[6] The device owned by the Italian was a tube with two lenses, and we may surmise that the image it produced was enlarged but fuzzy. At about the same time, the Italian poet,

[3]Eileen Reeves, *Galileo's Glassworks:The Telescope and the* Mirror. Harvard University Press, 2008. See also [3].

[4]William Bourne, *A treatise on the properties and qualities of glasses for optical purposes, according to the making, polishing, and grinding of them*, ca. 1580. See [4].

[5]See Willach, *The Long Road*, 93–97.

[6]*Journal tenu par Isaac Beeckman de 1604 à 1634, avec une introduction et des notes par C. De Waard*, (4 vols., The Hague: Martinus Nijhoff, 1939–1953), III: 376. See also Van Helden, *Invention*, p. 53.

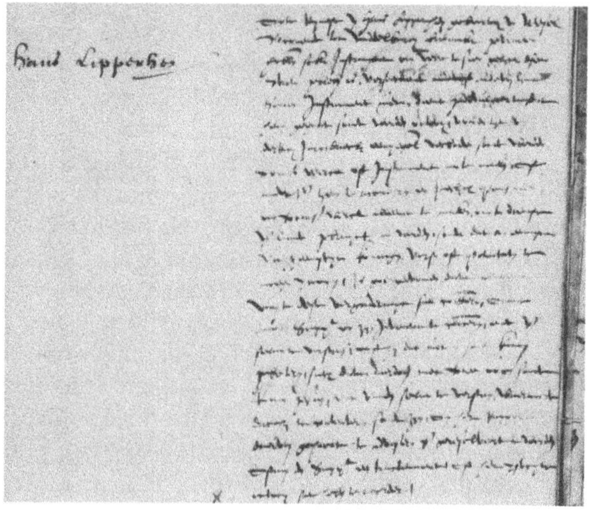

Fig. 2 Entry concerning Hans Lipperhey's patent application in the minutes of the meeting of the States General of the Netherlands, 2 October 1608. Courtesy Rijksarchief, The Hague

Raffaello Gualterotti, had a similar device but put it aside because it was "a feeble thing."[7] In Italy and the Netherlands—and presumably in other regions of Europe as well—devices such as these were present, but they were feeble things. What was wrong with them?

Although lens-grinding and polishing techniques had improved greatly in the sixteenth century, the area of the best spectacle lenses that was figured accurately remained small—a circle of perhaps a centimeter in diameter centered on the optical axis. Since these lenses were used as spectacles, there was no specific drive to make that area larger. In Middelburg, capital of the province of Zeeland, one of the "United Provinces," Italian glass-making techniques were practiced. And it was here that a counter-intuitive move was made: the objective lens of one of those early devices was stopped down drastically, from about 3 to 1 cm. The result was that the enlarged images seen through it were now reasonably clearly defined (Fig. 1). The spyglass was born, and the man who claimed it as his own was the spectacle maker Hans Lipperhey.

On 25 September 1608, Lipperhey set out from Middelburg with his spyglass and a letter of recommendation from the provincial government. Upon his arrival in The Hague, he presented his instrument to Prince Maurice of Nassau, commander-in-Chief of the Dutch armed forces. Maurice was impressed and sent the instrument to the States General, the national government, to decide whether to grant Lipperhey a patent (Fig. 2). Because the instrument could be easily copied, or even constructed after one heard a description of it, Lipperhey was not granted a patent; however, the gentlemen ordered two

[7] Raffaello Gualterotti to Galileo, 24 April 1610, Le Opere di Galileo Galilei, X: 341–342. Settle, "Danti, Gualterotti, Galileo;" and Van Helden, *Invention*, p. 45.

more spyglasses, this time binocular and with lenses made of rock crystal. For these he was handsomely rewarded.[8]

2 Galileo's telescope

Within 2 weeks of Lipperhey's patent application, the States General received a petition from Jacob Metius of Alkmaar, who claimed to have the same instrument, although, he admitted, not as good as Lipperhey's; and from Middelburg came news that there was another person there who claimed to know the secret of making spyglasses.[9] A newsletter printed in The Hague early in October was sent to the major European cities in diplomatic pouches,[10] and a French soldier departed from The Hague for Paris with knowledge of how to make spyglasses.[11] In November, the newsletter was received by Paolo Sarpi, who shared its content with Galileo.[12] We can get some idea of what Galileo's first, three-powered spyglass looked like from a sketch made in August 1609 of the first actual spyglass to arrive in Naples, by the internationally known polymath, Giambaptista della Porta (Fig. 3).

Della Porta wrote that the larger tube was one (Neapolitan) *palmo* in length (ca. 26.5 cm) and 3 in. (*oncie*, ca. 6.5 cm) in diameter; the smaller tube was 4 in. long (ca. 9 cm), and slightly smaller than the longer tube, so that it could slide into it. This instrument magnified three or our times. It is interesting to note that the objective aperture is very small, of the order of 1 cm.[13]

Over the next 6 months, Galileo steadily improved his skills as a lens maker, and progressed to instruments that magnified 20 times with reasonable quality by December of that year. It was at this point that he began to make his discoveries: mountains on the Moon, the difference in the telescopic appearances of stars and planets, and the satellites of Jupiter. His little book, *Sidereus Nuncius* came off the press in March 1610. Galileo became an international celebrity, and he moved back to his native Tuscany, where he was appointed "Philosopher and Mathematician to the Grand Duke."

In the meantime, he continued his observations. In July 1610 he discovered that Saturn is flanked by two bodies that do not move with respect to each other or the planet, and by the end of that year he was able to verify that Venus goes through phases like the Moon, and this proved that it, and by implication Mercury as well, goes around the Sun [5].

[8]Van Helden, *Invention*, 35–38, 42–43.

[9]Van Helden, *Invention*, 38–40

[10]Anonymous, *Ambassades du Roy de Siam envoyé à lÉxcellence du Prince Maurice, arrivé à la Haye le 10 septemb. 1608* (The Hague, October 1608). See *Embassies of the King of Siam sent to his Excellency Prince Maurits, arrived in The Hague on 10 September 1608*, ed. Huib Zuidervaart (Wassenaar, 2008); Van Helden, *Invention*, 40–42;

[11]Van Helden, *Invention*, 43

[12]Reeves, *Galileo's Glassworks*, 122–138

[13]Giambaptista della Porta to Federico Cesi, 28 August 1609, Galileo, *Opere*, X: 252

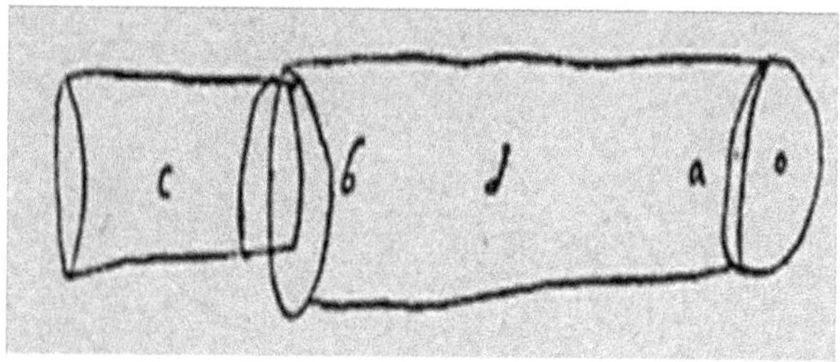

Fig. 3 Drawing of a spyglass by Giambaptista della Porta, 28 August 1609. *Le opere di Galileo Galilei*. Edizione Nazionale, vol. 10 (1968), p. 252

By that time, other observers were scrutinizing the heavens with telescopes of various powers and qualities. The study of sunspots was therefore initiated not by Galileo, although he had observed them earlier, but by Johannes Fabricius in Osteel, East Frisia, and by Christoph Scheiner. Galileo became involved in a controversy with Scheiner over the interpretation of the spots: Scheiner claimed that they were satellites of the Sun transiting the solar disc, while Galileo argued that they were like our clouds and on or near the surface of the Sun [6]. At stake was the perfection of the Sun and by extension the entire heavens. His arguments were convincing, even to Scheiner, and when *Istoria e dimostrazioni intorno alle macchie solari e loro accidenti* was published, in 1613, the arguments against the Ptolemaic–Aristotelian cosmology and supporting the system of Copernicus were so many that Galileo, although cautious in print, was defending the Copernican system openly in Florence. This was the point at which the Inquisition became involved. Three years later Copernicus's book, published in 1543, was put on the Index of Forbidden Books (until corrected), and Galileo was told not to hold or defend the heliocentric system. When he disobeyed this order, in 1632, the Inquisition condemned him to prison for life. He spent his remaining time under house arrest in his villa near Florence.

3 From the "Galilean" to the "astronomical" telescope

Galileo made his earth-shattering discoveries between 1609 and 1611. By the summer of 1611, the mathematicians of the Jesuit Collegio Romano had verified these discoveries. The phenomena were accepted as real; the attention now shifted to interpreting them.[14] Except for some minor additions, such as Simon Marius's discovery of the Andromeda nebula (1612), no further discoveries were made. What remained were elaborations: working out the parameters of the orbits of Jupiter's moons in the hope of solving the problem of longitude at sea; mapping the Moon, perhaps with an eye toward using

[14]The mathematicians of the Collegio Romano to Robert Cardinal Bellarmine, 24 March 1611, Galileo, *Opere*, XI: 92–93.

the method of lunar dichotomy (first described by Aristarchus of Samos in the third century BC) to determine longitude; and continuing observations of sunspots in order to determine the angle of the axis of rotation of the Sun. The telescope also made better observations of solar eclipses possible and opened the door to observing transits of Venus and Mercury.

But the instrument had its limitations. The small field of view, perhaps 15 arc min, limited magnification to about 20, and what modern astronomers call eye relief was very poor: the exit pupil was larger than the pupil of the eye, which meant a shifting field when the eye moved laterally to the optical axis. Because there was no fixed point for positioning the eye, the field was even affected by motion parallel to the optical axis.

Let us suppose that the apertures of Galileo's early telescopes were about 1.5 cm, and that his dark adapted pupil was half that. This means that the telescope gathered four times as much light as his naked eye, enough to show the satellites of Jupiter, about sixth magnitude, but not enough to reach much deeper into the solar system and reveal other satellites or surface features on the planets. We can thus say that by 1611 the potential of the Galilean telescope for discovery had been exhausted. Further discoveries had to wait for a different type of telescope.

As argued above, the combination of a strong concave and a weak convex lens produced obvious magnification. Combining two convex lenses, one strong and the other weak, did not, however, produce an obvious effect. The eyepiece has to be positioned by exactly its focal distance beyond the focal point of the objective. The "astronomical telescope" therefore came not out of the background of spectacle making, as the Galilean telescope did, but rather out of more theoretical considerations. In his *Dioptrice* of 1611, Johannes Kepler wrote that replacing the concave eyepiece by a convex one would also produce the telescopic effect, although the image would be inverted. To re-invert it, yet another convex lens would be necessary.[15]

There was no obvious advantage to this configuration, and one clear disadvantage: it showed objects on earth inverted. This type of telescope was not immediately put into practice. One application where it would have an advantage was projecting an image through the telescope. The Galilean instrument produced an inverted projected image whereas the Keplerian version produced an erect image. It was Christoph Scheiner who, after learning the projection technique from Galileo, continued his investigations of sunspots and around 1615 first used a Keplerian version for projecting the Sun's image. In his magnum opus on sunspots, *Rosa Ursina* (1630), Scheiner related that if this configuration was used in direct observation it would produce a much brighter image and a much larger field of view. The fact that the image was inverted did not matter in astronomical observation [8].

[15]Johannes Kepler, *Doptrice* (1611). See [7].

In the 1630s Francesco Fontana, a telescope maker in Naples, was the first to make astronomical telescopes with magnifications double those of the Galilean version, and his results slowly became known all over Europe through correspondence channels (His observations were not published until 1646.) By the early 1640s a hybrid terrestrial/astronomical telescope was offered for sale by the optician Johannes Wiesel in Augsburg. His telescopes had a removable erector lens. In 1645, the Capuchin monk Antonius Maria Schyrlaeus de Rheita published *Oculus Enoch et Eliae*, in which he spoke of yet a fourth lens, the field lens. (Rheita also introduced the terms *objective* and *ocular* into the language of telescopic astronomy.)[16] A new chapter in telescopic astronomy now began.

4 The long refractor and further discoveries

The substitution of the convex for the concave eyepiece made the problems of spherical and especially chromatic aberration worse. Thus, curvature had to be minimized, and therefore higher magnifications meant longer telescopes. Certainly a length of 3 m was already modest, when, in 1655, Christiaan Huygens in The Hague made his first research telescope. Its length was 12 ft, its field of view was 30 arc min, and its magnification was 50.[17] Recent test have shown that the objective of this instrument, preserved in the Utrecht science museum, was not very good (Fig. 4).[18] But Huygens turned his telescope to the heavens at an auspicious time, as we shall see.

As mentioned above, Galileo had observed Saturn's strange appearances for the first time in July 1610. He claimed he saw a central globe flanked by two smaller globes; in 1611, the mathematicians at the Collegio Romans claimed they saw an oval shape. In his third sunspot letter, printed in 1613, Galileo reported that since there had been no change in the positions of Saturn's "companions," he had lost interest, but that when he happened to look at the planet again in 1612 he was surprised to see the appendages gone. Had Saturn devoured his children [11]?

Since Saturn's appearance was neutral in the debate between the world systems, the problem was only occasionally discussed until the 1640s, when a new generation of telescopic observers usually armed with better telescopes took on the problem. By about 1655 the different appearances had been catalogued and some astronomers were beginning to formulate theories to explain them. What they were not looking for was a satellite of the planet.

---

[16] Rolf Willach [9]. See also [10].

[17] Christiaan Huygens *De Saturni luna observatio nova* (1656), *Oeuvres Complètes*, XV: 177; see also, 10–11.

[18] Rolf Willach, private communication.

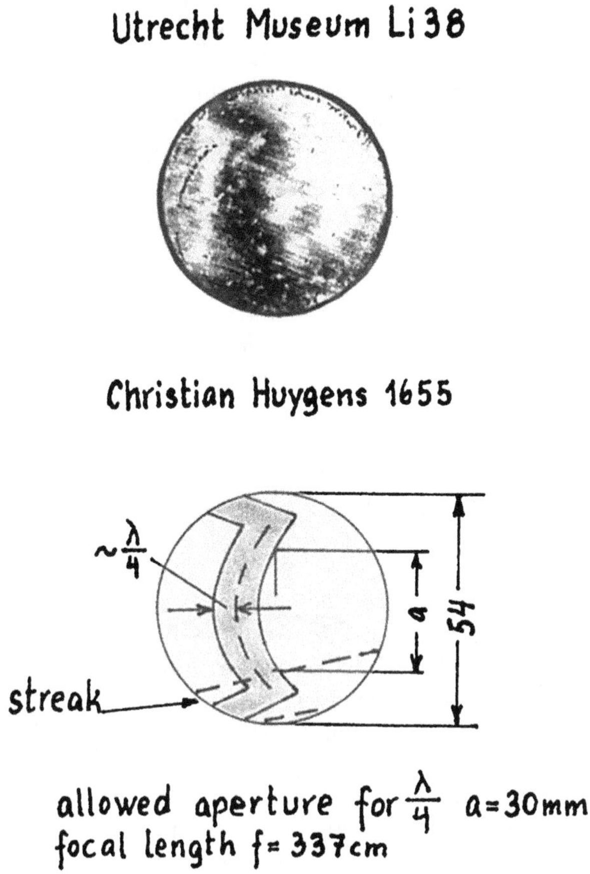

Fig. 4 Ronchi interference test of the objective of the telescope with which Huygens discovered Saturn's largest moon in 1655. The lens bears the inscription *Admovere oculis distantia sidera nostris* (from Ovid's *Fasti*), part of the anagram he used to hide his discovery of Saturn's moon, *Oeuvres Complètes de Christiaan Huygens*, vol. 15 (1967), p. 13

The Saturnian moon we now call Titan, $M = 8$, had been within the grasp of the best telescopes for about a decade, but its motion around the planet was a large ellipse (because its orbital plane more or less coincides with the ring plane). Observers such as Johannes Hevelius in Gdansk had seen it but considered it a fixed star. When the younger Huygens trained his telescope on the planet, early in 1655, he quickly noticed a bright little star moving back and forth on a straight line through Saturn. A new chapter of discoveries had begun.

The clue that led Huygens to his ring-theory was the fact that as Saturn's appendages became smaller they kept their lengths (Fig. 5). In England,

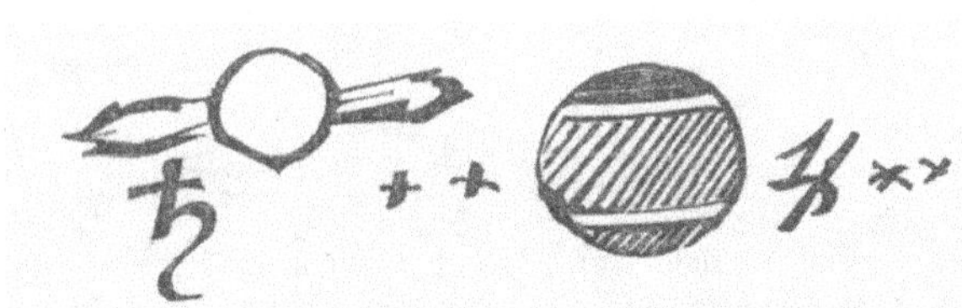

Fig. 5 Christiaan Huygens's first recorded observations of Jupiter and Saturn. *Oeuvres Complètes de Christiaan Huygens*, vol. 15 (1967), p. 35

Christopher Wren had observed the same and arrived at the conclusion that a very thin elliptical "corona" was attached to the planet in two places and that it turned or librated about this axis to produce the observed appearances. Huygens's solution certainly wasn't a matter of seeing a ring: he solved the problem late in 1655 or early 1656, during and shortly following the planet's conjunction with the Sun, and when the planet appeared its appendages were entirely gone. Rather, Huygens started from the supposition that Saturn was the center of its own (Cartesian) vortex. Since the newly discovered moon had a period of about 16 days and the planet was thought to rotate around its axis in a period not very different from the Earth on its axis, all matter in between had to revolve around the planet with periods less than 16 days. Presumably, the appendages or *anses* rotated around Saturn's axis in about a day, but their appearances changed much more slowly. A thin flat ring turning about the planet, and touching it nowhere was the solution: *Annulo cingitur plane, tenui, ad eclipticam inclinato* [12].

Huygens published a complete account of Saturn's moon and his ring solution in *Systema Saturnium*, 1659 (Fig. 6). This book was, however, more than a mere report about Saturn. Huygens also discussed his way of measuring angular distances in the field of view of his telescope by inserting thin rods into the focal plane of his telescope, and this led to the development of the screw micrometer in France, and the revival of William Gascoigne's micrometer in England. Having measured the angular diameters of the planets, Huygens made an educated guess of the Earth's apparent diameter as seen from the Sun, putting it halfway between those of Venus and Mars. His result was a solar parallax of 8.2 arc sec. *Systema Saturnium* thus also contained a measure of the solar system and everything in it [13].

Famous as he was because of his telescopic discoveries, Huygens's objectives were not as good as those made in Rome by Eustachio Divini and the new

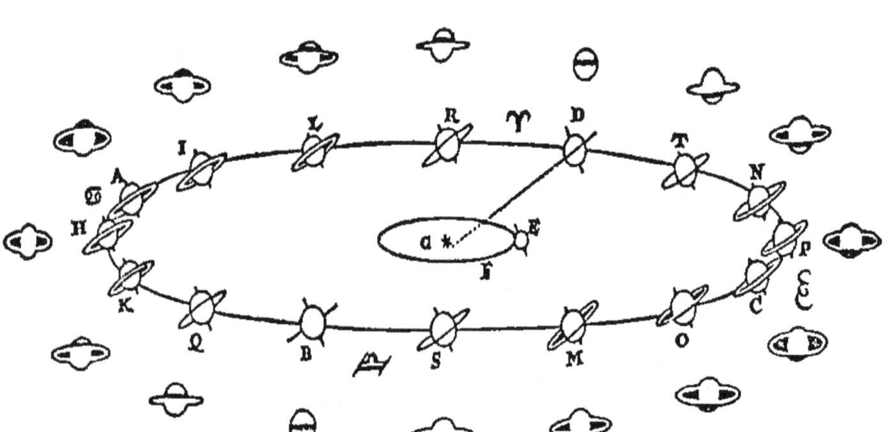

Fig. 6 Diagram demonstrating how a ring around Saturn can produce the various appearances. *Oeuvres Complètes de Christiaan Huygens*, vol. 15 (1967), p. 317

telescope maker, Giuseppe Campani. And whereas Huygens's instruments had clear apertures of a few inches, by 1665 Campani was producing biconvex objectives with a clear aperture of 10 cm for a focal distance of 10 m, with a peak-to-valley departure from the spherical of 0.14 wavelengths. Against these lenses, no one could compete [14].

No wonder that virtually all discoveries after that of Titan were made with Campani telescopes, especially by Giovanni Domenico Cassini, professor of mathematics at the University of Bologna and advisor to the Pope on engineering matters, and beginning in 1669 a member of the new Académie Royale des Sciences in Paris and head of the new royal observatory there. Cassini observed and published surface markings on Mars (earlier seen by Huygens but not published), determined the rotation periods of Mars and

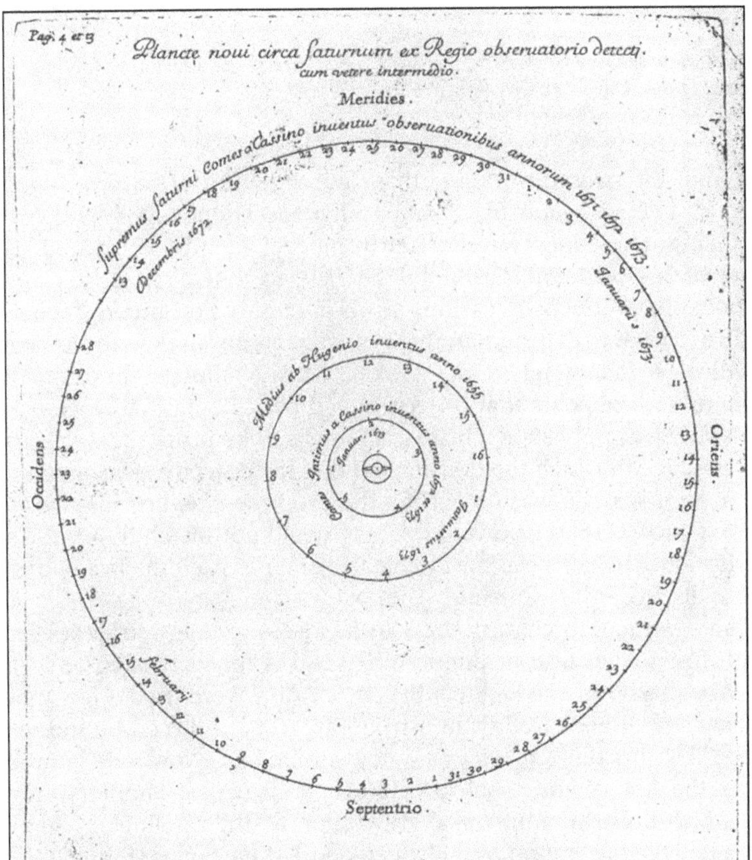

Fig. 7 Illustration from Giovanni Domenico Cassini, *Decouverte de deux nouvelles planetes autour de Saturne*. Paris 1673

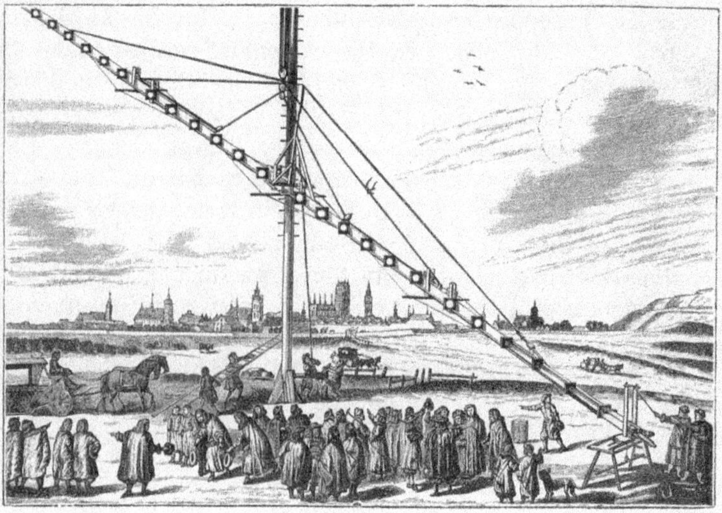

Fig. 8 40–45 m telescope erected on the beach near Gdansk by Johannes Hevelius. *Machinae Caelestis pars prior*. Gansk, 1673

Jupiter, and discovered no fewer than four satellites of Saturn, Iapetus and Rhea in 1671 (Fig. 7) and 1672,[19] and Tethys and Dione (1684) [12].

Campani's telescopes were of fairly modest lengths compared to those of some of his contemporaries. In the race to make telescopes more and more powerful, telescopes as long as 45 meters were already being raised around 1670 (Fig. 8). These were high-prestige items, useless for astronomical observation. With long telescopes, the only sensible thing to do was to do away with the tube. It appears that observers had been doing that since the 1660s. In 1684, Huygens published a little tract in which he proposed a design for an aerial telescope in which the housing holding the objective was mounted on a ball joint so that by means of a string the eyepiece could be aligned with the objective (Fig. 9) [15]. His own telescopes could not match those of Campani for clarity, and Huygens was never able to see for himself the last two satellites of Saturn discovered by Cassini.

As the seventeenth century drew to a close, the long refractors gradually fell out of favor because of diminishing returns for increasing unwieldiness. In the first quarter of the eighteenth century they were used a few times, and after that they disappeared from astronomy, to be replaced by the reflecting telescope. By this time, there was a new cosmology, Newton's infinite space ruled by the law of universal gravitation, and the astronomers' new solar system which illustrated universal gravitation by the satellites of Jupiter and Saturn and gave the system roughly its modern dimensions (Fig. 10).

[19]Giovanni Domenico Cassini, *Decouverte de deux nouvelles planetes autour de Saturne*. Paris 1673.

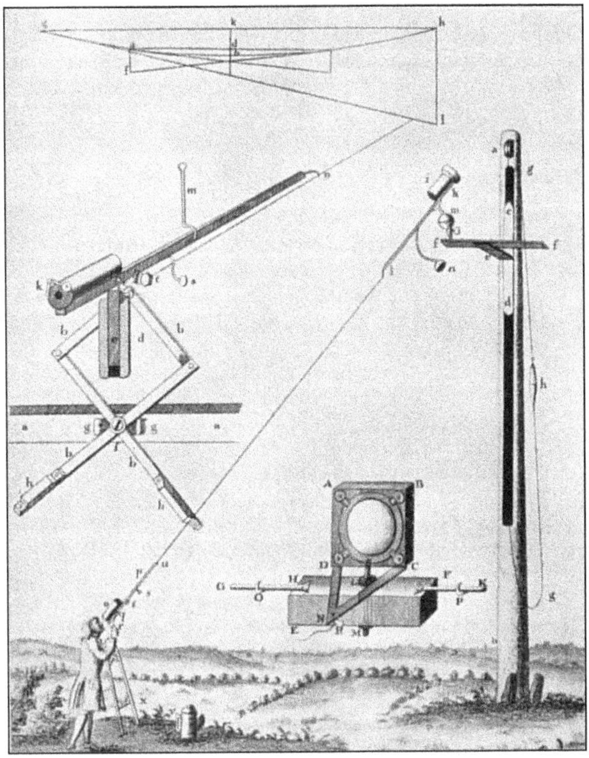

Fig. 9 Christiaan Huygens's design of a telescope without tube. *Astroscopia compendiarium*. The Hague, 1684

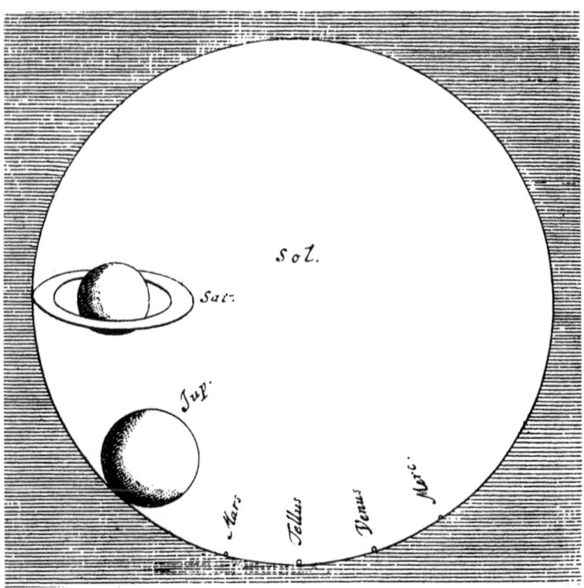

Fig. 10 Sizes of the planets compared with the Sun's. Christiaan Huygens, *Cosmotheoros*, The Hague 1698, p. 123

References

1. Settle, T.B.: Danti, Gualterotti, Galileo: their telescopes? Atti. Fond. Giorgio Ronchi **61**(5), 625–638 (2006)
2. Pelling, N.: Who invented the telescope? Hist. Today **58**(10), 26–31 (2008)
3. Dupré, S.: Ausonio's mirrors and Galileo's lenses: the telescope and sixteenth-century practical optical knowledge. Galilæana **2**, 145–180 (2005)
4. Van Helden, A.: The invention of the telescope. Trans. Am. Phil. Soc. **67**, 30–34 (1977)
5. Van Helden, A.: Sidereus Nuncius or the Sidereal Messenger, pp. 102–109. University of Chicago Press, Chicago (1989)
6. Reeves, E., Van Helden, A.: Galileo and Scheiner on Sunspots, 1611–1613. University of Chicago Press, Chicago (2009, in press)
7. Riekher, R. (ed.): Johannes Kepler Schriften zur Optik, 1604–1611, Ostwald Klassiker der Exakten Wissenschaften, vol. 198, pp 485–492. Verlag Harri Deutsch, Frankfurt a. M. (2008)
8. Van Helden, A.: The 'astronomical telescope,' 1611–1650. Ann. Ist. Mus. Stor. Sci. Fir. **1**, 13–36 (1976)
9. Willach, R.: Schyrl de Rheita und die Verbesserung des Linsenfernrohres in der Mitte des 17. Jahrhunderts. Sterne Weltraum **1**, 102–110, 186–192 (1995)
10. Thewes, A.: Oculus Enoch... Ein Beitrag zur Entdeckungsgeschichte des Fernrohrs, pp. 13–15. Verlag Isensee, Oldenburg (1983)
11. Van Helden, A.: Saturn and his anses. J. Hist. Astron. **5**, 105–121 (1974)
12. Van Helden, A.: Annulo Cingitur: the solution of the problem of Saturn. J. Hist. Astron. **5**, 155–174 (1974)
13. Huygens, C.: Systema Saturnium. Oeuvres Complètes **XV**, 343–351 (1659)
14. Miniati, M., Van Helden, A., Greco, V., Molesini, G.: Seventeenth century telescope optics of Torricelli, Divini, and Campani. Appl. Opt. **4**, 644–647 (2002)
15. Huygens, C.: Astroscopia compendiarium. Oeuvres Complètes **XV**, 201–231 (1684)

Exp Astron (2009) 25:17–31
DOI 10.1007/s10686-009-9142-0

HISTORICAL REVIEW

Galileo's telescope: history, scientific analysis, and replicated observations

Giorgio Strano

Received: 28 October 2008 / Accepted: 6 January 2009 / Published online: 24 January 2009

Abstract In the summer of 1609, Galileo Galilei started to improve upon the Dutch spyglass. Only a few months were needed to increase the magnifying power of the instrument up to 30 times. This transformation allowed Galileo to perform astronomical observations destined to change the traditional Universe. At present, the Institute and Museum of History of Science of Florence is conducting a research programme on the lenses of early telescopes. Historical research is combined with scientific analyses conducted by the National Institute of Applied Optics in Arcetri, the National Institute of Nuclear Physics in Florence, and the Glass Experimental Station in Murano. The shape and composition of the lenses are studied by applying different techniques. Moreover, the composition of early lenses and scientific glass objects are compared. Finally, Galileo's observations are checked with an optical replica of Galileo's telescope held at the Astrophysical Observatory in Arcetri.

Keywords History of the telescope · History of optics · History of astronomical observations · Galileo Galilei · Johann Kepler

1 Premise

Within a year of its appearance, the Dutch spyglass spread across Europe. After this, the so-called "Galilean" telescope dominated the scientific scene,

G. Strano (✉)
Istituto e Museo di Storia della Scienza, Piazza dei Giudici 1, 50122, Florence, Italy
e-mail: extraneus@imss.fi.it

especially in the Italian States, for about 30 years. Such a formidable impact raises the following problems concerning the early evolution of the telescope:

1. How did Galileo Galilei transform the Dutch spyglass into a powerful telescope?
2. What use did Galileo make of his improved telescope, and how was the instrument perceived in the scientific context of the time?
3. How can the functioning of Galileo's telescope be explained, especially considering the initial failure of a new type of telescope devised by Johann Kepler?

The historical discussion of these problems will be combined with a presentation of the provisional results of the scientific research still being conducted on the early telescopes of the Institute and Museum of History of Science (IMSS) in Florence.

2 Galileo's development of the telescope

Galileo began to recount his improvements upon the Dutch spyglass in a letter to the Venetian Doge Leonardo Donato, dated August 24, 1609. In it he claimed he would be glad to present the Doge with an "eyeglass invented by applying the most profound theories of optics", by means of which "something at a distance of nine miles appears as if it is only one mile far". This device could be very useful in detecting enemy ships or forces approaching.[1]

Galileo embellished this story in a letter to his brother-in-law Benedetto Landucci, dated August 29, 1609. He explained that between June and July 1609 he had received news that a Dutchman had shown Maurice of Nassau an eyeglass by which far objects could be seen as if close to the observer. Galileo had thought about the device, and worked out the principle of its functioning. As a result, he had made an instrument which he was called to present to the Senate of the Venetian Republic. On August 22 or 23, from one of the bell towers of Venice, he was able to show a ship at a distance of 50 miles as if it were just 5 miles away.[2]

Galileo gave a third report on the story of the telescope in the *Sidereus Nuncius* (Starry Messenger), printed in Venice on March 1610. In it he said that he had received confirmation of the invention of the Flemish optical device from one of his former pupils at the University of Padua. As a result, he had set out to develop the instrument by applying the principles of optics. He constructed an eyeglass with a magnifying power of only three times by inserting a plano-convex objective and a plano-concave ocular into a lead pipe. After this instrument, he was able to make a second one with a magnifying

[1] G. Galilei [7], v. 10, p. 250.

[2] Ibid., pp. 253–254. See also Antonio Priuli's report: ibid., v. 19, pp. 587–588, and Giovanni Bartoli's letter to Belisario Vinta, dated August 22, 1609: ibid., v. 10, p. 250.

power of eight times. Finally, he prepared a high-power instrument, by means of which far things could be observed about 30 times closer.[3]

Galileo's reports raise a series of interesting questions:

1. Did Galileo see one of the Dutch spyglasses before he began to develop his own telescope? In order to answer this question, I will mention a few examples of the early circulation of Dutch spyglasses through the Italian States.

In his *Telescopium* (The Telescope), published in 1618, Girolamo Sirtori, from Milan, claimed that in May 1609 a spyglass was delivered to the Count of Fuentes, in Milan, by a French gentleman.[4]

In a letter to Paolo Gualdo dated August 1st, 1609, the Paduan historian Lorenzo Pignoria wrote that one of those eyeglasses that Gualdo had mentioned in one of his former letters had just appeared in Padua. The instrument had been carried to the city by someone coming from beyond the Alps.[5]

In a letter dated August 28, the Neapolitan natural magician Giovanni Battista della Porta told Prince Federico Cesi, the founder of the *Accademia dei Lincei* in Rome, that he had finally seen a spyglass. He clamed it was "a hoax" that was taken from his *De refractione* (On Refraction), published in 1593. No reference to a spyglass can be found in this book. However, in his letter, della Porta explained that the instrument was made of a palm-size tin pipe into which a secondary smaller pipe could slide back and forth for focusing. Two spectacle lenses were fitted at either end of the tube. The first lens, to correct presbyopia, produced fuzzy magnified images; the other, to correct myopia, made the images clear.[6]

Finally, in a second letter to Gualdo dated August 31, Pignoria mentioned another spyglass that was sent from Flanders to Cardinal Scipione Borghese in Rome. This seems to be the same instrument that Gualdo had already mentioned to Pignoria as a sample, before August 1609. In addition, Pignoria stated that the spyglass market had appeared to meet with great success in Padua.[7]

These examples show that between May and August 1609 several Dutch spyglasses had been distributed in Italy. They probably came from France and had filtered down from the north to the south of Italy. In particular, many spyglasses had been sold in Padua, where Galileo had been professor of mathematics at the University since 1592. Therefore, although Galileo never claimed to have seen a Dutch spyglass, it is very hard to believe he had not.

[3]G. Galilei [6], pp. 6v–6r; see also, G. Galilei [7], v. 3-1, pp. 60–61. On Galileo's reports, see also M. Camerota [3], pp. 152–158.

[4]G. Sirtori [13], pp. 14–15.

[5]G. Galilei [7], v. 10, p. 250.

[6]Ibid., p. 252.

[7]Ibid., v. 10, p. 255. On the early diffusion of the Dutch spyglass, see also M. Biagioli [1], p. 84, n. 19.

2. Did Galileo improve upon the Dutch spyglass by applying the laws of optics? From Galileo's writings one can deduce that he had a limited knowledge of optics. The only extensive optical text in Galileo's handwriting is a copy of the *Theorica speculi concavi sphaerici* (Theory of the spherical concave mirror) by the mid-sixteenth century Venetian mathematician Ettore Ausonio.[8] Ausonio's manuscript was in the library of the nobleman Giovan Vincenzo Pinelli, in Padua. Galileo made a copy of the manuscript after his arrival to the city in 1592 and before leaving for Florence in 1610.[9]

Another confirmation of Galileo's not being particularly interested in optics until he heard of the Dutch spyglass can be found by comparing the drawings included in the Medieval and Renaissance works on optics to the drawings of the telescope included in the *Sidereus Nuncius*. As to the former, see for instance the formation of images in a convex mirror in the *Della Prospettiva* (On optics), by the fifteenth century physician and mathematician Giovanni Fontana. In order to investigate where the images of a ruler appear on a spherical mirror, Fontana marked on paper several positions of the eye. He then connected the different spots by a careful net of straight visual rays that locate the supposed real or virtual images of the ruler.[10]

Nothing of a similar accuracy can be found in Galileo's manuscript of the *Sidereus Nuncius* (Fig. 1a). The telescope appears as a slightly flaring quadrilateral, with two holes. The rough rendering of these holes gives the false impression that the instrument is made of two bi-convex lenses. Of course, this is not the optical structure of the telescope, which, in Galileo's explanation, has a plano-convex lens in *cd* and a plano-concave lens in *ab*. As for the light rays, only two parallel straight lines are drawn exiting on the side of the objective, and two divergent straight lines entering the eye in *e*.[11] This is not the accuracy that one can expect from an expert in optics, a fact which is confirmed by the printed version of the same drawing (Fig. 1b).[12] In any case, Galileo made no attempt to determine the image formation by geometric construction.

Although Galileo was not really interested in optics, he was not un-prepared to tackle the Dutch spyglass. As a natural philosopher, he was trying to solve the scientific problems by means of "sound reasoning", but also by means of "sensible experiences", that is, with the attention to the data arising from both observation and experimentation. Therefore, in addition to the mathematical components of his thought, Galileo needed to devise and make the instruments to analyze the world of reality.

Around 1597, he devised a new type of compass, which he produced, or had produced in several copies by his private craftsman, Marcantonio

[8]Biblioteca Nazionale Centrale di Firenze (BNCF), Ms. Gal. 83, f. 4r; see also: G. Galilei [7], v. 3-2, p. 869.

[9]S. Dupré [5], pp. 152 ff.

[10]Biblioteca Riccardiana, Florence, Ricc. 2110, f. 12r.

[11]BNCF, Ms. Gal. 48, f. 10; see also: G. Galilei [7], v. 3-1, p. 19.

[12]G. Galilei [6], p. 7r.

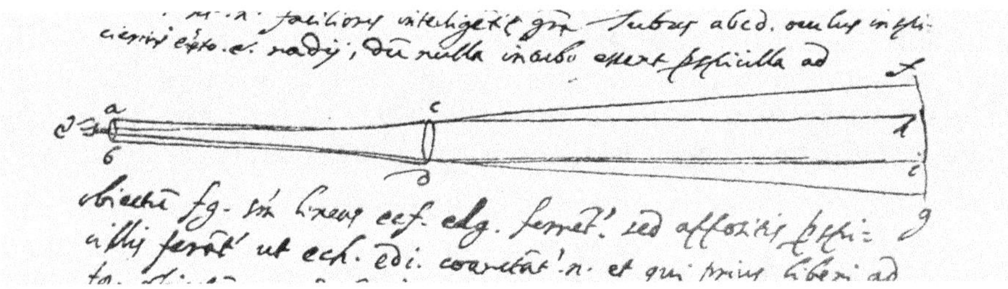

a

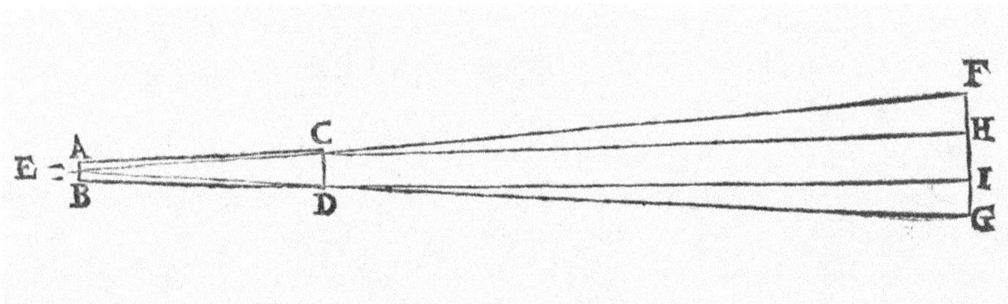

b

Fig. 1 Scheme of the telescope (G. Galilei, *Sidereus Nuncius*, 1610): **a** manuscript; **b** printed version

Mazzoleni. This compass could be employed to make calculations, and also for measuring angles, which was useful in surveying, cartography, ballistics, and even astronomy, as Galileo stated in his *Le operazioni del compasso geometrico et militare* (Operations of the Geometrical and Military Compass), published in Padua in 1606.[13]

This was not the only instrument that Galileo devised and made in Padua. After William Gilbert had published his *De magnete* (On the Magnet) in 1600, Galileo tried to determine how the power of suitable pieces of magnetite could be increased. In a few months he learnt how to transform pieces of mineral into powerful magnets. He sent one of them to his friend the nobleman Giovanfrancesco Sagredo, in Venice, and planned to sell another to the Grand Duke of Tuscany, Ferdinand I de' Medici.[14]

In addition, between 1602 and 1609 Galileo conducted studies on the mathematical law governing the motion of falling bodies. Once again, in order to accomplish reliable experiments he made suitable apparatus, including a sort of water clock to measure short time intervals. The invention of the thermoscope is also attributable to Galileo's Paduan period.[15]

[13]Several samples of the instrument are preserved in musea. See, for instance, Istituto e Museo di Storia della Scienza (IMSS) in Florence, inv. 2430. On the instrument, see also M. Miniati [12], p. 60, no. 6; M. Biagioli [1], pp. 8–13.

[14]Galileo's armour magnets are preserved at the IMSS, inv. nos. 6–11 and 2471. See also M. Miniati [12], p. 62, nos. 8–9.

[15]M. Camerota [3], pp. 130–132, 144–145.

In brief, the constant dealing with the making and improving of instruments, made Galileo well prepared for the news of the Dutch optical device. His ability in handcrafting instruments made it possible for him to grasp the secret of the spyglass more readily than most of his contemporaries.

3. How did Galileo improve upon the Dutch spyglass? The three stories that he recounted about the telescope are in agreement in that they distinguish three main steps. Initially, between July and August 1609, Galileo made a spyglass with a magnification power of three times out of a tube of lead. This instrument is only mentioned in the *Sidereus Nuncius*, and was just an imitation of what Galileo had heard or seen of the Dutch spyglass. Perhaps it was made by using two spectacle lenses.

As a second step, in August 1609, Galileo made another instrument capable of magnifying distant objects eight, nine or ten times. A magnification of nine is confirmed in a report by the Venetian nobleman Antonio Priuli, dated August 21, on Galileo's demonstration for the Senators of the Republic.[16] Apparently, Galileo mentioned the higher figure to his brother in law, and the lower figure in the *Sidereus Nuncius*, perhaps in order to emphasise his final achievement with the telescope. This second instrument was made of a tin tube, covered with reddish cotton frieze. It was about 60 cm long, and about 4 cm in diameter.[17] From the length of this telescope it is possible to argue that Galileo needed a spectacle lens for a moderately far-sighted individual as an objective, and a lens for an extremely near-sighted individual as an ocular. Since the time interval between the first and second attempt in telescope making was short, it is unlikely that Galileo was able to instruct any craftsmen to meet his special lens requests. We can only guess that, in order to obtain a suitable combination, Galileo searched in spectacle-makers' workshops for very well polished lenses.

As for the third step in telescope making, from the *Sidereus Nuncius* we can only obtain the information that Galileo's best instrument had a magnifying power of 30 times. Galileo spent no words on describing the construction of this telescope, perhaps to maintain secret a process that, when replicated, would have deprived him of scientific and material rewards.[18] Notwithstanding his silence, a precious private document casts light on the instrument. On November 23, 1609, Ottavio Brenzoni sent a letter to Galileo. On the occasion of a trip from Padua to Venice, Galileo used the blank part of the letter to jot down a list of items to look for while in Venice. The list begins with "Slippers and small hat for Vincenzo [Galileo's son]; The case with Marina's things [Galileo's lady friend]", and continues with: "Lentils, chick peas, rice, raisins, einkorn [...]". However, the list suddenly becomes a little bit disconcerting: "Artillery balls no. 2; Organ pipe made of tin; German flat glasses; [...]; Tripoli; [...]; Colophony; Felt, mirror for rubbing; [...]". The list also singles

[16] G. Galilei [7], v. 19, pp. 587–588.

[17] Ibid.

[18] Cf. M. Biagioli [1], pp. 81–86.

out shops where some of these items could be found: "The mirror-maker at the sign of the King; In Calle delle Aqque they make chisels". It also includes *desiderata* from the furnishers: "Deal for iron bowls, or to be made of stone, as for the artillery balls".[19]

This shopping list has great significance for the construction process of Galileo's best telescope. The list contains adhesives, abrasives, basic instruments for grinding and polishing, and the rough materials to produce lenses. Finally, it contains the moulds to be used to create the lenses. Indeed, for grinding concave lenses Galileo planned to use cannon balls. As for the convex lenses, he was wondering if it would be possible to use iron or stone bowls. Did Galileo have success in finding the items included in the list? As a document, the list only shows that, by the end of 1609, Galileo was not satisfied with the lenses from spectacle-making shops, and that he managed to set up his own lens-making workshop. However, the evidence that Galileo had some success is indirect: by the end of 1609 he completed his most powerful telescope.

The surviving documents allow us to outline the story of this telescope. Between December 1609 and January 1610, Galileo used it to study the Moon, the Milky Way, and to discover the four satellites of Jupiter. Immediately after the publication of the *Sidereus Nuncius*, he sent a still unbound copy of the booklet to the new Grand Duke of Tuscany, Cosimo II, along with the telescope. In the accompanying letters, Galileo pointed out that such a telescope was without aesthetic value, its importance consisting of the exceptional discoveries made with it.[20] The Grand Duke appreciated the gifts, and invited Galileo to come to Tuscany during the summer. The telescope became part of the Grand-ducal collection of scientific instruments. In less than 30 years, the tube and the ocular were lost, and the objective broke into several pieces. In 1677, the objective was salvaged, put back together, and inserted in a precious frame (Fig. 2).[21]

Recently, the lens has been examined with state-of-the-art scientific techniques. For instance, the interferometer map obtained at the National Institute of Applied Optics at Arcetri (INOA), Florence, by Giuseppe Molesini and his colleagues, has revealed Galileo's ability to shape very good lenses. Regrettably, the map also has revealed two different series of parallel interference fringes, one for each of the two major pieces of the lens. This implies that the individual who, in the past, had put together the pieces of the objective was not able to carry out a perfect restoration.[22]

In 2007, with the hope of revealing the glass recipe of the lens, a portable X-Ray Fluorescence-detecting apparatus was brought to the IMSS by a team of the National Institute of Nuclear Physics (INFN) in Florence, coordinated by Pier Andrea Mandò. No lead was found in the glass of Galileo's broken

[19]BNCF, Ms. Gal. 88, f. 16; see also; G. Galilei [7], v. 10, p. 270, n. 1.

[20]Ibid., v. 10, pp. 297–298, and p. 299.

[21]IMSS, inv. no. 2429. See also M. Miniati [12], p. 60, no. 1; A. Van Helden [16], p. 32, no. 3.

[22]V. Greco, G. Molesini & F. Quercioli [8], pp. 6219–6223; G. Strano [14], pp. 67–69.

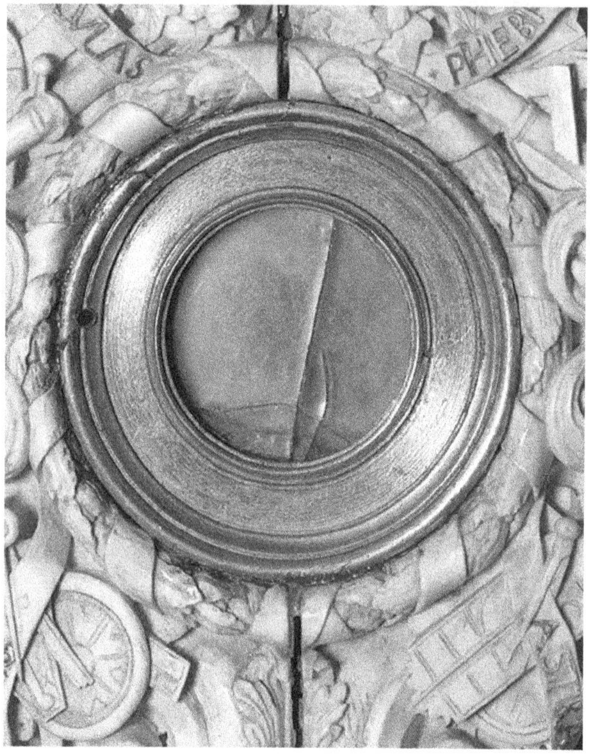

Fig. 2 Galileo's broken objective lens, detail (IMSS, inv. 2429)

objective. This confirms of course that Galileo did not use "optical lead glass", which was not developed until the eighteenth century. However, did Galileo really use the flat German glass he was looking for in Venice, as we know from his shopping list? Unfortunately, the study conducted by Marco Verità of the Experimental Glass Station of Murano, Venice, has led us to conclude that someone, in the past, may have varnished the surface of the lens in order to protect it from weathering, or that the lens is undergoing such a weathering process, and is in need of restoration. Further analyses will reveal the true problem and will aid in identifying the provenance of the glass.[23]

After the plain telescope he had sent to Florence in March 1610, Galileo had to maintain his promise to furnish the Grand Duke of Tuscany with ornate telescopes. He did this once he was nominated "first mathematician" at the University of Pisa and "first mathematician and philosopher" of the Grand Duke. Galileo's presentation telescope was carefully preserved in the grand-ducal collection of scientific instruments (Fig. 3). In order to make the tube, Galileo glued a series of strips of wood onto a piece of black fabric. Successively, he rolled the tube and covered it in brown and red leather. As a result, the tube was very strong, capable of resisting the pressure that was

[23] Ibid., pp. 75–78.

Fig. 3 Galileo's presentation telescope, detail (IMSS, inv. 2428)

needed to emboss it with a gilt decoration. Galileo stopped the objective, 37 mm in diameter, to an aperture of 15 mm with a diaphragm. This reveals that he learned very early the importance of eliminating the contribution from the rim of a lens to obtain sharper images. The magnification of the instrument, the original ocular of which was lost at the end of the nineteenth century, is 21 times. The field of view is only 15 arcminutes.[24]

What a plain Galilean telescope really looked like can be seen in another instrument at the IMSS. The tube is slightly conical and consists of two grooved wooden shells held together by copper threads and covered in paper. The housings of the objective and ocular are also made of wood, and can slide inside the main tube for focusing. Once again, the objective, 51 mm in diameter, is stopped to an aperture of 26 mm. The magnification is only 14 times.[25]

3 Galileo's discoveries

Galileo aimed the telescope towards the stars during the summer of 1609. Perhaps he used telescopes with a magnifying power that was intermediate between the instrument he presented to the Venetian Senators in August 1609 and his best telescope made in December 1609. This is confirmed by replicated observations made at the Arcetri Astrophysical Observatory, Florence, by a group co-ordinated by Francesco Palla. The astronomers used an optical replica of Galileo's presentation telescope that is capable of a magnification of 21. As a result, the Moon appears a little bit less detailed in Galileo's first watercolours (Fig. 4a) [26] than through the optical replica (Fig. 4b).

In 1609, Galileo also began to observe the Milky Way and the Ptolemaic nebulae. However, it was not until January 1610 that he realized the importance of his celestial discoveries. On the night of January 7, he observed three

[24]IMSS, inv. no. 2428. See: M. Miniati [12], p. 72, no. 2; A. van Helden [16], p. 30, no. 1.
[25]IMSS, inv. no. 2427. See: M. Miniati [12], p. 72, no. 1; A. van Helden [16], p. 32, no. 2.
[26]BNCF, Ms. Gal. 48, f. 28r–28v.

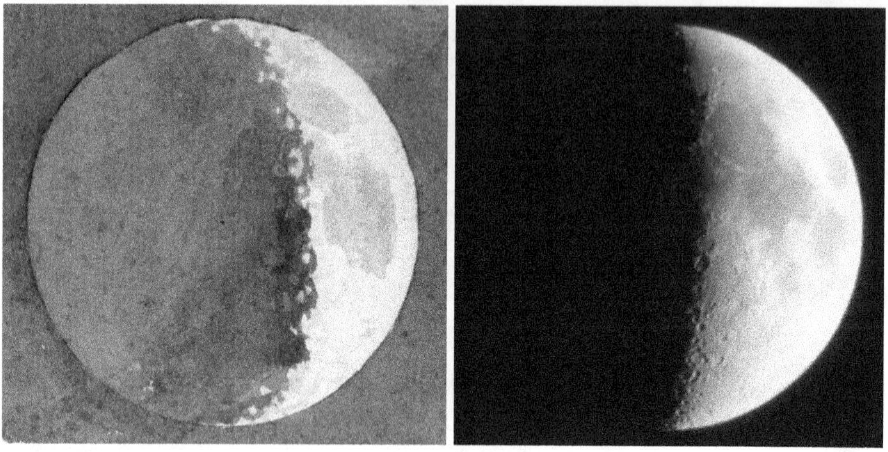

Fig. 4 The Moon before the first quarter as it appears: **a** in Galileo's watercolour, **b** through a replica of Galileo's presentation telescope

faint stars aligned with Jupiter. Additional observations led him to distinguish a fourth faint star on January 13, and to understand that, like the other three, it was orbiting Jupiter.[27] This was the impetus that caused Galileo to publish. Between February and March he continued to perform observations, worked upon the manuscript of the *Sidereus Nuncius*, and frequently travelled from Padua to Venice to meet the publisher and correct proofs. By the end of March, the booklet was printed, and sold out.

The first discovery which the *Sidereus Nuncius* communicated to the world was the terrestrial nature of the Moon. By using his telescope Galileo was able to distinguish that the lunar surface was not perfectly smooth as astronomers had usually believed. In particular, the observations contradicted the Aristotelian tradition that the celestial bodies, the Moon included, were made of perfect and incorruptible special matter, the so-called "quintessence". Like the Earth, the Moon had valleys and mountains. To emphasize this conclusion, Galileo calculated that the lunar mountains were about four Italian miles high.[28]

As a second discovery, Galileo described the stellar constitution of the Milky Way. He set aside any traditional notion of this celestial entity by claiming that its nebulosity was due to the eye of the observer being unaided. When the eye was aided by the telescope, the nebulosity disappeared, revealing an incredible number of faint stars. Such a conclusion was also valid for the Ptolemaic nebulae: the Head of Orion and the Heart of Cancer.[29]

[27] G. Galilei [6], pp. 17r ff.; see also G. Galilei [7], v. III-1, pp. 79 ff.; v. 10, p. 277, 280–281.

[28] G. Galilei [6], pp. 7r–16r; see also G. Galilei [7], v. III-1, pp. 62–75.

[29] G. Galilei [6], pp. 16r–16Bv; see also G. Galilei [7], v. III-1, pp. 75–79.

Apparently, the telescope was the means to reduce anything appearing fuzzy to the geometrical forms in which the book of nature was written.

Galileo's third discovery was the existence of four new stars orbiting Jupiter. In the *Sidereus Nuncius*, Galileo recounted the occasion of the discovery, and included daily observations of the planet up to March 2, 1610.[30] Galileo's desire to include up-to-date observations shows that the satellites of Jupiter formed the central topic of the work. Their discovery invalidated a classic objection to the Copernican world-system. In the Ptolemaic geocentric system, all the planets orbited the central immobile Earth. In the Copernican heliocentric system all the planets orbited the immobile Sun, excluding the Moon. The latter formed an exception, as it was the only celestial body to orbit the Earth. Jupiter's satellites provided evidence that other planets also had moons.[31]

The *Sidereus Nuncius* gave rise to conflicting opinions within the scientific community. Lenses were considered as magical attractions in the *wunderkammern*, the princely rooms destined to contain collections of natural, scientific and artistic items. Therefore, several detractors of the Copernican theory claimed that Galileo's celestial novelties were produced by the lenses. Since a single lens could produce fictitious images, a combination of two lenses could produce kaleidoscopic effects, and show mountains on the Moon, multiply the stars, and create moons of whatever planet.[32]

Galileo's celestial discoveries were welcomed by those individuals that had a firm Copernican belief. For example, immediately after the publication of the *Sidereus Nuncius*, on April 19, Johann Kepler, Imperial Mathematician of Rudolph II in Prague, wrote an enthusiastic letter to Galileo. He complimented the natural philosopher on the impact his discoveries had for supporting the heliocentric world-system.[33] Galileo appreciated Kepler's commendation and, as soon as the letter was published in Prague with the title of *Dissertatio cum Nuncio Sidereo* (Dissertation on the Starry Messenger), Galileo wanted it to be published also in Florence.[34]

As Galileo began to confront the different opinions the *Sidereus Nuncius* provoked, he also continued to make important celestial discoveries. On July 30, 1610, he wrote a letter to Belisario Vinta, Secretary of State of Cosimo II, announcing that Saturn was a compound of three stars. At first glance, this discovery provided additional evidence for the existence of satellites orbiting other planets. However, from night to night, Saturn's companions appeared to remain immobile.[35] In 1612, they temporarily disappeared,[36] and when they reappeared in 1616, they looked like two ears or handles on either side of the

[30] G. Galilei [6], pp. 17r–27v; see also G. Galilei [7], v. III-1, pp. 80–94.

[31] G. Galilei [6], pp. 28r–28v; see also G. Galilei [7], v. III-1, pp. 94–96.

[32] See M. Camerota, 2004, pp. 171–174.

[33] See G. Galilei [7], v. 10, pp. 319–340.

[34] See respectively: J. Kepler [9] and [10].

[35] G. Galilei [7], v. 10, p. 410. Galileo published the discovery in his *Discorso intorno alle cose che stanno in su l'acqua* (Discourse on Bodies on Water), printed in 1612; see ibid., v. 4, p. 63.

[36] Ibid., v. 5, pp. 237–238.

body of the planet.[37] These puzzling appearances caused Galileo to discard Saturn as another point in favour of the Copernican theory. However, Saturn gave Galileo the possibility of focusing on the importance of having very good telescopes. Only a perfect telescope could show Saturn's companions. A weak instrument would only have revealed an olive-shaped planet.[38]

On December 11, 1610, Galileo wrote to Giuliano de' Medici, Ambassador of Cosimo II in Prague, about a new discovery.[39] Galileo provisionally sent him an anagram that was solved on January 1, 1611. Venus revealed phases very similar to those of the Moon. In Galileo's mind, such phases formed additional evidence for the Copernican world-system.[40] In the Ptolemaic geocentric system, Venus orbited the Earth along two circles, the second of which always had its centre on the Earth-Sun segment. This implied that, at maximum illumination, the planet could be observed from the Earth as a crescent. Instead, in the Copernican heliocentric system, Venus orbited the Sun. This fact implied that the planet would display a complete series of phases when viewed from the Earth, as Galileo had confirmed.

Paradoxically, Jupiter's satellites and Venus' phases did not create any problem from the ecclesiastic point of view. In the spring of 1611, Galileo went to Rome to show his telescope to the Jesuit Fathers of the Roman College. On this occasion, the mathematician Christoph Clavius honestly confirmed the soundness of Galileo's discoveries,[41] and stated that it was indispensable to modify the architecture of the cosmos accordingly.[42] Of course, neither Clavius nor the other Jesuits gave weight to Galileo's conclusion on the validity of the Copernican theory. Instead, they accepted the system of the world that the Danish astronomer Tycho Brahe had expounded in 1588. In this system, the Moon and the Sun orbited the immovable Earth, and all the other planets orbited the Sun. It is clear that the phases of Venus fit into this system perfectly.

The Jesuits were more puzzled by the terrestrial nature of the Moon. This discovery effaced the cosmological gap between the super- and the sub-lunar worlds established by Aristotle.[43] This also explains the controversy that existed between Galileo and the Jesuit Father Christoph Scheiner on the nature of sunspots.[44] When Galileo observed the sunspots, he immediately interpreted them as evidence that, like the Moon, the Sun also had a terrestrial nature. The Sun was not perfect and incorruptible; it was a place of continuous mutations. In Galileo's opinion, sunspots might be dense clouds moving across its bright surface.[45]

---

[37] Ibid., v. 12, p. 276.

[38] Ibid., v. 10, p. 474; v. 5, p. 110.

[39] Ibid., v. 10, p. 483.

[40] Ibid., v. 11, pp. 11–12.

[41] Ibid., v. 10, pp. 484–485. See also ibid., v. 3-1, pp. 293–298.

[42] C. Clavius [4], p. 75.

[43] See S. E. Booth, A. Van Helden [2], pp. 470–473.

[44] See A. Van Helden [15], pp. 357–395.

[45] G. Galilei [7], v. 5, pp. 105–107.

4 Conclusions: early evolution of the telescope

The early evolution of the telescope reveals an interesting impasse. Galileo was able to improve upon the magnification of the Dutch spyglass, going from three to 30 times in a few months, and nobody else was able to improve upon Galileo's telescope for more than 20 years. Even more astonishing is the fact that, after the quick development of the telescope between 1609 and 1610, Galileo himself was not able to further improve the instrument. This suggests that, in 1610, he had at hand an optical device that could not be developed further.

Both the Dutch spyglass and the Galilean telescope were formed by combining a convex lens as an objective, and a concave lens as an ocular. The combination has the advantage of showing erect images. However, the combination also produces a small field of view, which progressively diminishes as the magnifying power increases. This drawback depends on the concave ocular, which has the property of sharpening the distant objects seen through the objective, but diminishing the apparent diameter of the objective.

The appearance of a new type of telescope had no effect on the evolution of astronomy until the new social figure of the specialized optician came into being. Note that, in order to observe the four satellites of Jupiter, Kepler had to search for a good telescope. When he finally found a suitable instrument on loan, it was the one that Galileo had made for the Elector of Cologne.[46] Nevertheless, on the basis of his thorough knowledge of optics, Kepler himself had conceived a new type of telescope that he expounded in the *Dioptrice*.[47] This was published in 1611, that is, the same year he received Galileo's telescope on loan...

From the theoretical point of view, Kepler's telescope, made of two convex lenses, had great advantages. Although the instrument created upside-down images, it was capable of a higher magnification and a larger field of view. However, for technical reasons this type of telescope became a reality only in the 1630s. Indeed, Kepler's combination of lenses amplifies optical aberrations more than Galileo's.

The transition to Kepler's telescope required more precise lens grinding and polishing techniques, especially by the use of appropriate metal moulds. A careful scientific study of the lenses made by Jacopo Mariani, Evangelista Torricelli, Eustachio Divini, and Giuseppe Campani is presently contributing to clarify the early evolution of the telescope. At present, micro-interferometric maps made by the INOA have revealed that the surface of Divini's objectives had a lower roughness. Interferometric maps of the lenses have also revealed that Campani's objectives had an overall good shape. However, micro-photographs of the surface of Campani's lenses show tracks left

---

[46] J. Kepler [11], v. 4, pp. 318–319.

[47] Ibid., pp. 387–388.

by abrasive grains. The presence of these tracks testifies that, after grinding, Campani used very soft polishing.[48]

In addition to controlled grinding methods, the improvement of the telescope was partly due to a careful selection of the glass. For instance, we know that Divini usually bought a large quantity of flat glass, but only few pieces successfully passed a careful examination with a microscope and became telescope objectives. Spectral transmittance analysis made by the INOA has revealed an overall similarity in the glass used by early telescope makers. In particular, the absorption curves obtained have revealed the presence of manganese. This element was employed to eliminate green coloration due to the presence of metal impurities in the glass.[49]

X-Ray Fluorescence analysis by the INFN has revealed that, as for Galileo's objective lens, the glass recipe for lens making did not include lead. Moreover, a comparison with a few glass instruments made for the Galilean Accademia del Cimento—which was active in Florence between 1657 and 1667—has revealed other interesting information. The white glass from the furnace installed in the garden of the Pitti Palace of Florence to produce blown glass items is very similar to the white glass used for lenses.[50] Such a similarity forms additional evidence that the evolution of early telescopes was based upon grinding and polishing methods. Varying the glass composition was not yet considered important.

Acknowledgements A condensed version of this paper has also been presented to the 27th Symposium of the Scientific Instrument Commission (SIC), Lisbon, September 16–21, 2008. I would like to thank Karen Lee Tomashavsky for her essential support in revising the English text of this article.

References

1. Biagioli, M.: Galileo's Instruments of Credit: Telescopes, Images, Secrecy. The University of Chicago Press, Chicago (2006)
2. Boot, S.E., Van Helden, A.: The Virgin and the telescope: the moons of Cigoli and Galileo. Sci. Context **5**, 463–486 (2000)
3. Camerota, M.: Galileo Galilei e la Cultura Scientifica nell'età della Controriforma. Salerno Editrice, Rome (2004)
4. Clavius, C.: Commentarium in Sphaeram Ioannis de Sacro Bosco. Sumptibus Antonii Hierat excudebat Reinhardus Eltz, Munich (1611)
5. Dupré, S.: Ausonio's mirrors and Galileo's lenses: the telescope and sixteenth-century practical optical knowledge. Galileana **2**, 145–180 (2005)
6. Galilei, G.: Sidereus Nuncius. Apud Thomam Baglionum, Venice (1610)
7. Galilei, G., Opere (A. Favaro ed.). Barbèra, Florence (1890–1909)
8. Greco, V., Molesini, G., Quercioli, F.: Telescopes of Galileo. Appl. Opt. **32**, 6219–6226 (1993)
9. Kepler, J.: Dissertatio cum Nuncio Sidereo nuper ad mortales misso a Galilaeo Galilaeo Mathematico Patavino. Typis Danielis Sedesani, Prague (1610a)

[48] G. Strano [14], pp. 71–72.

[49] Ibid., pp. 72–74.

[50] Ibid., p. 146, n. 3.2.1.

10. Kepler, J.: Dissertatio cum Nuncio sidereo nuper ad mortales misso Galilaeo Galilaeo. Apud Io. Antonium Canaeum, Florence (1610b)
11. Kepler, J.: Gesammelte Werke. C.H. Beck'sche Verlagsbuchhandlung, Munich (1937–)
12. Miniati, M. (ed.): Museo di Storia della Scienza: Catalogo. Giunti, Florence (1991)
13. Sirtori, G.: Telescopium, sive Ars perficiendi novum illud Galilaei visorium instrumentum ad sydera. Typis Pauli Iacobi, impensis Lucae Iennis, Frankfurt (1618)
14. Strano, G. (ed.): Galileo's Telescope: The Instrument that Changed the World. Giunti, Florence (2008)
15. Van Helden, A.: Galileo and Schener on sunspots: a case study in the visual language of astronomy. Proc. Am. Philos. Soc. **140**, 357–395 (1996)
16. Van Helden, A.: Catalogue of early telescopes. Giunti, Florence (1999)

Exp Astron (2009) 25:33–42
DOI 10.1007/s10686-009-9165-6

HISTORICAL REVIEW

The Era of Newton, Herschel and Lord Rosse

Jim Bennett

Received: 29 March 2009 / Accepted: 1 April 2009 / Published online: 22 April 2009

Abstract In the eighteenth century England was dominant in building telescopes and instrumentation. This paper describes the contributions of the most important opticians and telescope builders, from Newton's Opticks and the telescope design that bears his name, through various instrument makers who constructed 'popular' telescopes and published descriptions of mirror grinding (Smith, the Dollonds and their patent on achromatic lenses), to Herschel, who refined the description of his polishing procedures, and Lord Rosse, who attempted to communicate his through publication. The narrative of theory, practice and communication takes unexpected turns.

Keywords Reflectors · Refractors · Newton · Dollond · Herschel · Rosse

1 Telescope development in England

The period when English opticians and telescope builders made the most profound contributions to their subject might be characterised as 'the long eighteenth century', from the fundamental work of Newton in the late seventeenth century to the practical achievements and the visionary ambitions of Herschel extending into the nineteenth. Much of the technical, design and scientific development in telescopes over this period takes place in England. If we think of the successful designs of reflectors, their widespread adoption, the application of the refractor as a telescopic sight as part of the development of observatory measuring instruments, the achromatisation of the refractor, the introduction of the divided object-glass for measurement in both the

J. Bennett (✉)
Museum of the History of Science, Broad Street, Oxford, OX1 3AZ, United Kingdom
e-mail: jim.bennett@mhs.ox.ac.uk

refractor and the reflector, the extension of the aperture of the reflector and the development of giant instruments for the investigation of cosmological questions, much of the story takes place in England. That would change with the Munich School of optical instrument making and Fraunhofer as its principal representative, while a strand of development—the giant reflecting telescope—continues for a time in a 'British' context, a term not inappropriate for the scientific milieu of Lord Rosse, even though his telescopes were built in Ireland. The early nineteenth century marks the beginning of the end of British dominance in the design and manufacture of astronomical instruments. A claim could not be made for a similar standing in astronomy more generally over the same period, only in relation to telescopes and astronomical instrumentation, and it might be wondered why they were not more closely linked. This paper considers some of the central episodes from this period and, while the narrative is selective, it allows some general observations on relationships between theory and practice and between their respective practitioners.

2 Newton

Newton is mainly remembered in this context for a systematic mathematical treatment of optics and the establishment of a mathematical tradition of optical work in Britain. But it is important that it was not solely mathematical and theoretical, as Newton emphasises clearly in *Opticks* that he made his own reflecting telescopes, grinding and polishing the metal mirrors by hand. He goes through the process in detail, while making a point that would frequently be echoed in the eighteenth century: 'This Art of polishing will be better learn'd by repeated Practice than by my Description' [11].

In *Opticks* Newton offers two contexts for proposing a shift from the refractor to the reflector. One is more familiar, being the one generally emphasised: colours are not *created* by the process of refraction, they pre-exist in the incident light and their different refractive properties mean that refraction inevitably results in their dispersion. Dispersion into colours was unavoidable if one used lenses. Furthermore dispersion for a given degree of refraction was thought by Newton to be the same for all types of glass, so there could be no possibility of recombining the colours to give a resultant deviation without dispersion, and so a refracting telescope could never be perfected, so as to give a truly sharp image.

'Seeing therefore the Improvement of Telescopes of given lengths by Refractions is desperate; I contrived heretofore a Perspective by Reflexion, using instead of an Object-glass a concave of Metal' [11].

But Newton also presented the telescope construction that bears his name under the title 'To shorten telescopes', that being an obvious desideratum of contemporary telescopic astronomy. Slight curvature or long focal lengths were one way to reduce the effects of spherical and chromatic aberration, but the result was that telescopes were becoming unmanageably long. The combination of lenses and mirrors to shorten telescopes was also a concern of Robert Hooke. The designs of both Hooke and Newton were characterised as 'catadioptrical' instruments, rather than, at this stage, 'reflecting' telescopes, as they used combinations of mirrors and lenses.

Newton's legacy to optics in Britain was this peculiar combination of geometrical and artisan practice, together of course with issues in what we would call physics and they called philosophy or natural philosophy. After all, the original quarrel with Hooke in optics—their first quarrel—was over a physical hypothesis which Hooke detected and which Newton denied.

3 Robert Smith's *Opticks*

Given the character and content of Newton's *Opticks*, the structure and organisation of the great Newtonian optical treatise of the eighteenth century, Robert Smith's *Opticks* of 1738, is, for the most part, unsurprising. It is presented in four 'books', denominated 'popular', 'mathematical', 'mechanical' and 'philosophical' [12]. 'Mechanical' here refers to the design, manufacture and use of instruments while the 'philosophical' book contains what would be thought of today as the physics. The surprise comes from the book Smith called 'popular', this being a very early use of that adjective describing an account of science, meaning an introductory text for a general, non-specialised public. Newton's *Opticks* was certainly mathematical, mechanical and philosophical, but it was hardly 'popular', despite being the most accessible of his books. However, that aspect of optics had begun to develop in the early decades of the century and Smith's book would give it a further impetus. A great many serious amateurs in astronomy and telescope making began with Smith or encountered him at an early stage, while less committed participants increasingly had other resources to turn to in popular books, lecture series and instruments aimed at non-astronomers.

What we should properly call 'the natural philosophy of the heavens', rather than astronomy (for professional, observatory astronomy remained preoccupied with measurement and planetary theory), grew into a fashionable interest for the middling sort of clients of instrument makers. It was the instrument makers, rather than the mathematicians such as Smith, who promoted this popular culture and the commercial opportunities it presented. It was the makers who provided further and even more 'popular' books and who sold the instruments, almost exclusively Gregorian reflectors. The larger apertures possible with reflectors fitted them for enjoying greater detail in the moon or Saturn, or wondering at the variety of nebulae or clusters of stars, while the Gregorian had the intuitive advantage over the Newtonian that the observer was looking towards the object of interest. Rather than using the word 'popular', authors were inclined to signal a general interest by offering their work to ladies as well as gentlemen; examples would be James Ferguson's *Easy introduction to astronomy, for young gentlemen and ladies*, or Benjamin Martin's *The young gentleman and lady's philosophy* [7–9].

While instrument makers wrote textbooks and even devised courses and gave lectures, which we might think rather a literary or academic set of activities, the academic sorts rolled up their sleeves and applied themselves to grinding and polishing [4]. Smith describes these mechanical procedures in detail, so that they can be followed, presenting this as a polite occupation, appropriate to the rational gentleman [12]. His leading practitioners are John Hadley, Vice-President of the Royal Society, James Bradley, Savilian Professor at Oxford, and Samuel Molyneux, from a family of Irish gentry, a Member of Parliament and Privy Counsellor. These men cast metals and hand-grind and polish the resulting discs, while their exploits are written up in detail by the Plumian Professor in Cambridge, who will shortly become Master of Trinity College and Vice-Chancellor of the University. This episode in mirror making is not unlike the contemporary fashion among German aristocrats for turning delicately-shaped objects on a lathe. But of course the precedent and legitimation here is specifically Newtonian.

Smith goes on to tell his readers of a promising maker of reflectors in Edinburgh called James Short who is a protégé of Newton's mathematical disciple and collaborator Colin Maclaurin. Short was educated at the University of Edinburgh and already became a fellow of the Royal Society before moving permanently to London in 1738 [5]. As is well known, he made nothing but reflecting telescopes—something which sets him apart from the generality of makers, who usually sought to diversify for commercial reasons. Even if they were actually *making* only a small range of goods, they would retail more generally so as to attract trade. Short became an active fellow of the Royal Society, publishing in the *Philosophical Transactions* and serving on the Society's Council. He came close to being appointed Astronomer Royal, believing he had been denied the appointment on account of his support for the chronometer maker John Harrison. The trade description he adopted of 'optician solely for reflecting telescopes' has generally been understood as a mere statement of fact, but in a Newtonian culture where optical theory and experimental practice were combined in the reflector, Short's stipulation may have been an assertion of status as much as a business title.

4 John and Peter Dollond

If we look at the fortunes of the refractor, dismissed by Newton as 'desperate', we find a reversal in the intellectual and mechanical aspirations of telescope makers, with mechanics passionately seeking the status of theorists and engaging in sustained squabbles over their theoretical credentials. In John Dollond's well-known portrait (Fig. 1), there is not a telescope in sight, even though making and selling telescopes became his trade. He presents himself as a student of Newton's *Opticks* and, holding experimental prisms rather than lenses or telescopes, as an experimental natural philosopher. He began as a silk-weaver in a Huguenot family in the unfashionable district of Spitalfields in London, but with optics becoming a field of general interest, open to all comers—the first

Fig. 1 Portrait of John Dollond

'popular' science as presented by Smith—Dollond encouraged his son Peter to take up this more profitable trade and also followed his own investigations [6]. Naturally he began with complete confidence in Newton's conclusions but he came to doubt the reliability of the assumption that the dispersion of light for a given degree of refraction was the same for all types of glass. His subsequent experiments showed that this was not the case and that high-lead 'flint' glass had a significantly greater dispersive power than the more common optical 'crown' glass, and this opened the possibility for an 'achromatic' combination of two lenses that could produce overall deviation without dispersion. Newton had been wrong.

Dollond explained his technique in a paper to the Royal Society in 1758 and was awarded the prestigious Copley Medal. In the same year he also took out a patent on his new lenses in partnership with the optical instrument maker Francis Watkins, who contributed his financial support. Dollond was elected a fellow of the Royal Society in May 1761 but died suddenly in November and his share in the patent passed to his son Peter.

The subsequent story is well known. John Dollond's motives for patenting the achromatic lens are not entirely clear but his son Peter quickly set about exploiting the commercial advantage it gave him over other instrument makers, buying the share owned by his father's partner Francis Watkins in 1763 and in the same year taking legal action against Watkins for infringement, and in the following year against the widow of the optician James Ayscough. Disgruntled and alarmed at this, 35 London opticians petitioned the Privy Council for the patent to be revoked on the grounds that the achromatic lens had been invented previously by someone else, an amateur optician called Chester Moor Hall, and had been widely sold before the patent application. They said that, for example, the optician George Bass had made such lenses and indeed that it was in conversation with Dollond that the latter had learnt about them. The petition was not upheld and Dollond continued with successful court actions against other makers to secure his position.

The instrument maker Jesse Ramsden married Peter Dollond's sister Sarah in 1766 and it is thought that he benefited by his family connection to the patent holder, but after his separation from Sarah and the expiry of the patent, Ramsden too attacked Dollond; indeed he was probably the most effective critic of Dollond's position. Peter, ever anxious to defend his father's reputation, in 1789 presented a paper on the subject to the Royal Society. It was read to a meeting but not published, whereupon Dollond himself arranged its publication. Ramsden responded with an account, also read before the Society, where he offered his personal experience of the matter and attributed the invention of the achromatic lens, not to his father-in-law John Dollond, but to Chester Moor Hall, who had been championed previously by the London opticians in their dispute with Peter. Peter then drafted a rebuttal, which was never published but survives in manuscript [3]. It is from this document that we learn for the first time that John Dollond had actually met Chester Moor Hall. According to Peter, none of this mattered, as his father had independently reached his practical technique on the basis of his theoretical command of Newtonian optics. It was central to Peter's cause that his father had not simply been a jobbing optician, but a natural philosopher:

'In the beginning of the year 1757 Mr Dollond having tried the ... 8th Expt. of the 2d part of the first book of Newton's Opticks and by that means having discovered the new principle which was that the dissipation of the different coloured rays was not in the same proportion to the mean refraction, in water as in glass...'

Peter even admitted that there had been an early conversation between his father and George Bass, but that Bass himself had not understood the significance of what he had been saying: 'He did not know that such a difference in the quality of two kinds of glass could be applied to any advantage. Mr Bass was a practical optician; not a theorist.'

There are three themes worthy of attention in this episodic account of the period and each occurs in a form of dialectic. The first is the relationship between theory and practice, which may not seem a very original opposition but here there is a curious inversion of ambitions, where not only do makers want to be seen as theorists and even compose synthetic popular accounts of natural philosophy, the elite mathematicians and natural philosophers—from Newton on—present their mechanical credentials. They design and manufacture instruments themselves and they have troubled and unsatisfactory relations with the real mechanics, whose practice they are sure they can improve and reform. Again this begins with Newton, who writes in *Opticks* about his problems with the London artificers and his reliance on his own mechanical skills. Looking forward briefly to the end of this account, Lord Rosse, like Newton, becomes President of the Royal Society, but he is absorbed in the mechanical detail of his telescopes, now on a vastly greater scale. Rosse's contributions are in engineering design as well as mechanical production, but of all the ingredients that make up his extraordinary telescopic enterprise, it could be argued that he is least interested in the astronomy.

5 William Herschel

Our next builder of reflectors, William Herschel, was a product in the first place of 'popular' astronomy. He was a musician who was a student of Smith and followed his instructions first in the making of specula, an amateur whose enthusiasm meant that he was forever pushing the limits of the metal speculum and the reflecting telescope, a man who in relative obscurity and isolation created telescopes more powerful that any in the world of professional astronomy and who elaborated a set of goals in cosmology that would scarcely have been recognised by the professionals as astronomy at all, but that perfectly matched his instrumental ambitions.

Herschel could easily have been ignored—would have been ignored—in a sense there was nothing to ignore since, properly speaking, he was not an astronomer, had it not been for the discovery of Uranus—a discovery so astonishing that it gave Herschel a secure platform to pursue and insist on his real programme of large-scale cosmology. Jérôme Lalande rightly said that this extraordinary and unprecedented discovery had been made, not by an astronomer at all, but by a musician. But now, however outlandish Herschel's future plans for building telescopes, however imaginative and even embarrassing his speculations, he could not be ignored.

Although it was an 'amateur' vision in the eighteenth century, Herschel had imagined giant reflectors looking deep into space and back in time, detecting unknown galaxies, discovering the physical arrangement of the entire universe and tracing its evolution over time. It was a vision that seemed deranged to some of his contemporaries and Herschel himself 'fit for Bedlam', but it closely resembles the programme of modern astronomy.

Herschel was not simply a lone visionary but was troubled and challenged by a problem that beset all of our telescope makers. Our second theme of oppositions is between acquired or tacit skill and its codification. We have seen it already present in Newton, when he said that 'This Art of polishing will be better learn'd by repeated Practice than by my Description.' Smith set out the known descriptions of polishing procedures in 1738 in great detail. The exclusive manipulative skills of James Short and their reputation, and to some extent their mystery, were what allowed him to command much higher prices that his competitors. The physician John Mudge sought to publish the methods that were understood to be Short's in a substantial paper in the *Philosophical Transactions* and such was the importance attached to this that he—like Herschel and John Dollond—was awarded the Copley Medal.

But was such codification possible at all? Mudge admitted that his account could be only a partial one:

> ... I will describe it in the best manner I can; though many little circumstances which will be unavoidably omitted (and which at the same time are frequently essential to the success of a mechanic process) can only be supplied by actual experience.

Throughout the whole account, Mudge is detailed and loquacious, yet he doubts whether his literary technology is adequate for the purpose:

> ... for though I have endeavoured to be as particular as I can (I am almost afraid too much so) it is yet scarcely possible to supply a want of dexterity, arising from habit only, by the most laboured and minute description [10].

In Herschel's case we know that he attempted to be systematic in carefully recording his polishing experiments and seeking to develop a vocabulary of the differently named 'strokes' he used. There is even something of his former profession as a music teacher in the type of system he tried to develop.

Much of the private record of Herschel's polishing experiments has survived. A day-by-day journal fills four folio volumes preserved at the Royal Astronomical Society, and there have been attempts in the Herschel household to systematise the results into a treatise on polishing. The individual results have, for example, been copied on to many slips of paper, arranged into bundles, and recopied in a rational order as 'Results of Experiments on the Construction of Mirrors for Telescopes' in eight folders. There is also an essay 'On the Construction of Specula' in five folders. Nothing of this was published, and it is clear that Herschel struggled to synthesise and codify the skill and experience he acquired through many years of practice.

Herschel tried many combinations of materials in his alloys and a multitude of different polishing techniques, disciplining his manipulative routines according to a personalised mechanical syntax that allowed him to record his progress and not to lose any advance he might discover. For example, on 14 November 1778, when he was polishing an objective mirror for his 7-ft telescope, he noted: 'I used the divided reducing stroke of the 170th experiment, and in a very short time made it a most capital speculum.' It was when using this mirror that he would later discover Uranus [1].

Herschel was criticised for not publishing his polishing methods but the manuscript evidence suggests that he struggled and failed to achieve a secure codification that could have communicated his skills. One of the critics was Lord Rosse, who complained that polishing specula was an area where each individual had to begin afresh, struggling, as he put it, in a dark labyrinth. But although he resolved from the beginning of his work to change all that, his methods were not much more effectively communicated by text.

6 Lord Rosse

Rosse introduces our final dialectic, that between originality and credibility. Expanding Herschel's vision—still mainly the province of amateurs—led to telescopes so special and outrageous that they could not be replicated or their results repeated. If their observations depended to such an extent on individual mechanical skill and even using them required practice to develop individual observational acuity, how could the results and conclusions based on them maintain credibility far from their origins? Herschel might not have been able to describe to you how to make one of his telescopes but at least there was the possibility of buying one, and he did make a great many for sale, though it has to be said that observing with one was such a particular skill that very few of his clones were effective with other observers.

Rosse's policy, as well as describing his methods in detail, was to keep open house for observers and to invite whoever wanted to come to try his telescopes for themselves and to witness his results [6]. Being relatively inaccessibly located in the middle on Ireland, he was not overwhelmed with visitors, but worse, those who did come often found that they could not see what had been reported. This was either because the weather was poor or because they had no appropriate experience—none of what might be called the tacit skill for observing with a particular type of instrument, gained again only by practice.

Thomas Romney Robinson sought to bolster Rosse's credibility by other arguments. The telescope's builder and owner was a nobleman, who did not need commercial success as earlier makers had, since honour was the only currency of any importance in his economy. His word may have been reliable, but what of his hand and eye? Rosse himself stressed that there were no secrets here: he had had recourse to no commercial or even experienced makers, the only people who had worked on building his telescopes and making his mirrors being simple Irish peasants with no knowledge of optics. There was nothing to suspect: everything was open and evident.

Eventually, however, the fourth Earl, who continued the enterprise, had to admit that replication was quite impossible. Every time the great speculum was re-polished, its figure was completely new and optically the former telescope was replaced by a different one [2]. For all Rosse's spectacular success, it was clear that further progress would depend not only on design and technical originality, and on enthusiasm, commitment and determination, but also on more structural and organisational elements, on new systems of manufacture and protocols of research in a reformed telescope culture. We have reached the end of the line for the individual telescopic polymath: funder, designer, builder, mechanic and astronomer.

References

1. Bennett, J.A.: On the power of penetrating into space: the telescopes of William Herschel. J. Hist. Astron. **7**, 15–108 (1976)
2. Bennett, J.A.: A viol of water or a wedge of glass. In: Gooding, G., Pinch, T., Schaffer, S. (eds.) The Uses of Experiment, pp. 105–114. Cambridge University Press, Cambridge (1989)
3. Bennett, J.A.: Peter Dollond answers Jesse Ramsden. Sphaera **8**, 4–5 (1998)
4. Bennett, J.A.: Catadioptrics and commerce in eighteenth-century London. Hist. Sci. **44**, 247–278 (2006)
5. Bryden, D.J.: James Short and His Telescopes. Royal Scottish Museum, Edinburgh (1968)
6. King, H.C.: The history of the Telescope. Griffin, London (1955)
7. Millburn, J.R.: Benjamin Martin: Author, Instrument-make and 'Country Showman'. Noordhoff, Leiden (1976)
8. Millburn, J.R.: Wheelwright of the Heavens: The Life and Work of James Ferguson, FRS. Vade-Mecum, London (1988)
9. Millburn, J.R.: Adams of Fleet Street; Instrument Makers to King George III. Ashgate, Aldershot (2000)
10. Mudge, J.: Directions for making the best composition. Philos. Trans. **67**, 296–349 (1777). doi:10.1098/rstl.1777.0017
11. Newton, I.: Opticks. The Warnock Library, London (1704)
12. Smith, R.: A Compleat System of Opticks. Cornelius Crownfield, Cambridge (1738)

Exp Astron (2009) 25:43–61
DOI 10.1007/s10686-009-9154-9

HISTORICAL REVIEW

The great nineteenth century refractors

James Lequeux

Received: 4 October 2008 / Accepted: 24 February 2009 / Published online: 13 March 2009

Abstract The first modern refractor was built in 1824 by Fraunhofer; its diameter was 23 cm. The sizes of the largest refractors grew linearly with time, culminating in the 102-cm Yerkes telescope (1897) and the 125-cm horizontal refractor with siderostat for the 1900 Universal Exhibition in Paris. This was the end of the giants; however long-focus photographic or visual refractors with diameters 60-70 cm were still being constructed until 1972. In parallel, many short-focus, wide field astrographs were built after 1893. We recall the scientific uses of the large refractors, and discuss why these cumbersome and expensive instruments were not dethroned by the silvered-glass reflecting telescopes until late, in spite of the advantages of the latter.

Keywords Telescope · Refractor · Reflector · Astrograph · History of astronomy

1 The first modern refractors

During the whole nineteenth century, the main work of astronomical observatories consisted in meridian observations of stars, planets and satellites. However, refractors of various sizes were also built for discovering and observing solar-system objects, and for observations of visual binaries and parallaxes of nearby stars; small refractors equipped with heliometers were also used for determining parallaxes; other small instruments specialized in searching for comets. Any observatory of some importance needed one or several refractors. Reflecting telescopes were little used before 1860, with the

J. Lequeux (✉)
LERMA, Observatoire de Paris, 61 Av. de l'Observatoire, 75014 Paris, France
e-mail: james.lequeux@obspm.fr

very notable exceptions of those of William Herschel (1738–1822) and of Lord Rosse (1800–1867).

The first modern refractor was built in 1824 by Joseph von Fraunhofer (1787–1830) for the Russian Imperial Observatory in Dorpat, now Tartu in Estonia. Fraunhofer made his own flint and crown glass for the 23-cm achromatic objective, ground the lenses and designed the German-type equatorial mount driven by a clockwork. The result (Fig. 1) was of outstanding quality, so that most later equatorials are essentially scaled versions of Fraunhofer's refractor. This telescope was timely, because astronomers were realizing that observations of the orbits of visual binary stars could yield their masses, provided that their distances were known: in 1827, Félix Savary (1797–1841) had shown how to calculate their orbits. Wilhelm Struve (1793–1864) used the Dorpat refractor to observe many visual binaries, and attempted to measure the distance of stars through their visual parallaxes. He also obtained accurate values for the diameters of Jupiter's satellites.

Another early equatorial, the Northumberland telescope (Fig. 2), with a good 30-cm objective by Cauchoix, was installed in Cambridge (England) in

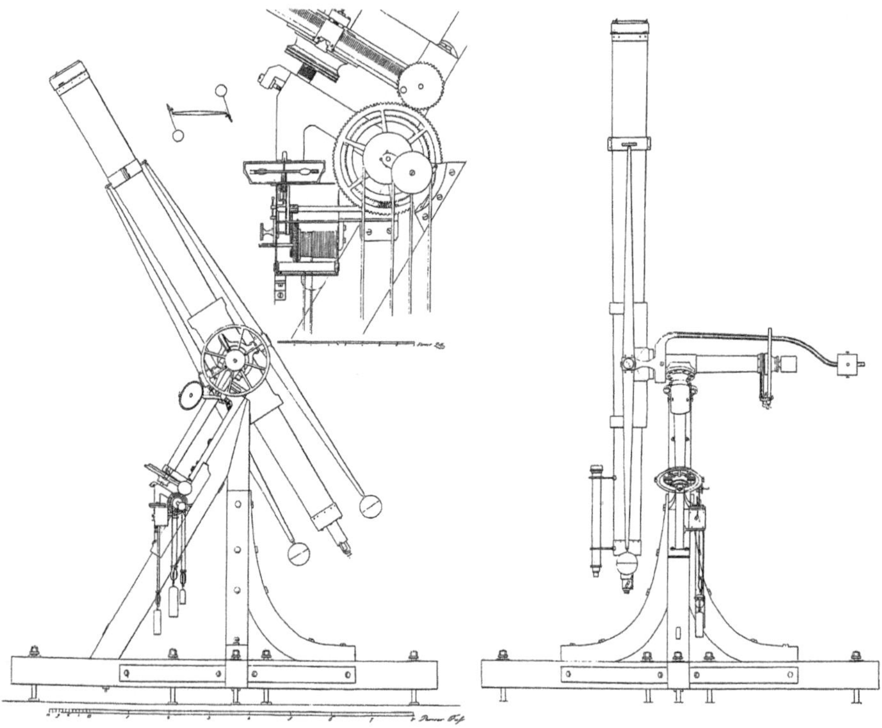

Fig. 1 The Fraunhofer refractor in Dorpat (1824, diameter 23 cm, $f/d = 18$). Details of the driving sidereal time clockwork with fan blades governor are shown at the *top of the left image*. The levers with spherical weights on the sides of the tube compensate for its flexion when the inclination changes. This instrument has been recently restored

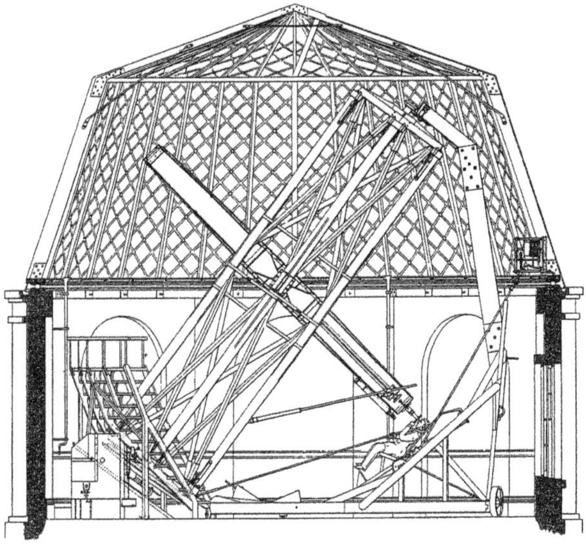

Fig. 2 The Northumberland telescope in Cambridge, England (1835, diameter 30 cm, $f/d = 20$)

1835 on a mount designed by George Bidell Airy (1801–1892). This English-type mount, in spite of clear mechanical advantages, had little success outside England because the polar regions cannot be observed.

A few years later, in 1839, the German firm Merz & Mahler, successor of Fraunhofer & Utzschneider, built a 15-inch (38-cm) refractor very similar to Fraunhofer's for the brand-new observatory of Pulkovo, near St-Petersburg in Russia. This was such a good instrument that the Harvard Observatory ordered a similar one, which was delivered in 1847.

2 The race for giant refractors

Some people were already dreaming of larger refractors. In 1852, a wealthy Englishman, Rev. Mr. Craig, installed a 61-cm refractor on a very primitive mount;[1] this was too large for the time, the objective was of very poor quality and the site terrible, so that the telescope was completely useless and was dismantled after a few years. Four years later, one could see in Paris a 52-cm refractor which was not a success either (we will comment later on this instrument).

However, the time was ripe for large professional refractors. Table 1 gives a list of 72 refractors with diameters larger than 40 cm, ordered by date of installation, that we believe to be reasonably complete. We can see that while increasingly large refractors were constructed in Europe, Americans were not late in the competition, essentially due to the skill of Alvan Clark (1804–1887).

[1]See for an image [1, p. 225]

Table 1 List of large refractors

Initial location	Institution	Date	Diameter (cm)	F/d	Mount	Glass	Objective	Note
Dorpat = Tartu, Estonia	Imperial Obs.	1824	23	18	Fraunhofer	Fraunhofer	Fraunhofer	
Cambridge, UK	Observatory	1835	30	20	Airy (English)	Guinand	Cauchoix	1
Pulkovo, Russia	Observatory	1839	38	18	Merz & Mahler	Merz	Merz	
Harvard, USA	Observatory	1847	38	18	Merz & Mahler	Merz	Merz	
Wandsworth , UK	Private	1852	61	?	Rennie	?	Slater	2
Paris, F	Private	1856	52	29	Porro	Guinand	Porro	3
Walworth, UK	Private	1858	54	16	Buckingham	Common, Wray	?	4
Moscow, Russia	Observatory	1859	49	~ 40	Merz	Merz	Merz	5
Chicago, Ill.	Dearborn Obs.	1862	47	17	Clark	Chance	Clark	6
Buffalo, N.Y.	Private	1862	41	16	Fitz	Chance	Fitz	
Newcastle, UK	Private	1869	64	14	Cooke	Chance	Cooke	7
Cambridge, UK	Solar Phys. Obs.	1871	63	14	Grubb	Chance	Grubb	
Washington DC	Naval Obs.	1877	66	15	Warner & Swasey	Chance	Clark	8
Madison, Wis.	Washburn Obs.	1879	40	16?	Clark	Chance	Clark	
Strasbourg, F	Obs. de l'Univ.	1880	49	14	Repsold	Merz	Merz	
Vienna, Austria	Univ. Sternwarte	1880	69	15	Grubb	Feil	Grubb	
Daramona, Ir	Private	1881	61	?	Grubb	Feil	Grubb	9
Milan, I	Oss. di Brera	1881	49	14	Repsold	Merz	Merz	
Princeton, N.J.	Halstead Obs.	1882	58	16	Clark	Feil	Clark	10
Rochester, N.Y.	Private	1882	41	16?	Clark	Feil	Clark	11
Pulkovo, Russie	Observatory	1884	76	19	Repsold	Feil	Clark	
Charlottesville, Va.	Mc Cormick Obs.	1887	66	15	Clark	Feil	Clark	
Nice, F	Observatory	1887	76	24	Gautier	Feil	Henry brothers	
Mount Hamilton, Cal.	Lick Obs.	1888	91	19	Warner & Swasey	Feil	Clark	12
Paris, F	Observatoire	1890	60 vis.	30	Gautier (coudé)	Feil	Henry brothers	
			60 phot.	30		Mantois		

La Plata, Argentina	Univ. Obs.	1890	43	22	Gautier (coudé)	Mantois	Henry brothers	
Northfield, Minn.	Goodsell Obs.	1891	41	16	Warner & Swasey	Mantois	Brashear	
Nice, F	Observatoire	1892	40	23	Gautier (coudé)	Mantois	Henry brothers	
Arequipa, Peru	Harvard station	1893	61 phot.	5.7	Clark	Mantois	Clark	13
Manille, Philippines	Weather Bureau	1893	51	?	Saegmuller	Merz	Merz	14
Greenwich, UK	Royal Obs.	1893	71	12	Airy, 1859 (English)	Mantois	Grubb	15
Denver, Co	Chamberlin Obs.	1894	51	17	Saegmuller	Mantois	Clark	
Flagstaff, Ariz.	Lowell Obs.	1894	46	?	Clark	Mantois	Clark	
Heidelberg, D.	Landessternwarte	1894	41phot.x2	5	?	Mantois	Brashear	16
Greenwich, UK	Royal Obs.	1896	66 phot.	10	Grubb	Mantois	Grubb	17
Upper Darby, Pa.	Flower Obs.	1896	46	?	Warner & Swasey	Mantois	Brashear	18
Meudon, F	Observatoire	1896	83 vis.	20	Gautier	Mantois	Henry brothers	
			62 phot.	26		Feil ?		
Berlin, D	Popular Obs.	1896	68	31	Hoppe	Schott	Steinheil	
Flagstaff, Ariz.	Private	1896	61	16	Clark	Mantois	Clark	19
Williams Bay, Wis.	Yerkes Obs.	1897	102 vis.	19	Warner & Swasey	Mantois	Clark	
Zo-Se, China	Jesuit Obs.	1897	40 phot.	17	Gautier	Mantois	Henry brothers	
			40 vis.	17				
Athens, Gr	National Obs.	1899	40	12	Gautier	Mantois	Henry brothers?	
Potsdam, D	Astroph. Sternw.	1899	80 phot.	15	Repsold	Schott	Steinheil	
			50 vis.	25				
Paris, F	Univ. Exhib.	1900	125 phot.	46	Gautier (cœlostat)	Mantois	Gautier	20
Cape Town, S. Afr.	Royal Obs.	1901	61 phot.	11	Grubb	Mantois	Grubb	
			46 vis.	15				
Oxford, UK	Radcliffe Obs.	1901	61 phot.	11	Grubb	Mantois	Grubb	21
			46 vis.	15				

Table 1 (continued)

Initial location	Institution	Date	Diameter (cm)	F/d	Mount	Glass	Objective	Note
Amherst, Mass.	Wilder Obs.	1903	46	17	Clark	Mantois	Clark	
Cincinnati,Ohio	Univ. Obs.	1904	41	16	Clark	Mantois	Clark	
Swarthmore, Pa.	Sproul Obs.	1906	61	18	Warner & Swasey	Schott	Brashear	
Rio de Janeiro, Br	National Obs.	1907	46 phot.	?	Cooke	Mantois	Cooke	
Harvard, Mass.	Oak Ridge Obs.	1909	41 phot.x2	5.2	Metcalf	?	Metcalf	
Berlin-Babelsberg, D	Univ. Sternw.	1912	65	16	Zeiss	Schott	Zeiss	
Madrid, Spain	Obs. Astr. Nacion.	1912	41	?	Grubb	Mantois	Grubb	
Pittsburg, Penn	Allegheny Obs.	1912	76 phot.	19	Warner & Swasey	Schott	Brashear	22
Sétif, Algeria	Private	1913	52	12	Schaer	?	Schaer	23
Oakland, Cal.	Chabot Obs.	1914	51	17	Warner & Swasey	Schott	Brashear	
Berlin-Babelsberg, D	Sternwarte	1915	40 phot.	?	Toepfer?	Schott	Steinheil	
Bergedorf, D	Hamburg Obs.	1916	60 vis.	15	Repsold	Schott	Steinheil	
			60 phot.	15				
Potsdam, D	Astroph. Obs.	1924	60 phot.	23	Zeiss (coelostat)	Schott	Zeiss	24
Middeltown, Conn.	Van Vleck Obs.	1925	51	17	Warner & Swasey	Schott	Clark	
Johannesburg, S. Afr.	Union Obs.	1925	67	16	Grubb	Mantois	Grubb	25
Johannesburg, S. Afr.	Union Obs.	1925	67 phot.	16	Schlesinger	Corning	McDowell	26
Santiago, Chile	Observ. Nacional	1925	61	?	Grubb	Mantois	Grubb	
Bloemfontein, S. Afr.	Michigan U.	1928	69 vis.	18	Zeiss	Schott	Zeiss	27
Lembang, Java	Bosscha Sternw.	1928	60 vis.	18	Zeiss	Schott	Zeiss	
			60 phot.	18				
Belgrade, Serbia	Univ. Obs.	1929	65	16	Zeiss	Schott	Zeiss	
Mitaka, Japan	Tokyo Obs.	1930	65 phot.	16	Zeiss	Schott	Zeiss	
			38 vis.	27				
Stockholm, Sweden	Univ. Obs.	1931	61 phot.	13	Grubb-Parsons	Schott	Grubb-Parsons	
			46 vis.	16				

Stockholm, Sweden	Obs. Ac. Sci.	1932	40 phot.	5	Zeiss	Schott	Zeiss	
Nice, F	Observatoire	1933	40 phot. x2	5	Zeiss	Schott	Zeiss	28
Uccle, B	Obs. Royal	1933	40 phot. x2	5	Zeiss	Schott	Zeiss	28
Johannesburg, S. Afr.	Union Obs.	1934	41 phot. x2	5.5	Grubb-Parsons (English mount)	Schott	Grubb-Parsons	
Castel Gandolfo, I	Vatican Obs.	1935	40	5	Zeiss	Schott	Zeiss	28
Mount Hamilton, Ca	Lick Obs.	1942	61 phot.x2	6,7	Warner & Swasey	Schott	Warner & Swasey	29
Llano del Hato, Ven.	Observatory	1955	65 vis.	16	Zeiss	Schott	Zeiss	
Kamitakara, Japan	Kyoto Univ. Obs.	1972	65 vis.	16	Zeiss	Schott	Zeiss	

Diameter larger than 40 cm, with the exception of the first 4

1 Northumberland telescope; 2 Craig refractor, useless; 3 Ignazio Porro's private observatory. No significant scientific result; 4 Moved to Edinburgh; 5 New objective in 1866; 6 Installed only after civil war, moved to Evanston (Ill) in 1889; 7 Newall telescope. Moved to Cambridge (UK) in 1891 then to Athens (National Observatory, Penteli station) in 1957; 8 Initially mount and objective by Fitz, later re-worked by Clark; 9 Radcliffe refractor, moved to the Ratcliffe Observatory, Oxford, then to the University of London Observatory at Mill Hill in 1929; 10 Original mount replaced; 11 Later, Warner Observatory. Telescope moved to Mount Lowe in 1894; 12 84-cm correcting lens for photography, glass by Mantois, figured by Clark; 13 Bruce doublet astrograph. First large short-focus, wide field photographic refractor with guiding telescope. Installed in 1896 and moved to Bloemfontein in 1927 until 1950; 14 Lens destroyed during WW2; 15 Moved to Herstmonceux in 1957, back to Greenwich in 1971; 16 First triple astrograph: two photographic refractors for two different colors and guiding visual refractor; 17 With a 76-cm Common reflector on the same German mount; 18 Moved to New Zealand in 1962, not in use; 19 Now Lowell Observatory; 20 Only used for demonstrations; 21 Moved to the University of London Observatory at Mill Hill in 1935; 22 30-cm Brashear lens at the middle of the tube for visual observations; 23 Jarry–Desloges Observatory; 24 Einstein tower, mostly for solar observations. Five solar coronagraphs with single-lens objectives 53 cm in diameter have been installed in Eastern countries after WW2: Pulkovo Observatory (Kislovodsk station, Russia), Sayan Observatory (Izmiran, Russia), Crimean National Observatory (Nauchny, Ukrainia), Georgian National Observatory (Mt. Kanobili, Georgia) and Debrecen Observatory (Hungary). The first one was constructed at Pulkovo by V.G. Schreiber near 1966; 25 Innes telescope; 26 Yale telescope. Moved to Mount Stromlo in Australia and destroyed by fire in 2003; 27 Lamont–Hussey refractor. Additional lens for photography; 28 Similar double astrographs. There are others in the former Soviet Union : Kitab (now Uzbekistan), Ussurik (Russia), Abatsumani (Georgia), Golossevo-Kiev (Ukraine), Crimean astrophysical observatory (Ukraine); 29 Carnegie double (actually triple) astrograph. Refractor for the blue in 1942, for the yellow in 1961

Built from various sources, mainly: [1–9]; Repsold and Andrews give an abundant iconography

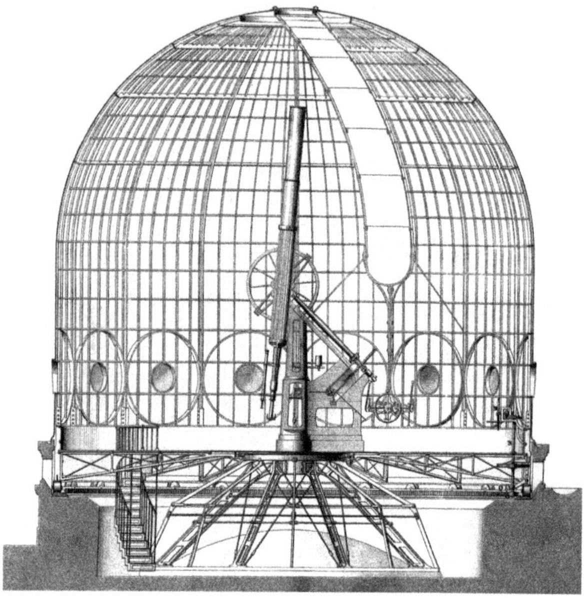

Fig. 3 The dome of the Paris Observatory, built by Travers in 1847

However they depended on Europe for the supply of glass for the objectives until late in the twentieth century.

We now comment on some landmarks in the series.

The Newall telescope installed by a rich amateur (1812–1889) in Newcastle in 1869 (64 cm, $f/d = 14$), although built in England by Cooke & Sons, had a German mount.[2] This was the largest refractor at this time, but the site was very bad. This telescope was moved in 1891 to Cambridge where it enjoyed a second life, then to Athens in 1957. Like most refractors of the second half of the century, it was installed inside a hemispherical dome. The first such large dome was built in 1847 on a tower of the Paris Observatory (Fig. 3) in order to house a 38-cm refractor (Brunner/Lerebours, 1857) which was a failure because of a damaged objective.

The Lick 91 cm, $f/d = 19$ telescope is a very impressive Warner & Swasey/Clark giant installed in 1888. For photography, a 84-cm lens made by Clark could be inserted. For the same purpose, a 30-cm lens could be placed midway inside the tube of the Allegheny refractor (1912, 76 cm, $f/d = 19$), and another one in the 69-cm Lamont–Hussey refractor of Bloemfontein in 1928. These lenses were little used after orthochromatic plates and yellow filters became available.

The Greenwich 71 cm, $f/d = 12$ equatorial was installed in 1893 in replacement of a 32-cm Merz of 1859 on the original English mount designed by Airy—a good testimony for his engineering skill. It can still be seen in a new dome. This was the second giant refractor which could in principle be

[2]For a contemporary description (in French, with figure) see [10]

Fig. 4 The Meudon double equatorial (1896, 83 cm visual, $f/d = 15$; 62 cm photographic, $f/d = 26$)

used for visual and photographic observations, by reversing and repositioning the objective lens (a similar system was implemented the following year on the Van Vleck refractor of the Chamberlin Observatory). This system proved flawed in practice, at least at Greenwich, so that the telescope was only used for visual double-star observations. Another 66-cm refractor with a shorter f/d of 10, specifically designed for photographic observations, was installed at Greenwich by Grubb in 1896.

The Meudon double equatorial of 1896 (Fig. 4; Gautier/Henry brothers,[3] visual objective 83 cm, $f/d = 20$; photographic objective 62 cm, $f/d = 26$) has been mainly used visually and for photographic spectroscopy, rarely for direct photography because of its extremely large focal ratio.[4]

[3]Paul (1848-1905) and Prosper (1849-1903) Henry always worked together.

[4](For details on this instrument see [11, 12])

Fig. 5 The Yerkes equatorial (1897)

The Yerkes telescope (Fig. 5), built like the Lick refractor by Warner & Swasey with a Clark objective (102 cm, $f/d = 19!$) is the largest of all giant refractors. Completed in 1897, it probably would never have existed if two crown and flint 42-inch blanks had not been available at Clark & Sons: they were ordered from Mantois in Paris by the University of Southern California, which could not pay for them.[5]

One can judge the optical quality of some refractors by looking at Fig. 6, which shows the results of Hartmann tests secured for several objectives, in particular for the excellent Yerkes one which has been studied in much detail [13].

Figure 7 shows the evolution of the size of large refractors. One sees that this size grows linearly with time up to the Yerkes refractor: presumably builders like the Clarks scaled up the same design a few inches next time to make a larger telescope.

There were controversies in the 1880s about the use of giant refractors. With small refractors, observers saw objects like the satellites of Mars which were invisible with bigger instruments and claimed that the giants were useless. A controversy raged over the channels of Mars that the Italian astronomer Giovanni Schiaparelli (1835–1919) "discovered" in 1877 with a 21-cm refractor

[5]See http://astro.uchicago.edu/yerkes/history.html

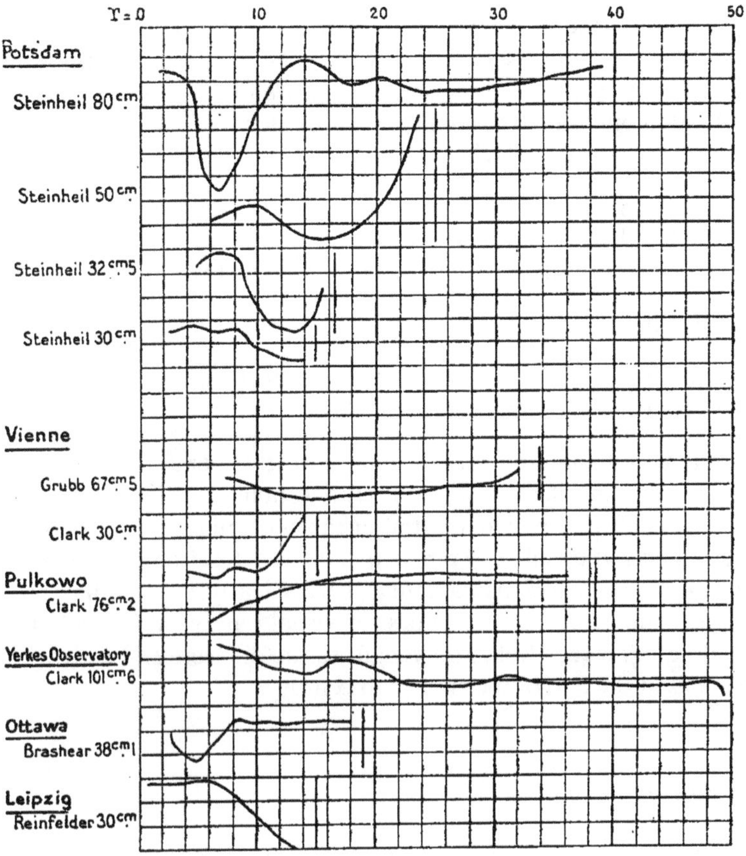

Fig. 6 Results of the Hartmann test for several objectives. The deviations of the focal length (*1 square* = 2 mm) are plotted for zones of different radii (abscissae in centimeters) [14]

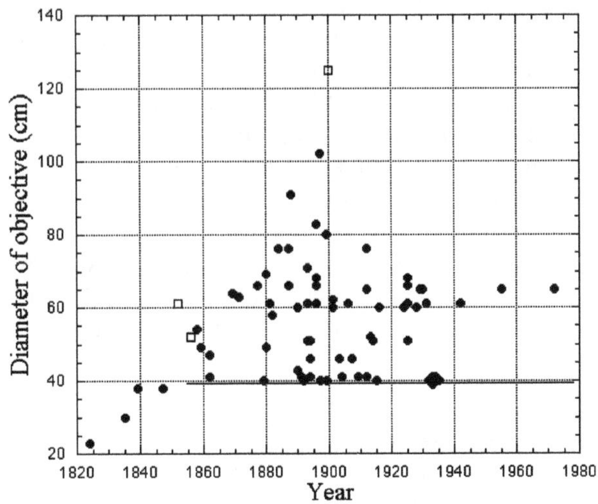

Fig. 7 The size of refractor objectives as a function of date of completion. After 1847, only refractors with diameters larger than 40 cm are plotted. There were many smaller telescopes. The *empty squares* correspond to instruments which did not produce significant results. This plot does not include several 40-cm astrographs and 53-cm diameter, single-lens solar coronagraphs built after 1940 in Eastern countries

(he even claimed them to be double in 1881). Percival Lowell (1855–1916) also pretended to have seen them with the Flagstaff refractors in 1894. But observations with the best giant refractors like the 91-cm Lick and the 83-cm Meudon ones failed to confirm their existence. As written by one of the Meudon observers, G. Millocheau,

> "The channels are made of small dark spots with jagged edges, separated from each other, and which easily seem to form a single line if the resolving power of the instrument is insufficient [15]."
>
> Anton Pannekoek (1873–1960) writes [16]:
>
> "Intricate complexes of objects cannot be distinguished by low powers and are conceived in simplified forms."

It is often said that the era of giant refractors ended in 1912 with the 76-cm Allegheny one. This is not really true. A number of long-focus refractors with diameters 60–67 cm were built after this date, in particular by Zeiss. This firm even managed to sell a 65 cm visual, $f/d = 16$ refractor to Venezuela in 1955 and another one to Japan as late as 1972. Clearly, even at that time, many astronomers still preferred refractors for visual binary work, while reflectors can also do the job very well. Many giant refractors were still actively used by professionals at the end of the twentieth century.

3 Curiosities and oddities

Several giant refractors did not conform to the usual designs. It is of interest to discuss a few of them briefly.

In 1856, an Italian builder of scientific instruments then living in Paris, Ignazio Porro (1801–1875), erected on his own premises an enormous refractor, 52 cm in diameter with a 15-m tube ($f/d = 29!$) on a simple mount.[6] The objective was said to be good, photographs of lunar and solar eclipses were obtained in 1856 and 1858 respectively, but the telescope was certainly difficult to use and it is said that its objective broke after a few years.

Another extremely-long focus monster was installed in 1896 in Berlin[7] Its 68-cm Steinheil visual objective was placed at the end of a 20-m tube on a strange equatorial mounting built by Hoppe (Fig. 8). Obviously they were looking for the largest possible image scale; an advantage of the design was that the observer was standing at the crossing of the axes and did not have to move during and between observations. A 110-cm, $f/d = 6.4$ objective was planned for photography but was never constructed.

[6] Fuentès [17], with a picture of the telescope.

[7] Scholl [18], with a photograph.

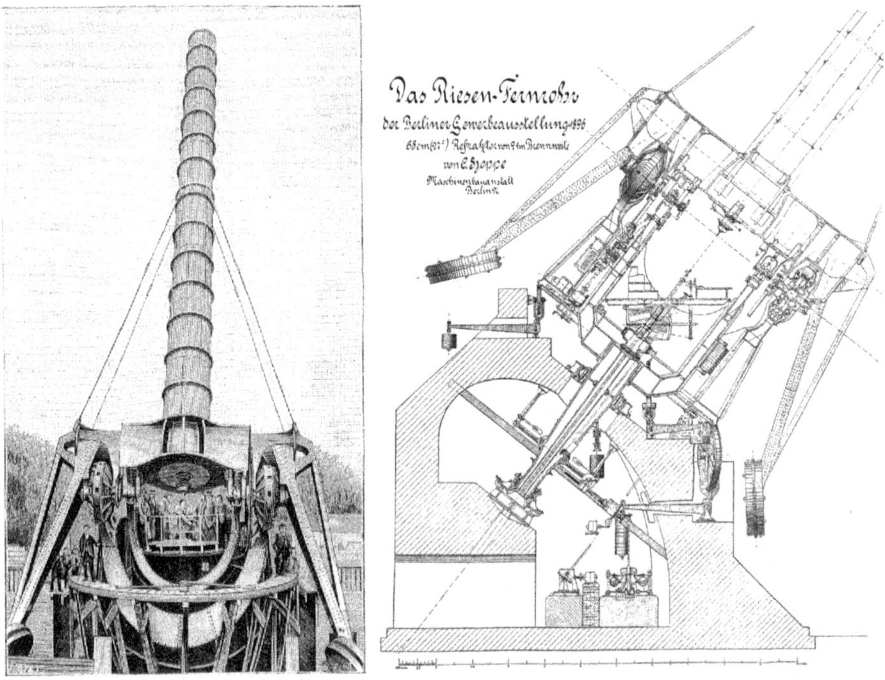

Fig. 8 The great Berlin equatorial of 1996. *Left*, a contemporary photograph. *Right*, a cut of the equatorial mount (*bottom scale* in metres). The observer stood on the platform at the middle

Another instrument designed for the comfort of the observer was the coudé équatorial of Maurice Lœwy (1833–1907).[8] After the success of his 27-cm diameter prototype, a large instrument was built in Paris in 1890 (Fig. 9). It had two interchangeable 60-cm objectives, for visual and photographic observations respectively. Strangely enough, the coudé équatorial was originally conceived as an astrometry instrument for building a star catalogue with the possibility "to study the zones on the right and the left of the meridian". It soon proved unsuitable for this purpose and was mainly used for photography or for searching for comets. The principle had much success in France and in Algeria where seven coudé equatorials were built in total; there were two others in Vienna and in La Plata (Argentina) respectively.

The largest refractor ever built was another French instrument (the French seem to have had a special taste for unorthodox refractors). It was shown at the Paris Universal Exhibition of 1900.[9] It was not an initiative of astronomers, but of an influential political man. This monster consisted of a Foucault-type

[8]For a description see [19, 20]. Pictures of these instruments can be found easily through google.
[9]Launay [21]. Also [22, 23], with a picture.

Fig. 9 The large coudé équatorial of the Paris Observatory (1890, 60 cm, $f/d = 30$; Gautier/Henry brothers). The observer sits in a fixed position and looks along the polar axis. The objective is visible in front of the cube at the end of the shorter tube. Inside this cube, a flat mirror at 45° sends the light into the tube; another 45° flat mirror at the intersection of the two tubes receives the light and sends it to the eyepiece or to the photographic plate. The whole instrument rotates around the polar axis, and the cube can rotate upon itself. In other models, the objective is behind the cube

siderostat with a 2-m diameter flat mirror feeding a fixed horizontal, 58 m long and 125 cm diameter refractor (Fig. 10). This should have been a double refractor, but only the photographic part was completed. Some photographic and even visual observations were secured with this instrument before it was dismantled, everyone judging it too cumbersome for practical use. The mirror and the two lenses of the photographic objective survive at the Paris Observatory.

4 Photography with large refractors; the astrographs

Even with photographic objectives corrected for the blue, or later with visual objectives, yellow filters and orthochromatic plates, the great equatorials were not well suited to photography because of their very large focal lengths. Only high-brightness objects like planets could be photographed, so that photographic objectives were mostly used for spectrography. It is thus not surprising to see that visual observations, mostly of double stars, dominated the use of these giant instruments. Reducing the f/d ratio as for the 1896 66-cm photographic equatorial at Greenwich helped to some extent, but chromatic aberration was then enhanced. In order to remedy this situation, short-focus,

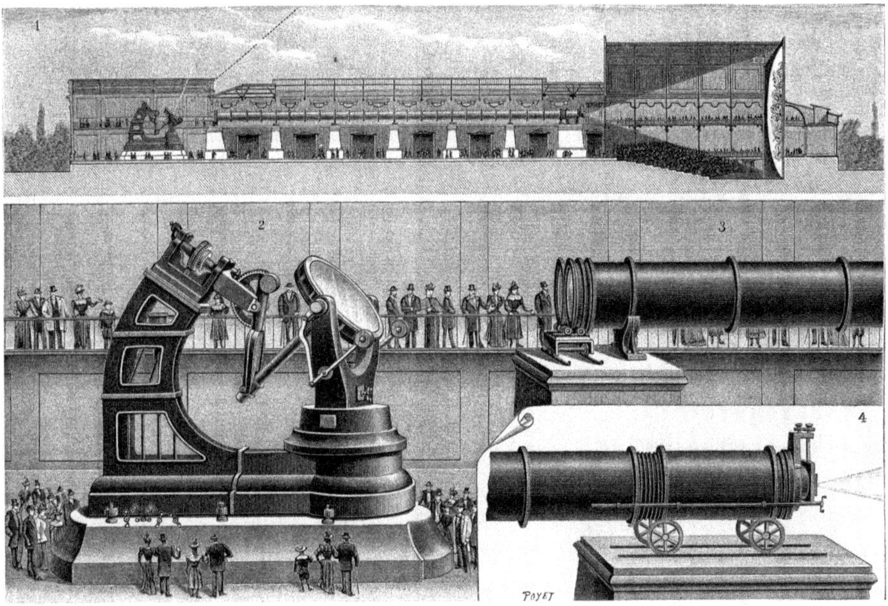

Fig. 10 The great horizontal refractor of the Paris Universal Exhibition of 1900. Built by Gautier including the optics, it consisted of a 2-m diameter siderostat feeding a 125-cm diameter, $f/d = 46$ refractor. This picture represents the instrument as planned, with an enormous image of the Moon projected on a screen by the visual telescope. However, only the photographic part was built, with a single objective and tube. The mechanics of the siderostat were said to be reliable, and the optics fairly good

wide-field photographic lenses were constructed after 1888, preferably, but not always, used with filters to minimize chromatic aberration: these are the astrographs. Reflectors could not compete with astrographs in spite of their lack of chromatism because the optical combinations allowing wide fields were not available: the first Ritchey–Chrétien reflector was only built in 1927 although the principle dates from 1910, and the Schmidt camera was invented in 1930. In astrographs, the photographic refractor is coupled to a visual guiding refractor, often of large diameter so that one speaks of double astrographs. Some even have two photographic refractors adapted to two-color observations and are called triple astrographs.

The first sizeable, and actually the largest astrograph ever made together with the Lick astrograph of 1942 (61 cm, $f/d = 5.5$ and 6.7 respectively), was the Bruce doublet built by Clark in 1893 for the Harvard station in Arequipa, Peru: plates taken with this instrument were used by Henrietta Leavitt for her famous study of cepheids in the Small Magellanic Cloud. In spite of the advent of wide-field instruments like Schmidt or Maksutov telescopes, many astrographs continued to be constructed until late in the twentieth century. We know of 14 astrographs with diameters larger than 40 cm, including many astrographs built by Zeiss for the former Soviet Union (see the Table and its Note 28). The list is perhaps still incomplete.

5 Refractors versus reflectors

The first "serious" reflecting telescope with a silvered glass mirror was built in 1862 by Léon Foucault (1819–1868) for the Marseilles Observatory. Its 80-cm diameter, $f/d = 5$ mirror was excellent, and in general the instrument can be considered as nearly perfect:[10] it is the reflecting equivalent of the Fraunhofer refractor. It was used for a century as a visual instrument with little modification, in very innovative ways. Édouard Stéphan (1837–1923) used it from 1871 to 1884 to find 800 high-brightness galaxies. They include his famous quintet and are listed in Dreyer's *New General Catalogue*. From 1906 to 1962 the telescope was used by Robert Jonckheere (1888–1974) to discover 3,350 new binary stars, a proof that reflectors can indeed be used for double-star studies. In 1873, following an idea of Hippolyte Fizeau (1819–1896), Stéphan attempted to use it to measure interferometrically the diameter of a number of stars, and in 1914 Charles Fabry (1867–1945) and Henri Buisson (1873–1944) obtained with this telescope the first astronomical Fabry-Pérot interferogram on the Orion Nebula.

One could imagine that silvered-glass reflecting telescopes would have rapidly dethroned refractors of the same size, the more so because they are particularly suited for photography (the Foucault telescope was too early to have been equipped for it). But this was not the case. It seems that the contemporaries did not know or did not acknowledge the outstanding qualities of the 80-cm reflector, and several of the following reflectors were failures. One of these was the great Melbourne telescope of 1869, with its 122-cm speculum mirror; the speculum was one of the reasons for its failure, but more important was the unfortunate choice of an extremely large f/d of 41 for its Cassegrain combination, which precluded photography and even spectroscopy [25]. George W. Ritchey (1864–1945) wrote in 1904:[11]

> "I consider the failure of the Melbourne reflector to have been one of the greatest calamities in the history of instrumental astronomy: for by destroying confidence in the usefulness of great reflecting telescopes, it has hindered the development of this type of instrument, so wonderfully efficient in photographic and spectroscopic work, for nearly a third of a century."

Another failure was that of the Edinburgh 61-cm silvered glass mirror reflector. Built in 1872, this instrument was placed in a too small, very crowded dome and was abandoned in 1879. The Toulouse Observatory had an 83 cm reflector installed in 1875, with a good silvered mirror by the Henry brothers, but the staff was new and inexperienced so that little science was produced. In Paris, a 120 cm reflector was erected in 1877 but Foucault, who had been expected to polish the mirror, was dead: his successor did not have his skills

[10] For details, see [24].

[11] Ritchey [26]; cited by King, *op. cit.*, p. 267.

and the silvered glass mirror was very bad. Apparently these successive failures overwhelmed the success of the 80-cm reflector. The first really successful reflector after it was the refurbished Crossley telescope at Lick, in 1896. The photographs taken with it are famous.[12] But as long as photography was not developed, in spite of some exceptions,

> "The trouble with reflectors as used visually was that within his own lifetime William Herschel had virtually exhausted their possibilities... All that remained was for John Herschel to extend his father's surveys to the southern sky, and for Rosse to discover spiral structure" (Sidney Gascoigne[13]).

Moreover, the builders of reflecting telescopes themselves acknowledged some advantages of the refractors:

> "Contrary to my expectations, I had to admit that, in certain aspects, the refractor was manifestly superior to the reflector." (Foucault 1860)

William Tobin comments[14] that, for extended objects like planetary surfaces, residual chromatic aberration in refractors produces changes of tint which are easier to detect than changes of brightness.

On their side, the Henry brothers noticed that image instabilities are smaller in refractors because of their closed tube; open-air reflectors are good in this respect but only in the absence of wind. In order to remedy this problem they proposed a catadioptric telescope closed by a thin meniscus [28].

These are minor defects, however. The difficulty of construction and use of reflectors, and above all conservatism, are the main reasons for the long survival of refractors, as evidenced by the next two texts:

> "One should be aware that the use of these large silvered-glass mirrors is very delicate and requires very skilled and experienced observers... The smallest temperature inequality, the slightest condensation of water on the surface, the lightest asymmetry in the way the mirror is supported and squeezed suffice to affect images to the extent that observation becomes impossible... Furthermore, it will probably be necessary to re-silver the mirror every year after the bad season."
>
> Ernest Mouchez, director of Paris Observatory, 1879[15]

> "Refractors have always been found better suited than reflectors to the ordinary work of the observatories. They are, so to speak, of a more robust, as well as of a more plastic nature. They suffer less from

[12]For the early development of reflecting telescopes in the USA, see [27]

[13]Gascoigne, *op. cit.,* p. 106.

[14]Tobin, *op. cit.,* p. 263.

[15]Mouchez [29]. He was obviously trying to find excuses for the bad quality of the 120-cm reflector of the Observatory, but he is nevertheless right.

vicissitudes of temperature and climate. They retain their efficiency with fewer precautions and under more trying circumstances. Above all, they co-operate more readily with mechanical appliances, and lend themselves with far greater facility to purposes of exact measurements."

Agnes Clerke, scientific journalist, 1887.[16]

The last sentence is clearly an expression of conservatism, as there is no reason why refractors should be better than reflectors for these "exact measurements" whatever they are.

References

1. King, H.C.: The History of the Telescope. Dover, New York (1979)
2. Repsold, J.A.: Vol. 1: Zur Geschichte der astronomischen Messverkzeuge: von Purbach bis Reichenbach, 1450 bis 1830; Vol. 2: von 1830 bis um 1900, Reinicke, Leipzig, reed. (2004) Köln, Gerhard Sauer (1908, 1914)
3. Harper, W.E.: J. R. Astron. Soc. Can. **23**, 351–355 (1929)
4. Danjon, A., Couder, A.: Lunettes et Telescopes. Blanchard, Paris (1979)
5. Andrews, A.D.: Cyclopaedia of telescope makers. Ir. Astron. J. **20**, 102–183 (1992)
6. Andrews, A.D.: Cyclopaedia of telescope makers. Ir. Astron. J. **21**, 1–82, 167–250 (1993–1994)
7. Andrews, A.D.: Cyclopaedia of telescope makers. Ir. Astron. J. **22**, 43–96 (1995)
8. Andrews, A.D.: Cyclopaedia of telescope makers. Ir. Astron. J. **23**, 57–117, 215–242 (1996)
9. Andrews, A.D.: Cyclopaedia of telescope makers. Ir. Astron. J. **24**, 125–192 (1997)
10. de Fonvielle, W.: La lunette de M. Newall, La Nature 2^{d} sem., p. 216–218 (1874)
11. Dollfuss, A.: La Grande Lunette de Meudon. CNRS Éditions, Paris (2006)
12. Launay, F.: Un Globe-Trotter de la Physique Céleste, L'astronome Jules Janssen, p. 134–137. Vuibert, Paris (2008)
13. Fox, P.: An investigation of the forty-inch objective of the Yerkes telescope. Astrophys. J. **27**, 237–253 (1908)
14. Baillaud, B.: Rapport Annuel Sur l'état de l'Observatoire de Paris en 1912, p. 59. Gauthier-Villars, Paris (1913)
15. Fuentès, P.: Flammarion, Martien d'adoption. L'Astronomie **122**, 32–39 (2008)
16. Pannekoek, A.: A History of Astronomy, p. 378. Dover, New York (1989)
17. Fuentès, P.: L'affaire Porro. L'Astronomie **111**, 270–272 (1997)
18. Scholl, A.: Le grand équatorial de l'exposition de Berlin. La Nature 1st semester, 209–210 (1897)
19. Fraissinet A.: Le grand équatorial coudé de l'Observatoire de Paris. La Nature 1st semester, 305–307 (1891)
20. Loewy, M.: Description d'un nouveau système de telescope. C. R. Acad. Sci. **118**, 1295–1299 (1894)
21. Launay, F.: The great Paris exhibition telescope of 1900. J. Hist. Astron. **38**, 459–475 (2007)
22. Gautier, P.: Annuaire du Bureau des longitudes, C1–C26 (1899)
23. Barré, L.: La grande lunette de 1900. La Nature 1st semester, 167–170 (1899)
24. Tobin, W.: The Life and Science of Léon Foucault, chap. 12. Cambridge University Press, Cambridge (2003)
25. Gascoigne, S.C.B.: The great Melbourne telescope and other 19th-century Reflectors. Q. J. R. Astron. Soc. **37**, 101–128 (1996)
26. Ritchey, G.W.: On the modern reflecting telescope. Smithsonian Contributions to Knowledge **34** (1904)

[16]Cited by King, *op. cit.*, p. 246.

27. Osterbrock, D.E.: The Quest for more photons: How reflectors supplanted refractors as the monster telescopes of the future at the end of the last century. Astron. Q. **5**, 87–95 (1985)
28. Henry P. & P.: Nouveau télescope catadioptrique. La Nature 1st semester, 307–308 (1879)
29. Mouchez, E.: Rapport annuel sur l'état de l'Observatoire de Paris. Gauthier Villars, Paris (1879)

Miscellaneous

Acknowledgements Publications available through asdwww.harvard.edu are [3, 5–9, 13, 25, 27], that through gallica2.bnf.fr is [20] and those through cnum.cnam.fr are [10, 18, 19, 21, 23, 28].

Exp Astron (2009) 25:63–77
DOI 10.1007/s10686-009-9146-9

HISTORICAL REVIEW

In the grip of the big telescope age

David H. DeVorkin

Received: 19 January 2009 / Accepted: 20 January 2009 / Published online: 5 February 2009

Abstract George Ellery Hale was a man of many dreams. One of his most persistent was to find the means to collect as much light as possible, but there is another element in his designs for the modern astrophysical observatory that has even greater significance, as it defines and distinguishes the practice of astrophysics from that of classical astronomy. Here we examine factors that either impeded or drove the acceptance of reflectors over refractors around the turn of the twentieth century at the outset of what may best be called the "Hale era." This commenced in the late nineteenth century, when the first large multi-focus photographic reflectors emerged during the reign of the great refractors. It lasted through to the onset of World War II when astronomical practice was dominated by ten reflectors with mirrors between 60 and 100 in., and was about to add one more whose surface area would almost double that of all the rest combined. We will touch upon how design choice reflected both scientific priorities and technological limitations.

Keywords Hale era · Telescope · Reflector · Refractor · Coudé

1 Introduction

There are numerous excellent surveys of the development of the reflector in the first half of the twentieth century [7, 10, 27, 28, 31, 33, 41, 51, 56]. A few individual telescopes have received solid attention, like the 200-in. [17, 59], and the 100-in. at Mount Wilson [40, 58], the 84-in. McDonald reflector [16],

D. H. DeVorkin (✉)
National Air and Space Museum, Smithsonian Institution, Washington, DC, USA
e-mail: devorkind@si.edu

to name a few. Most of these works emphasize four elements that play a role in telescope building in the twentieth century:

1. Means of achieving maximum effective light-gathering power (making bigger mirrors with fewer reflections to the focal plane)
2. Mechanical solutions for mounting structures and focal plane accessibility
3. Modes of patronage (private and corporate philanthropy, state support)
4. Site testing (accessibility, political stability, institutional identity)

These all played a role in the twentieth century, to be sure, but there is a fifth element that I feel has not gained as much attention as I think it deserves:

5. Flexibility through multi-use (creating efficient multi-focus designs)

Of all the elements above, I feel that this one bears closer scrutiny. To do that, we return to the 1890s, and first ask: why did reflectors supplant refractors when they did, and not earlier?

Albert Van Helden has described the latter half of the nineteenth century as the "golden age of the great refractor" ([51], p. 134). Indeed, in his authoritative encyclopedic essay "Building Large Telescopes, 1900–1950," as well as in many synthetic reviews and compilations before and since, including Henry King's classic history, we find that the 1880s and 1890s saw the largest of all astronomical refractors constructed by the US as well as by Great Britain and France. But this was also a time when the limitations of this form of telescope were being felt the most: problems with mechanical flexure, the intrinsic difficulties of lens production, absorption of light by thick lenses, chromatic aberration, and transparency for photography and for spectroscopy all posed serious problems. In addition, the goals of astronomy were changing. The spatial stability of the optical field offered by the great refractor was still very important for positional work, but there were new, quickly growing, lines of research that called for, in Van Helden's words, a "continuous quest for greater light-gathering power" ([51], p. 134).

One can, of course, claim that this started with William Herschel's plan to increase his power of penetrating into space in order to reach the ends of the visible system of stars and thus deduce its form and structure ([8], p. 75). One can also find elements of this quest in the great speculum reflectors of the third and fourth Earls of Rosse, as well as in the valiant efforts by William Lassell, James Nasmyth and the hapless builders of the great Melbourne telescope led by Thomas Romney Robinson and Thomas Grubb.

H. C. King has suggested that Rosse's engineering achievements paved the way for great reflectors, demonstrating that "the building of large telescopes could no longer be regarded as 'a perilous adventure, in which each individual must grope his way'" ([31], 217). If King's assertion is correct, why didn't the reflector supplant the refractor before the twentieth century? Here we will explore the reasons why, and will show how, by the direction of George Ellery Hale, the first truly astrophysical reflector was built in the first years of the twentieth century.

2 Problems of the reflector: part I

In the late nineteenth century, one of the most important questions astronomers asked was: how do things change over time? In astronomy, most things beyond the solar system change rather slowly: e.g. the positions of the stars, and of double stars in their orbits. For a growing subset of astronomers, the question narrowed down to whether stellar systems were static or changed over time, a question at the heart of what was regarded as the "sidereal problem," or the *agenda* of what Robert Paul has called statistical cosmology [30, 42].

One of the best examples I've found of the failure of reflectors to become competitive in this new arena comes from an exchange between Edward S. Holden and Lawrence Parsons, the fourth Earl of Rosse. In the late 1870s, Holden was intent on collecting evidence that nebulae changed form over time and so wrote to (of all people) the Earl for advice. Predictably, the Earl was highly skeptical that any real changes had been recorded visually over time, even over some 30 years, and the main reason for this was, as he expressed it, that "the condition of the mirrors so affect the appearance of the object and I should of course add the power of the telescope employed." One could not, advised the Earl, merely compare the drawings of the "Herschels & Lassells & Bonds & Laporinovs..." and the reason for this is that they were all using different telescopes, and:

> It is one of the drawbacks to a reflector used in the open air & in a damp climate that you cannot say that you have examined an object to the very fullest advantage unless the speculum has [not] been dewed since leaving the polishing room [34].

The Earl's insight into the limitations of his and his father's reflectors, as builders and users of these cantankerous devices, well illustrate one reason why the refractor remained favourite among those seeking large apertures. In the same year as the Earl's letter to Holden, Howard Grubb tried to argue that the "Great Telescopes of the Future" would be reflectors. But at the 1893 Columbia Exposition, Alvan Clark passionately countered Grubb, claiming the refractor was still the telescope of the future. Large reflectors, he argued "have never accomplished much except in the hands of the opticians who made them..." ([13], see also [54]), such as Herschel, Rosse, Nasmyth, Lassell, or even A. A. Common. Clark's claim was applauded by visual observers like S. W. Burnham and by other astronomers; the only voices in quiet opposition were a few physicists and the yet young George Ellery Hale [57, 58].

3 Campaigning for the reflector: Hale's vision

In 1897, on the occasion of the opening of his new Yerkes Observatory, which boasted the largest refractor in the world, for reasons that will soon become clear Hale felt he not only had to revive, but must amplify Grubb's arguments to make his point once again, viz: that reflectors possessed:

1. Perfect Freedom from chromatic aberration
2. Relatively small absorption for large apertures
3. Possible large angular apertures [fast f-ratio]
4. Small cost of speculum, mounting and dome.
5. Possible large linear aperture [22].

Hale felt that the main reason for not only repeating, but strengthening Grubb's original arguments was because by 1897, the continued technological refinement of the silver on glass method and the casting of large glass disks had stabilized the reflecting telescope. These arguments alone, he felt, should be enough to make the astrophysicist conclude that reflectors are preferable. A fully equipped observatory still needed both refractors and reflectors, Hale diplomatically asserted, "in order that all classes of work can be prosecuted in the most advantageous manner possible" ([22], p.130). But to all this, Hale added the conclusion that advantages for the astrophysicist are even more enhanced by adopting one new element: the fixed eyepiece, or as he called it, the "*equatorial coudé* reflector" ([22], p.130). This hybrid design had been suggested by his British friend Arthur Cowper Ranyard, and elaborated by his Yerkes assistant F. L. O. Wadsworth, who had the previous year argued for this design, and maintained that the future of astrophysics lay with the large equatorial reflector, over the popular siderostat [53].

If there is one characteristic of Hale's vision that can be claimed to be truly modern, it is that, without exception, on the eve of the completion of any one of his major achievements, whether it be the instruments of the Kenwood, the Yerkes, or the Mount Wilson Observatories, Hale was already well along planning his next big step. Accordingly, Hale reported at the dedication of Yerkes that a new 60-in. reflector that was now being built by his optician George Willis Ritchey, "will be provided with a mounting of this type [Ranyard's design], and devoted exclusively to investigations in astrophysics" ([22], p.131) using devices far too heavy and dynamically fragile to be sitting on a moving telescope in a drafty dome.

It is this last feature of Hale's telescopes that, I feel, best characterizes the "Hale era" of building big telescopes—it was a time when large reflecting telescopes emerged to dominate professional astronomy. But it was also when large multi-focus photographic reflectors emerged that could easily and faithfully accommodate many diverse applications, and ancillary equipment useful to the wider research interests of astrophysicists. This era extends to the onset of World War II, and one might argue, well into the 1960s, as Richard Learner, Aden B. Meinel and others argue, when the mechanical "heavies" were phased out in favor of new design philosophies that led to vast increases in light-gathering power without exponentially proportionate physical expansion of the framework ([33], p.198).

4 Problems of the reflector: part II

Van Helden has provided the key for tracing out the origins of the multi-focus reflector, and in effect the acceptance of the reflector as the dominant instrument of the astronomer's tool kit. By 1900, despite Hale and Grubb's assertions, there was still "considerable opposition to large reflectors among observatory astronomers..." ([51], p.134). Reflectors were considered imprecise and, moreover, tediously and even treacherously awkward. Indeed, in 1901, describing what he called the revolution of the last quarter century in astronomy, H. H. Turner argued that while the reflector was a better light collector, size wasn't everything. Rosse's 6-ft could not compare to Common's 5-ft due to the superior reflectivity of silver on glass over speculum. However, he added in his engaging way, large reflectors have not been as accessible as large refractors, and he wondered rhetorically why: refractors are "handier and more trustworthy" maybe, the reflector "being subject to curious moods which render it sometimes difficult to work with satisfactorily." According to Turner, even Common was known to compare the reflector to "the female sex for uncertainty" ([49], p.46). Fickleness could certainly describe the large reflectors; I. S. Glass notes that, though the failure of the Melbourne reflector could be seen in its design, it was mainly due to the incompetence of its users that it did not meet expectations. To this one might add, its notorious inconvenience [20].

Much of the problem was indeed inconvenience, a constant annoyance from the 1860s through the turn of the century that we need to appreciate in assessing the rise of the reflector and the use of multiple foci [5]. Indeed, as H. C. King relates, Common's magnum opus, his 5-footer, was first used as a Newtonian in 1891, "but one night he nearly had a serious fall from the observing stage and there and then decided never to mount it again" ([31], p. 277). H. H. Turner once again made this point clearly: as telescopes get larger, access gets more difficult: "The most noteworthy recent improvements are mainly concerned with the comfort of the observer, which, though in some aspects a mere detail, is of the greatest importance in delicate observation, when the nerves and muscles should be as free from strain as possible" ([49], p. 124). As an example of an advance in improving comfort he cited Grubb's rising floors, and then the fixed eyepiece of the Coudé as developed by Loewy for refractors. Louis Bell repeats Turner's conclusions, and his are in turn echoed by more derivative writers like G. Edward Pendray and other popular writers in the 1930s. Contemporary writers, such as Fred Watson, speak of the quest for "comfort and joy" as a motivation for big telescope builders from Nasmyth on, striving to create mountings that made the eyepiece not only accessible but as fixed as possible, making, as he described it, "Gigantic telescopes at once easy and comfortable" ([55], p. 211).

Thus, early efforts to search for alternatives to the Newtonian focus were based upon convenience and safety for the observer, and not upon problem choice, until Hale made his entrance. Hale's goal was to create a fixed eyepiece so that instruments like radiometers and spectrographs could be mounted under highly stable conditions, both thermal and mechanical. Hale's argument anticipates the priorities set by modern visionary builders like Aden Meinel, who noted in 1960 that beyond the quest for unlimited aperture, there was the fact that reflectors could be designed with multiple foci: Prime, Newtonian, Cassegrain, and Coudé. For Meinel, this versatility in f-ratio was a hallmark of the new era of the giant reflector ([37], p.25).

5 The multi-focus reflector

The earliest suggestion thus far found for multiple foci is by the Reverend Thomas Romney Robinson, the third Director of the Armagh Observatory and a good friend and confidant of the Earl of Rosse. Robinson believed the reflector to be the telescope of the future, but believed there were serious problems with its construction, mainly its speculum mirror but also with the inaccessibility of its Newtonian focus. Robinson envisioned what became the Melbourne reflector as a combination 48-in. Newtonian-Cassegrain in the early 1850s, and by 1869 Grubb built it, complete with a spectrograph, as the great southern telescope ([20], p. 42).

Those who designed large reflectors, from Nasmyth to Draper to Grubb and Common, all expressed the need to make the telescope accessible and conducive to delicate work. This problem became critical with the introduction of photography, and especially the dry plate process, which made long exposures possible, and highly sought after. King has argued, in fact, that the introduction of the dry emulsion plate into astrophotography greatly increased the need for precision mountings for tracking and guiding. In some form, telescopes had to be mounted equatorially, yet they had to be guided, and therefore accessible ([32], 305).

As Glass and others have recounted, Newtonians were far from dead. In the tradition of the Herschels and Rosses, builders and observers like Smyth, Lassell and Nasmyth opted for the Newtonian focus, accessed by elaborate "aerial chambers." In the 1860s, Lassell chose to build a gigantic fork-mounted Newtonian reflector accessed by an elaborate tower, and advocated a Newtonian fork mount, whereas others, like Robinson, stuck to the composite English/German equatorial advocated by Grubb. There was constant "danger to the observer from falling" of course, with any of these designs ([20], p.40 and 44).

An ingenious and very promising alternative design that became popular worldwide was the coudé developed by Maurice Loewy of the Paris Observatory; the first example was open for business in 1882. By the end of the decade at least six existed in Paris, Lyons, Bescancon, and Algiers and two more were soon bound for Paris and Vienna. In 1884 David Gill and Norman Lockyer both inspected the Loewy coudé and raved about it. Its great rigidity and unparalled comfort for the observer made it yet another "instrument of the future" ([6], p. 139). In 1889, when Loewy was awarded the Gold Medal of the RAS for his design, the Astronomer Royal, W. H. M. Christie hailed it as comparable to the use of the microscope in a natural history laboratory ([12], p.47). Turner's greatest approval was for a modification of Loewy's design by Grubb, as embodied in the Sheepshanks at Cambridge ([49], p. 124).

Loewy's coudé was a refractor, but a few reflecting designs were entertained. The earliest was by the German maker Steinheil who proposed an equatorially mounted reflector utilizing a siderostatic plane mirror as early as 1858. The primary mirror would look in the direction of the pole, and a large plane mirror with a hole in it would be set in its way. The observer would look through a hole in the plane mirror, which would also move in declination. Among several disadvantages noted by commentators, seeing very far north was impossible. But it is clear from Steinheil's own conceptual drawing that one could reverse the entire system, having the mirror looking down onto the flat [12].

Most of these alternate designs dealt with accessibility, but some included rigidity. A few suggested multiple use, but it was not a first priority Not until Hale do we see multiple use come to the fore. Hale wanted his telescopes to be versatile, multi-tasking devices that could collect, sift and sort light as if in a physicist's laboratory. His observatories were akin to 'test benches' as he promoted a life in astronomy at Mount Wilson to young promising physicists he tried to lure away from campuses to mountain tops ([25], cited in [14]). He often spoke of versatility. In his 1908 monograph, *The Study of Stellar Evolution,* he argued that a telescope like his newly opened 60-in. was well suited to feed the "most powerful spectrographs" under controlled conditions, and "such a telescope is also adapted for many other classes of work..." ([24], p. 238) By 1915, with over 6 years of experience with the 60-in. in operation, Hale argued that "the initiator of new methods almost necessarily becomes an instrument maker." The modern astronomer, contrary to popular practice, had to be ready to "take immediate advantage of those new points of view and new possibilities of attack which his investigations are certain to disclose" ([23], p.5). Thus poised, Hale then described the essence of Mount Wilson as:

> A laboratory or observatory like that of Mount Wilson, planned for the exploration of unfamiliar fields, can thus possess no fixed and standard equipment. Its mode of attack and its means of progress must grow with its work and develop with the disclosure of new and unexpected possibilities ([23], pp.5–6).

Highly unorthodox for his era, thinking along these lines, Hale, as well as Harvard's E. C. Pickering and a few other observatory directors, were attracted to experimenting with some form of coudé or heliostatic design. Hale tried horizontal heliostats at Yerkes. But the great drawback to all these designs, especially Steinheil's, was the need for producing large flats, equivalent to, or larger than, the primary objective. Even so, variations on the coudé remained sufficiently popular to suggest competition with the reflector, enough so that F. L. O. Wadsworth and others in the late 1890s felt it was necessary to show why it was not as efficient or effective as an equatorially mounted reflector.

In so doing, they found a compromise between the collecting power of an equatorial reflector and the versatility of the fixed coudé eyepiece.

Wadsworth managed Hale's shop staff at Yerkes. Trained in physics and mechanical engineering at Ohio State and Clark University, he worked for a succession of visionary tinkerers: first for A. A. Michelson at Clark, then for Samuel Langley in the early 1890s at the Smithsonian and then for Hale through the rest of the decade, running the instrument shop [11]. He was inventive, and opinionated, much like, but to a lesser degree, than his optical associate, George Willis Ritchey. At first Wadsworth focused on the design of spectrographs and their incorporation into telescopic mountings. As such, he became intimately involved with Hale's plans for improved means of mounting very large sensitive spectrographs and spectrohelioscopic equipment. In fact, soon after he arrived at Yerkes, Hale gave him a set of drawings and a description of a new form of telescope mounting sent by a friend in England.

How Hale came to know Arthur Cowper Ranyard is not known, but it must have been through Hale's visits to England and the RAS. In 1891, Ranyard accompanied Hale on a hurried trip to Paris to inspect Deslandres' new device for observing the disk of the Sun in isolated wavelengths, something disturbingly similar to Hale's own spectroheliograph. The two found they had much in common ([58], pp. 82–83; [36], p. 225). Ranyard, a barrister by profession, was an avid student of the Sun, and attended eclipse expeditions with relish and creativity. So he naturally befriended the 23-year-old American, and even built a Hale spectroheliograph, lending it to Hale for his expedition to Etna in 1894.

Like Hale, Ranyard was not satisfied with having to attach heavy but delicate devices to moving telescope tailpieces. According to Wadsworth, he looked for a way to combine the virtues of the reflector with Loewy's coudé design, but was prevented from carrying out his plan by his death in 1894. At the end of his life, he sent Hale his designs, and Hale celebrated his memory by lavishing credit at every opportunity ([47], p. 42; [52], pp. 132–134; [45], p. 221; [21]).

In 1901, Ritchey claimed that the design of the Yerkes 2-ft reflector was in every way seen as a test bed for the design of the ultimate 5-ft, with one notable exception: "The plan for the five-foot is adapted to include, if desired, the *coudé* arrangement of mirrors which was suggested by Mr. Ranyard to Professor Hale in 1894" ([45], p. 221). As Donald Osterbrock has noted, Ritchey considered Ranyard's design for the 5-ft, adding a tertiary mirror just above the main mirror to direct light up a polar axis and into a fixed and thermally isolated room north of the telescope chamber. Wadsworth, however, modified it to send the light down a hollow polar axis, for greater mechanical and thermal stability (see Fig. 1). As the 2-ft was a dual-focus German equatorial, the ultimate triple-focus arrangement remained to be tested on the 5-ft ([40], p. 68; [52]).

Hale's goal for the 60-in. was high-dispersion stellar spectroscopy of stars of solar type and later, stars that would reveal the subtleties of temperature classification that he felt sure held the key to the course of stellar evolution.

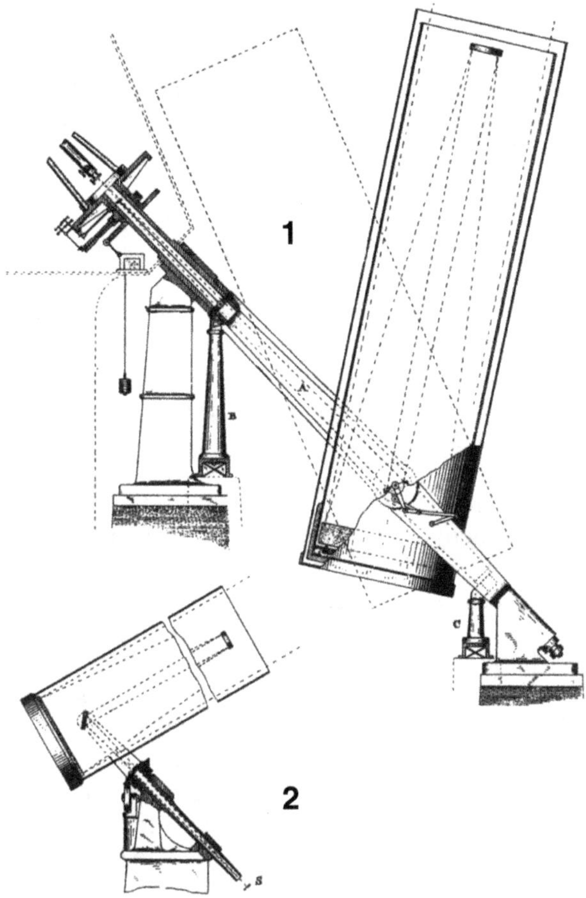

Fig. 1 Plate IV from Wadsworth [52] depicting Ranyard's original design concept (Fig. 1) and Wadsworth's modification (Fig. 2). Ranyard's concept included a double-walled telescope tube and an enclosed optical beam, thus obviating the need for a protective dome. It in some ways anticipated the double yoke design of the 100-in. reflector

His goal, to extend solar astrophysics to the stars, studying the Sun as a star, led to a wide range of instruments first at Yerkes and then on Mount Wilson. But Hale, in need of funds to complete his dream, needed support in order to convince his superiors at the University of Chicago to invest in his latest dream. He thus resorted to a tactic new to astronomers: polling his peers.

In a world-wide appeal to 23 observatory directors in late 1901, Hale asked what instrument could best solve the most pressing problems of astronomy. The first priority on his list was photography of nebulae, followed by photography for stellar parallaxes and proper motions, spectroscopic observations of both bright and faint stars at higher dispersions than presently possible, studies of the heat from celestial objects, using sensitive radiometers, and photographic studies of variable stars "beyond the reach of the 40-in. telescope" ([18], p. 4). Hale then asked each of the astronomers and physicists to comment on how suitable a 60-in. reflector was for these tasks, and if a better technology existed. He described the state of production of the mirror, and the form of mounting which would provide three focal points, each maximized for one or more of the tasks enumerated earlier. The radiometer, in particular, required both thermal and mechanical stability, and very high dispersion spectroscopy required large instruments, both best suited to the coudé focus.

Nineteen astronomers agreed that the 60-in. multi-focus design could perform this wide range of tasks. Four of them, however, felt that the astrometric work could be better done with a refractor. Several remarked that Ritchey's recent observations of changes in Nova Persei were spectacular and warranted a larger telescope. It was striking, frankly, how frequently the successful nebular photography recently performed with reflectors was mentioned by each of these astronomers. [One cannot help but reflect on his how this factor reappeared recently in the form of intense public pressure that convinced NASA to mount a final servicing mission to extend Hubble's vision].

The story of how Hale finally succeeded in building the 60-in. is well known [38]. Allan Sandage is one of the latest to provide a detailed accounting, arguing that the 60-in. "broke new engineering ground. Its three focal positions were achieved by using three different secondary and tertiary mirror arrangements. The 60-in. was the first to boast this versatility of changeable mirror trains." It was this feature, and the fact that the 60-in. mounting allowed "sustained observations in all positions of the sky" taking advantage of the cumulative properties of the photographic process, which distinguished the 60-in., in Sandage's opinion, from all predecessors ([46], p. 161). Indeed, within its first decade of life, it provided the means for observations as varied as Harlow Shapley's studies of variables in globular clusters, which led to his galactocentric revolution [9, 19, 50]. The 60-in. also made it possible for the first time to examine efficiently the high-dispersion spectra of stars, equivalent to those available for the Sun and for laboratory standards [2, 4]. It also produced the observations that led to the recognition of the technique of spectroscopic parallaxes [1, 15]. All three foci, and eventually the fourth, the prime focus, were equipped with spectrographs that saw heavy use, and also a significant degree of experimentation, especially at the coudé, where Sinclair Smith tried out highly precise means of radiometrically recording intensities in high-dispersion stellar spectra [3].

In the wake of the Mount Wilson 60-in., all large reflectors until 1948 were multi-focus, save for a 20-in. built at Trieste in 1925, a 40-in. at Geneva in 1927, one in Bloemfontein, and a 36-in. at Edinburgh in 1928. Up to 1948, the four largest telescopes in the world were triplets: Newtonian (or Prime), Cassegrain and coudé, and among those of 60-in. and above six were triple, four double and one Newtonian only, at Bloemfontein. The Harvard telescope at Bloemfontein was based upon a reworked 60-in. Common mirror that was re-mounted by Fecker, and was intended for routine spectroscopic programming.

With the exception of Hale's efforts, and those of his staff who continued to build larger telescopes on new designs, the majority of professional astronomical institutions of the first half of the twentieth century concentrated on conventional activities, building large enough telescopes to be competitive in areas of stellar and galactic astronomy that could be approached with conventional instrumentation [27]. What innovation did take place centered upon new designs for focal plane instrumentation (see Patrick McCray's essay, this volume).

The emergence of the multi-focus reflector is a milestone of twentieth century telescope design that warrants more historical attention, specifically to how it changed observatory practice, as more and more observatories concentrated on problem-based observing programs rather than open-ended surveys in the Post World War II era.

6 But doesn't size matter?

Despite most claims by historians and astronomers alike that astronomers always strive for larger apertures, in the late nineteenth century and into the first decade of the twentieth there was a debate over the most efficient size for a telescope. Bigger aperture was not universally regarded as better for visual work. A number of factors contributed to this debate:

First, those most vocal against the large telescope were amateurs, who had enjoyed a long tradition, in Victorian Britain especially, of making substantial contributions to the science. This was the end of the era of the grand amateur, protesting the rise of the professional class, as John Lankford has suggested. Very vocal and highly influential amateurs, from William Frederick Denning to Percival Lowell, raised the question: Is bigger better? As Lankford shows, only with the establishment of the professional class, "The conflict of the kind witnessed in the 1880s and 1890s during the debate over telescope size ... was virtually eliminated ..." ([32], p. 11).

Not completely, of course, as the title of my talk suggests; a few voices still called out well into the twentieth century, and even up to recent times. In the February 1935 issue of the journal "Southern Stars," the house organ of the New Zealand Astronomical Society, the editor, I. L. Thomsen, editorialized about feeling relevant in an age when a 200-in. telescope was being built, when people associated astronomy with big telescopes. "Let it not be forgotten" he cried, "that many factors other than the telescope itself enter into the successful use of any instrument." Then he added, quoting an anonymous astronomer, "It is the man behind the gun that counts." He reassured his brethren amateurs that "Large instruments have not and will not destroy extensive fields of usefulness," citing, ironically, astronomers at Mount Wilson like Paul Merrill. Thomsen also quoted Hale, that "there is a splendid field for any man who wishes to accomplish results, wherever he may be situated, and however simple his means of research may be ([48], p. 239). We still hear assurances like this today, as in how the new gargantuan optical surveys being assembled will in fact provide far more data than can be exploited by professionals alone [35].

It is indeed true that professionals like Merrill and Hale were always eager to encourage the amateur, and even today there is a definite sentiment that one does not have to enjoy access to the largest telescope to do perfectly acceptable and welcome work in astronomy. The debate has continued among astronomers in recent times [29, 39], but to my knowledge no professional had voiced public doubts of larger and larger telescopes to the extent E. C. Pickering did as late as 1909, where, speaking on the future of astronomy, he proclaimed that:

> Much more depends on other conditions, especially those of climate, the kind of work to be done and, more than all, the man behind the gun. The case is not unlike that of a battleship. Would a thousand feet long always sink one of five hundred feet? [44], p. 105).

The second factor was rivalry, couched in conservatism. Pickering, one of the most powerful astronomers of his day, and then the president of the American Astronomical Society, was here getting worried that his west coast competitor, George Ellery Hale, would continue to run away with more prizes, dwarfing his Harvard-centric east coast establishment aspirations ([14], pp. 147–148). Pickering therefore told his Cleveland audience that "...we have reached the limit in size with the 60, 72, and planned 100 inchers" ([44], p. 105).

A few years earlier, Pickering had purchased two problematic 60-in. mirrors from A. A. Common, and was desperate to keep Harvard in the game, in the face of the great success of Hale's recently opened 60-in. Only 8 years earlier, however, Pickering was more sanguine. He was one of the astronomical world leaders who responded to Hale's appeal to advise the University of Chicago, and ultimately after 1902 the Carnegie Institution, that spending money on completing his 60-in. telescope was the best course to take. His response to Hale was guarded, of course, cautioning that:

> The views of astronomers have changed during the last few years regarding the use of large telescopes, especially of reflectors. This is largely due to the results obtained by Keeler with the 3-foot reflector, and the work, under your direction, with the 2-foot reflector. Instead of being a luxury, a large reflector is now a necessity in many departments of astronomy, if work of the highest grade is to be done [43].

Pickering's attitude may well be the basis for a third significant factor, well beyond rivalry. In the twentieth century, to say simply that the push was constantly toward larger telescopes oversimplifies the situation. The need for large reflectors was established, of course, and they were seen as the telescopes of the future, but the question of what "large" was—what indeed was a sufficient aperture to be professionally competitive—remained debatable. Local, regional and national pride more than once added a few inches to aperture, but there was also need for economy and efficiency.

If one looks at the most common form of telescope mounting, 60-in. and above, following Richard Jarrell's perceptive analysis of the steps Plaskett went through to decide upon a design for his telescope at the Dominion Astrophysical Observatory, one finds that the Mount Wilson 60-in. was not the prototype for the standard "conventional large telescope" of the first half of the twentieth century [27]. As Jarrell points out, Plaskett was not a radical builder; he needed only medium-dispersion spectra for his radial velocities. He was an adept but conventional spectroscopic observer who wanted more photons to observe more radial velocities for fainter and fainter classes of stars. He was neither a Ritchey nor a Hale, both of whom, to the mainstream, were radical risk takers. Those who adopted the 72-in. DAO design were similarly mainstream data collectors, as opposed to the experimental or exploratory approach Hale advocated. No one person, institution or nation challenged Hale's 100-in., and certainly not Hale's 200-in. Why? Jarrell's observations hint at the answer, but indeed, so critical a question deserves further inquiry someday.

It is safe to say that Hale was, indeed, an aberration, but a happy one for astronomers. Among astronomers, especially among the elite circle of observatory directors who defined the discipline in the first half of the century, he was the most progressive, professionally and socially. Despite what some historical writers have recently suggested, his staff enjoyed greater professional freedom and more autonomy than staff at other major observatories, and part of this was due to the fact that Hale had fostered so many highly versatile instruments, most without peer, and fostered a wide range of research interests. It was an unbeatable combination.

References

1. Adams, W.S., Kohlschûtter, A.: Some spectral criteria for the determination of absolute stellar magnitudes. Contr. Mt. Wilson Obs. **89**, 1–14 (1914)
2. Adams, W.S.: Some results of a study of the spectra of Sirius, Procyon, and Arcturus with high dispersion. ApJ **33**, 64–71 (1911) doi:10.1086/141829
3. Adams, W.S.: Summary of the Year's Work at Mount Wilson. PASP **43**, 394–403 (1931). doi:10.1086/124172
4. Adams, W.S.: Sunspot lines in the spectrum of Arcturus. ApJ **24**, 69–77 (1906). doi:10.1086/141374
5. Anon: Reflecting telescopes. Astron. Regist. **4**, 318–319 (1866)
6. Lockyer, N.: Meeting of the royal astronomical society. Astron. Regist. **22**, 139–142; 142 (1884)
7. Bell, L.: The Telescope. McGraw Hill, New York (1922)
8. Bennett, J.A.: On the power of penetrating into space: the telescopes of William Herschel. J. Hist. Astr. **7**, 75–108 (1976)
9. Berendzen, R.: Geocentric to heliocentric to galactocentric to acentric: the continuing assault to the egocentric. Vistas Astron. **17**, 65–83 (1975). doi:10.1016/0083-6656(75)90049-5
10. Blank, D., Andersen, G.: The telescope. Its history, technology, and future. J. Astron. Hist. Herit. **10**, 247 (2007)
11. Cattell, J.M. (ed.): Frank Lawton Olcott Wadsworth. Am. Men of Sci., 5th edn., p. 1157. The Science Press, New York (1933)
12. Christie, W.H.M.: Award of the Gold Medal to Maurice Loewy. Mon. Not. R. Astron. Soc. **49**, 247 (1889)
13. Clark, A.G.: Great telescopes of the future. A&A **12**, 673–678 on 674–675 (1893)
14. DeVorkin, D.H.: Henry Norris Russell: Dean of American Astronomers. Princeton University Press, Princeton (2000)
15. DeVorkin, D.H.: The discovery and exploitation of spectroscopic parallaxes. In: Philip, A.G.D., Van Altena, W.F., Upgren, A.R. (eds.) *Anni Mirabiles*: A Symposium Celebrating the 90th Birthday of Dorrit Hoffleit, pp. 17–24. L. Davis Press, Schenectady, NY (1997)

16. Evans, D.S., Derral Mulholland, J.: Big and bright: a history of the McDonald observatory. University of Texas Press, Texas (1986)
17. Florence, R.: The Perfect Machine: Building the Palomar Telescope. Harper-Collins, New York (1994)
18. Forgan, J.B., et al.: Second Annual Report of the Visiting Committee of the Yerkes Observatory. University of Chicago Press, Chicago (1902)
19. Gingerich, O., Welther, B.: Harlow Shapley and the Cepheids. Sky Telesc. **70**, 540 (1985)
20. Glass, I.S.: Victorian Telescope Makers: The Lives and Letters of Thomas and Howard Grubb. IOP (Institute of Physics Press), Bristol (1997)
21. Hale, G.E.: Arthur Cowper Ranyard. ApJ **1**, 168–169 (1895). doi:10.1086/140031
22. Hale, G.E.: Note on the Ranyard mounting for reflecting telescopes. ApJ **5**, 148–149 (1897)
23. Hale, G.E.: Ten Year's Work of A Mountain Observatory. Carnegie Institution of Washington Publ 235, Washington, DC (1915)
24. Hale, G.E.: The Study of Stellar Evolution. University of Chicago Press, Chicago (1908)
25. Hale, G.E.: T. Mendenhall, 1904. Hale Papers, Huntington Library and Archives, AIP Microfilm Edn.
26. Hale, G.E.: On the comparative value of refracting and reflecting telescopes for astrophysical investigation. ApJ **5**, 119–131 (1897). doi:10.1086/140317
27. Jarrell, R.J.S.: Plaskett and the modern large reflecting telescope. J. Hist. Astr. **30**, 359-390 (1999)
28. Jarrell, R.: The Telescope (1950). Encycl. Astron. Astrophys. **4**, 3293–3298 (2001)
29. Johnson, H.L., Richards, W.L.: Optimum size of infrared photometric telescopes. ApJ **160**, L111–L116 (1970). doi:10.1086/180540
30. Kapteyn, J.C.: The plan of selected Areas. In: Council Notes, MNRAS 74, pp. 348–352, on 348 (1914)
31. King, H.C.: H.C. King's book: The History of the Telescope, 1959. Griffin, London. Reprinted, Dover Publishing, New York (1979)
32. Lankford, J.: Amateurs vs. professionals: the controversy over telescope size in late Victorian science. Isis **72**, 11–28 (1981). doi:10.1086/352648
33. Learner, R.: Astronomy though the Telescope. Van Nostrand, New York (1981)
34. Lord Rosse to E. S. Holden, January 6, 1877. Holden Papers, Mary Lea Shane Archives, University of California, Santa Cruz
35. MacRobert, A.: Death of amateur astronomy... or new birth? Sky Telesc. **116**, 33 (2008)
36. Meadows, A.J.: Meadows's book: Science and Controversy: A Biography of Sir Norman Lockyer. MIT Press, Cambridge, MA (1972). Reprinted, MacMillan, London (2008)
37. Meinel, A.B.: Design of reflecting telescopes. In: Kuiper, G., Middlehurst, B. (eds.) Stars and Stellar Systems, vol. 1, pp. 25–42 (1960)
38. Misch, A., Sheehan, W.: Pioneering telescope turns 100. Sky Telesc. **116**, 38–41 (2008)
39. Oke, J.B.: On the optimum size of optical telescopes. ApJ **162**, L77–L78 (1970). doi:10.1086/180628
40. Osterbrock, D.E.: Pauper & Prince; Ritchey, Hale & Big American Telescopes, p. 68. Arizona, Tucson (1993)
41. Panek, R.: Seeing and Believing: How the Telescope Opened our Eyes and Minds to the Heavens. Viking, New York (1998)
42. Paul, R.: The Milky Way Galaxy and Statistical Cosmology 1890–1924. Cambridge University Press, Cambridge (1993)
43. Pickering, E. C. to G. E. Hale, nd circa December 1901. Reprinted in Report of the Visiting Committee of the Yerkes Observatory, October 1902, private circulation. p. 20
44. Pickering, E.C.: The future of astronomy. Pop. Sci. Mon. **75**, 105–116 (1909)
45. Ritchey, G.W.: The two-foot reflecting telescope of the Yerkes Observatory. ApJ **217**, 221 14 (November 1901)
46. Sandage, A.: The Mount Wilson Observatory, Breaking the Code of Cosmic Evolution. Volume 1 of the Centennial History of the Carnegie Institution of Washington, p. 161. Cambridge (2004)
47. Steinheil, C.A.: On the advantages to be derived from the use of silver mirrors for reflecting telescopes, and on a novel mode of mounting such instruments. MNRAS **19**, 56–60 (1858). Translated from AN 1138

48. Thomsen, I.L.: In the grip of the big telescope age. JRASC 29 **238**, 238–240 (1935). Reprinted from Southern Stars (February 1935)
49. Turner, H.H.: Modern astronomy, being some account of the revolution of the last quarter of a century, pp. 46–47, 124–129. Constable, Westminster (1901)
50. Van de Kamp, P.: The galactocentric revolution, a reminiscent narrative. PASP **77**, 325–335 (1965). doi:10.1086/128228
51. Van Helden, A.: Building large telescopes, 1900–1950. In: Gingerich, O. (ed.) Astrophysics and Twentieth-century Astronomy to 1950 part A, pp. 134–152. The General History of Astronomy 4A. Cambridge University Press, Cambridge (1984)
52. Wadsworth, F.L.O.: On a new form of mounting for reflecting telescopes devised by the late Arthur Cowper Ranyard. ApJ **5**, 132–134 (1897). doi:10.1086/140318
53. Wadsworth, F.L.O.: Review of G. Johnstone Stoney. On the equipment of the astrophysical observatory of the future. ApJ **4**, 238–243 (1986). doi:10.1086/140275
54. Warner, D.J., Ariail, R.: Alvan Clark & Sons, Artists in Optics, 2nd edn. Willmann-Bell, Inc, Richmond (1995)
55. Watson, F.: Stargazer: The Life and Times of the Telescope, p. 211. DaCapo, Cambridge (2004)
56. Woodbury, D.O.: The Glass Giant of Palomar, revised edn. Dodd, Mead, New York (1970)
57. Wright, H., Warnow, J.N., Weiner, C. (eds.): The Legacy of George Ellery Hale, p. 42. MIT, Cambridge (1972)
58. Wright, H.: Explorer of the Universe: A Biography of George Ellery Hale, revised edn. American Institute of Physics, Woodbury, NY (1994)
59. Wright, H.: Palomar, The World's Largest Telescope. MacMillan, New York (1952)

Exp Astron (2009) 25:79–89
DOI 10.1007/s10686-009-9138-9

REVIEW ARTICLE

"Beautiful and cantankerous instruments": telescopes, technology, and astronomy's changing practice

W. Patrick McCray

Received: 5 January 2009 / Accepted: 6 January 2009 / Published online: 3 February 2009

Abstract Between the dedication of the 200" Hale Telescope in 1948 and the completion of today's 8–10 m behemoths, astronomers' most iconic symbol, the telescope itself—its design, its technology, and its use—was transformed as a research tool. The importance of this is deceptively simple: in astronomy, technological innovations have often led to new discoveries. Driven by the need to get as much observing time as possible and the desire to take advantage of the best observing conditions, modern observatories have experimented with new technologies and modes of collecting images and spectra. This entailed a re-casting of the telescope by astronomers and science managers as a factory of scientific data. At the same time, contemporary astronomers express considerable unease and apprehension about how these technological changes have altered, in ways subtle and profound, the nature of astronomical observing and what it meant to be an astronomer. This short essay addresses the issues associated with these recent changes in astronomical practice and their connections to astronomers' desire for ever larger and more complex telescopes.

Keywords Telescope history · Practice of astronomy · Giant telescopes

1 Introduction

Eight years ago, I visited Cornell University. There, in the Astronomy Department, I found a sign in a hallway that read: "Know Thyself. Know Your Telescope." This adage, paraphrased from Socrates, says a good deal about

W. P. McCray
Department of History, University of California, Santa Barbara, Santa Barbara, CA, USA
e-mail: pmccray@history.ucsb.edu

the connection between astronomers and the telescope—to do good research, one presumably should be aware of the connection between instrument and scientist. Albert Whitford, the director of Lick Observatory, 50 years ago in an oral history interview called telescopes "beautiful and cantankerous instruments" and said that using them demanded what he called "high artistry." Such work was challenging; it could be cold, tiring, boring, or simply frustrating. Nonetheless, time spent in the observatory dome fostered an intimate bond between scientist and machine.

Over the last four centuries, astronomers did indeed form a personal relation with telescopes, which are, after all, their central research tools. However, in the last half-century, astronomers' relationship with telescopes has changed profoundly. These interactions have become more productive, yet more? impersonal and remote. The classic image of an astronomer working alone under the night sky is now mostly a romantic anachronism. Efficiency, improved performance, comfort, and access to more data are goals that the science community has chosen to pursue.

In studying the history of the telescope, historians have often focused on their size and light-collecting ability. To be fair, this emphasis is understandable. In 1970, only three optical telescopes in the world had mirrors larger than 3 m. Today, sixteen telescopes have mirrors bigger than 6 m. Given the funding and technical challenges scientists faced 30 years ago—recall that in 1978 no one had yet demonstrated that they could even build a telescope mirror 8 m or bigger—today's global menagerie of telescopes seems quite remarkable, even unexpected.

I want to, therefore, approach the last half-century of telescope history from a perspective other than size and light collecting ability. This essay considers how technological changes since 1945—specifically, the adoption of computers and other electronic devices—have altered the nature of what it means to do astronomy and to be an astronomer.

2 Historical parallels

Many changes in the astronomy community parallel earlier ones that high-energy physicists experienced, debated and worried over. In the 1950s and 1960s particle accelerators and bubble chambers grew enormously in size, complexity, and cost as well as the amount of data generated. For many scientists, the data itself became THE experiment. Physicists at Berkeley, for instance, relied on machines like the Franckenstein that could automatically measure particle tracks much faster than a human operator. As Peter Galison has noted, experts in data-processing told physicists that they could envision eliminating people from the data analysis process, function by function through computerization. Berkeley physicist and future Nobel winner Luis Alvarez defended these techniques as a necessary and pragmatic part of modern physics, where high efficiency came from production line organization. The opinion of the physics community itself was divided over these new tools. At CERN—where the rush to automation was even more fast-paced than in the USA—one European physicist lamented that "in a few years... one would not go to start a new experiment but would just go into the data archives, get a few magnetic tapes, and... that would be the experiment.... This point of view frightens me," he said. Three decades later, many astronomers voiced similar ambivalence.

Another historical parallel that helps us understand the changing relation between astronomers and the telescope comes from the early days of the manned space program. As David Mindell discusses in his recent book *Digital Apollo*, hands-on flying was central to the test pilot's professional identity, making questions about the pilot's role a central debate in early spaceflight missions. As one pilot, a young man named Neil Armstrong wrote in 1960: "this hypersonic vehicle [he was talking about the X-15] is an instrument of the pilot" and depends on him "for control and flight success." Armstrong's viewpoint conflicted with that of NASA engineers who described the Mercury spacecraft as "an automatic system which, if all components work correctly, can complete an unmanned flight." The primary task of the astronaut, in their view, was "systems management" and not actual stick and rudder flying. These striking similarities speak not just to the increasing complexity of research tools after 1945 but also go the heart of scientists' professional identity.

3 Palomar nights

To get a sense of what has changed since 1948, consider how a typical night at the 200-in. at Palomar might have unfolded six decades ago. While the telescope operator opened the dome shutters and checked the weather, the astronomer climbed into the prime-focus cage or down to the coudé focus. "Each time you go up," one long-time Palomar observer recalled, "you carry some photographic plates in a tiny little box... a whole set of dreams of what the object you're going to work on is going to turn out to be." The telescope operator steered the telescope from the main control desk on the dome floor where analog dials showed the telescope's position. Throughout the night, the operator pointed the telescope to where the astronomer requested through an intercom or simply by shouting.

Many of the observation runs with the 200-in. were carried out at the telescope's prime focus station. There researchers sat in the cramped "observing cage" and rode with the telescope all night long while collecting data. A single exposure might last all night. The astronomer looked through an eyepiece almost continually to keep the object centered properly on the photographic plate or spectrograph slit. The work could be quite tiring and even inefficient.

A sense of the personal and physical experience of using a large telescope is revealed in an audio tape (surreptitiously made by a night assistant) of

Rudolph Minkowski talking to himself during a night's observing run on the Palomar 200-in.:

> "Where is that thing? I think that's it over there. [sound of slow motion motor] Damn! Wrong button. [motor again] There it is... now where's that little double to the left?... No, that's not it... [to the night assistant] Try a little west... Stop! [To himself] Here it comes [slow motion motor]. Yes, I think that's it. Pull it down a little [slow motion motor again] Ah! Too much! [motor sound] Now I've got it! [sound of camera opening; then, to the night assistant] Start the exposure! I'll take three hours on this one. You can rest awhile."

A long sigh followed as Minkowski began to hum Beethoven's "Ode to Joy" before the tape ended.

Using a big telescope required skill one acquired through time and perseverance. Astronomers had their own techniques for collecting data; telescope operators even claimed they could tell who was in the observing cage simply by the types of instructions that were being shouted out to them. But, as Maarten Schmidt said in an interview, the astronomer had the profound, perhaps humbling, experience of "sitting with his back to the whole universe."

4 Was "technology supplanting artistry"?

As early as 1945, Princeton's Henry Norris Russell predicted that "in another decade or so, those of us who slaved looking through eyepieces... will be pitied by the folks who use the new devices." After World War Two, astronomers gradually replaced photographic plates with electronic devices that could capture images and spectra more efficiently and from different focal points. These new tools slowly helped re-define the astronomer's interaction with the telescope. Things like auto-guiders and image intensifiers made observations easier and more efficient. But the telescope's complexity increased because now astronomers had to modify all the things that they used to do by hand.

While new tools made observations more efficient, they also began to separate the person from the machine. In some cases, this was literally the case: television-based guiding systems brought the astronomer in from the cold observing cage to warm, well-lit data rooms. The output of electronics systems was displayed in a separate control room and could also be used to control the telescope automatically.

By 1968, astronomers had managed to operate telescopes remotely. Scientists in Tucson, for instance, used punched paper tape and a computer to remotely collect data with a small telescope on Kitt Peak. Not all astronomers appreciated such technological feats. Some experienced observers, one scientist said, still preferred "manual controls, manual guiding, and fairly extensive human interaction with the instruments." At Lick Observatory, scientist George Herbig, however, cautioned his colleagues "not [to] hold too conservative a philosophy and be accused," he said, "of always planning to wage the next war with the weapons that won the last." Nonetheless, many scientists wondered whether "technology [was] supplanting artistry" and whether the craft of traditional observing would vanish.

One cause of astronomers' apprehension was the rapid appearance of computers in the observatory. In the 1950s, astronomers began using computers for routine tasks such as data reduction. These were large and costly machines centrally located on university campuses. The introduction of mini-computers, like the IBM 360 and Raytheon's 703 in the mid-1960s, changed this completely. As computer equipment became more common in astronomy, astronomers saw that it had the potential to affect other areas of science practice other than data processing. For example, in 1966, astronomers at Northwestern University described the use of a computerized information retrieval system that could provide answers to technical inquiries. By processing up to eight questions a minute, the system analyzed queries posed in English about stellar astronomy, searched a star catalog, and provided numerical answers. Once modern observatories installed powerful yet relatively inexpensive computers, astronomers could link them to other electronic equipment or even the telescope itself. As this happened, observatories were obliged to hire scientists and engineers with the skills to make the new devices work.

Edwin W. Dennison was one such person. After getting his PhD in astronomy at the University of Michigan, he was hired as a researcher at Caltech. He soon turned his attention toward the development of electronic systems for telescopes. In 1966, Dennison directed what came to be called the Astro-Electronics Laboratory at Palomar. By 1969, Dennison's laboratory employed a dozen people. One of their main projects was adapting mini-computers for direct use at the telescope. Dennison's devices found a welcome reception at Palomar, and by 1970 astronomers at the 200-in. used a digital data-recording system on one out of every three observing runs.

The degree to which technological enthusiasts believed that modern electronic instruments could change astronomers' nightly work is reflected in an article Dennison published in *Science*. Entitled "Electronic Optical Astronomy: Philosophy and Practice," Dennison's 1971 paper outlined two principles that guided his lab's efforts: to enable the astronomer to make observations that would otherwise be difficult or impossible and to improve the "operating efficiency of the instruments."

Besides describing the computer and electronic systems at the Hale Observatories, the article offered an overall philosophy that guided the development of electronic instrumentation. Dennison, while recognizing that no machine could have "the flexibility or ingenuity of the experienced observer," expressed concern that a fatigued and uncomfortable astronomer could accidentally make mistakes during the night, wasting increasingly valuable observing time. As a result, all telescope and electronic controls should be as simple and self-explanatory as possible. Only those the observer would need during the night should be available. The observer, according to Dennison, should be free to monitor the data as they were collected in order to "eliminate the feeling that he is being manipulated by the dictates of the equipment." But, to avoid systematic errors, the observer "must never be a link in the data-collection chain."

5 Change we need?

Not all astronomers felt at ease with the new technologies invading the observatory dome. At an ESO conference on telescopes in 1971, Dennison showed a diagram of the Hale Observatories' new computing system. On one side of the drawing, set apart from boxes labeled "teletype" and "future large computer" he included an icon identified simply as "Telescope." One astronomer, seeing this picture, remarked that in this new technology regime, the community's most treasured tool might just become what he sarcastically called "a big computer with a large optical analog-input at its periphery."

We can see the clash between these two cultures of doing astronomy at another conference ESO and CERN jointly organized in 1974. This was an especially critical time because astronomers throughout the world were in the process of building the next generation of large telescopes like the Anglo-Australian Telescope and the 4-m at Kitt Peak and Cerro Tololo. Debate over how these new facilities would be used was therefore of central importance. An entire morning session was devoted to the "Philosophy of Telescope Use." Speakers like Caltech's Jesse Greenstein assured the large international gathering that the telescope was still their central research tool. Nonetheless, Greenstein told colleagues that, while he was impressed and comforted by the extraordinary gains in data-gathering ability afforded by new electronic devices, they often caused him to "return from observing in a state of personal rage."

The 1974 ESO-CERN meeting also featured a lively panel discussion (chaired by Adriaan Blaauw, ESO's director) that reflected the state of unease and transition in the astronomy community. The astronomers who spoke that spring morning in Geneva reflected two different visions for the future of large telescope use. Harry van der Laan, a future ESO director, suggested that it might be better if the astronomer "doesn't go near the [observatory]...he communicates by telephone and telex and gets his data back." Of course there would be exceptional cases where interaction between the astronomer and the observatory staff might be necessary, but "big modern optical telescopes are evolving...where it should be mostly a hands-off policy as far as the using astronomer is concerned."

Another ESO astronomer, Jaap Tinbergen, made even more pointed remarks, saying that optical astronomers were "just too conservative, and too lazy in some cases, to try it any other way." He then proceeded to tell a story about a neophyte astronomer who wasted a night's worth of observing because he had not properly aligned his target on the spectrograph's slit. "Now that wouldn't have happened," Tinbergen concluded, "if he had allowed his spectra to be taken by an assistant."

Greenstein reacted angrily to this debate. He warned his colleagues not to make the mistake of believing that "they will be able to send messages" over the phone to the observatory and get reliable data. "I wouldn't trust anybody else," he contended, "Either you believe that the 'Establishment' has been a bunch of idiots... who like to be up on a freezing mountain, or you must believe that there was some point in it."

For tradition-minded astronomers like Greenstein, who had worked at telescopes of all sizes for nearly 40 years, the idea that the astronomer might not directly participate in data collection was anathema. For other astronomers, many of whom lacked California scientists' unparalleled access to big telescopes, there was not nearly enough telescope time to satisfy the needs of the community and any new technology that made observing more efficient was worth considering.

6 Reconsidering telescope design

Despite the reservations of Greenstein and his colleagues, these computer tools were incorporated directly into the design and construction of new and existing telescopes. Previously, large telescopes were understood by engineers as a system of separate parts—mirror, truss, dome. Yes, they interacted but their design was largely done independently. By the 1970s, telescope engineers started using computers to assist with engineering analyses. Prior to this, computer-aided design and analysis programs were not widely used, and telescope design was as much art as science. But, by separating the telescope into different "analytical segments," computer-assisted modeling, finite element analysis, and CAD programs let engineers model the mechanical behavior of telescopes before construction. This prompted one astronomer in the early 1970s to remark, "Now we have got rid of the observer, do we get rid of the engineer too by putting everything into a computer?"

Of course, as computer and electronic systems for the telescope became more sophisticated, there were changes in how the telescope was actually operated. For example, in the 1970s, project engineers working on the sophisticated Multiple Mirror Telescope (MMT) on Mt. Hopkins in southern Arizona built a facility that required a complex and connected system of lasers, electromechanical servos, and optical sensors.

Unlike conventional telescopes of the time, the MMT had no observer cage where the astronomer could ride and guide the telescope while collecting data. From the beginning, the MMT's designers envisioned that astronomers would operate the telescope remotely from a control room and move the telescope using a television system while monitoring the output of data on a computer display.

The sophisticated optics system of the MMT, for example, carefully combined images from each of the six 1.8-m primary mirrors and depended on three different electronic systems, linked together, to manage the movement of the telescope, the co-alignment of the mirrors, and its scientific instruments. Its ability to point and move quickly aided its efficient use. The astronomer was indeed becoming much more of a systems manager. MMT engineers actually went so far as to describe the telescope itself as a "large and unique peripheral device" for a computer system.

7 New modes of practice

Since 1945, scientists and engineers transformed the large optical telescope into the most visible part of a much bigger astronomical data-collecting network. Modern observatories like Gemini and the VLT are tools in a transoceanic, decentralized, data-oriented and collaborative research environment. They are linked in real-time to engineers and astronomers by high-speed data networks, fiber optic cables, video conferences, and the World Wide Web. This shift has entailed a re-casting of the telescope by astronomers and science managers as a factory of scientific data and scientists as customers who order up astronomical data that is delivered to them electronically while they monitor the process through Internet links.

Building giant new telescopes like Gemini and the VLT involved more than manipulating glass and steel to construct complex scientific instruments. For centuries, using a telescope meant going to the observatory at night, often alone, and collecting data. The advantage of classical observing was that control of the telescope rested with the astronomers. If they desired, scientists could always point the telescope at someplace new and hope for a discovery based on serendipity and instinct rather than extensive pre-planning. According to Alan Dressler, an astronomer at the Carnegie Institution of Washington, "A certain degree of cowboy-like behavior made a difference," Dressler explained, "People made their reputations based on how they performed at the telescope."

Ground-based astronomers who observed in the classical style, of course, took risks. If the weather was poor or cloudy they would collect few data. If an astronomer had an important observing program that called for the best possible seeing conditions and these did not materialize, then he had to fall back on other programs for that night. Classical observing was, in other words, something of a gamble when it came to getting data. There were more serious risks as well. In 1987, Marc Aaronson, a promising young scientist and skilled observer, was fatally injured while using the 4-m telescope at Kitt Peak.

As the new generation of giant telescopes saw first light in the last 15 years or so, astronomers developed new strategies for doing science with them. Many of the world's modern observatories, for instance, use some form of "queue observing." Adapted from the space telescope world, this mode enables observatory staff to collect scientific information and provide it in a standard format to researchers. The same staff uses computer algorithms to control scheduling and the order of executing research projects, taking into consideration weather conditions and the facility's time allocation agreements.

Astronomers' traditional concept of an "observatory" came to embrace a panoply of interdependent elements—the telescope's enclosure, instruments, mirror supports and control systems, data archives, and electronic linkages for remote observing. Tying these myriad systems together at Gemini or the VLT were miles of cables, wires, and fibers. Further complexity was added by the integration of new tools such as videoconferencing and Internet access which would allow astronomers to monitor the telescopes and participate in data collection remotely. Even local weather conditions on Mauna Kea and Cerro Pachón constituted part of the "observatory system." To optimize the thermal control of a thin primary mirror, for example, engineers needed to keep it close to the predicted temperature of the night air. Successfully doing this meant that staff needed extensive and accurate information on how the weather on the mountains varied seasonally as well as a system for predicting nighttime observing conditions from real-time meteorological data.

Advocates of queue observing insisted it enhanced efficiency. Unless they were familiar with the telescope and its instrumentation, astronomers might waste several hours in calibrating equipment before collecting any data. When the Very Large Telescope was completed in 2000, ESO managers had thousands of hours of telescope time to allocate annually. Each telescope cost about a dollar per second around the clock. "If a staff member's work can save two nights," one European scientist noted, "their salary is already accounted for." In many ways, observatories' emphasis on flexibility, adaptation, and streamlined efficiency resembled Japanese "just in time" manufacturing practices admired by American and European business leaders in the 1980s.

This style of research challenged some traditional views of what being an astronomer was about. Like Alan Dressler, Matt Mountain (former director of the Gemini Observatory and now head of the Space Telescope Science Institute) also opted for a cowboy analogy. "I do not believe discoveries are made at the telescope riding the weather like a cowboy riding a bucking bronco," he claimed, "Discoveries emerge at a computer terminal while trying to reconcile an awkward data set with preconceived models."

8 Data and more data

One of the things that strikes me when I read plans for the next generation of giant telescopes and sky surveys is the attention given to data flow and management—"science-oriented systems engineering." Just as Dennison located the telescope on the periphery of his chart in 1971, today's astronomer has become one node in the data collection chain, their expertise augmented by a whole host of specialists and systems. The data pipeline has become a central organizing idea in designing new telescopes and the telescope itself was transformed, like the atom smashers of the 1950s and 1960s, into a data factory. Astronomers' new lexicon included "productivity," "efficiency," and "mission statement," terms more common in the corporate boardroom than the telescope dome. Observatory managers began to speak of their colleagues as "customers" who actively shopped for telescopes equipped with the right instruments to make particular measurements.

Determining the catalyst for this shift in astronomers' thinking is difficult. The scientists themselves attributed the changes primarily to technological progress—bigger mirrors, more ambitious telescopes, and much greater computing power. Since the 1970s, the amount of information collected by telescopes worldwide had far outpaced gains made in light collecting power alone. While the total area of telescope mirrors around the world had increased by a factor of 30 or more since 1975, the number of pixels in a typical CCD detector grew by a factor of at least 3,000. The introduction and maturation of new communications tools—the World Wide Web, the Internet, videoconferencing, and high-bandwidth links connecting mountain peaks in Hawaii and Chile to astronomers in the mainland USA—also motivated astronomers to adopt new patterns of work and consider changes in their nighttime use of telescopes. Moreover, well-publicized science results from the Hubble Space Telescope encouraged scientists to maximize the capability of their new tools.

Not all changes occurred as a result of technological innovations. Additional impetus came from observatory directors and science managers who argued that the cost of building and operating new telescopes demanded greater efficiency. Some in the community recognized that even with smaller (and supposedly less complex) telescopes, the first night astronomers spent observing was often unproductive. In today's modern observatories, software engineers and observatory directors give considerable attention to the most efficient manner in which to examine, manage, and archive the vast amounts of data produced. To date, the Hubble Telescope has amassed some 120 terabytes of data. Future survey facilities like the Large Synoptic Survey Telescope are expected to produce that amount of data every week, a glut that has prompted some science writers recently to wonder whether the flood of data will presage an era where theory is less important, even unnecessary. Is today's astronomer someone who spends time at the telescope or has this occupation shifted to become much more focused on data management and mining?

9 Conclusion

Think back to that sign at Cornell that I described at the start of this essay. What does the injunction "know your telescope" mean in such a data-centric universe?

To some astronomers, first light at new observatories like the Very Large Telescope represented the loss of the romance they associated with telescopic observing and with it of? a certain degree of control. But with this loss came revitalized powers to observe the universe more efficiently.

Over the past half-century, a new generation of eyes –flesh and blood, glass and steel—has turned to the cosmos. If we look more closely at the practice of astronomy and how telescope technology has affected it, it seems to me that these people and their instruments increasingly appear as cyborgs—a blend of person, machine, computer working together. Since 1945, the telescope has merged with the computer, the web, the database into a hybrid instrument.

But there was nothing inevitable about this change. Computers, chips, circuits, and databases have not re-made astronomy. Rather, astronomers chose these new tools and systems and integrated them into their nightly work routines. And while scientists use them in a manner radically different from yesteryear's artistry, telescopes—still beautiful and still cantankerous—still produce new and exciting knowledge about the universe.

Bibliographic note For this paper, I drew upon two books that were helpful for setting up comparative examples; these included several chapters in Peter Galison, *Image and Logic: A Material Culture of Microphysics* (Chicago: University of Chicago Press, 1997) and David A. Mindell's *Digital Apollo: Human and Machine in Spaceflight* (Cambridge, MA: The MIT Press, 2008). Other than the discussion of particle physics and test pilots (which come from Galison and Mindell's books), all direct quotes are taken from my 2004 book *Giant Telescopes: Astronomical Ambition and the Promise of Technology* (Cambridge, MA: Harvard University Press, 2004); readers interested in the original sources of quotes used are directed to it.

Exp Astron (2009) 25:91–106
DOI 10.1007/s10686-009-9159-4

REVIEW ARTICLE

Astronomical spectroscopy in the last four decades: survival of the fittest

Guy J. Monnet

Received: 24 February 2009 / Accepted: 25 February 2009 / Published online: 20 March 2009

Abstract Spectroscopic astronomical instrumentation has much evolved in the last 40 years. Long-slit grating spectrographs with a photographic plate as the detector working in the 0.3–1 μm range were prevalent up to the early 1970s. The replacement of photographic plates by two-dimensional digital detectors provided gains in sensitivity of two orders of magnitude and much better photometric and radial velocity precision, and opened the 1 to 25 μm infrared domain. Another gain in speed by up to two orders of magnitude was then obtained through the development of various spectroscopic systems, each optimized for a subset of astronomical objects. This development was underpinned by a number of technological advances, in particular the development of automatic data reduction pipelines using sophisticated algorithms. With ever larger and more complex instrument systems for the present 8–10 m diameter telescopes—and soon even more for the next generation of Extremely Large Telescopes, the development of an instrument is now a big enterprise, ranging all the way from long-term enabling technology efforts to management of large teams for construction and deployment over typically 7–8 years.

Keywords Astronomy · Instrumentation · Spectroscopy · Detectors · Image analysis · Optical · Near-infrared

G. J. Monnet (✉)
European Southern Observatory (ESO), Karl-Schwarzschild Straße 2,
85748 Garching beï München, Germany
e-mail: gmonnet@eso.org

1 Scope and content

This paper covers the fast development of night-time ground-based optical spectrometric instruments and their associated detectors over the last 40 years. Section 2 presents their venerable *ancestor*, the long-slit spectrograph cum photographic plate, and its few competitors circa 1970. The next two sections cover the 1975–1995 'golden decades', with the advent of 2-Dimensional (2-D) digital detectors (Section 3) and in parallel of whole new spectrometric *phyla* (Section 4). This rapid evolution has been and still is accompanied by inexorable increases in size and complexity, from the nimble instruments of the 1970s to their huge offspring presently in operation on 8–10 m class telescopes, and even more with the behemoths currently planned for the new generation of Extremely Large Telescopes (ELTs). The resulting challenges and possible solutions to avoid overweight/complexity collapse are addressed in Section 5. Conclusions are given in Section 6.

2 The ancestors

2.1 Photographic-plate detector

For more than a century, the detector of choice for astronomical imaging (through filters) and spectroscopy (with a slit spectrograph) in the optical domain (0.3–1 μm range) has been the photographic plate or less often film. In its post-1900 form, this was an excellent 2-D detector with a huge number of resolving elements (pixels, or px in brief), 10^8 to 10^9 for a 30×30 cm plate depending on photon flux. Very long integration times (tens of hours) were made possible by suitable pre- and/or post-exposure priming. It was also light, easy to use (no cryogenics, no electronics) and doubled as a write-once read-many mass memory. On the negative side, Quantum Efficiency (Q.E.) was very low, a few 10^{-3} at best, dynamical range small, data calibration difficult and of low precision, and data extraction highly cumbersome.

Interestingly enough, digital detectors with much better Q.E., dynamics, and more science-ready data output were already fully available by the 1960s, viz. photomultiplier tubes for the optical and photodiodes for the infrared, wetting the astronomer's appetite for digital as opposed to analog data. These single-pixel detectors were then used intensively for photometric measurement of point-like (e.g. stars or galaxy centers) objects. They were also used for infrared imaging for lack of a 2-D or even 1-D detector in that wavelength range, as well as for spectroscopy of single lines in galaxies when coupled to a scanning grating spectrometer. These rather simple instrument types quickly became obsolete with the availability of digital 2-D detectors (see Section 3 below). In another vein, there was also an alternative 2-D detector with much better Q.E. and a somewhat higher dynamics than the photographic plate, the Lallemand electronographic camera, but it was extremely difficult to operate and was ultimately abandoned when digital detectors went into full gear.

2.2 Long-slit spectrography (LSS)

At first coupled with a photographic plate, later with digital 2-D detectors, (see Section 3), the Long-Slit Spectrograph has been the workhorse for astronomical spectrometry over most of the twentieth century, essentially due to its appealing simplicity: the long slit is a natural match to the prism, or more often ruled grating, light disperser and the resulting data format quite easy to handle, e.g. to subtract airglow emission lines, a most important point in that largely pre-computer era. As illustrated in Fig. 1, LSS provides a 1-D sky coverage along the narrow slit and 1-D wavelength (λ) coverage in the orthogonal direction, with usually 1 octave (i.e. a factor 2) in wavelength span. Spectral resolution $\Re = \lambda/\delta\lambda$ (where $\delta\lambda$ is the resolved wavelength width) could range from a few 10^2 to 10^5, depending on the scientific aim of the instrument from distant galaxies to bright stars.

By providing *simple* spectroscopic data for any type of astronomical object from solar system bodies to galaxies to quasars while requiring only a *modicum* of data reduction and analysis, the final output in terms of number of refereed papers per total observing plus data reduction effort was neatly optimized. While this mode is no longer the dominant spectrographic species in large telescope instrumentation complements, it is still quite current on large telescopes.

2.3 2-D spectroscopic imagery

Imagery with at first wide-band color filters, then increasingly since the early 1960s through wide or intermediate-band interference filters (with respectively $\Re \sim 5$ and $\Re \sim 50$), provides full 2-D sky coverage on a single exposure but only 1 wavelength bin at a time and thus does not really qualify as a bona

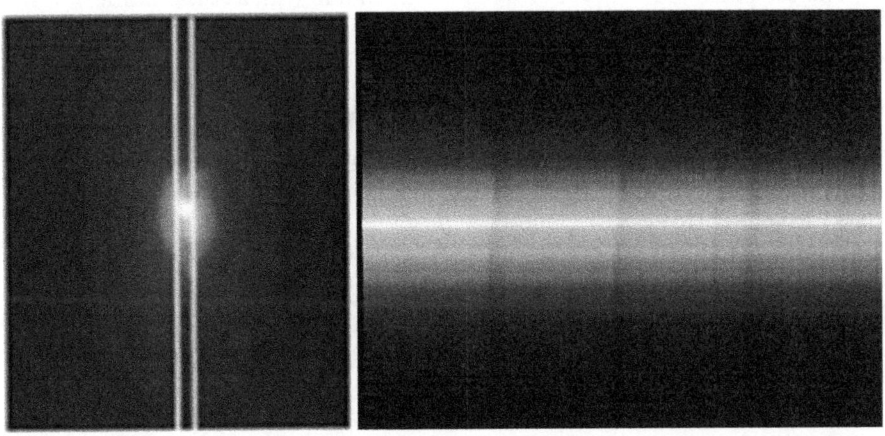

Fig. 1 *Left* K-band image of the central part of the active galaxy NGC 1808 with ISAAC at the VLT. *Right* K-band ISAAC infrared spectrum with the (vertical) slit positioned along the (vertical) galaxy major axis. CO bands are clearly seen; their strong tilts correspond to velocity shifts of ±110 km s^{-1} across the nucleus of the galaxy. ESO Press Release 18/01, August 2001

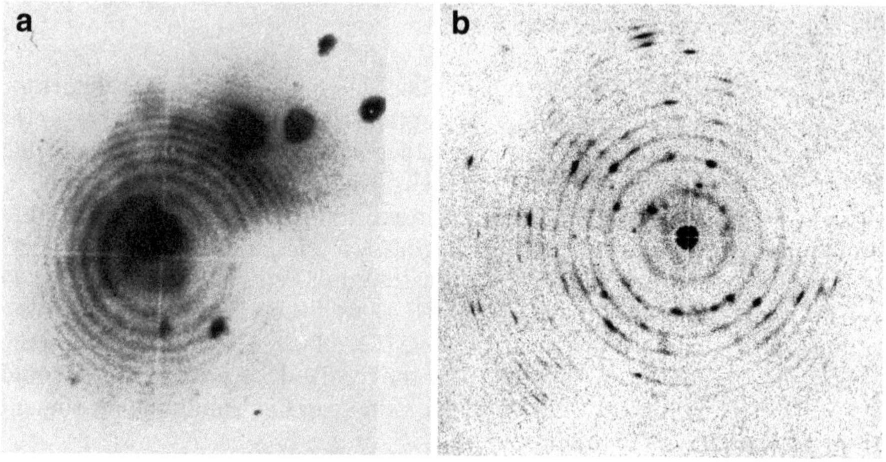

Fig. 2 **a** First astrophysical use of a Fabry–Perot etalon on the Orion nebula in the [OII] 3,727 Å 'nebulium' doublet at the Observatoire de Marseille 80 cm telescope by H. Buisson et al. [6]. **b** First use of a Fabry–Perot etalon on an external galaxy (M33; Ha emission line; [9]). From this single 6-h exposure at the OHP 193 cm telescope, radial velocities of some 300 HII regions were measured in the 15′ diameter field

fide spectroscopic technique. In practice, a few (<10) wavelength bins can be covered with successive exposures, each with a different filter. This scanning technique has always played and still is playing a major role in observational astronomy, in particular through massive multi-color imaging surveys of large fractions of the sky. An early example (circa 1900) is the huge Carte du Ciel project, even if it ultimately collapsed due to the unmanageable weight of data reduction.

The use of a fixed Fabry–Perot etalon in tandem with a color or interference filter to typically reach $\Re \sim 10^4$ over a large field did, if barely, qualify as a spectroscopic technique. With only one wavelength bin per spatial pixel, but ~8–15 in a very small spectral range across the field, it was strictly limited to kinematical studies of extended emission-line objects, viz. HII regions in our Galaxy and nearby gas-rich galaxies (Fig. 2a, b). With a rather heavy data extraction process and the need to get matching filters for basically every object/emission line combination, quite understandably this technique remained a connoisseur's item limited to a handful of team instruments.

3 The golden decades (1975–1995): I. The advent of 2-D digital detectors

Circa 1970, a flurry of different types of 2-D image intensifiers (developed for military purposes) coupled first with photographic plates, and then to various brands of television tubes (developed for the entertainment industry) became available. This culminated soon with Photon Counting Tubes (PCT) where photoelectrons are individually counted. The rapid rise of these detectors in the decade and the concomitant fall of the photographic plate can be seen from the left half of Fig. 3. The fully digital PCT featured much better Q.E. (~10% peak value), a fairly good λ range (typically 0.4 to 0.8 μm) and extremely low-noise, a few counts per pixel per hour, giving a huge boost to the detection of faint objects On the other hand, formats were small (~256 × 256 px) and dynamics severely limited at the high flux limit (<1 count/px/s).

By 1980, a digital version of the photographic plate (i.e. an integrating device, not a photon counting one), the Charge Coupling Device or CCD in short was developed for the consumer market, giving another huge Q.E. boost, initially up to about 50% in the optical domain. Early devices had small formats ($<10^5$ pixels) and huge read-out noises ($>100e^-$ rms). All image intensifier brands, including photon-counting ones, were nevertheless quickly replaced, as seen from the right half of Fig. 3, for two main reasons, (1) owing to their higher Q.E. and much higher dynamics, even early CCDs were already much better at detecting small light fluctuations over a strong background, by far the most usual case in astrophysics whenever imaging the faintest galaxies over the night-sky background, taking low-resolution (a few 10^2) spectra of faint galaxies or very high signal-to-noise-ratio high-spectral-resolution ($>10^5$) spectra of bright stars/quasars; (2) CCDs are extremely rugged (idiot/astronomer-proof) devices while PCTs could be permanently damaged in a blink by the occasional bright star in the field. As of now, the modern CCD mosaics with its huge format (10^9 px for the just-commissioned Pan-STARRS camera), nearly 100% peak Q.E. and very small read-out noise ($\sim 2e^-$ rms), is reigning supreme in the optical range. CCD mosaics sizes have actually followed a Moore-like law from 1983 to 2008 with a doubling in the number of pixels every 33 months.

Progress has been even more impressive in the near-Infrared (NIR) 1–5 μm domain. Single pixel detectors were used up to 1983 where the first 2-D digital Charge-Injected Devices (developed for military purposes) were made available. Q.E. also approaches the 100% level and read-out noise is now routinely lowered to $\sim 5e^-$ rms by clever reading strategies, while close to 10^8 pixel arrays are currently being deployed. NIR array mosaic sizes have also

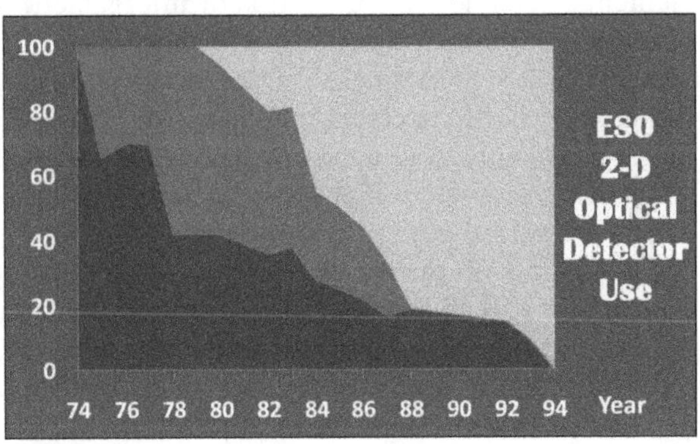

Fig. 3 The percentage use of the three main types of detectors (from left to right: photographic plate, image intensifiers, CCDs) at ESO from 1974 to 1994. The significant photographic plate tail after 1985 comes from the completion of the southern sky imaging survey at the 1-m ESO Schmidt Telescope

followed a Moore-like law from 1983 to 2008, with a doubling in the number of pixels every 24 months. 2-D arrays with up to 10^6 pixels are now also available for the 8–25 μm domain. Owing to this strong detector development across the board, instrumentation complements on Large Telescopes now usually cover the whole available range from the UV atmospheric cutoff at 0.3 μm to the IR one around 25 μm, or more than 6 octaves in wavelength.

4 The golden decades (1975–1995): II. Whole new spectroscopic phyla

Astrophysical objects feature highly diverse 2-D spatial shapes with sizes ranging from slightly under 1″ (e.g. a stellar seeing disk) to a few arc second (e.g. intermediate-redshift galaxies) to 1′ or more (e.g. stellar components in nearby galaxies). Their detectable 1-D wavelength shapes can range all the way from one emission line only (e.g. Lyα in high-z proto-galaxies) to a continuum with thousands of absorption lines over many octaves. Complete spectroscopic coverage on an object requires squeezing a 3-D data 'cube' (two dimensions of space plus one in wavelength) on detectors which presently offer two dimensions only. The impetus to develop the very diverse types of spectrographs in operation today came from this need to efficiently obtain data 'cubes' of highly different space-wavelength shapes; the opportunity came from the development of sophisticated data extraction and analysis software for the digital data given by the new generation of detectors presented in Section 3. Through such a substantial set of specialized species, observing efficiency can usually be boosted by up to a factor 100 or even more depending on the science targets, compared to the much simpler long-slit spectrography.

4.1 Point-like targets; wide wavelength coverage (one spatial super-px; $>10^5$ spectral px)

High-signal-to-noise-ratio ($S/N \geq 10^3$), high-resolution ($\Re \sim 10^5$) spectroscopic studies of stars or quasars over a huge wavelength range have played a major scientific role for e.g. determination of primordial abundances of light elements, detection and evolution of Intergalactic Matter and detection of the oldest stars in our Galaxy. The conundrum of fitting more than 10^5 resolved spectral pixels on a single 2-D detector is elegantly solved with the cross-dispersed echelle spectrograph: light from the $\sim 1''$ diameter seeing-enlarged target is usually fed to a narrower slit (e.g. $0.4 \times 2.5''$) via an image slicer; the slit is dispersed with a high-order echelle grating while the many overlapping orders are sorted out in the orthogonal direction by a lower-dispersion single-order grating or prism. The meters-long total spectrum, usually covering 1 octave (e.g. 0.45 to 0.90 μm), is therefore cut into say ~50 slices covering the full 2-D detector. Figure 4a shows one early-1970 example using an image intensifier, but it is only in the late 1980s, with relatively large-format CCDs (2×2 k px or more) and, most importantly, sophisticated data extraction and

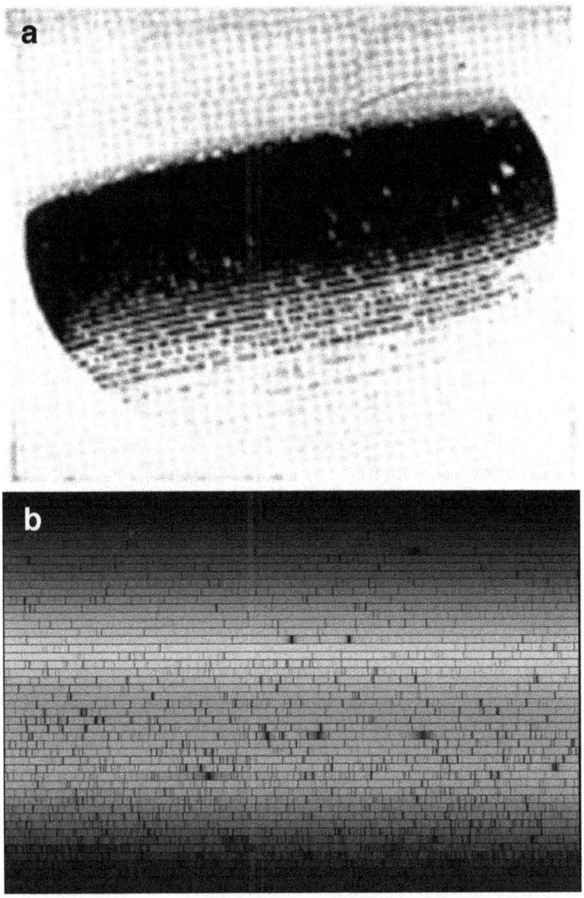

Fig. 4 **a** Cross-dispersed spectrum of Arcturus on an image intensifier plus photographic plate with the Harvard 61″ telescope [17]. **b** A true-color stellar spectrum made with HIRES at the Keck 10-m telescope. Courtesy W.M. Keck Observatory; http://www2.keck.hawaii.edu/inst/hires/

reduction algorithms, that the technique has become commonplace. See Fig. 4b for a modern (circa 1994) example with the 10-m Keck HIRES spectrograph.

The same approach is now also being used in the 1 to 5 μm NIR domain, where it has essentially outlived the Fourier Transform Spectrometer, a clever scanning technique using the single-pixel IR detectors available in the 1960s to early 1980s. Nor is the cross-dispersed technique now limited to the very high resolution range. One example is the moderate-resolution ($\mathfrak{R}$ ranging from 4,000 to 14,000) cross-dispersed 'X-shooter' spectrograph currently being put into operation at one ESO 8-m telescope (http://www.eso.org/sci/facilities/develop/instruments/xshooter/). With three different channels fed with dichroics, it covers a wide wavelength range—0.3 to 2.4 μm—in one shot and can be considered in terms of instrumentation ecological space as a modern avatar of the long-slit spectrograph.

4.2 Extended targets with small wavelength coverage (10^7 spatial px; ~20 spectral bins)

The main science application is to measure the 2-D velocity field of ionized gas in galaxies, including our own. The scanning Fabry–Perot technique follows closely the fixed Fabry–Perot technique (see Section 2.3); the crucial twist invented by Brent Tully (Fig. 5) is to scan the etalon over one full order—usually in 12 to 25 resolved spectral bins—and extract a full (if small) spectrum at each of the detector pixels. This results in single-line spectroscopic data over the full field of view covered by the detector (usually a few arc minutes), at the cost however of huge total integration times. This technique was well established by the early 1980s with sophisticated data extraction and reduction algorithms and photon counting detectors. It can easily reach the needed spectral resolutions between 10^3 and 10^4.

Nevertheless, this approach has always been limited to a handful of mostly team instruments (as opposed to general-use ones) and is still very much an endangered species. The quick demise of PCTs as observatory-maintained detectors (Section 3) did not help as their zero read-out noise and time-resolving ability made them ideal for such a scanning technique, in particular to beat atmosphere transmission and seeing fluctuations by scanning the full spectral range in typically 30 s. and repeating for up to tens of hours. However the most important genetic deficiency lies, as already pointed out by H. Buisson et al. [6], in having to use a fixed filter to select one of the many etalon orders.

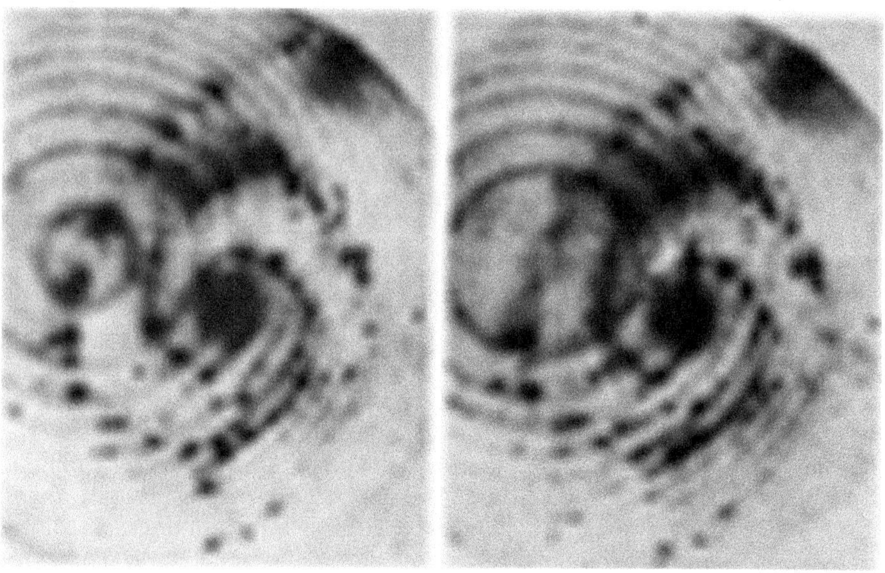

Fig. 5 First application of the scanning Fabry–Perot technique to M51 in Hα 6,563 Å light by B. Tully [23] at the KPNO 84″ telescope. Two shots on the image intensifier plus photographic plate detector with different etalon spacing are shown

The scanning etalon output on each spatial pixel is thus modulated by the filter band-pass, making it all but impossible to study individual absorption lines in galaxies, a truly huge loss in ecological space. To add insult to injury, the technique still requires a large filter bank to move from one object/line combination to another, a rather unappetizing fixture for a general-use instrument.

There are potential solutions. In the NIR, a spectrographic variant of the scanning Fourier Transform Spectrometer, using a 2-D IR array, invented by Jean-Pierre Maillard, has been successfully used at the 3.6 m CFHT telescope for more than a decade, including on absorption lines [22]. It is however no longer in operation essentially due to the 'non-standard' complexity of the instrument. Another approach using two Fabry–Perot (one high-order, one low-order) scanning synchronously in tandem is currently being developed [18]. As long suggested, this should also open the rich pastures of the absorption lines, but the Jury is still out.

4.3 Many point-like targets; wide patrol field (>10^3 objects; ~4×10^3 spectral px)

A long-slit spectrograph can usually gather the spectra of two objects at once by orienting the slit; the interest of getting a much higher multiplex (number of objects studied simultaneously) by using multiple slitlets, each pointed on a separate object in the instrument field of view, was already pointed out in the mid-1960s within the KPNO preliminary concept for the 15-m New generation telescope (Ngt) project. The key application envisioned was the study of clusters of galaxies with hundreds of targets in moderately large fields (a few arc minutes). The concept was soon verified (e.g. the author of this paper made a successful if unpublished test on 24 HII regions in M33 at the OHP 193 cm telescope in 1971) but it took the 1980s with the high QE, high dynamics CCDs, almost real-time mask-making devices and efficient data extraction and reduction algorithms for the technique to take off, using either multi-slits or multi-fibers (Fig. 6a, b).

In the following three decades, this phylum has been highly successful for the study not only of galaxy clusters but also of the Universe Large Scale Structure through huge surveys of field galaxies. Multi-object spectrographs, as both the multi-slit and multi-fiber variants are collectively called, are now ubiquitous on the 4-m, 8–10-m, as well as on survey facilities, e.g. 2DF and Sloane, including a few working in the NIR. One NIR version will actually relatively soon be in operation on board of the James Webb Space Telescope (JWST).

4.4 Moderate-size target; wide wavelength coverage (~60^2 spatial px; 4×10^3 spectral px)

This is arguably the most common individual astronomical object format, from solar system bodies, resolved gaseous and stellar components in galaxies, all

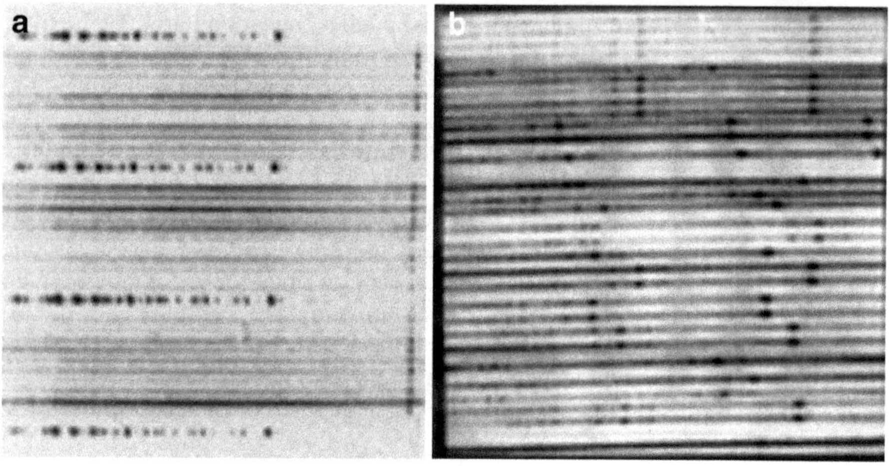

Fig. 6 **a** First light in the multi-fiber mode (3″ diameter fibers on 26 individual galaxies with all fibers realigned on a single entrance slit) with the MEDUSA instrument on the Abell 754 Cluster by J.M. Hill et al. [14] at the Steward 2.3 m telescope. **b** First light of a multi-slit spectrograph on the 0054 + 1654 Cluster by H. Butcher [7] at the Kitt Peak 4-m Telescope

the way up to the highest redshift proto-galaxies. Actually, nearly all astrophysical extended object shapes, except edge-on disks of SO/Spiral galaxies, match a square spatial format much better than a long-thin slit. The basic trick of what its pioneer Georges Courtès [11] christened Integral Field Spectrography (IFS) is to slice a small squarish field at the telescope focus and reconstruct either a long-slit or regularly spaced multi-slits feeding a classical grating spectrograph. There are three main slicer variants in operation, using either fibers, lenses, or mirrors (see e.g. [8]). A fourth, using micro-slices on an anamorphic image of the field, has been proposed by S. Morris et al. [19].

While the IFS concept was first successfully demonstrated on-sky by Christian Vanderriest in 1980 (Fig. 7a), it took more than a decade before it began to be operated regularly, first with team instruments (see first lights of the three variants respectively in Fig. 7a–c), then also as general use facilities, essentially because of the crucial need for providing sophisticated user-friendly data extraction, reduction and analysis pipelines to often at first skeptical potential consumers.

A hybrid multi-object/integral field technique uses multiple deployable fiber-based Integral Field Units in a large patrol field to get coarse spatial information ($<10^2$ spatial pixels) on a moderate ($<10^2$) number of targets at once. It was first operated at ESO on the FLAMES instrument in 2003 (see Fig. 7d).

The fully flexible 'HONEYCOMB' concept able to cover a number of arbitrarily-shaped regions in the field has been developed by J. Bland-Hawthorn et al. [4].

IFS presently operated or under development come in all three flavors, with the mirror slicer version most popular as the only one that offers a nearly

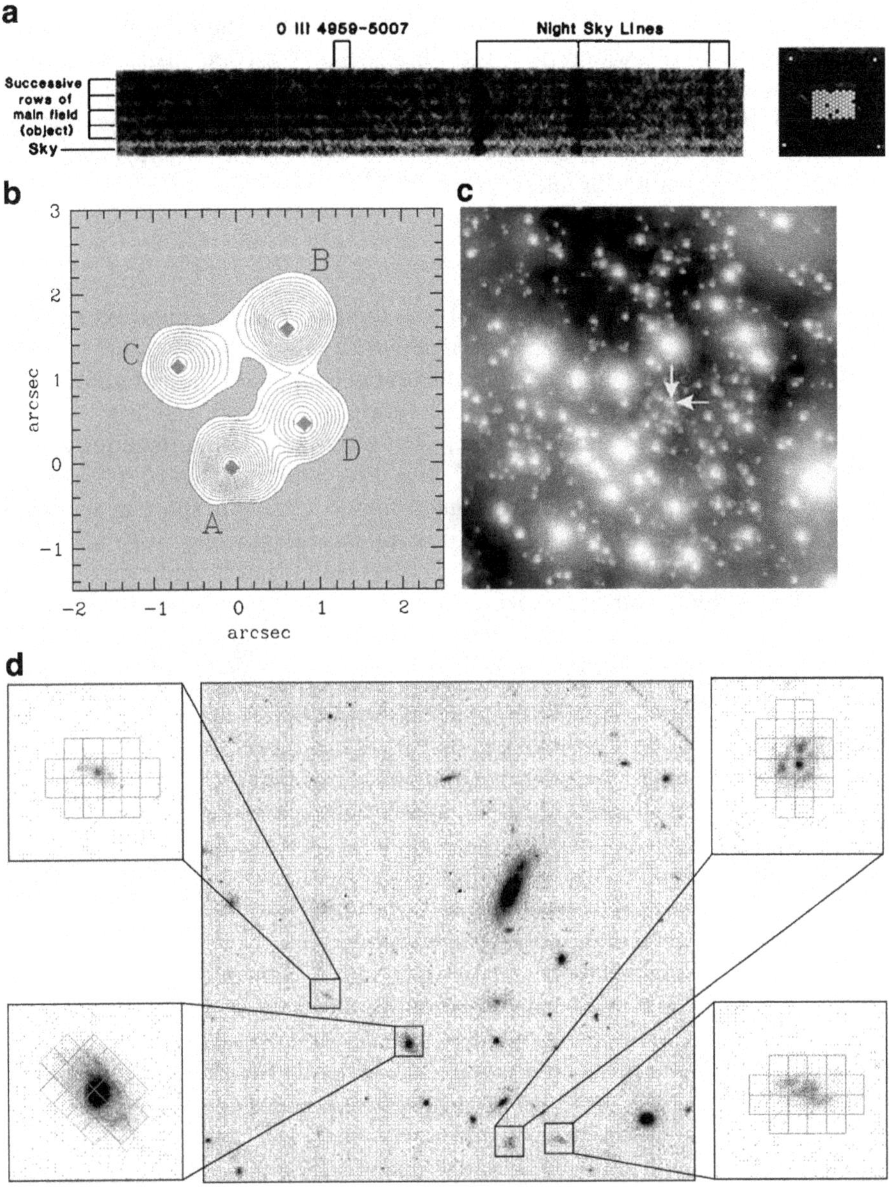

Fig. 7 **a** First light of the fiber-based SILFID instrument at the Mauna Kea UH 2.2 m UH telescope with multiple spectra across the active galaxy 3C 120 [24]. **b** First light of the lens-based TIGER instrument, showing the CIII] 1,909 Å emission line intensity map of the Einstein Cross [1]. **c** 8 × 8″ near-IR image of the Galactic center (NACO instrument at the VLT). In an early observation with the MPE 3-D cryogenic spectrograph, L. Weitzel et al. [25] simultaneously got 12 K-band stellar spectra in this field at the 2.2 m La Silla telescope. This first IFS working in the near-IR used an innovative mirror slicer concept, which now exists in various avatars (see e.g. [10]). **d** Principle of the deployable Integral Field Units technique [13]

100% packing ratio of the data cube on the precious (and costly) detector pixels. They feature widely in large optical-IR Telescope instrumentation. This will relatively soon include the space-based JWST which holds both near and mid infrared IFS. IFS also features prominently in presently envisioned instrumentation for the upcoming ELTs, as even the most distant galaxies are not point-like anymore for telescope sizes of 25 m or more.

4.5 Conclusions

The many different ways to squeeze a three-dimensional data cube onto a two-dimensional-only detector summarized above all suffer from the same 'dimensional curse': all methods are equally efficient in generating the same data cube volume using the same number of detector pixels at the same time (see e.g. [2]). The universal spectroscopic instrument that would cover simultaneously both a wide wavelength range and a huge field is thus essentially out of reach.

One 'easy' way out is to use a scanning technique, either in space by scanning the 2-D field with a 1-D long-slit [26] or in wavelength (e.g. with a Fabry–Perot, see Section 4.2 above). However, all on-ground scanning techniques are vulnerable to temporal variations in atmospheric transmission, background emission and turbulence seeing, actually the main reason why integral field spectroscopy which by nature is largely immune from these temporal variations is becoming so strong. Besides, it would just take forever to build a huge data cube in that way.

The brute force way, as exemplified by the MUSE and VIRUS projects in Section 5.1 below, is to build non-scanning instruments with a huge number of spectroscopic pixels—10^8 and more versus the classical $\sim 10^7$—at the cost of a big surge in size, cost and if to a lesser extent complexity. There are thus some practical limits and none of the instruments being contemplated now would fully cover say a half a degree field on an 8-m telescope over an octave or more in wavelength.

5 Rise of the giants

The increase of telescope size over the last 40 years from the 4-m to the 8–10 m class has been accompanied by a large parallel increase in the size and complexity of spectroscopic instruments. There are quite a number of reasons, in particular the following: (1) the availability of large detector arrays makes it scientifically attractive to design instruments with large field and wavelength coverage, albeit with a concomitant cost and size impact; (2) the now nearly perfect detector sensitivity makes it possible to effectively address simultaneously thousands of targets with multi-object spectrographs, hence requires large fast robotic positioners; (3) matching the decreasing detector angular pixel size to the basically constant angular seeing disk requires large gratings and camera optics; (4) IR instruments coupled to large digital arrays need to be housed in huge cryogenic vacuum vessels and (5) some new science goals require ultra-high precision measurements on the detector, e.g. getting spectral line positions at the Å level and extracting minute light fluctuations ($<10^{-6}$) for respectively exo-planet indirect and direct detection.

Conversely, there is a significant mitigating factor in the relatively new development of adaptive optics systems feeding the instruments. They offer the often scientifically attractive mode of delivering images down to the telescope diffraction limit, e.g. 30 milli-arc second (mas) at 1 μm for an 8-m and 5 mas for a 40-m telescope, both well below the best turbulence seeing at any site on Earth. Matching these small seeing disks to a detector pixel is much easier, requiring more reasonable size instruments. On the other hand, the added adaptive optics systems are large and highly complex in their own rights and need to be connected to the instruments with micrometer stability over the full integration time (tens of minutes).

Besides the two deciding factors—availability of large digital detector hardware (detector chip and electronics controller) and development of sophisticated software pipelines for data handling and analysis—many other technological developments contributed to make modern instrumentation possible. This includes large size refracting and reflecting optics with efficient anti- or super-reflection coatings, large efficient reflective and transmission gratings, and a host of smart focal plane paraphernalia like multi-mode fibers, laser drillers, robotic cryogenic positioners and mini-optics assemblies (fiber dissectors, lens arrays, mirror slicers). As essentially always since at least Galileo's telescope, most of these technological advances came from civil and/or military industrial developments, very often tuned further for optimal astronomical performance through combined industry and research institutes development and testing work.

The ever increasing ambitions for new generations of instruments for the 8–10 m and the formidable challenge just looming over the horizon for the ELTs are good reasons to worry about a possible 'dinosaur demise', where future instruments (and the people that design, build, integrate and operate them) would just collapse under excessive weight and unbearable complexity. New avenues in spectroscopic instrument concepts are currently being explored across the globe to evade such a gloomy outcome. Three notable tracks, viz. multiple small instruments, huge numbers of micro-instruments and detector-only spectroscopy are briefly addressed below.

5.1 Multiple small instruments

It is conceptually easy to replace a single huge instrument by a moderate number (a few tens) of smaller replicated ones, paving either the field or the telescope pupil (or both). One example is the MUSE project [3] with 24 simple identical spectrometers-cum-image slicers making up for a survey-type integral field system, giving in a single exposure 9×10^4 spectra, 1 octave long, over a $1 \times 1'$ field on an 8-m ESO Unit Telescope. An even more ambitious project features 145 small replicated VIRUS spectrometers [15] at the Hobby–Eberly Telescope. It is too early to pass final judgment on these pathfinders to ELT instrumentation; current outcome is that the approach looks very attractive including in cost, while requiring quite elaborate distribution systems for vacuum, cryogenics and detector control.

5.2 Micro-instruments

Integrated photonic techniques, developed by industry for telecommunication via optical fibers, offer very attractive opportunities with the promise of cheap mass-produced adjustment-free innovative optical components, all the way up to turn-key full spectrometers complete with detectors. Bragg single-mode photonic fibers have been developed by J. Bland-Hawthorn [12] to suppress the airglow lines in the J and H-band: faint object spectrometers between 0.6 and 1.75 μm currently are forced to operate at a relatively high spectral resolution (>3,000) in order to detect faint-object spectra between the many strong emission lines emitted by the Earth upper atmosphere. With such pre-filtering with a photonic fiber, a much smaller and more sensitive spectrometer, with, say, a spectral resolution of 750 only, could be used instead. J. Bland-Hawthorn and A. Horton [5] have also tested an H-band commercial photonic disperser using phased fiber arrays which could replace an entire spectrograph. E. LeCoarer [16] is developing the integrated photonics counterpart of a Fourier Transform Spectrometer, complete including the detector.

This field is picking up momentum and clearly offers much hope of pulling out of our present size-cost-complexity quagmire. There is one strong proviso however. By design, these single-mode components cover only one diffraction core, hence a few mas on the sky. This is no problem for small-field (one diffraction-limited core only) interferometric observations using pairs of telescopes. For more conventional large telescope instrumentation, this is a major challenge as huge numbers of individual components must then be assembled to cover the full field, from typically 10^4 components for spectrometry at the diffraction limit to 10^7 for diffraction-limited imagery, and with another factor of 10^{2-3} for maximum-light-gathering instruments.

5.3 3-dimensional (3-D) detectors

None of the diversity of complicated and bulky spectrographic systems used or planned at this time would actually be needed if we could get that Holy Grail of astronomical instrumentation, a large noiseless 3-D detector giving both the position and the energy (wavelength) of every photon impinging on its surface (as an aside an ever holier detector would also provide the photon direction, then making telescopes unnecessary!). While commonly used in the X-ray domain, such detectors are still very much prototypes in the optical to very near IR region. Examples are the Superconducting Tunnel Junction (STJ) detector [20] and the Transition Edge Sensor (TES) spectrophotometer [21]). One strong limitation is that they have, at least presently, far too few pixels ($<< 10^3$) to be able to compete with 1 billion pixels classical imaging cameras even if the latter give only one spectral pixel per exposure versus about 10–20 with these 3-D detectors. An even more limiting—and possibly quite fundamental—factor is that their spectral resolution ($\Re << 10^2$) is far too small for them to be able to significantly encroach on the hunting grounds of classical faint-object spectrographs.

6 Conclusion

Instrumentation is a vital component of large telescope systems, with a major impact on their scientific breadths and successes. Conception, development and deployment of maybe 15 different instruments on any large telescope over its lifetime is now a big enterprise involving many research institutes and industries over the globe. As an example, even ESO's preliminary conceptual studies for its 42-m diameter ELT project instrumentation complement already involve directly some 30 research institutes from nine different European nations. Much ingenuity is being spent on adapting the most advanced technological components and systems for astronomical instrumentation needs, while at the same time the ever-growing size and complexity of the instruments require careful management of large and diverse teams over nearly a decade in time. How well these quite contradictory requirements will be met will bear much weight in the successes—and pitfalls—of observational astrophysics in the next decades.

In the long term, the evolution of night-time astronomical instrumentation will continue to depend much on technological developments, often prompted by the evolving scientific needs, from small evolutionary steps to possibly giant leaps thanks to 'disruptive' approaches, like 3-D detectors if they ever come to full fruition. In any case, the evolutionary race experienced in the last four decades where many instrument species—many more actually than could be accounted here in a few pages—have struggled for survival is bound to continue unabated.

Acknowledgements The author is deeply indebted to the many individuals and organizations which have fueled the rapid development of astronomical instrumentation in the four decades covered here. I would like to apologize for having been forced to skip many clever and highly useful innovations in the field for lack of space (and no doubt in some cases from lack of insight). I have tried hard to give proper credit to the pioneers but again would like to apologize in advance for any glaring mistake or omission.

References

1. Adam, G., Bacon, R., Courtès, G., Georgelin, Y., Monnet, G., Pécontal, E.: Observations of the Einstein cross 2237+030 with the TIGER integral field spectrograph. A&A **208**, L15–L18 (1989)
2. Allington-Smith, J.: Integral field spectroscopy for panoramic telescopes. RevMexAA (Serie de Conferencias) **28**, 17 (2007)
3. Bacon, R.: Probing unexplored territories with muse: a second-generation instrument for the VLT. ESO Messenger **124**, 5 (2006)

4. Bland-Hawthorn, J., McGrath, A.J., Saunders, W., Haynes, R., Gillingham, P.: Honeycomb: a concept for a programmable integral field spectrograph. Proc. SPIE **5492**, 242 (2004). doi:10.1117/12.550291
5. Bland-Hawthorn, J., Horton, A.: Instruments without optics: an integrated photonic spectrograph. Proc. SPIE **6269**, 21 (2006)
6. Buisson, H., Fabry, C., Bourget, H.: An application of interference to the study of the Orion nebula. Ap. J. **40**, 241 (1914). doi:10.1086/142119
7. Butcher, H.: Multi-aperture spectroscopy at kitt-peak. Proc. SPIE **331**, 296 (1982)
8. Cameron, M., Weitzel, L., Krabbe, A., Genzl, R., Drapatz, S.: 3D: the new MPE near-infrared field imaging spectrometer. B.A.A.S. **25**, 1468C (1993)
9. Carranza, G., Courtès, G., Georgelin, Y., Monnet, G., Pourcelot, A.: Interferometric study of ionized hydrogen in M 33: new kinematical and physical data. Ann. Astrophys. (Paris) **31**, 63 (1968)
10. Content, R.: New design for integral field spectroscopy with 8-m telescopes. Proc. SPIE. **2871**, 1295 (1997). doi:10.1117/12.269020
11. Courtès, G.: An integral field spectrograph (IFS) for large telescopes. In: Humphries, C.M. (ed.) Instrumentation for Astronomy with Large Optical Telescopes, vol. 92, p. 123. Dordrecht, Reidel, ASSL (1982)
12. Ellis, S.C., Bland-Hawthorn, J.: The case for OH suppression at near-infrared wavelengths. Mon. Not. R. Astron. Soc. **386**, 47 (2008). doi:10.1111/j.1365-2966.2008.13021.x
13. Flores, H., Puech, M., Hammer, F., Garrido, O., Hernandez, O.: GIRAFFE multiple integral field units at VLT: a unique tool to recover velocity fields of distant galaxies. A&A **420**, L31–L34 (2004)
14. Hill, J.M., Angel, J.R.P., Scott, J.S., Lindley, D., Hintzen, P.: Multiple object spectroscopy - The Medusa spectrograph. Astrophys. J. **242**, L69–L72 (1980). doi:10.1086/183405
15. Hill, G.J., MacQueen, P.J., Adams, J., Tufts, J., Blanc, G., Smith, M.P., Roth, M.M., Kelz, A., Segura, P., Gebhard, K., Good, J., Drory, N.: VIRUS-P: a powerful integral field spectrograph designed for replication. B.A.A.S. **39**, 747 (2007)
16. LeCoarer, E., Blaize, S., Benech, P., Stefanon, I., Morand, A., Lérondel, G., Leblond, G., Kern, P., Fedeli, J.M., Royer, P.: Wavelength-scale stationary-wave integrated Fourier-transform spectrometry. Nature Photonics **1**(8), 473 (2007)
17. Liller, W.: High dispersion stellar spectroscopy with an echelle grating. Appl. Opt. **9**(10), 2332 (1970). doi:10.1364/AO.9.002332
18. Marcelin, M., Amram, P., Balard, P., Balkowski, C., Boissin, O., Boulesteix, J., Carignan, C., Daigle, O., de Denus Baillargeon, M.-M., Epinat, B., Gach, J.-L., Hernandez, O., Rigaud, F., Vallée, P.: 3D-NTT: a versatile integral field spectroimager for the NTT. SPIE Conf., vol. 7018. Marseille (2008)
19. Morris, S., Content, R., Sharples, R., Bower, R., Davies, R., Baugh, C.: A million element integral field unit (MEIFU). Proceedings of the ESO conference on scientific drivers for ESO future VLT/VLTI instrumentation, p. 99 (2002)
20. Perryman, M.A.C., Favata, F., Peacock, A., Rando, N., Taylor, B.G.: Optical STJ observations of the crab pulsar. A&A **346**, L30 (1999)
21. Romani, R.W., Miller, A.J., Cabrera, B., Figueroa-Feliciano, E., Nam, S.W.: First astronomical application of a cryogenic transition edge sensor spectrophotometer. Astrophys. J. **521**, L153–L156 (1999). doi:10.1086/312202
22. Simons, D., Clark, C.C., Smith, S., Ken, J., Massey, S., Maillard, J.P.: CFHT's imaging Fourier transform spectrometer. Proc. SPIE. **2198**, 185 (1994). doi:10.1117/12.176746
23. Tully, R.B.: The kinematics and dynamics of M51. 1. The observations. Astrophys. J. Suppl. Ser. **27**, 415 (1974). doi:10.1086/190304
24. Vanderriest, C.: A fiber-optics dissector for spectroscopy of nebulosities around quasars and similar objects. PASP **92**, 858 (Dec. 1980–Jan. 1981)
25. Weitzel, L., Krabbe, A., Kroker, H., Thatte, N., Tacconi-Garman, L.E., Cameron, M., Genzel, R.: 3D: The next generation near-infrared imaging spectrometer. A&AS **119**, 531 (1996)
26. Wilkinson, A., Fosbury, R., Wallace, P., Sharples, R.: Stellar Dynamics of Cen A. B.A.A.S. **15**, 921 (1983)

Exp Astron (2009) 25:107–124
DOI 10.1007/s10686-009-9140-2

REVIEW ARTICLE

The history of radio telescopes, 1945–1990

Woodruff T. Sullivan III ·

Received: 27 November 2008 / Accepted: 6 January 2009 / Published online: 5 February 2009

Abstract Forged by the development of radar during World War II, radio astronomy revolutionized astronomy during the decade after the war. A new universe was revealed, centered not on stars and planets, but on the gas between the stars, on explosive sources of unprecedented luminosity, and on hundreds of mysterious discrete sources with no optical identifications. Using "radio telescopes" that looked nothing like traditional (optical) telescopes, radio astronomers were a very different breed from traditional (optical) astronomers. This pathbreaking of radio astronomy also made it much easier for later "astronomies" and their "telescopes" (X-ray, ultraviolet, infrared, gamma-ray) to become integrated into astronomy after the launch of the space age in the 1960s. This paper traces the history of radio telescopes from 1945 through about 1990, from the era of converted small-sized, military radar antennas to that of large interferometric arrays connected by complex electronics and computers; from the era of strip-chart recordings measured by rulers to powerful computers and display graphics; from the era of individuals and small groups building their own equipment to that of Big Science, large collaborations and national observatories.

Keywords Radio telescopes · History · Radio astronomy

1 Introduction

The advent of radio astronomy was far more than simply the introduction of another interesting technique. In today's astronomical world it is difficult

W. T. Sullivan III (✉)
Department of Astronomy, University of Washington, Seattle, WA, USA
e-mail: woody@astro.washington.edu

to appreciate the magnitude of the effect caused by discoveries made at wavelengths a million times longer than those of (optical) astronomy. There had been infrared observations of the sun, moon, and a few stars, but these were not revolutionary in character and were conducted on familiar (optical) telescopes. But starting at the time of World War II, radio physicists and engineers (not astronomers) used radio antennas of many kinds to reveal a wholly different universe (Fig. 1). And they could observe during the daylight and during rainstorms, too! The sun was no longer the strongest object in the sky, although at times of radio bursts it could become a million times stronger. Especially at the lowest frequencies, the band of the Milky Way dominated the sky. Several discrete radio sources were also stronger than the radio sun, and although they were called "radio stars", most of them did not correspond in position with any optical objects. By the early 1950s a few did have suggested optical identifications, however, and these were remarkable: the Crab nebula (called Taurus A), the elliptical galaxy Messier 87 (Virgo A), and a faint, distant peanut-shaped galaxy (or colliding galaxies?) (Cygnus A). The last was particularly significant, for it implied that some radio sources were being detected from distances far beyond the reach of the largest optical telescope (the Palomar 200-in. [5.1 m]), and in fact could be used to investigate the large-scale structure of the universe.

It took 10–15 years for radio astronomy to establish itself and become accepted as a part of astronomy. In the process the radio physicists and engineers became "radio astronomers" and invented the need to now call the traditional astronomers "optical astronomers". Furthermore, as it turned out, the expansion of the observed electromagnetic spectrum into the radio range was only the first of many later expansions, mostly made possible by the advent of space platforms (X-ray, ultraviolet, infrared, gamma-ray). This

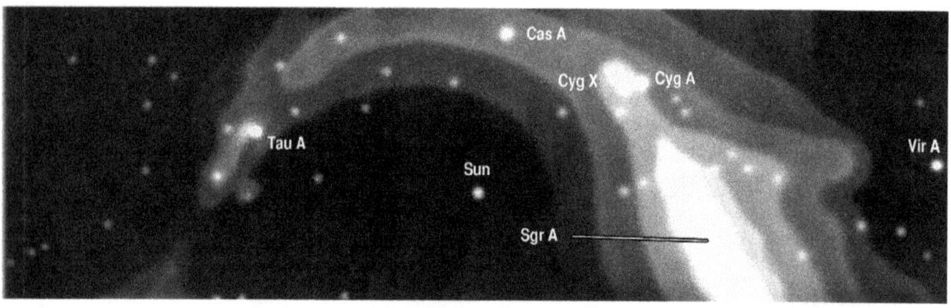

Fig. 1 The northern sky as seen with "radio eyes." The map extends from −40° to +70° in declination; north is *up* and east to the *left*. This artistic rendering is based on a map, observed with a beam of 1° × 8° at a frequency of 250 MHz (wavelength of 1.2 m), taken with a helix array at Ohio State University in the mid-to-late 1950s [10]. The *broad curving band*, defining the radio Milky Way, corresponds in position to the optical Milky Way (the galactic plane). The position of the radio sun varies through the year and is shown for the spring equinox; its intensity varies greatly on various time scales, but normally is far fainter than the radio Milky Way. The strongest radio stars are labelled; Cyg X is a large, diffuse region and Sgr A, the galactic center, is lost in the surrounding region of strong emission (modified version of Fig. 8–2 in [10])

overall opening of the electromagnetic spectrum was to become the defining characteristic of twentieth-century observational astronomy.

In this paper I focus on the first decade of radio astronomy after World War II, and then briefly continue the story until about 1990. In accord with the 400th anniversary of the invention of the (optical) telescope, I emphasize the role of *radio telescopes*, especially those most important to basic discoveries; there is unfortunately not room to mention every major radio telescope. I do not cover the history of lunar and planetary radar, nor of millimeter-wavelength radio astronomy. Details of the history of radio astronomy through ~1953–54 can be found in my *Cosmic Noise: A History of Early Radio Astronomy* [1]. There are also many other books on the history of radio astronomy [2–9].

2 Phase 1: the beginnings of radio astronomy (through 1953)

Although the first detection of extraterrestrial radio waves was not until 1932, in the period 1894–1901 there were several attempts in Europe to detect Hertzian (radio) waves from the sun. Yet in the ensuing four decades, as radio technology steadily improved, no one made serious attempts to detect the sun. This was because of a lack of sensitive, directional radio equipment, discipline specialization, and a reliance on Planck's blackbody theory (1900) to calculate expected signal levels. So it was that Karl Jansky, a radio physicist working at the Bell Telephone Laboratories in New Jersey, only serendipitously discovered extraterrestrial radio emission in 1932—and it came from the Milky Way, not the sun. With his effective beam size of ~30° at a frequency of 20 MHz, he established that the emission (he called it "star static") was strongest toward the center of our Galaxy, yet appeared all along the galactic plane.

The hard times of the economic Depression and the bizarre nature of Jansky's discovery meant that, although his results were generally known, little serious follow-up occurred throughout the 1930s. The lone exceptions were investigations carried out over a decade by radio engineer Grote Reber. Living in a Chicago suburb, in his spare time and with his own funds, Reber took it upon himself to build a 31-ft (9.4-m) diameter paraboloid dish in his backyard in order to check Jansky's findings at much higher frequencies. He ended up making two maps over the northern sky of what he called "cosmic static". Along the way, Reber also detected solar radio emission (but did not know of earlier unpublished work by others) and established working relations with the astronomers at Yerkes Observatory. Yet despite this pioneering and ingenious work through 1947, Reber had little influence on the postwar development of radio astronomy, especially outside of the US.

The man who was seminal for the beginnings of radio astronomy in England was Stanley Hey. As a radar-operations researcher during the war, in 1942 he deduced that apparent jamming of British coastal radar installations was in fact radio radiation picked up from the sun. This was the earliest conclusive detection of the radio sun (and kept secret until war's end), although before the war amateur radio operators and ionospheric physicists had garnered some evidence of same. Later during the war, the sun was independently studied by George Southworth of Bell Telephone Labs at centimeter wavelengths, as well as detected during wartime operations in Germany and in New Zealand.

Before the end of the war, Hey also encountered and investigated both the Jansky/Reber galactic noise and anomalous echoes that he determined originated from the ionized trails of meteors. Until the Cold War heated up in 1948, Hey was able to hold together his small Army group and conduct significant research in all aspects of radio astronomy, most importantly establishing the existence of the first discrete "radio star," later known as Cygnus A (Fig. 2).

Two other major postwar groups also began in England at this time as an outgrowth of radar research at the wartime Telecommunications Research Establishment. Martin Ryle started at the Cavendish Laboratory of Cambridge University in an ionospheric research group, but soon was investigating radio waves from the sun and radio stars, of which by 1949 his group had found 23 (including Cassiopeia A). In the process he invented many techniques fundamental to radio astronomy, such as the Michelson interferometer and the phase switch, and assembled the core of a powerful research group featuring Graham Smith and Antony Hewish. The second British group, led by Bernard Lovell, studied radar echoes from meteor trails at Jodrell Bank, a field station of the Physics Department of Manchester University. Over the postwar decade

Fig. 2 64 MHz antenna of Hey's group at Richmond Park, near London. A reflecting screen is supported just off the ground; the vertical structure is a vestige of the wartime radar configuration. This setup (with only two Yagis) was used in 1945 for a first survey of galactic noise and then again in 1946 (with four Yagis as pictured) for a second survey, during which fluctuations led to the deduction of the discrete source Cyg A

Lovell's group developed techniques to measure the trajectories and velocities of meteors and dominated the world in this field. For instance, new techniques enabled study of *daytime* meteor showers. In 1949 Lovell was joined by Robert Hanbury Brown, another leading wartime radar veteran. Hanbury Brown used a huge fixed dish of diameter 218 ft (66 m), and with his student Cyril Hazard detected for the first time radio waves from an external normal galaxy, the Andromeda nebula.

The largest of all the postwar groups was led by Edward "Taffy" Bowen and Joseph Pawsey of the Radiophysics Laboratory (RP), Sydney. This Australian laboratory was one of the world's premier wartime radar labs and remained intact after the war, switching its electronic know-how to investigation of cosmic noise. With Bowen's management skills and nose for good projects, and Pawsey's inspired scientific leadership of teams of radio engineers and physicists at field stations scattered around Sydney, RP developed techniques such as the sea-cliff interferometer (Fig. 3) and Fourier-synthesized maps. Basic discoveries in the early years included a hot radio solar corona at a temperature of 1×10^6 K (Pawsey and David Martyn); angular sizes for solar radio burst regions of $<10'$ (Pawsey, Lindsay McCready, and Ruby

Fig. 3 Pairs of 100 MHz (*left*) and 60 MHz Yagi antennas used for monitoring the sun and measuring its circular polarization at Dover Heights (1947). The same pairs, with the Yagi elements parallel and pointing toward the horizon, were used as a sea cliff interferometer for the first surveys of discrete sources by John Bolton (*pictured*) and his group

Payne-Scott); motions and frequency–drift structure of solar bursts (Payne-Scott; Paul Wild); suggested identifications of three radio sources with optical objects: Taurus A with the Crab nebula (remnant of a supernova explosion), Centaurus A with a weird nebula catalogued as NGC 5128, and Virgo A with the distant elliptical galaxy Messier 87 (John Bolton, Gordon Stanley and Bruce Slee); and pioneering measurements of radio source properties and distributions across the sky (Bernard Mills). The success of the RP group was all the more remarkable given that Australia at this time was geographically and scientifically isolated, and not known for its science. The esteem of RP's colleagues can be gauged by the decision to hold the 1952 meeting of the International Union of Radio Science (URSI) in Sydney, the first time *any* international scholarly society had met in Australia.

The story of early Dutch radio astronomy is anomalous in that it was not driven by radar veterans and in fact was directed by (optical) astronomers. Upon learning of Reber's results, astronomers Jan Oort and Hendrik van de Hulst, at Leiden University in occupied Holland, as early as 1944 considered the possibilities of using radio spectral lines to study the structure of the Milky Way, a specialty of Oort's since the 1920s. Once the war ended, resources were scarce and radar veterans were nonexistent, but Oort persisted and eventually succeeded when he found Alexander Muller, a talented radio engineer fresh from undergraduate study, and adapted an abandoned German Würzburg radar dish for astronomical observations (Fig. 4). Muller built a receiver that

Fig. 4 The 7.5-m Würzburg reflector (at Kootwijk, in central Holland) that was used for early work on the 21 cm line by Muller and Oort

in 1951 detected a spectral line of atomic hydrogen at a wavelength of 21 cm (closely following its discovery in the US [see below]). This technique made accessible an entirely new component of the interstellar medium. There quickly followed the first of many Galaxy surveys exploiting the line's Doppler shifts in order to detect spiral arms composed of cold hydrogen. An "international Dutch school" of 21 cm Milky Way studies was soon established.

Besides the above five major groups in the postwar decade, many other efforts sprung up around the world. In the United States Robert Dicke, a physicist at the Radiation Laboratory run by the Massachusetts Institute of Technology, the largest of the American wartime radar labs, developed his eponymous microwave receiver and with it in 1945 was the first to observe the moon and a solar eclipse. At the US Naval Research Laboratory (Washington, DC), another leading wartime radar lab, John Hagen's group, starting in 1946, undertook many solar projects at microwavelengths, including eclipse expeditions, and also built a 50-ft (15-m) diameter dish. Smaller groups were at Cornell University and Stanford University. Finally, a one-time effort (a Ph.D. thesis project) represents the most significant US contribution to radio astronomy in the postwar decade. This was the discovery at Harvard University in 1951 of the 21-cm hydrogen line by physicists Harold Ewen and Edward Purcell (both former wartime radar experts).

Radio astronomy in the USSR was strongest in theoretical studies and dominated from the start by physicist Vitaly Ginzburg (Lebedev Institute [FIAN], Moscow) and astronomer Iosif Shklovsky (Moscow State University). In 1946 both made seminal contributions to solar coronal radio theory. From 1951 onwards, Ginzburg argued in detail for a connection between cosmic rays and the still-unexplained galactic background radiation, namely that relativistic electrons in the Galaxy's magnetic field copiously produced radio waves via the synchrotron mechanism. Shklovsky worked out a theory for interstellar radio lines of hydrogen, hydroxyl (OH) and CH in a 1949 paper, and in 1953 suggested that *both* the enigmatic radio and optical radiation from the Crab nebula arose from the synchrotron mechanism.

In France, the specialty at the Ecole Normale Supérieure in Paris and Meudon Observatory was solar observations both at home and on eclipse expeditions to exotic locales. Jean-François Denisse was a leader in solar radio theory and recipient in 1949 of the first doctorate awarded for a radio astronomical topic. Canada produced important observations of the sun and of meteor echoes and in Japan several isolated solar groups made contributions.

Meteor radar was a fruitful subfield during the postwar decade, although unique in early radio astronomy in that by the mid-1950s it had peaked and gone into decline. There had been several studies before the war suggesting that the ionized trails of meteors could effectively reflect radio waves, but the effect was not confirmed until 1945 when Hey's group undertook a three-station triangulation experiment on anomalous echoes that had often been seen during wartime operations. The October 1946 Giacobinid meteor shower, spectacular (as predicted) for visual observers, also represented a milestone for the many radar teams who at its peak were inundated with echoes. The major scientific question to which meteor radar contributed was the issue of whether there existed a component of meteors arriving from interstellar regions, completely outside the solar system. Astronomers had been debating this issue for decades, and in the end statistics from huge numbers of echoes gathered at Jodrell Bank and in Canada convinced most that very few, if any, meteors had hyperbolic orbits indicative of interstellar origin.

Studies of solar noise constituted fully ~70% of the efforts of early radio astronomers, and yet this was the subfield where the fewest fundamental astronomical discoveries were made. Radio observers did, however, contribute important evidence for the million-degree corona that astronomers had been suspecting since the 1930s. Tremendous efforts went into monitoring radio burst activity and correlating it with various optical properties—the main outcome was the recognition of a "slowly varying component", a long-term oscillation in total microwave intensity that followed the 11-year sunspot cycle. In other studies, expeditions were mounted to eight solar eclipses over the period 1945–52 in order to gain higher angular resolution by using the moon as a moving knife edge. Perhaps the greatest contribution of the sun, so to speak, to radio astronomy was as a handy test bed for new high-resolution mapping techniques developed by Ryle at Cambridge (Section 4) and Wilbur "Chris" Christiansen in Sydney. By 1953 both groups had made two-dimensional maps of the sun using Fourier analysis of data gathered with interferometric portable antennas and grating arrays.

Radio stars, or radio sources as they eventually became known, were the central puzzle in early radio astronomy. Four major surveys were carried out by the groups in Cambridge (Fig. 5), Manchester and Sydney over 1949–53, but only about 5% of the ~250 known radio stars correlated with any optical object, i.e., had what became known as an "optical identification". Furthermore, different surveys in areas of overlap often seriously disagreed. Two astronomers with access to the world's largest (optical) telescopes, Walter Baade and Rudolph Minkowski, were central to taking the deep photographs required to make identifications, which then bolstered the degree of reality of any associated source. On the radio side, positions accurate to a few arcminutes or better were required for any chance at an identification; one particularly pivotal effort was the precision interferometry of Smith at Cambridge. Positional accuracy of ~1′ finally allowed the two strongest sources in the sky, which had stubbornly resisted previous attempts, finally to be identified with (a) a strange network of fast-moving filaments that was perhaps a supernova remnant (Cas A), and (b) a faint, distant peanut-shaped object, taken to be a pair of galaxies in collision (Cyg A). The latter especially was significant, for it implied that if Cyg A's intrinsic properties were at all typical, then the bulk of the *un*identified radio sources were at distances beyond what (optical) telescopes could explore. Thus did radio astronomy become relevant to cosmological inquiries at an early stage. Together with the earlier identifications of Tau A, Vir A and Cas A, it was also clear that radio astronomers were revealing violent, high-energy phenomena hitherto unsuspected by (optical) astronomers.

Fig. 5 One element of the Long Michelson (1C) interferometer (ca. 1950). It comprised an array of forty 81 MHz halfwave dipoles mounted 1/4 wavelength above a reflecting screen

Another desired measurement for a radio source was its angular size. The intrinsically poor resolution of most early antennas and interferometers, however, made this a difficult task except for the strongest few sources. These turned out to have sizes of a few arcminutes, which argued that they were not radio *stars* at all, but *radio nebulae.* One novel technique for measuring sizes, the intensity interferometer, was developed at Jodrell Bank by Hanbury Brown and Richard Twiss. With it Roger Jennison and M. K. Das Gupta not only measured the angular extent of Cyg A, but found that it mystifyingly consisted of two blobs separated by 1.5′, one on either side of the optical object [11].

3 Where did radio astronomy find itself in 1953?

The year 1953 was a watershed for radio astronomy for several reasons. By that time the bulk of the radio astronomers were no longer astronomical novices, but at a mid-career stage in their 30s and some even recipients of significant professional honors. Furthermore, integration with astronomy as a whole was gaining strength; for instance, to aid communications, the radio astronomers at this time officially voted to discourage use of terms like *solar noise* and *cosmic noise* in favor of *solar radio emission* and *galactic radio emission.* In addition, the field's most important sectors had migrated from solar to galactic and extragalactic questions (although solar research still dominated quantitatively). The first two textbooks of radio astronomy also appeared about this time [12, 13]. Finally, the field was moving from empirical exploration to a more programmatic approach, entailing enormous antennas whose correspondingly larger budgets meant that they could no longer be built in-house by the individual groups. Radio astronomy was moving into a Big Science phase.

We can recognize three major historical themes in early radio astronomy: (1) World War II and the Cold War; (2) Technoscience; and (3) A "visual culture".

3.1 World War II and the Cold War

The first theme is how World War II and the ensuing Cold War spawned and shaped the field of radio astronomy, albeit with important national differences. In the first place, every major postwar group (except Holland) was filled with war veterans of radar research or operations, often skilled in ionosphere research or other aspects of radio communications. Especially in England and Australia, these researchers not only had know-how and ample surplus equipment, but also a can-do attitude and camaraderie that continued to permeate their postwar groups. Also forged during the war were management skills vital for the success of the new field's leaders. Furthermore, programs after the war to support research and train new scientists were funded at much higher levels than before.

The strong military postures of the Cold War, especially in the United States and the Soviet Union, also influenced the type and amount of radio astronomy that was done. The highly successful activities in radio astronomy of one military group, Hey's Army Operational Research Group in England, were in fact terminated in 1948 due to Cold War pressure. In the US money flowed from military agencies to university physicists and engineers, but the direction of their research in radio astronomy (toward shorter wavelengths) was inappropriate for the most fruitful results at that time. US groups thus long lagged behind Australia and the two university groups in England (at Cambridge and Manchester), who, despite much smaller budgets, achieved pioneering discoveries at longer wavelengths. Postwar US radio astronomy provides another example of historian Paul Forman's thesis that military funding during the Cold War substantially warped the contents of seemingly "pure" university research [14]. A topic such as radio emission from sources thousands of light-years distant may at first seem of no military consequence, but in fact the sensitive radio technology necessary to find and track such sources was of direct use to the military. In contrast, the Australian and English groups' relative isolation from "Cold War science" contributed to the success of their postwar radio astronomy.

3.2 Technoscience

The second principal historical theme deals with the role of instrumentation and technology in early radio astronomy—was science or technology in the lead? The practitioners themselves had a mixture of engineering and scientific skills, continually switching between building radio instrumentation and making calculations about the sun or Milky Way. The field itself grew out of military radio and radar engineering, and new technical capabilities (especially those that improved angular resolution or sensitivity to weak signals) often led to serendipitous scientific discoveries as investigators took a raw empirical approach and tried out the latest equipment at hand. However, for other episodes (especially later), astronomical goals dictated experimental design. And in many other cases it is hopeless to ask whether the technology led or followed the science, or sometimes even whether the two can be separated. A more useful formulation may be that radio astronomy was *technoscience*, an amalgam activity in which technology and science are intimately related. For example, based on cost, experimental design and happenstance, different antenna systems and receivers were favored by the various groups conducting radio star surveys. Largely because of these choices, serious discrepancies arose between these surveys and serious arguments between the surveyors, casting doubt on the legitimacy of any but a handful of their putative sources. This in turn led to revisions of the equipment and follow-up surveys.

3.3 A "visual culture"

The third theme is that radio astronomers ironically drove themselves to more and more of a "visual culture" even as they continued to record their data as very "un-visual" jiggly ink lines on rolls of strip chart paper. By *visual culture* I mean concerns with (a) finding optical galaxies and nebulae that were also radio emitters, (b) creating radio images that one could begin to compare with photographic images, and (c) seeking the expertise, company and advice of (optical) astronomers. Regarding (a), a continual aim of radio astronomy was for better angular resolution in order to pinpoint more finely the locations and sizes of discrete sources. And the main reason for wanting better angular resolution was the hope that an unambiguous optical correlate at the same sky location could be found, a so-called *optical identification*. Such an optical identification, carrying the imprimatur of (optical) astronomy and appealing to epistemically superior visual information, conferred on a radio source a degree of legitimacy and reality that it could not otherwise have. As Bolton wrote to the optical astronomer Minkowski in 1952: "Photographs are much more satisfying than the evidence we get out of our machinery". His Sydney colleague Mills agreed and was prompted to produce "radio pictures", pseudo-photographs representing radio emission. Regarding (b), by 1953 two groups had worked out the details of how in principle to create a radio image, i.e., a map of high resolution (by sampling its Fourier components and then mathematically calculating the image), although it would be another decade before enhanced computer power would make the technique tractable (Section 5).

The final aspect of the visual culture of radio astronomers was their desire *not* to set up a new discipline, but rather to become part of the very old discipline of astronomy. Although their backgrounds differed immensely from those of astronomers and their research style was much more hurried and involved more teamwork, by 1953 radio astronomers had learned enough astronomy and gained enough legitimacy that they were welcomed at astronomy meetings, often collaborated with astronomers, and were viewed as making significant contributions to astronomy. In the process they invented the research specialty of *radio astronomy* (the term was first used in 1948), and made astronomers comfortable with the notion that, for example, what looked like a field full of clothes lines (Fig. 5) was in fact worthy of being called a telescope, albeit by the new term *radio telescope*. It was also *radio* astronomy that created the obverse category of *optical* astronomy, which until then had not been necessary. Radio astronomers became amalgamated into an existing discipline (astronomy), and did not form their own discipline, although full integration was not complete until two decades later.

4 Phase 2: radio telescopes 1953–70

As we move into the "early adulthood" of radio astronomy, the field grows significantly and this account must be much sketchier. Furthermore, I will emphasize the development of the radio telescopes themselves, and minimize the science. During the late 1950s and 1960s radio astronomy entered the era of Big Science, by which I mean that the antennas could not be built in-house any longer and required contracts, soon for millions of dollars, with large engineering firms. Moreover, the concept of a national radio observatory arose in several countries, with researchers from many institutions granted access to large radio telescopes that they had not in the least designed. In the quest for better angular resolution and with improving microwave receivers, radio astronomers also pushed to shorter wavelengths and longer-baseline interferometers. This meant that paraboloidal reflectors, "dishes", became more common and that baselines stretched across the countryside and eventually even crossed national borders.

Perhaps the best known dish in the world is the Jodrell Bank 250-ft (76-m) diameter reflector, now called the Lovell Telescope, in the Cheshire countryside near Manchester, England [2, 15]. Placed into service in 1957 after many delays and cost overruns, it turned out to have many abilities besides radio astronomy—in fact, its first high-profile tasks were tracking Russian space satellites, starting with Sputnik 1's rocket in late 1957. Its collecting area dwarfed that of previous steerable dishes of ~25-m diameter. It soon became a British icon and in fact is *still* in service 51 years later, albeit with many improvements over the years to the accuracy of its surface and to its pointing. Over the past few decades it has served as the anchor for the MERLIN interferometer array (Section 5) scattered about England.

Another audacious dish, also still in service and still the largest in the world, is the 1,000-ft (305-m) diameter spherical reflector at Arecibo, Puerto Rico (Fig. 6). Built with military money in 1963, its original purpose was as much to study the ionosphere through low-frequency radar as to do planetary radar astronomy and passive radio astronomy. Desiring some degree of steerability despite the huge size, its designers placed it in a natural hollow (a karst depression) and made it spherical, topped by a huge superstructure supporting a feed antenna at the focus movable as far as 20° from the zenith. By placing the dish in the tropics, the accessible 40°-wide strip of sky amounted to ~40% of the celestial sphere and also allowed access to the moon and planets [9]. Finally, huge line feeds had to be developed in order to correct the spherical aberration. Probably its greatest astronomical contributions have involved pulsars; in fact, Joseph Taylor and Russell Hulse received the 1993 Nobel Prize in Physics for their work demonstrating the likely existence of gravitational waves emanating from a decaying binary pulsar system (B1913+16) discovered at Arecibo in 1974 and monitored thereafter for decades. The Arecibo dish is also the radio telescope most associated with searching for extraterrestrial intelligence (SETI), in particular via Project Phoenix during the 1990s.

Fig. 6 The 1,000-ft (305-m) diameter spherical reflector at Arecibo, Puerto Rico (photo ca. 1970s). The 800 metric ton feed support platform is suspended 152 m above the dish surface

In the southern hemisphere the iconic role has been played by the 210-ft (64-m) diameter Parkes dish in Australia, even christened *The Dish* by Hollywood in a 2000 comedy film. Commissioned in 1961, early-on it contributed importantly to the discovery of quasars and has been a workhorse to this day. The longevity of all these antennas is remarkable and rivals optical telescopes such as the Palomar 200-in. (5.1 m), completed in the late 1940s.

The last and largest of the fully steerable Big Dishes of this era was the 100-m diameter reflector operated by the Max Planck Institute for Radio Astronomy at Effelsburg, Germany. Starting in 1970, it reigned at microwavelengths for three decades (until the Green Bank Telescope came online around 2000). Its main technical innovation was its homologous surface design, meaning that the reflector's support structure was designed such that the surface gravitationally deformed into a succession of paraboloids as the dish swung down in elevation angle; software could then automatically adjust the pointing and focus of the dish without loss of gain.

Other types of radio telescopes were also important in this era. Bernie Mills built a large cross-type antenna in the mid-1950s, which created a narrow pencil beam by multiplying the outputs of an east–west dipole array and a

Fig. 7 The rotating paraboloidal horn at Bell Telephone Labs, Holmdel, New Jersey, USA with which Arno Penzias and Robert Wilson (*pictured*) discovered in 1965 an isotropic signal with a brightness temperature of about 3 K, soon explained as the cosmic microwave background

north–south array. His group's catalogs of radio sources soon came into conflict with those of Martin Ryle's group at Cambridge, who employed large parabolic cylinders as the elements in a Michelson interferometer. Well into the 1960s bitter battles ensued over observing techniques, definitions of a "source", and cosmological implications of the source counts [4]: was it a steady-state universe or an evolving (Big Bang) universe? Ironically, despite all the energy thrown into these battles, most astronomers in the end abandoned this line of evidence altogether, and instead adopted the hot Big Bang on the basis of another, serendipitous radio discovery—made with the strange-shaped, sugar-scoop horn shown in Fig. 7. In 1965 Arno Penzias and Robert Wilson of Bell Telephone Labs (the very place where Jansky had made *his* discovery 30 years before) found an unexplained isotropic signal with a brightness temperature of about 3 K. This was soon confirmed to be blackbody emission, consistent with expected relict radiation from an early hot, dense universe [4, 16]; the discovery was good enough for the Nobel Prize in Physics for 1978.

Although radio work trended towards microwaves during 1953–70, important research continued at lower frequencies. In particular, the Culgoora Radioheliograph in Australia, completed in 1967, was the culmination of Paul Wild's work on solar bursts begun two decades earlier. A 3-km-diameter ring of 96 small dishes was connected to form a high-resolution 80 MHz radio image of the sun every second; movies of burst motions were now possible. Later this array was used for nonsolar radio astronomy, too. And in Cambridge Antony Hewish built the "4-acre array", a field of dipoles spread over 470 m × 45 m that looked like nothing so much as a structure waiting to support hops vines whose product was destined for Guinness (Fig. 8). Designed in 1967 as a transit telescope to study low-frequency source scintillations due to interplanetary plasma, this array produced strip-chart recordings on which Jocelyn Bell recognized the puzzling signals that turned out to be the first pulsars.

5 Phase 3: radio telescopes 1970–90

Interferometry reigned during this phase of the history. The first type was aperture synthesis, whose roots extended back primarily to Ryle's group in Cambridge throughout the 1950s. By the early 1960s phase stability could be achieved to a sufficient degree over 12 h that one no longer had to physically move antennas from place to place in order to record fringe visibilities at various baseline orientations, but rather could let the rotation of the earth effectively do it for you. In 1962 Ryle and Ann Neville [17] published the seminal map proving this technique, which by 1964 led to the One Mile Telescope: three 18-m dishes, one of which was movable on an 800-m-long track. The other vital factor enabling aperture synthesis was the increase in computational power needed for all of the Fourier transforming required to turn fringe amplitudes and phases into sky maps. And what maps! Their sen-

Fig. 8 The "4-acre array" of 2,000 full-wave 81.5 MHz dipoles spread over 470 m × 45 m, built in 1967 by Antony Hewish (Cavendish Lab, Cambridge) as a transit telescope to study low-frequency radio source scintillations due to interplanetary plasma. Hewish's student Jocelyn Bell soon recognized puzzling signals from this array, which turned out to be the first pulsars

sitivity, angular resolution, and dynamic range were far superior to anything before. Ryle was awarded the 1974 Nobel Prize in Physics largely for this work (with Hewish sharing the prize for leading the pulsar discovery).

The next giant leaps in earth-rotation aperture synthesis, aided by larger budgets and ever faster computers, were in Holland and the US. In 1970 the Dutch inaugurated the very sensitive Westerbork Synthesis Radio Telescope, consisting of ten fixed 25-m dishes and two movable dishes. Then over the late 1970s, on the Plains of San Augustine in New Mexico, began sprouting an array of twenty-seven 25-m dishes of far greater power (Cassegrain foci with shaped surfaces, operable at shorter wavelengths, observing wavelength instantaneously switchable, broader bandwidth, longer baselines, etc.). The Very Large Array, officially completed in 1980 with its three 20-km-long arms, allowed radio astronomy to explore well below the millijansky limit and has remained in many senses the world's premier radio telescope now for almost three decades.

The other important type of radio interferometry was very-long-baseline interferometry (VLBI), for which Jodrell Bank researchers led the way starting in the mid-1950s with ever-longer microwave-linked baselines; by 1980 this work had developed into a dedicated array of six dishes with longest baseline of 130 km, the MERLIN array (Multi-Element Radio-Linked Interferometer Network). But the first independent antenna observations, i.e., without a realtime link, were achieved in 1967 in Canada and the US. Signals at each antenna were directly recorded on fast-moving 7-track computer tapes and correlated later at a central location. The key enabling technology was the new Varian rubidium oscillator frequency standard, good to one part in 10^{11}. By 1972 a VLBI baseline at 1.35 cm wavelength between Crimea and the US had achieved an 0.0002 arcsec fringe spacing. VLBI observations became routine only when the Very Long Baseline Array, a dedicated, multi-wavelength array of ten 25-m dishes, was built during 1986–93 across North America (actually the Virgin Islands to Hawaii). In sum, at least for small sources with extremely high brightness temperatures, radio astronomy had overcome its old nemesis of poor angular resolution and now far surpassed what optical astronomers could achieve!

6 A twentieth-century "new astronomy"

How should we view radio astronomy and its telescopes in the 400-year-long history of the telescope? Radio astronomy caused a transformation, and created what I call the twentieth-century's *New Astronomy*. The radio sky was astoundingly at variance with the familiar optical sky, including new aspects of previously known objects, wholly new phenomena, and novel ways to attack old problems (Fig. 1). This transformation of the sky by a New Astronomy was not unlike earlier ones brought on by the introduction of novel technology. These included Galileo with his telescope, William Herschel with his large reflectors, and the late-nineteenth-century astrophysicists with their spectroscopes and photography.

In fact, the definition of the twentieth century's New Astronomy should be expanded beyond just radio astronomy. Radio astronomy should be viewed as the harbinger of a broader twentieth-century phenomenon: the openings to astronomy, mostly made possible by access to outer space, of the infrared, ultraviolet, X-ray and γ-ray windows. In particular, the first years of X-ray astronomy in the 1960s had many close analogies with early radio astronomy [18]. The emergence of radio astronomy thus heralded the era of this broader New Astronomy, namely the opening of the *entire* electromagnetic spectrum over the period 1930–80.

Each of these New Astronomies—Galileo, Herschel, astrophysics, and radio astronomy (or more broadly, the opening of the electromagnetic spectrum)—radically changed their contemporary astronomical worldview. Each was enabled by innovative technologies and came to fruition through a combination of technoscience, faith in empiricism, and key intellectual players. Moreover, each encountered and overcame difficulties with the establishment of the day, eventually insinuating itself into and redefining mainstream astronomy. Such a broader context provides a gauge of the significance of this New Astronomy of the twentieth century: I conclude that, even within the compass of the millennia-long history of astronomy, the emergence of radio astronomy must be ranked as a major event.

References

1. Sullivan, W.T.: Cosmic Noise: A History of Early Radio Astronomy. Cambridge University Press, Cambridge (2009)
2. Lovell, A.C.B.: The Story of Jodrell Bank. Oxford University Press, London (1968)
3. Hey, J.S.: The Evolution of Radio Astronomy. Elek Science, London (1973)
4. Edge, D.O., Mulkay, M.J.: Astronomy Transformed: The Emergence of Radio Astronomy in Britain. Wiley, New York (1976)
5. Kellermann, K., Sheets, B. (Eds.): Serendipitous Discoveries in Radio Astronomy. National Radio Astronomy Observatory, Green Bank, West Virginia (1983)
6. Sullivan, W.T. III: Classics in Radio Astronomy. Reidel, Dordrecht (1982)
7. Sullivan, W.T. III (Ed.): The Early Years of Radio Astronomy: Reflections Fifty Years after Jansky. Cambridge University Press, Cambridge (1984)
8. Robertson, P.: Beyond Southern Skies: Radio Astronomy and the Parkes Telescope. Cambridge University Press, Cambridge (1992)
9. Butrica, A.J.: To See the Unseen: A History of Planetary Radar Astronomy. Pub. SP-4218. NASA, Washington, D.C. (1996)
10. Kraus, J.D.: Radio Astronomy. McGraw-Hill, New York (1966)
11. Jennison, R.C., Das Gupta, M.K.: The measurement of the angular diameter of two intense radio sources. II. Diameter and structural measurements of the radio stars Cygnus A and Cassiopeia A. Phil. Mag. **1**, 65–75 (1956)
12. Lovell, A.C.B., Clegg, J.A.: Radio Astronomy. Chapman and Hall, London (1952)
13. Pawsey, J.L., Bracewell, R.N.: Radio Astronomy. Oxford University Press, Oxford (1955)
14. Forman, P.: Behind quantum electronics: national security as basis for physical research in the United States, 1940–1960. Historical Stud. Physical Sciences **18**, 149–229 (1987)
15. Lovell, A.C.B.: Out of the Zenith: Jodrell Bank 1957–1970. Oxford University Press, London (1973)
16. Kragh, H.: Cosmology and Controversy: The Historical Development of Two Theories of the Universe. Princeton University Press, Princeton (1996)
17. Ryle, M., Neville, A.C.: A radio survey of the North Polar region with a 4.5 minute of arc pencil-beam system. Mon. Not. Roy. Astr. Soc. **125**, 39–56 (1962)
18. Hirsh, R.F.: Glimpsing an Invisible Universe: The Emergence of X-ray Astronomy. Cambridge University Press, Cambridge (1983)

Exp Astron (2009) 25:125–141
DOI 10.1007/s10686-009-9148-7

HISTORICAL REVIEW

History of infrared telescopes and astronomy

G. H. Rieke

Received: 18 November 2008 / Accepted: 21 January 2009 / Published online: 20 February 2009

Abstract The first attempts to measure the infrared outputs of stars preceded by nearly a century the permanent establishment of infrared astronomy as an important aspect of the field. There were a number of false starts in that century, significant efforts that had little impact on the astronomical community at large. Why did these efforts fizzle out? What was different in the start that did not fizzle, in the 1960s? I suggest that the most important advances were the success of radio astronomy in demonstrating interesting phenomena outside of the optical regime, and the establishment virtually simultaneously in the United States of a number of research groups that could support each other and compete against one another in their approach to infrared astronomy.

Keywords Infrared astronomy · History of astronomy

1 Introduction

The infrared spectral range has moved to the center of our goals for astronomy. Future facilities including Herschel, the James Webb Space Telescope (JWST), the Space Infrared Telescope for Cosmology and Astrophysics (SPICA), the Stratospheric Observatory for Infrared Astronomy (SOFIA), and 30-m-class groundbased telescopes will finally let us reach the potential of this field. However, infrared astronomy began fitfully and continued that way until less than 50 years ago, when it suddenly began to grow rapidly.

In this article, I discuss this early, fitful development of galactic and extragalactic infrared astronomy, and then briefly touch on the beginnings of

G. H. Rieke (✉)
Steward Observatory, The University of Arizona, Tucson, AZ 85721 USA
e-mail: grieke@as.arizona.edu

modern infrared astronomy in the 1960s and the rapid growth that ensued. In this way, I can complement other material relevant to the history of this field. For example, Lequeux (2008, private communication) describes the very early stages of infrared astronomy; Connes [15] describes the development of Fourier Transform infrared spectroscopy; Sinton [82] puts emphasis on the development of planetary infrared astronomy; Price [72] describes infrared surveys, including a number that never received broad recognition; Jennings [37] and Storey [86] emphasize the growth of infrared astronomy in England and Australia, respectively; Dolci [17] discusses the history of airborne astronomy in general, which includes its major role in the infrared; Rieke [76] describes the development of the Spitzer Space Telescope; and Low et al. [50] describe the rapid emergence of infrared astronomy in the 1960s. Although a number of holes remain (e.g., for most infrared space missions), the combination of these articles starts to yield a reasonably complete description of the emergence of the field.

2 False starts

2.1 Nineteenth century

Infrared astronomy had a faltering beginning. Huggins [34] built a relatively simple thermopile with a galvanometer readout, for an 8-in. refractor. Similar instrumentation had been applied by Luigi Magrini to detect the solar corona during the eclipse of 1836 and by the fourth Earl of Rosse (son of the builder of the 72-in. "Leviathan" Telescope) to study the thermal radiation of the Moon [74, 79]. Despite his modest instrument, Huggins reported detections of Sirius, Pollux, Regulus, and Arcturus. Stone [85] improved the reliability of thermal measurements by using a matched pair of thermocouples at the telescope focus, to compensate both for thermal drifts in the instrument and for fluctuations in the emission of the sky (a measure also adopted by Lord Rosse). Stone reported detections of Arcturus and Vega. Only Lord Rosse seems to have made an effort to demonstrate that he was observing predominantly in the infrared, by putting a piece of glass in the beam and observing the resulting reduction in the signal. The others evidently assumed this result, given the broad response of their thermopile detectors.

The American inventor Edison excelled at discovering new applications for capsules of finely divided carbon powder. One example was an infrared detector consisting of a pellet of this material with a galvanometer to record its resistance as it reacted to changes in temperature. As a form of sophisticated vacation, science experiment, and public relations exercise, Edison observed the 1878 solar eclipse with his "tasimeter." To calibrate the instrument, he observed and may have detected Arcturus [18]. However, the tasimeter was a very finicky and unstable instrument, probably usable only in the hands of its inventor, and the small size of the telescope and absence of thermal compensation in the instrument calls the detection into some question [64]. In any case, Edison had grander interests, and once the extensive publicity died down, there was no further effort.

The publicity for Edison's feat did bring a rejoinder from Huggins, staking his claim for priority [35]. However, an attempt by Boys [6] to confirm the results of Huggins and Stone was not successful. Boys used a new detector, which he called a radiometer. It consisted of two matched thermocouples suspended in a current loop and inside a magnet. A small mirror attached to this delicate torsional pendulum (oscillation period of 10 s) allowed measuring its position with a light beam. Energy absorbed by one of the thermocouples modulated the current in the loop and caused a torque that moved the pendulum. Boys used a government grant of 50 lbs Sterling to build a telescope specifically around the requirements of this detector and those of measuring infrared radiation. He demonstrated that his detector was far more sensitive than those of Huggins and Stone. Nonetheless, he was unsuccessful at detecting any stars, but had to be satisfied by confirming Lord Rosse's results on the heat radiation of the moon. He rather politely points out the issues (including Huggins' use of a refractor) in describing the discrepancy [6]:

> Dr. Huggins did not obtain any effect from the moon, but as he used an object glass which would absorb nearly all the moon's heat or, at any rate, a large proportion, and as it had a smaller aperture and a greater focal length than the mirror which he lent me, so that the image of the moon was larger, and on this account also had a smaller heating power, it is not surprising that he did not obtain a large effect from the moon; but that he obtained none is, in the face of these measures, strong evidence that the deflections observed, and which he attributed to the stars, were spurious.

Detecting a star appears to have been a challenge taken up to demonstrate the performance of new types of infrared detector. The next attempt was by Nichols, who used a high performance version of Crookes' Radiometer. This device survives as a novelty item: a set of vanes blackened on one side is placed inside a partially evacuated glass bulb, arranged so they can turn easily. When exposed to sunlight, the vanes in the novelty device spin rapidly due to the interaction of the heated surface with the gas in the bulb. Nichols used very light vanes of mica with a delicate suspension to improve over Boys' device by about a factor of five in sensitivity [28]. Small signals were observed from Arcturus and Vega, in a ratio of about 2.1:1 in favor of Arcturus. Nonetheless, Hale [28] reports that:

> In view of the smallness of the deflections, and the uncertainty which arises from rapid fluctuations in the atmosphere, Professor Nichols does not greatly rely upon the quantitative value of the results.

Thus, the nineteenth century saw the first efforts at detection of a star in the infrared (by Huggins, Stone, and Edison), the first government grant for infrared astronomy and the first infrared-optimized telescope (both by Boys), but ended with some finality that no astronomy had been accomplished.

3 A real beginning is ignored

But was that negative conclusion correct? Nichols [64] mulled over his measurements and decided he had achieved a detection after all. Indeed, if we reject the one measurement he describes as questionable, he finds a difference between Vega and Arcturus of 0.9 magnitudes, which can be compared with the value of 1.1 magnitudes measured 30 years later by Pettit and Nicholson [68] using equipment with similar spectral response. Nichols concludes:

> The ratio greater than 2 to 1, of the total radiation of Arcturus to that of Vega (stars which by most observers are estimated to be of nearly equal photometric magnitude) indicates a proportionately more intense infrared spectrum for the former than for the latter star. ... The photosphere of Arcturus may be at a lower temperature than that of Vega, but the star be of sufficiently greater angular diameter, as seen from the Earth, to equal Vega in light intensity and surpass it in total radiation.

We therefore give Nichols credit for the first stellar astronomical measurement in the infrared. However, his detector was clearly inadequate to open up a large field of investigation. Coblentz [8] solved this problem by dramatic improvements in a thermocouple-based detector. He realized that the best performance depended on achieving the smallest possible heat capacity along with good thermal isolation—the basic principles used to optimize thermal detectors in general. To achieve the first goal, he made single-junction detectors of only 0.3 to 0.4 mm size; placing the detector in a vacuum was an essential step toward the second goal. He also introduced a cell of water as a way to remove the infrared signal and deduce its strength through the change relative to the total signal with it out of the beam [13]). Thus, he provided a complete solution to the problems that had prevented much progress in the nineteenth century. Already by 1914 he had measured more than a hundred celestial objects, including a more accurate ratio of the outputs of Arcturus and Vega [8, 9]. He continued to improve detectors and filters (e.g., [10–12]), and determined a reasonable temperature scale for stars of different spectral types [14]. He clearly deserves the title of first infrared astronomer. Moreover, he set the scene for further advances that brought the field to a high state of development.

Pettit and Nicholson carried out an ambitious infrared astronomy program from about 1920, building on the technical foundation established by Coblentz. They began with a thorough optimization of the thermopile detectors [67]. They mounted a pair of these detectors on the Newtonian plate holder of the Mt. Wilson 100-in. Telescope, oriented so the observer could switch the stellar image between the thermocouple receivers by turning a guiding screw.

The signals were conveyed by long wires to the basement of the telescope dome, where there was a sensitive galvanometer well-protected from any disturbances. In addition to a visible "real time" observer to monitor the readings and convey the results to the instrument operator at the telescope focus, the signals were permanently recorded on a photographic plate. As with Coblentz, the problem of isolating true infrared radiation was solved with a water cell; with 1 cm of water in the beam, virtually all radiation at wavelengths longer than ~1.2 μm was blocked. The instrument had adequate sensitivity to reach hundreds of stars. It was sufficiently stable that the photometric data would be judged of reasonable quality by the standards of modern photometers (see the signals recorded in [67]). Comparison with modern commercial thermopiles (e.g. [81]) indicates that the devices employed by Pettit and Nicholson were an order of magnitude more sensitive. As we show below, their photometry was highly repeatable and quite accurate, although the photometric system had obvious shortcomings. For example, although the Harvard visible photometry indicated a difference between Sirius and Vega of 1.72 magnitudes, Pettit and Nicholson [68] measured a radiometric difference of only 1.37 magnitudes. The modern values are 1.46 magnitudes at V and 1.31 magnitudes in the near infrared.

Their masterwork addressed the then topical issue of stellar temperatures [68]. The paper is remarkable for the meticulous description of the derivation of accurate photometry, including corrections for effects such as the tarnishing of the telescope silver coatings. They also made observations with a specially prepared detector with a salt window to allow them to correct to total infrared flux from the cutoff near 3 μm imposed by the glass or quartz windows of the devices they normally used. Finally, their photographic recordings of the galvanometer output allowed precise reductions that could, for example, compensate for slow instrument baseline drifts, and could combine multiple photometric readings in a consistent way (they typically took eight for each observation).

From their measurements, Pettit and Nicholson [68] derived reasonably accurate stellar temperatures. A significant error source was the visible magnitudes, which were taken from the revised Harvard magnitudes [70]. These values represented a huge effort (more than a million individual measurements), obtained by direct visual observation without bandpass-defining filters. The effective wavelength was therefore uncertain. A comparison with modern V photometry indicates that a value significantly shorter than 0.555 μm is appropriate. Therefore, if we take the temperatures derived from the Heat-Index of Pettit and Nicholson and an assumed effective wavelength of 0.529 μm, we find that the agreement with modern values is of comparable accuracy as for other contemporary determinations (e.g., the tabulation they provide from Russell, Dugan, and Stewart). For cool stars, it is better.

In parallel with the work of Pettit and Nicholson, Abbott [1] used a low-dispersion prism to obtain infrared spectra of stars. His detector was based on the Crookes' Radiometer; Nichols assisted him in the construction of early versions. To make the moving parts of very low mass and thereby boost the sensitivity, he substituted the wings of house flies for the mica plates Nichols had used as the radiation receivers. Although his published spectra show a general trend toward larger infrared emission for cooler stars, there is no clear pattern from star to star, even among the spectra he classifies as the best. It appears that the experimental difficulties were too great for his approach.

Pettit and Nicholson wrote a second very impressive paper, comparing the infrared and visible variability of cool stars [69]. Their two main works represented a high point of early infrared astronomy, with accurate and sensitive measurements, very careful analysis, and application to important astrophysical problems. Kuiper [42] used their work in computing what became the standard reference on stellar temperatures and bolometric corrections. Otherwise, neither paper seems to have received much attention until the 1950s. A few additional papers by others reported further radiometric measurements over the following decade, similar to the work of Coblentz and of Pettit and Nicholson [20, 80]. They had little impact. Most practicing astronomers considered infrared observations irrelevant.

4 Starting over again

Fifteen years after the work of Pettit and Nicholson, Whitford [89] obtained near-infrared measurements of a few stars to extend the interstellar extinction curve. The thermopiles had been replaced with a lead sulphide photoconductive cell, a spinoff of World War II. The galvanometer and photographic recording of its readings had also been replaced with vacuum-tube-based electronics. Bandpass definition remained a challenge; in this case, a dye-based filter isolated a band that merged the modern H and K bands. Whitford [90] later reported some additional measurements with the same apparatus. This work, although promising, was not extended beyond his specific interest in the form of the reddening curve. Kuiper et al. [43] also used a lead sulphide photocell in an infrared spectrometer and published stellar spectra, but thereafter turned the instrument to planetary studies.

Fellgett [21] also started an initiative in infrared astronomy. Despite the use of a lead sulphide photocell and vacuum-tube electronics, Fellgett remarks that the sensitivity of the photometer was only modestly improved over that of Pettit and Nicholson. The definition of photometric bands was also similar; a crude filter of mica blocked the visible but transmitted in the infrared to the cut-off wavelength of the detector. The accuracy of both systems is attested by Fellgett's comparison of measurements of stars in common; it shows an average deviation between them of only 0.08 magnitudes. A surprising shortcoming of Fellgett's photometer is the absence of any sky reference beam that would allow differential rather than total-power measurements. A differential approach had been found useful by Stone [85] and had been used in all the following photometers—including Boys, Nichols, Coblentz, and

Pettit and Nicholson. Fellgett shows fluctuations in the signal from a star that he attributes to variations in atmospheric extinction, but he also claims that "it was often impossible to detect even considerable unsteadiness of the extinction by visual inspection of the sky." It is likely that these fluctuations had a significant contribution from emission variations not compensated because his measurements did not include a simultaneous sensing of the sky.

However, unlike the compelling efforts to overcome all obstacles that infuses the papers of Pettit and Nicholson, Fellgett's style suggests that his paper was intended to be a swan song. He rationalizes shortcomings in the equipment by saying they would be addressed only if a "first programme had shown whether a worthwhile infrared-red sensitivity would be realized." On this issue, he proclaims, "there is little chance of detecting, in the infra-red, stars that are too cool to be visible." He never published a sequel to his paper, and it is cited only for its historical significance.

Lunel [51] describes an extensive effort using an instrument similar in concept to Fellgett's. However, an important improvement is that she utilized filters that cleanly isolated the H and K photometric bands, and also a band that roughly combined I and J. She measured 61 stars in this system and showed that the data correspond well with the observations of Pettit and Nicholson [68] and Fellgett [21]. Although the instrument and measurements are described in detail, her paper lacks the careful physical interpretation that is a strength of the work of Pettit and Nicholson [68]. Posterity has ignored it.

Moroz [57–60] made the next start in infrared astronomy. His photometer was based on lead sulphide and germanium photocells, with spectral band definition not significantly advanced from the water cell of Coblentz. The most important innovation was his choice of non-stellar targets: the Crab Nebula, Galactic Center, and the Orion Nebula. The success of radio astronomy and the leadership of Russia in the theory of the newly discovered radio emissions had provided new classes of object for observation. However, Moroz's efforts were largely overlooked in the rest of the world, part of a widespread indifference toward Russian observational astronomy as well as the indifference to infrared astronomy in general. Moroz had broad interests, and about the time his single-man infrared program began to be overrun by efforts in the USA, he found himself attracted into planetary exploration, where he spent the rest of his career.

Connes [16] developed high-spectral-resolution infrared spectrometers. A few stellar observations were obtained (e.g., [15]), which provided detailed lists of absorption features in cool stars [56]. However, the primary applications for these instruments were for planetary studies.

5 The beginning that took root

Modern infrared astronomy grew from a new start in the USA in the 1960s. The story is told in detail in Low et al. [50]. The first step was the lead-sulphide-based stellar photometry by Johnson [38], followed by expansion to the 10 μm atmospheric window by Wildey and Murray [91] and Low and Johnson [46]. Having seen so many false starts, it is important to consider what made this one different:

1. Despite what one might assume, it was NOT due to advances in sensitivity. Johnson's [38] original set of measurements was in the same magnitude range, generally even a bit brighter, than the stars measured by Pettit and Nicholson [68]. The modest gains in photometer performance were nearly exactly cancelled by the use of smaller telescopes (as had previously been remarked by Fellgett [21]). Even Low's [44] bolometer initially operated at similar detection levels to the previously available mid-infrared photoconductors (compare the results of Wildey and Murray [91] with those of Low and Johnson [45, 46]).
2. There were a number of important technical advances, nonetheless: (a) electronic integration allowed long accumulation of signals with full realization of the intrinsic system signal-to-noise potential; and (b) two decades of improvements in filter technology had led to high-performance infrared interference filters [52]), critical to isolate well-defined photometric bands matched to the atmospheric windows. Although Johnson originally used total-power chopping, the huge sky emission at 10 μm required differential measurements ([45, 91]), leading to rediscovery of their value originally established a century earlier ([79, 85]). Improved modulation capabilities were the first critical innovation introduced as part of the growth of the field; they let the improving detectors actually function at their full performance levels.

There was a third and far more important element. The brilliant success of radio astronomy demonstrated the scientific potential for astronomy in new spectral regions. The new start in infrared astronomy was led almost entirely by experimental physicists, not by astronomers [50]. They found infrared astronomy to be an open playing field where a bit of ingenuity would yield many exciting discoveries. As with Moroz, they were strongly influenced in their selection of targets by discoveries in the radio regime.

The development of infrared astronomy primarily by physicists, even in the face of indifference and antipathy in traditional astronomy departments, also had a precedent in radio astronomy. Jarrell [36] summarizes the backgrounds of 95 early radio astronomers, finding there were 61 physicists, 21 engineers, two mathematicians, and only 11 astronomers. Publication in the new field was often in physics or neutral journals (such as Nature). Citations of these papers by proper astronomers was rare; Jarrell [36] summarizes a study of citation rates: "To be sure, several important optical and theoretical astronomers....showed interest in the early radio work. But, for most of their colleagues, radio data were simply irrelevant to their research. In fact, radio results were as likely to be of as much value to atmospheric scientists as to astronomers." The situation led to the following editorial:

> Radio astronomy is a rapidly developing branch of astronomy; and any schism between this newer branch and the older branch of optical astronomy is to be greatly regretted. For this reason, the Astronomical Journal, already established as one of the principal astronomical journals, would like to see that papers on radio astronomy, commensurate with its growing importance, will be represented in the Astronomical Journal."—Brouwer and Smith [7]

A parallel and similarly worded editorial also appeared in the Astrophysical Journal.

X-ray astronomy also grew out of physics and other efforts not linked to traditional astronomy departments ([87]; Giacconi 2008, private communication). There was evidently a general resistance to initiatives into new spectral regimes on the part of most twentieth century optical astronomers. Thus, we find a fourth key element to the successful growth of infrared astronomy from the new start in the early 1960s. By chance, there were enough centers of activity for a community to be established—to be sure a community split by intense rivalries, but still where one's accomplishments would be appreciated, sometimes envied, and rapidly incorporated into other scientific and technical advances.

In the decade from the early 1960s to early 1970s, virtually the entire foundation for infrared astronomy was put in place, such as:

1. Establishment of the photometric system, its calibration, and application of it to large numbers of stars [39]
2. Completion of an all-sky survey that successfully disproved Fellgett's discouraging prediction that "there is little chance of detecting, in the infrared, stars that are too cool to be visible." [62]
3. Detection of the internal energy of Jupiter and the other giant planets [3]
4. Identification of oxygen-rich and carbon-rich interstellar dust, the former in a range of environments [53, 84, 92]
5. Discovery of the prototypical dust-embedded very young star [4]
6. Finding the huge and ubiquitous far-infrared outputs of regions where young and massive stars are forming [33, 48]
7. Discovery of the center of the Milky Way, both in stellar emission [5] and in thermal re-emission of the stellar energy by interstellar dust [31]
8. Detection of the strong infrared outputs of star-forming galaxies[41] and discovery of Luminous Infrared Galaxies and Ultraluminous Infrared Galaxies (LIRGs and ULIRGs) [75]
9. Demonstration of the infrared excesses associated with active galactic nuclei [47, 65]

In parallel, the prime approaches toward infrared observations were established:

1. Precepts were laid out for optimizing ground-based telescopes for good infrared performance [49]
2. Airborne infrared astronomy got under way [17]
3. The advantages of operating cold telescopes in the space environment were demonstrated (e.g. [33, 72])

6 Building on the foundation

The sky had been super-saturated with potential discoveries in the infrared. That they had not been made earlier, when some of the technical means to do so were already in place, runs contrary to our usual assumption that discoveries follow quickly after we have the capability to make them [30]. Human imagination had, for once, come up short. However, the explosion of discoveries in the 1960s and early 1970s attracted the attention of the international astronomy community. The instrumentation required to join in was quite modest. Therefore, infrared programs rapidly sprang up in the UK [37], Europe, Australia [86], and South Africa. The discoveries also attracted the interest of advisory bodies (e.g., the National Academy "Greenstein" report) and government funding agencies. Soon, there were plans for substantial upgrades in all three main observing approaches. These upgrades were fuelled by incredible technical progress that occurred, ironically, *after* the field had taken root rather than before.

6.1 Groundbased telescopes

A key step in the USA was a meeting in July, 1975, of an ad hoc advisory group convened by Nancy Boggess (who oversaw the interests of infrared astronomy for NASA) at the Snowmass Colorado ski resort. This group urged NASA to proceed with a large, national infrared telescope. The resulting Infrared Telescope Facility (IRTF) was completed in 1979. In the interim, Gehrz and Hackwell [24] had persuaded the Wyoming legislature and NSF to fund a 2.3-m infrared optimized telescope. Both efforts were trumped by the UK, which completed the UK Infrared Telescope (UKIRT) also in 1979. Thus, by 1980 the astronomical community had access to a number of infrared telescopes at excellent sites (low water vapor).

In 1975, InSb photodiodes were introduced with performance far better than lead sulphide cells [29]. Otherwise, the detectors, optics, and instrumentation evolved slowly from those used in the 1960s. With telescopes virtually as large as possible at the time, it appeared that the rapid growth of the field might level off. Instead, it accelerated due to the introduction of infrared detector arrays, plus the success of the Infrared Astronomical Satellite (IRAS).

Rudimentary arrays became available in the early 1980s. The challenges in the mid-infrared, with the immense signal from the thermal background, proved intractable for sensitive operation, although a few images of bright objects were produced [2]. However, in the near infrared even the very early arrays worked reasonably well. The first example was a 32 × 32 pixel InSb array provided through Alan Hoffman of Santa Barbara Research Center to the University of Rochester, where Hoffman had been an Assistant Professor before joining SBRC [22]. A similar array but with HgCdTe detectors and from Rockwell Science Center was deployed at the University of Arizona shortly thereafter [78]. In both cases, the array readout circuits were charge-coupled-devices (CCDs). CCDs are fundamentally incompatible with high-performance infrared detectors; the detectors require operating temperatures so low that the CCDs must operate in surface channel mode, and hence the read noises of these devices were high (>1,000 electrons). Nonetheless, the performance of individual pixels was comparable to that of the best single detectors. The readout limitations were soon removed by development of devices that had a readout amplifier hidden behind each detector and hence were not subject to the limitations of charge transfer. The first widely applied array of this type, from SBRC, used a readout of 58 × 62 format fitted with a variety of detector types.

All of these initial detector arrays were developed for other (usually military) customers. They were followed by arrays financed by the NSF and NASA and optimized for astronomical applications–slow readout, low noise, large format, and high quantum efficiency. These developments were fostered by the enthusiasm of the array manufacturers. Most of their other customers operated on a principle of compartmentalization of information, so it was not clear what use, if any, was intended for the arrays. Astronomy was wide open; the intended uses were public and interesting, and the customers highly interactive on the design details. A number of astronomers even migrated into array manufacturing. As a result of this partnership, the funding from NSF and NASA was used very effectively to yield the current state of the art devices [77]. A single pixel of one of these megapixel arrays is two to three orders of magnitude more sensitive than the single detectors used 30 years ago.

Soon after the first astronomy-optimized arrays became available, the first 6–10 m telescopes were completed. With the application of adaptive optics, increasingly advancing to a general-purpose capability, these telescopes routinely provide diffraction-limited infrared images. Compared with capabilities 40 years ago, the net gain across the 1–25 μm spectral range is factors of 1,000 to more than 10,000 in sensitivity, a gain in multiplex capability of up to four million, and enhancement of the angular resolution (compared with typical aperture photometers) of two orders of magnitude.

7 Airborne astronomy

Frank Low's initiation of airborne infrared astronomy was a natural step from: (1) his previous measurement of the mm-wave output of the sun from a Navy

Douglas A3-B bomber; and (2) his position at the University of Arizona Lunar and Planetary Laboratory, where director Gerard Kuiper and his group made use of the CV-990 Galileo I Airborne Observatory for planetary observations. Low and Carl Gillespie fitted a 12-in. telescope into the emergency exit of a Lear Jet at NASA Ames Research Center. On their first flight series, the telescope was not well stabilized and there were only two accessible sources bright enough for guiding—Arcturus and Jupiter. The observations revealed the internal energy of the planet–airborne astronomy started with a fanfare! The Lear Jet Observatory continued to fly until 1984 [17]).

However, a much larger telescope was desired. Plans were developed for a 36-in. (0.9 m). The C-141 cargo airplane to carry it was delivered in early 1972 and research flights started 2 years later. In May, 1975, the facility was named the Kuiper Airborne Observatory (KAO). Observing with the Lear Jet had been a barebones adventure, wearing an oxygen mask and squeezed between the back of the telescope and the opposite wall of the fuselage. It conjured up self-images of having the Right Stuff. The KAO ushered in the shirt-sleeve era of airborne infrared astronomy, an environment more conducive to demanding observing. It flew about 70 flights per year for more than 20 years, both for infrared astronomy and for other applications where a highly mobile observatory was required (e.g., the discovery of the rings of Uranus [17, 19]).

Balloons have been intermittently popular to lift infrared telescopes above the atmosphere. Early near-infrared spectra were obtained with the Stratoscope II telescope [93]. We have already mentioned the discovery of the far-infrared emission of the Galactic Center by Hoffmann and Frederick [31]; this group made a number of additional flights to map the region in detail (e.g., [32]). Shortly thereafter, a 1-m balloon was deployed (e.g. [94]). Subsequent efforts have led up to the BLAST, a 2-m telescope with a high performance bolometer imaging array [66]. However, in general the far-infrared community has had a preference for airplane-borne telescopes, in part because such observatories have been more successful at establishing a long-term infrastructure than have balloons.

The ultimate airborne observatory would use the largest possible airplane. NASA approved a joint project with Germany to build just that in 1996. Named the Stratospheric Observatory for Infrared Astronomy (SOFIA), the new observatory would utilize a 2.5-m telescope in a Boeing 747. In part to avoid splitting the infrared astronomy community, SOFIA and Spitzer were sold as a package, and in an era of very tight funding. The plans for SOFIA also further enhanced the shirt-sleeve environment of the KAO. Perhaps lulled by the very rapid deployment of the KAO, astronomers agreed to decommission that observatory to free funds for the development of SOFIA.

The modifications to the 747 airframe proved more complex to implement than expected. In addition, the desire to certify the entire observatory to passenger transport standards added substantial complexity to the instruments and other aspects of the project. The schedule and costs grew until 2006, when after a period of review, a brief halt, and threatened cancellation, work eventually was resumed. Science observations are now expected to begin with

SOFIA in 2009–2010. At that time, the observatory will offer the far-infrared community the major advantage of a readily accessible telescope, with the ability to update instrumentation regularly to keep the observatory at the state of the art and to introduce and demonstrate new technologies [26]. SOFIA is planned for a 20 year service life.

8 Space infrared astronomy

The advantages of operating a cold infrared telescope above the atmosphere were realized in early sounding-rocket experiments at Cornell under Martin Harwit and Jim Houck [50], and through a variety of missions funded by the military and described by Price [72]. They included HISTAR, operated by Price from the Air Force Geophysical Laboratory (AFGL) and the first publicly available mid-infrared survey (a satellite-borne effort, the Celestial Mapping Program, had been released previously but in an irregular fashion in the form of a photographic reproduction of an all-sky infrared source locator).

Besides a large groundbased telescope, the Snow Mass meeting recommended an ambitious all-sky survey to cover the mid and far infrared. In parallel, the Netherlands had studied the feasibility of a far-infrared satellite telescope. NASA issued an Announcement of Opportunity that attracted a number of proposals. They were combined with a down-selection of members and an agreement was negotiated in which the Dutch would provide the spacecraft, the United States the telescope, data system, and dewar, and the United Kingdom the ground station. The resulting IRAS mission is described by Neugebauer et al. [63]. It was one of the most successful space astronomy missions ever conducted and remains a fundamental resource more than 20 years after it was completed.

The success of IRAS led to efforts in both Europe and the United States to build a follow-up mission in the form of a pointed cryogenic telescope that could concentrate on individual sources. A proposal had already been prepared in Europe in 1979, and was accepted as a cornerstone European Space Agency (ESA) mission as soon as the success of IRAS became apparent. Construction started in 1988, with a launch in 1995 and a mission that lasted for more than 2 years. ISO (the Infrared Space Observatory) was the first infrared astronomy mission to use true detector arrays. Its suite of instruments included many capabilities, such as high resolution spectrometers, to enable detailed astrophysical studies. An overview of the mission is given by Kessler [40], and a review of the extragalactic studies by Genzel and Cesarsky [27]).

The US effort ran into repeated political setbacks [76] and was finally launched in 2003, and renamed *Spitzer*. It had been descoped into a much simpler mission than originally envisioned, based on a pioneering concept of developing the design around a small number of defining science programs. It also introduced a new cooling concept that placed heavy reliance on radiative cooling with only a supplemental capacity to reach the final telescope operating temperature and cool the detectors; the mission cryogenic lifetime has been more than 5 years. Fortunately, the development of very high performance infrared detector arrays allowed a mission with superb performance and that, like ISO, has yielded many science breakthroughs. The mission design is described by Gehrz et al. [25], and early reviews of the science accomplished are provided by Werner et al. [88] and Soifer et al. [83].

The Akari Mission was developed by the Japan Aerospace Exploration Agency (JAXA) and completed a double-pass all-sky survey in a number of bands from 1.7 to 180 μm, over its 18 month cryogenic mission (see PASJ [73]). It also had a program of pointed observations both with imagers and with additional instrument capabilities, primarily spectroscopy. The success of this mission underlies the proposal from Japan for a much more ambitious far-infrared cold telescope, the Space Infrared Telescope for Cosmology and Astrophysics (SPICA) [55, 61]. A more immediate advance in far-infrared astronomy will be provided by Herschel, due to be launched by the European Space Agency in mid-2009. It is a large-aperture (3.5 m) radiatively cooled (to about 100 K) far-infrared mission with three instruments providing imaging and a range of spectroscopic capabilities from 60 to 670 μm [71]. Herschel will be the first long-duration far-infrared observatory with a telescope aperture greater than a meter! Also planned for launch in 2009, the Wide-field Infrared Survey Explorer (WISE) will map the sky at 3.3, 4.7, 12, and 23 μm [54].

However, the realization of a dream for infrared astronomers will be the James Webb Space Telescope (JWST), of 6.6 m aperture and with four powerful focal plane instruments [23]. It will cover the 0.7 to 28 μm wavelength range and due to the radiative cooling of its primary to about 40 K will be natural-background limited over most of this wavelength region. Upon the launch of this mission in 2013, the astronomical capability in the infrared—number of pixels/(detection limit squared)—will have increased by a factor of 10^{20} over 50 years, a doubling time of 9 months sustained over this entire time span.

Acknowledgements Preparation of this article was partially supported by contract 1255094 from Caltech/JPL and by contract NAG5-12318 from NASA/Goddard, both to the University of Arizona.

References

1. Abbott, C.G.: Energy spectra of the stars. ApJ **69**, 293 (1929)
2. Arens, J.F., Lamb, G.M., Peck, M.C., Moseley, H., Hoffmann, W.F., Tresch-Fienberg, R., Fazio, G.G.: High spatial resolution observations of NGC 7027 with a 10 micron array camera. ApJ **279**, 685 (1984)
3. Aumann, H.H., Gillespie, C.M., Low, F.J.: The internal powers and effective temperatures of Jupiter and Saturn. ApJL **157**, L69 (1969)
4. Becklin, E.E., Neugebauer, G.: Observations of an infrared star in the Orion nebula. ApJ **147**, 799 (1967)
5. Becklin, E.E., Neugebauer, G.: Infrared observations of the galactic center. ApJ **151**, 145 (1968)

6. Boys, C.V.: On the heat of the moon and stars. Proc. R. Soc. **47**, 480 (1890)
7. Brouwer, D., Smith, H.J.: Publication of papers in radio astronomy. AJ **64**, 36 (1959)
8. Coblentz, W.W.: Note on the radiation from stars. Publ. Astron. Soc. Pac. **26**, 169 (1914)
9. Coblentz, W.W.: A Comparison of Stellar Radiometers and Radiometric Measurements on 110 Stars. Government Printing Office, Washington (1915)
10. Coblentz, W.W.: Report on instruments and methods of radiometry. J. Opt. Soc. Am. **5**, 259 (1921)
11. Coblentz, W.W.: The measurement of solar, sky, nocturnal, and stellar radiation. J. Opt. Soc. Am. **5**, 269 (1921)
12. Coblentz, W.W.: A portable vacuum thermopile. J. Opt. Soc. Am. **5**, 356 (1921)
13. Coblentz, W.W.: New measurements of stellar radiation. ApJ **55**, 20 (1922)
14. Coblentz, W.W.: The effective temperature of 16 stars as estimated from the energy distribution in the complete spectrum. Proc. Natl. Acad. U. S. A. **8**, 49 (1922)
15. Connes, P., Connes, J., Bouigue, R., Querci, M., Chauville, J., Querci, F.: Sur les spectres à étoiles rouges M et C entre 4000 et 9000 cm^{-1} (2,5 et 1,1 μ). I. Description génélrale. ANAP **31**, 485 (1968)
16. Connes, P.: Early history of Fourier transform spectroscopy. IR Phys. **24**, 69 (1984)
17. Dolci, W.W.: Milestones in Airborne Astronomy: From the 1920's to the Present. Am. Inst. Aeronautics and Astronautics, Reston, Virginia (1997)
18. Eddy, J.A.: Thomas A. Edison and infrared-red astronomy. J. Hist. Astron. **3**, 165 (1972)
19. Elliot, J.L.: The rings of Uranus. Nature **267**, 328 (1977)
20. Emberson, R.M.: Radiometric magnitudes of some of the brightest stars. ApJ **94**, 427 (1941)
21. Fellgett, P.B.: An exploration of infrared stellar magnitudes using the photo-conductivity of lead sulphide. MNRAS **111**, 537 (1951)
22. Forrest, W.J., Moneti, A., Woodward, C.E., Pipher, J.L., Hoffman, A.: The new near infrared rray camera at the University of Rochester. Publ. Astron. Soc. Pac. **97**, 183 (1985)
23. Gardner, J.P. et al.: The James Webb space telescope. Sp. Sci. Rev. **123**, 485 (2006)
24. Gehrz, R.D., Hackwell, J.A.: Exploring the infrared universe from Wyoming. Sky Telesc. **55**, 466 (1978)
25. Gehrz, R.D., et al.: The NASA Spitzer space telescope. Rev. Sci. Inst. **78**, 011032 (2007)
26. Gehrz, R.D., Becklin, E.E.: The stratospheric observatory for infrared astronomy (SOFIA). Proc. SPIE **7012**, 61 (2008)
27. Genzel, R., Cesarsky, C.J.: Extragalactic results from the infrared space observatory. ARAA **38**, 761 (2000)
28. Hale, G.E.: Heat radiation of the stars. ApJ **9**, 360 (1899)
29. Hall, D.N.B., Aikens, R.S., Joyce, R., McCurnin, T.W.: Johnson noise limited operation of photovoltaic InSb detectors. Appl. Opt. **14**, 450 (1975)
30. Harwit, M.W.: Cosmic Discovery. Basic Books, New York (1981)
31. Hoffmann, W.F., Frederick, C.L.: Far-infrared observation of the galactic-center region at 100 microns. ApJL **155**, L9 (1969)
32. Hoffmann, W.F., Frederick, C.L., Emery, R.J.: Far-infrared observation of the galactic-center region at 100 microns. ApJL **170**, L89 (1971)
33. Houck, J.R., Soifer, B.T., Pipher, J.L., Harwit, M.: Rocket-infrared four-color photometry of the galaxy's central regions. ApJL **169**, 31 (1971)
34. Huggins, W.: Note on the heat of the stars. Proc. Royal Soc. **17**, 309 (1869)
35. Huggins, W.: Heat of the stars. Astron. Regist. **16**, 309 (1878)
36. Jarrell, R.: Radio astronomy, whatever that may be. In: Orchiston, W. (ed.) The Marginalization of Early Radio Astronomy. Astrophysics and Space Science Library, vol. 334. Springer, Dordrecht (2005)
37. Jennings, R.E.: History of British infrared astronomy since the second world war. QJRAS **27**, 454 (1986)
38. Johnson, H.L.: Infrared stellar photometry. ApJ **135**, 69 (1962)
39. Johnson, H.L.: Astronomical measurements in the infrared. ARAA **4**, 193 (1966)
40. Kessler, M.: The infrared space observatory (ISO) mission. Adv. Space Res. **30**, 1957 (2002)
41. Kleinmann, D.E., Low, F.J.: Observations of infrared galaxies. ApJL **159**, L165 (1970)
42. Kuiper, G.P.: The magnitude of the sun, the stellar temperature scale, and bolometric corrections. ApJ **88**, 429 (1938)

43. Kuiper, G.P., Wilson, W., Cashman, R.J.: An infrared stellar spectrometer. ApJ **106**, 243 (1947)
44. Low, F.J.: Low-temperature germanium bolometer. J. Opt. Soc. Am. **51**, 1300 (1961)
45. Low, F.J., Johnson, H.L.: Stellar photometry at 10 microns. ApJ **139**, 1130 (1964)
46. Low, F.J., Johnson, H.L.: The spectrum of 3C273. ApJL **141**, L336 (1965)
47. Low, F.J., Kleinmann, D.E.: Infrared observations of seyfert galaxies, quasistellar sources, and planetary nebulae. AJ **73**, 868 (1968)
48. Low, F.J., Aumann, H.H.: Observations of galactic and extragalactic sources between 50 and 300 microns. ApJL **162**, 79 (1970)
49. Low, F.J., Rieke, G.H.: The instrumentation and techniques of infrared photometry. In: Carleton, N. (ed.) Methods of Experimental Physics, vol. 12, part A. Academic Press, New York and London (1974)
50. Low, F.J., Rieke, G.H., Gehrz, R.D.: The beginning of modern infrared astronomy. ARAA **45**, 43 (2007)
51. Lunel, M.: Recherches de photométrie stellaire dans l'infra-rouge au moyen d'une cellule au sulfure de plomb. Ann. D'Astrophysique. **23**, 1 (1960)
52. Macleod, H.A.: Thin-Film Optical Filters, 3rd edn. Bristol, Philadelphia (2001)
53. Maas, R.W., Ney, E.P., Woolf, N.J.: The 10-micron emission peak of comet Bennett 1969I. ApJL **160**, L101 (1970)
54. Mainzer, A.K., et al.: Preliminary design of the wide-field infrared survey explorer (WISE). Proc. SPIE **5899**, 262 (2005)
55. Matsumoto, T.: Large-aperture cooled telescope (SPICA) for mid- and far-infrared astronomy. Proc. SPIE **5487**, 1501 (2004)
56. Montgomery, E.F., Connes, P., Connes, J., Edmonds, F.N.: The infrared spectrum of Arcturus. ApJ **19**, 1 (1969)
57. Moroz, V.I.: The radiation flux from the Crab Nebula at 2 microns and some conclusions on the spectrum and magnetic field. Astron. Z. **37**, 265 (1960)
58. Moroz, V.I.: An attempt to observe the infrared radiation of the galactic nucleus. Astron. Z. **38**, 487 (1961)
59. Moroz, V.I.: Radiation emission from the Orion Nebula in the 0.85–1.7 micron wavelength region. Astron. Z. **40**, 788 (1963)
60. Moroz, V.I.: Infrared observations of the Crab Nebula. Astron. Ah. **40**, 982 (1963)
61. Nakagawa, T., Murakami, H.: Mid- and far-infrared astronomy mission SPICA. Adv. Space Res. **40**, 679 (2007)
62. Neugebauer, G., Leighton, R.B.: Two-micron sky survey. A preliminary catalogue. NASA Spec. Pub. **3047** (1969)
63. Neugebauer, G., et al.: The infrared astronomical satellite (IRAS) mission. ApJL **278**, L1 (1984)
64. Nichols, E.F.: On the heat radiation of Arcturus, Vega, Jupiter, and Saturn. ApJ **13**, 101 (1901)
65. Oke, J.B., Neugebauer, G., Becklin, E.E.: Absolute spectral energy distribution of quasi-stellar objects from 0.3 to 2.2 microns. ApJ **159**, 341 (1970)
66. Pascale, E. et al.: The Balloon-borne large aperture submillimeter telescope: BLAST. ApJ **681**, 400 (2008)
67. Pettit, E., Nicholson, S.B.: Application of vacuum thermocouples to problems in astrophysics. ApJ **56**, 295 (1922)
68. Pettit, E., Nicholson, S.B.: Stellar radiation measurements. ApJ **68**, 279 (1928)
69. Pettit, E., Nicholson, S.B.: Measurements of the radiation from variable stars. ApJ **78**, 320 (1933)
70. Pickering, E.C.: Revised Harvard photometry. Harvard Annals **50**, 1 (1908)
71. Pilbratt, G.: Herschel mission: status and observing opportunities. SPIE **5487**, 401 (2004)
72. Price, S.D.: The infrared sky: a survey of surveys. Publ. Astron. Soc. Pac. **100**, 171 (1988)
73. Publ. Astron. Soc. Jpn. (Akari Special Issue) **59**, SP2 (2007)
74. Ranyard, A.C.: Edison's tasimeter. Astron. Regist. **16**, 309 (1878)
75. Rieke, G.H., Low, F.J.: Infrared photometry of extragalactic sources. ApJL **176**, L95 (1972)
76. Rieke, G.H.: The Last of the Great Observatories. University of Arizona, Tucson (2006)
77. Rieke, G.H.: Infrared detector arrays for astronomy. ARAA **45**, 77 (2007)

78. Rieke, M.J., Rieke, G.H., Montgomery, E.F.: Rockwell HgCdTe arrays as imagers. In: Wynn-Williams, C.G., Becklin, E.E. (eds.) Infrared Astronomy with Arrays, p. 213. University of Hawaii Institute for Astronomy (1987)
79. Rosse, E.O: On the radiation of heat from the moon, the law of its absorption by our atmosphere, and its variation in amount with her phases. Proc. R. Soc. **21**, 241 (1873)
80. Rust, C.F.: Spectral types and radiometric observations of stars with large infrared index. ApJ **88**, 525 (1938)
81. Schilz, J.: Thermoelectric Infrared Sensors (Thermopiles) for Remote Temperature Measurements. Perkin Elmer Optoelectronics GmbH, Wiesbaden, Germany (2000)
82. Sinton, W.M.: Through the infrared with logbook and lantern slides: a history of infrared astronomy from 1868 to 1960. Publ. Astron. Soc. Pac. **98**, 246 (1986)
83. Soifer, B.T., Helou, G., Werner, M.W.: The Spitzer view of the extragalactic universe. ARAA **46**, 201 (2008)
84. Stein, W.A., Gillett, F.C.: Spectral distribution of infrared radiation from the trapezium region of the Orion Nebula. ApJL **155**, L197 (1969)
85. Stone, E.F.: Approximate determinations of the heating-powers of Arcturus and Alpha Lyrae. Proc. R. Soc. **18**, 159 (1870)
86. Storey, J.W.V.: Infrared astronomy: in the heat of the night. Publ. Astron. Soc. Aust **17**, 270 (2000)
87. Tucker, W., Giacconi, R.: The X-ray Universe. Harvard University Press, Cambridge (1985)
88. Werner, M.W., Fazio, G.G., Rieke, G.H., Roellig, T.L., Watson, D.M.: First fruits of the Spitzer space telescope: galactic and solar system studies. ARAA **44**, 269 (2006)
89. Whitford, A.E.: An extension of the interstellar absorption-curve. ApJ **107**, 102 (1948)
90. Whitford, A.E.: The law of interstellar reddening. AJ **63**, 201 (1958)
91. Wildey, R.L., Murray, B.C.: Stellar photometry from B8 to M7 in the 8–14 micron window. AJ **68**, 300 (1963)
92. Woolf, N.J., Ney, E.P.: Circumstellar infrared emission from cool stars. ApJL **155**, L181 (1969)
93. Woolf, N.J., Schwarzschild, M., Rose, W.K.: Infrared spectra of red-giant stars. ApJ **140**, 833 (1964)
94. Wright, E.L., Fazio, G.G., Low, F.J.: A high-resolution far-infrared survey of the W31 region. ApJ **217**, 724 (1977)

Exp Astron (2009) 25:143–156
DOI 10.1007/s10686-009-9139-8

REVIEW ARTICLE

History of X-ray telescopes and astronomy

Riccardo Giacconi

Received: 5 January 2009 / Accepted: 6 January 2009 / Published online: 24 January 2009

Abstract The development of X-ray telescopes over the last 45 years is briefly summarized. The first applications to the study of solar X-ray emission are described up to the 1973 Skylab mission. The rather convoluted path that led to the first stellar X-ray orbiting telescope Einstein (1978) and later to Chandra (1999) are discussed. During this 45 years period the angular resolution improved from 20 to 0.5 arc sec and the sensitivity by ten billions. X-ray observations have discovered new types of stellar systems (binary X-ray sources containing neutron stars and black holes) and intergalactic high temperature plasmas containing most of the baryonic mass of the universe. They have become an indispensable tool to study the role of energetic phenomena in the creation and dynamic evolution of cosmic structures. The methodology introduced by X-ray astronomy has influenced all of astronomy.

Keywords History · X-ray telescopes · Astronomy · Stellar systems · Universe

1 Beginnings

Soon after the end of World War II, a Group at Naval Research Laboratories led by Herbert Friedman started the observation of solar X-ray emission. Using Geiger counters carried aloft by captured V-2 German rockets they were able to discover X-rays from the Sun and to establish the variability of the emission over an 11-year period. They were unable to discover any X-ray radiation from stars.

R. Giacconi (✉)
Johns Hopkins University, Baltimore, MD, USA
e-mail: rgiacconi@comcast.net

However, when considering the scientific potential of space observations in the late 1950s, several Committees of the National Academy of Sciences realized that X-ray observations could yield important new astronomical information. X-rays were the lowest-energy radiation in the high-energy domain that could reach Earth unimpeded over cosmological distances and therefore they would provide the highest photon flux in the emission from high-energy processes.

When in 1959 George Clark, Bruno Rossi and I started to think about X-ray astronomy we quickly realized that the fluxes we could predict from extra-solar sources were quite small [1]. While the Sun produced at Earth a flux of 10^6 photons/cm^2 s, well above the cosmic-ray background of about 1/cm^2 s in the Geiger counters used at the time, the fluxes expected from other celestial sources were much smaller. A Sun-like star at a distance of one light year would produce 2.5×10^{-4} photons/cm^2 s.

We considered emissions from a number of potential sources including: flare stars, peculiar A stars, Supernova remnants and the Moon. Their fluxes were difficult to estimate on the basis of available knowledge. Lunar fluorescence from solar X-rays appeared as possibly the strongest source to be expected, at 0.4 photons/cm^2 s. This was at least 50 times below the upper limit quoted by Friedman. Although this limit was only for a small portion of the sky, it still was a daunting prospect.

I had been working in cosmic ray research with Beppo Occhialini in Milano and with Robert Thompson at the University of Indiana prior to joining AS&E (American Science and Engineering), and I was well aware of the difficulty of working in a field with very low fluxes and therefore with poor statistic.

I started thinking of ways to concentrate the X-ray radiation from a large area collector onto a small detector, thereby increasing the signal to noise in proportion to the ratio of the area of the collector to that of the resolution element. In the book on X-rays by Compton and Allison and in the Encyclopedia of Physics by Flugge I found all I needed to design the first X-ray "Telescope" with an improvement in sensitivity of orders of magnitudes. When I first discussed this idea with Bruno Rossi he suggested the concept of nesting. Improvements in sensitivity of 4×10^5 could be anticipated for angular resolutions of 20 arc sec (10^{-4} rad), [2].

My group at AS&E started the development of grazing-incidence telescopes in 1960, before the discovery of Sco X-1. The early laboratory tests were quite necessary to establish the scientific and technological basis for grazing-incidence mirrors. As to the optics design it became immediately clear that a real imaging telescope could be built using conics. Wolter had earlier studied several configurations for X-ray telescopes, which however could not be built because of the very high precision required in small dimensions [3]. Using double reflections we could obtain real imaging optics satisfying Abbe's conditions with a simple design (Fig. 1). We tried several materials and techniques before building the first flight-worthy telescope, by electrodepositing nickel on a steel mandrel. The first flight took place in 1963, on a rocket instrumented with a solar pointing control by Ball Brothers Corporation. The experiment worked

Fig. 1 An imaging X-ray telescope

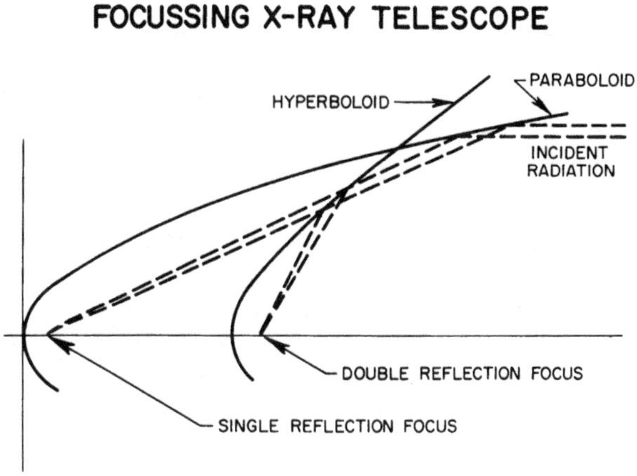

but poorly; we did not obtain the first scientifically useful image until 1965 (Fig. 2) [4].

2 Counter techniques and stellar X-ray astronomy

I now must take a wide detour to bring up to date the progress made in stellar X-ray astronomy using a simple counter technique. In June 1962 we had flown an Aerobee rocket carrying the first successful counter experiment which discovered Sco X-1 and the isotropic X-ray background [5]. This discovery

Fig. 2 The X-ray picture of the Sun obtained in 1965

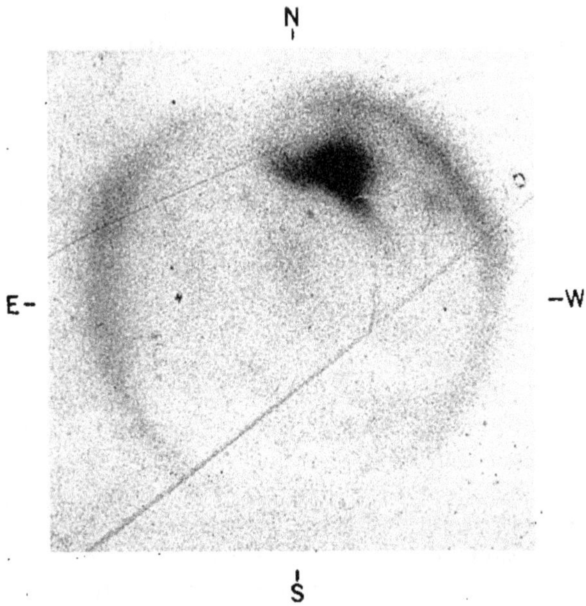

completely revolutionized X-ray astronomy. We had discovered a celestial source with an X-ray luminosity a thousand times that of the Sun at all wavelengths. The ratio of X-ray to visible-light luminosity was 1,000. This implied a new type of stellar system and a new process for X-ray emission. The background was also very interesting, because it could carry cosmological information.

In 1963 Herbert Gursky and I proposed to NASA a telescope of 1.2-m diameter (the same size as *Chandra*, but much simpler) which could image and detect this background with an angular resolution of 1 arc min [6]. But there was a great deal of resistance at NASA Headquarters and also in the scientific community to the development of telescopes for stellar astronomy. To give an example: Philip Morrison of MIT had declared that all X-ray sources were likely to be extended and therefore there was no point in high-resolution studies.

In the 1960s we continued the study of stellar sources with counter experiments and we were joined by many groups in the world. These efforts culminated in the launch by our group at AS&E of the first orbiting X-ray observatory UHURU in 1970 [7]. It is with UHURU that binary X-ray sources containing neutron stars and black holes were discovered. It was found that the energy for the abundant X-ray emission from these sources was provided by the infall of gas from the companion onto the collapsed objects. It was also with UHURU that the intergalactic plasma concentrations in clusters of galaxies were discovered. These plasma concentrations were found to contain many times the mass of all galaxies in the cluster.

During the same period we concentrated on the use of X-ray telescopes in solar X-ray astronomy, where there was an understood need to improve the angular resolution and where we had the support of John Lindsay of Goddard Space Flight Center. This was not only interesting scientifically, but allowed us to continue developing better and better telescopes which eventually could be used for stellar astronomy.

The group at AS&E now included Leon Van Speybroek and Giuseppe Vaiana, the first an outstanding optical designer and builder, the second a brilliant solar and later stellar astrophysicist. They designed a much finer solar X-ray telescope with thick beryllium walls (kanogen-coated), which allowed the achievement of a much better optical figure. In 1968 our group obtained the picture of a solar flare with 5″ resolution [8]. The X-ray image is shown in Fig. 3.

In 1968 we competed for a solar experiment to be flown on board of SKYLAB. SKYLAB was the first US manned space station, with three planned visits by astronauts. Our experiment consisted of a 30-cm X-ray telescope which imaged the sun on photographic film. The film was to be retrieved and brought to the ground by the astronauts. I should emphasize that this was not just a matter of convenience but the only way at the time to obtain high resolution over a wide field.

Always thinking ahead to stellar astronomy we designed the SKYLAB mirror to consist of two nested mirrors. Although not strictly required for solar

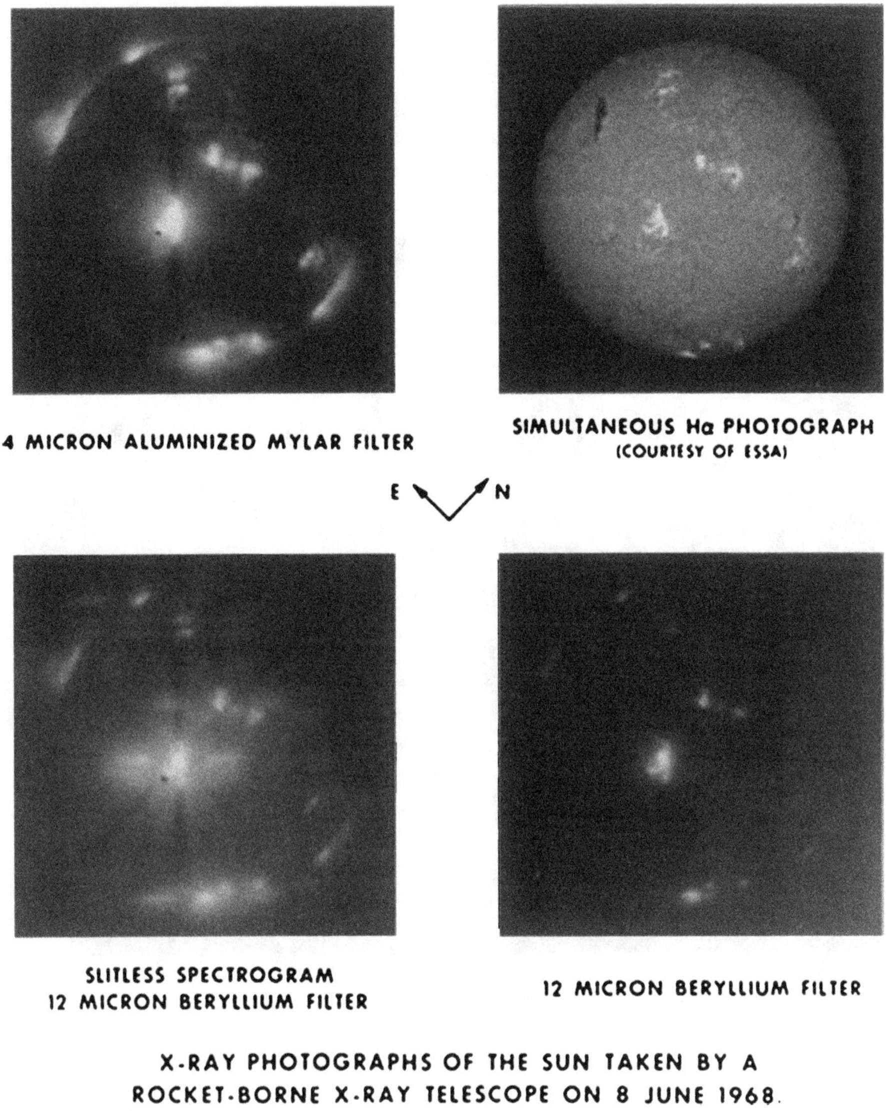

Fig. 3 X-ray photograph of the Sun 8 June 1968

physics this approach increased the sensitivity, something that was absolutely necessary in stellar astronomy, and tested the nesting concept.

There was another new feature in the AS&E SKYLAB experiment design: a transmission-grating spectrometer, which was invented by Herbert Gursky and Ted Zehnpfennig of our group and tested within a week of the proposal deadline.

SKYLAB flew in 1973, and the astronauts were essential not only for the exchange and retrieval of film, but also for repairing a problem with the solar

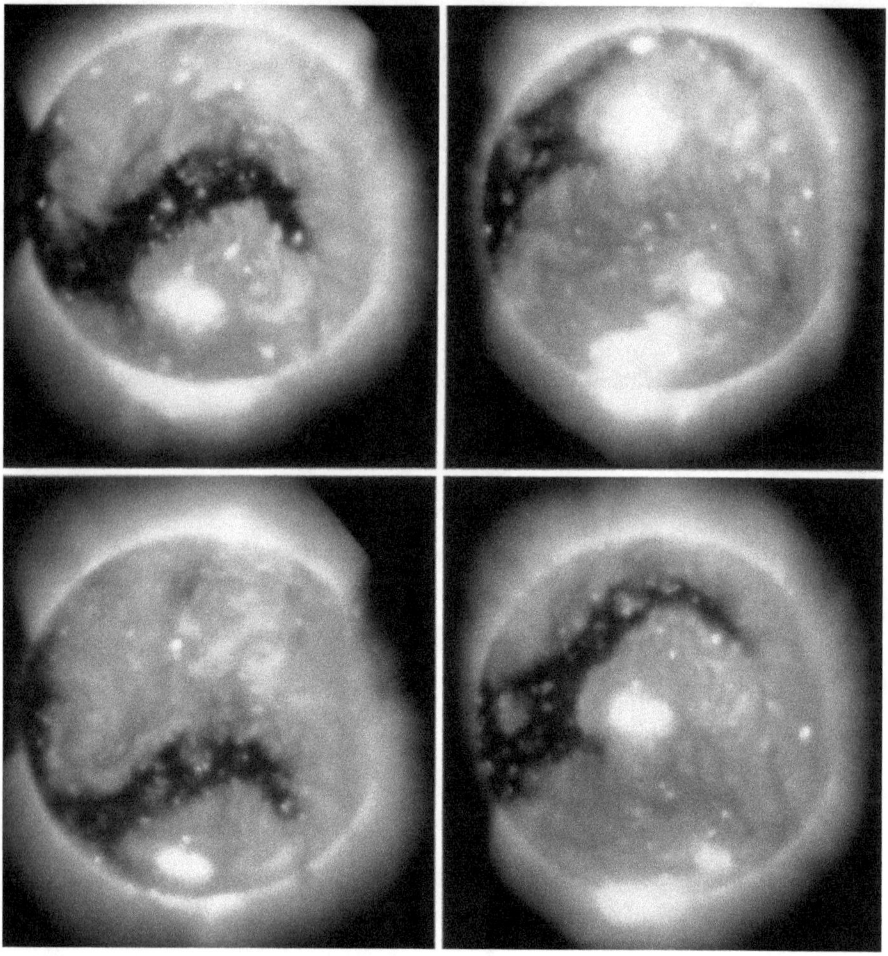

Fig. 4 X-ray pictures of the Sun during a solar rotation

cell deployment and for instrument malfunctions. Scientifically the mission was extremely successful. Solar X-ray emission was studied in detail over several solar rotations (Fig. 4) and the results led to a substantial change in our understanding of the heating and confinement of plasmas in the corona [9].

3 Einstein

We continued to urge NASA to support the development of stellar X-ray telescopes, the main difference with solar telescopes being the requirement for higher efficiency and for electronic rather than film detectors. Finally, following the recommendations of the National Academy of Sciences Summer

Study at Woodshole in 1967 and the demonstration in 1968 of the power of imaging X-ray telescopes, NASA was prepared to proceed with a stellar X-ray telescope as part of the High Energy Astronomical Observatory mission. The HEAOs were huge spacecraft intended to be large but inexpensive.

A consortium of institutions including AS&E, Columbia University, Goddard Space Flight Center, Massachusetts Institute of Technology and Marshall Space Flight Center proposed to NASA an observatory called the "Large Orbiting X-ray Telescope" or LOXT. LOXT was selected for the third mission; it consisted of two telescopes: the first was a 1.2-m parabola–hyperbola design with four nested surfaces for arc second imaging, the second a 1 × 1-m Kirkpatrick–Baez design as a collector for spectroscopy and polarimetry. In essence LOXT incorporated the capabilities that were later realized separately by the *Chandra* and XMM missions. The LOXT program was approved in September 1970 and we proceeded with the design and development for 3 years, until it was abruptly cancelled by NASA to pay for overruns on the Viking planetary program. In a reconstituted HEAO program of half the initial size we were offered to fly only one telescope, of 0.6 rather than 1.2 m diameter, on a smaller spacecraft. This reduced version of LOXT became what was ultimately known as *Einstein*. *Einstein* was the first orbiting observatory to carry a high resolution X-ray telescope [10].

The *Einstein* mirrors were of fused quartz to obtain higher surface smoothness and therefore efficiency. A set of four co-focal mirrors was assembled in which the outside mirror had a diameter of 0.6 m. For this mission a new type of

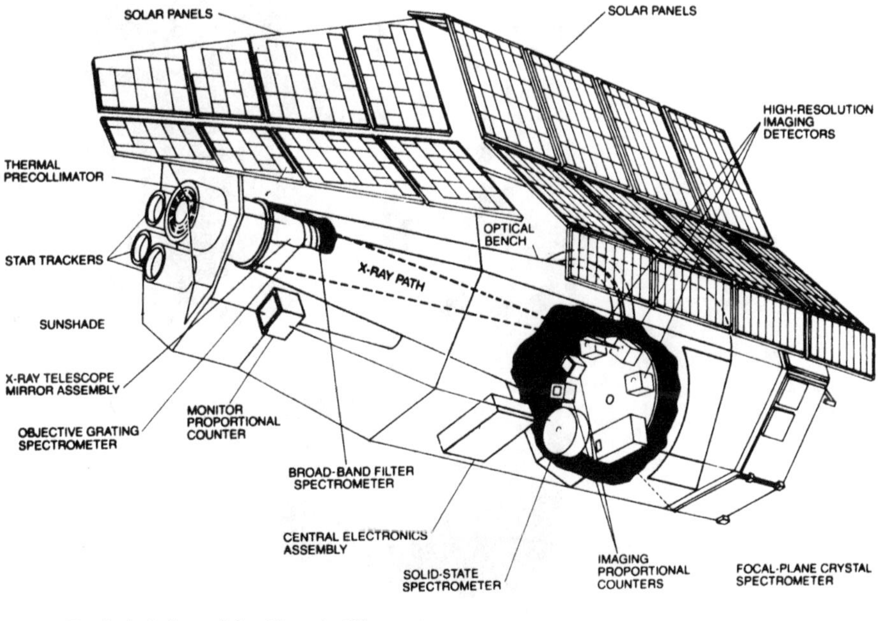

Fig. 5 Exploded view of the *Einstein* Observatory

detector (the High Resolution Imager or HRI) was developed by Steve Murray of our group at Harvard. The detector was based on the use of channeltron multipliers read out by a simple grid. The position and time of arrival of each photon was measured and transmitted to the ground by telemetry.

The coordinates of the detector were translated into sky positions by an on-board self-calibrating system. In this system optical fiducial lights on the detector front surface were projected through the X-ray mirror and a corner reflector onto the visible-light star tracker, which was read out every second. This system had first been developed for LOXT and was used successfully on *Einstein*. It had the great advantage that it freed the X-ray observatories from the need for accurate pointing. They only required stability over timescales of seconds, which is easy to achieve. This approach is made possible by single photon counting and the low fluxes of X-ray astronomy, and gives a great advantage in simplicity compared with optical and infrared space telescopes.

The *Einstein* observatory contained a number of imaging cameras and spectrometers on a rotary table which could be placed in the focus on command (Fig. 5). A special test facility was constructed at MSFC and used to test the fully integrated spacecraft with a quasi-parallel beam produced by a point-like source at a distance of 1,000 ft (Fig. 6). The test took place over a period of only

Fig. 6 The X-ray test facility at MSFC. The *Einstein* Observatory telescope and instrument module is fully integrated

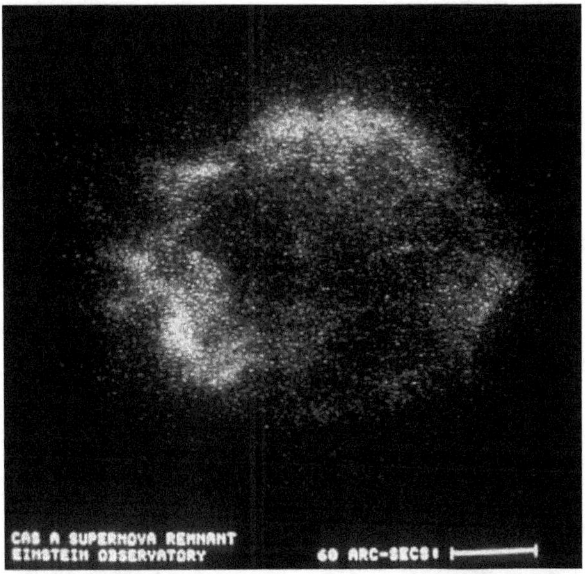

Fig. 7 An X-ray picture of the Cas A Supernova

30 days because of programmatic requirements. This forced us to a computer-driven plan of operations for efficient execution of the more than 1,000 tests we had planned to carry out over a 6-month period. The data rate was comparable to that expected in orbit and the data were analyzed with the same software that would be used after launch.

Einstein was launched on November 13, 1978. The results were equal or superior to our expectations. I will show only three pictures to illustrate the advances that *Einstein* brought to X-ray astronomy. Figure 7 shows a picture of the Supernova remnant CAS A [11]. The temperature distribution and composition of the gas was studied with the on-board spectrometers. Figure 8 shows a time-phased photography of the Crab Nebula and its pulsar folded over the pulsar period [12]. The time resolution required has not been duplicated even with *Chandra*. Figure 9 shows the stellar sources, individual binaries or supernovas, in the center of M31 at ten times the distance to our own galaxy center [13]. M87 and its barely resolved jet were also detected. Here the greatly superior (by a factor of 10) angular resolution of *Chandra* became apparent in 1999.

But the most important advance that *Einstein* brought to X-ray astronomy is that it made it relevant to all of astronomy.

This forced us to a significant effort to make the data available to astronomers in all disciplines in an immediately usable form, and promptly enough to allow follow-up observations. This included on-line reduction and calibration of the data as well as data archiving and distribution. I believe this was the first end-to-end system for astronomy.

Although the *Einstein* project was a Principal Investigator project which typically involved proprietary data, from the start we had also planned to make

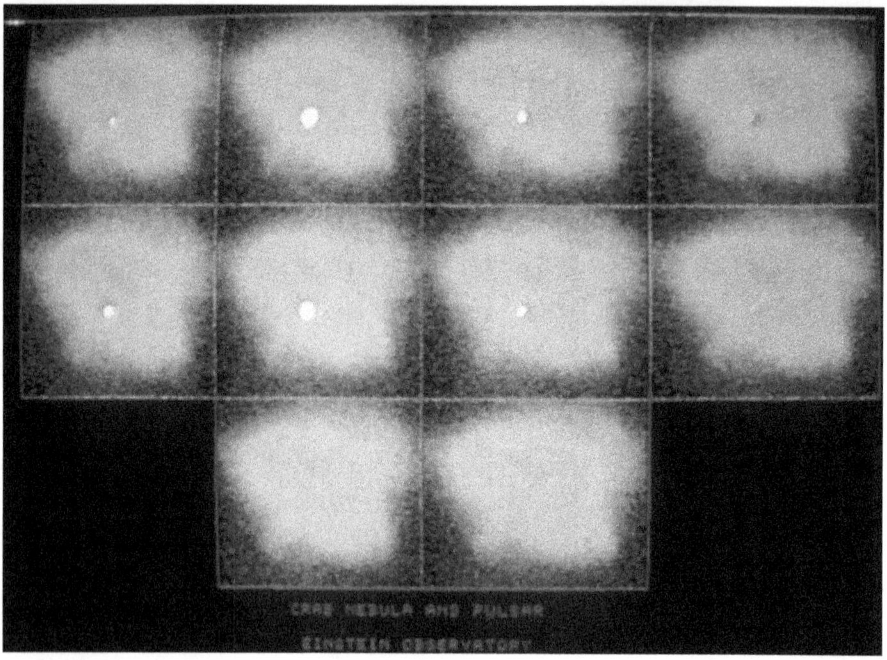

Fig. 8 The Crab Nebula and its Pulsar in X-rays

observing time available to guest investigators competitively selected by peer review. In practice some 2,000 out of 5,000 total observations were carried out on behalf of 150 guest observers around the world. Some 1,000 papers appeared in refereed journals in the first 10 years after the *Einstein* launch.

Fig. 9 Stellar X-ray sources in M-31

Einstein was a truly international observatory and the program yielded an important heritage of methodology for future observatories both in space and on the ground. This heritage has been used in the execution of the HST, the ESO VLT, and *Chandra* and has been almost universally adopted since then.

I have discussed so far almost entirely work in which I was personally involved. I want to explicitly recognize the fact that groups from many nations joined us in the use of focusing optics after the launch of *Einstein* in 1978. All the major X-ray missions conducted after 1990 used X-ray telescopes. Particularly important was the launch in 1990 of the satellite Rosat by the Max Planck Institute group led by Joachim Trümper. Rosat had a larger mirror than *Einstein* (0.8 rather than 0.6 m), a wider field of view, and about the same angular resolution. It proved invaluable for filling in the gap in worldwide X-ray observational capabilities, between the demise of *Einstein* in 1980 and the launch of *Chandra* and of XMM in 1999.

4 Chandra

In conclusion I should like to discuss *Chandra*. In 1976, even before the launch of *Einstein*, Harvey Tananbaum and I joined in re-proposing a 1.2-m

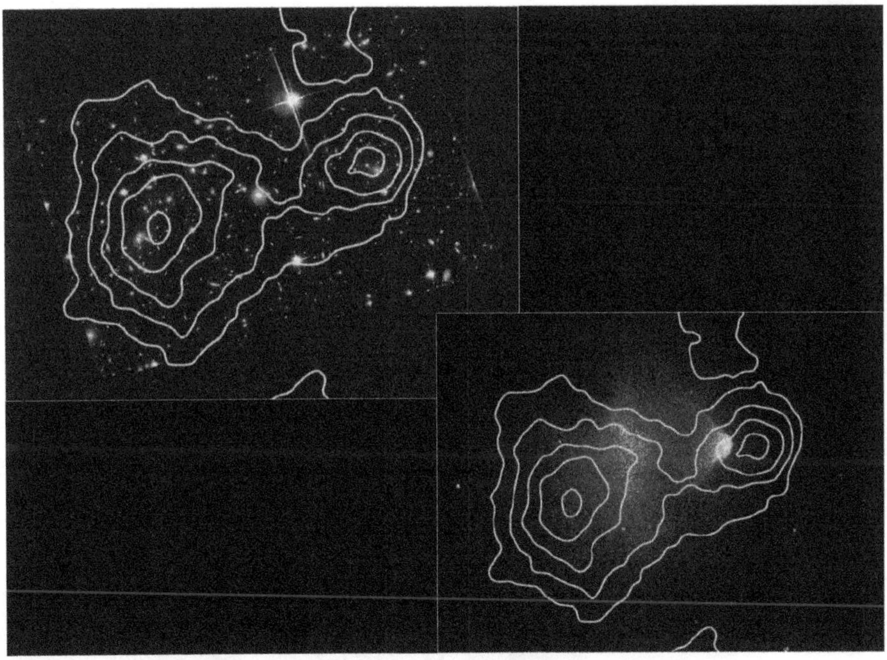

Fig. 10 The Bullet Cluster in X-rays, optical and with dark energy isopotentials

telescope, fearing a 20 years hiatus in high resolution X-ray astronomy capability. Unfortunately these fears were well-founded: there was such a hiatus.

After I took on the first directorship of the Space Telescope Science Institute in 1981, which gave me, together with Ethan Schreyer and Rodger Doxsey, the opportunity of transferring some of the X-ray experience to the HST, Harvey Tananbaum and Martin Weiskopf led the consortium that brought *Chandra* to launch in 1999. Currently the new ESA-NASA satellite, the International X-ray Observatory is not expected to be launched any earlier than 2020. Hopefully a number of smaller missions will keep the field alive until then.

The science based on *Chandra* observations is superb. One of the most significant of the *Chandra* achievements is the study of the interacting properties of dark matter in the Bullet Cluster (Fig. 10) [14].

I myself was finally able to image at least 90% of the 0.5 to 6.0 KeV background in the one million second exposure in the *Chandra* Deep Field South (CDFS) with a source density of about one source per square arc minute [15], (Fig. 11). I used HST for optical identifications and VLT for spectroscopy of the sources, which are by now 99% identified. As suggested by Setti and Woltjer in 1973 the sources are predominantly Active Galactic Nuclei.

I was also able to observe a faint distant group of galaxies in the CDFS. This is the type of study that might be extended to large regions of the sky by use of wide-field X-ray telescopes, particularly in view of the cosmological significance of the study of distant clusters.

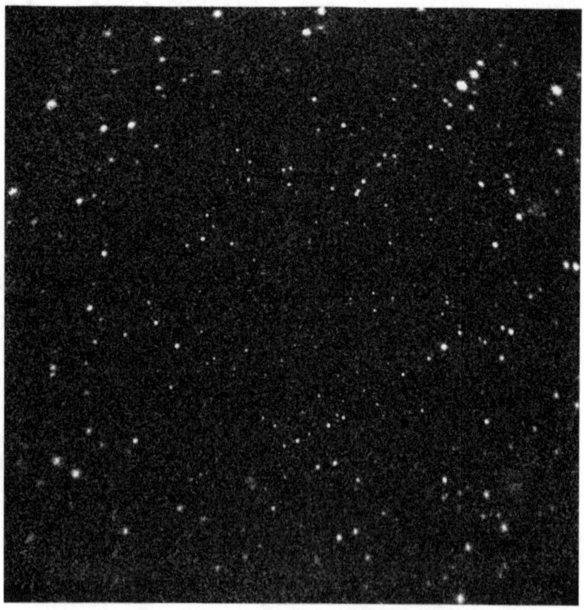

Fig. 11 X-ray image of the Chandra Deep Field South

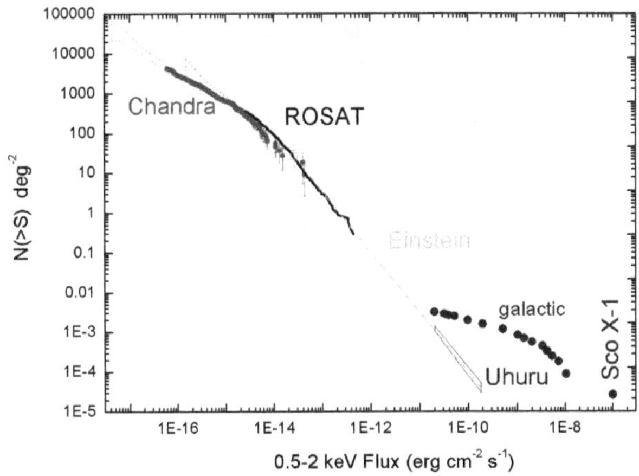

Fig. 12 The increase in sensitivity from the rocket observation of Sco X-1 in 1962 to the observations by the *Chandra* Observatory in CDFS

The development of X-ray astronomy in the last 40 years has been very rapid, comparable to going from the naked eye to the VLT (Fig. 12); one might well feel overwhelmed by the abundance of new data in the field.

References

1. Giacconi, R., Clark, G.W., Rossi, B.B.: "A brief review of experimental and theoretical progress in X-ray astronomy". Document AS&E TN-49. American Science and Engineering, Cambridge, MA (1960)
2. Giacconi, R., Rossi, B.B.: A "telescope" for soft X-ray astronomy. J. Geophys. Res. **65**, 773 (1960). doi:10.1029/JZ065i002p00773
3. Wolter, H.: Spiegelsysteme streifenden Einfalls als abbildende Optiken fur Rontgenstrahlen. Annalen Phys. **445**(1), 94–114 (1952). doi:10.1002/andp.19524450108
4. Giacconi, R., Reidy, W.P., Zehnpfennig, T.F., Lindsay, J.C., Muney, W.S.: Solar X-ray image obtained using grazing incidence optics. Astrophys. J. **142**, 1274 (1965). doi:10.1086/148404
5. Giacconi, R., Gursky, H., Paolini, F., Rossi, B.B.: Evidence for X-rays from sources outside the solar system. Phys. Rev. Lett. **9**, 442 (1962). doi:10.1103/PhysRevLett.9.439
6. Giacconi, R., Gursky, H.: "An experimental program of extrasolar X-ray astronomy". September proposal by AS&E to NASA. Document AS&E-449 (1963)
7. Giacconi, R., Kellog, E.M., Gorenstein, P., Gursky, H., Tananbaum, H.: An X-ray scan of the galactic plane from Uhuru. Astrophysical J. Lett. **165**, L27–L35 (1971). doi:10.1086/180711
8. Vaiana, G.S., Reidy, W.P., Zehnpfennig, T.F., Giacconi, R.: X-ray structures of the Sun during the importance 1N flare of 8 June 1968. Science **161**, 564–567 (1968). doi:10.1126/science.161.3841.564
9. Vaiana, G.S., Rosner, R.: "Recent advances in coronal physics". Annual Review of Astronomy and Astrophysics, Palo Alto, CA: Annual Reviews, pp. 393–428 (1978)
10. Giacconi, R., Branduardi, G., Briel, U., Epstein, A., Fabricant, D., Feigelson, E., Forman, W., Gorenstein, P., Grindlay, J., Gursky, H., Harnden, F.R., Henry, J.P., Jones, C., Kellog, E., Kock, D., Beeker, R.H., Bolt, E.A., Serlemitzos, P.J., Clark, G., Canizares, C., Markert, T., Novik, R., Helfand, D., Long, K.: The Einstein-HEAO-2 X-ray observatory. Astrophys. J. **230**, 540–550 (1979). doi:10.1086/157110
11. Murray, S.S., Fabbiano, G., Epstein, A., Giacconi, R., Fabian, A.C.: High resolution X-ray observations of the Cassiopea A Supernova remnant with the Einstein Observatory. Astrophys. J. **234**(L), 69 M (1979)

12. Harnden, F.R., Jr., Buehler, B., Giacconi, R., Grindlay, J., Hertz, P., Shreier, E., Seward, F., Tananbaum, H., Van Speybroeck, L.: X-ray observations of the Crab Nebula with the Einstein Observatory. BAAS **11**, 789 H (1979)
13. Van Speybroeck, L., Epstein, A., Forman, W., Giacconi, R., Jones, C., Liller, W., Smarr, L.: Observations of X-ray sources in M31. Astrophys. J. **234**(L), 45 V (1979). doi:10.1086/183106
14. Markevitch, M., Gonzales, A.A., Clowe, D., Vikhlinin, A., Forman, W., Jones, C., Murray, S., Tucker, W.: Direct constraints on the dark matter self-interaction cross section from the merging galaxy cluster 1E 0657-56. Astrophys. J. **606**, 819 M (2004). doi:10.1086/383178
15. Giacconi, R., CDFS Team: The Chandra deep field south one million seconds catalog. Astrophys. J. (Supplement 139), 369–410 (2002)

Exp Astron (2009) 25:157–171
DOI 10.1007/s10686-009-9143-z

REVIEW ARTICLE

History of gamma-ray telescopes and astronomy

Klaus Pinkau

Received: 30 December 2008 / Accepted: 6 January 2009 / Published online: 17 March 2009

Abstract Gamma-ray astronomy is devoted to study nuclear and elementary particle astrophysics and astronomical objects under extreme conditions of gravitational and electromagnetic forces, and temperature. Because signals from gamma rays below 1 TeV cannot be recorded on ground, observations from space are required. The photoelectric effect is dominant <100 keV, Compton scattering between 100 keV and 10 MeV, and electron–positron pair production at energies above 10 MeV. The sun and some gamma ray burst sources are the strongest gamma ray sources in the sky. For other sources, directionality is obtained by shielding / masks at low energies, by using the directional properties of the Compton effect, or of pair production at high energies. The power of angular resolution is low (fractions of a degree, depending on energy), but the gamma sky is not crowded and sometimes identification of sources is possible by time variation. The gamma ray astronomy time line lists Explorer XI in 1961, and the first discovery of gamma rays from the galactic plane with its successor OSO-3 in 1968. The first solar flare gamma ray lines were seen with OSO-7 in 1972. In the 1980's, the Solar Maximum Mission observed a multitude of solar gamma ray phenomena for 9 years. Quite unexpectedly, gamma ray bursts were detected by the Vela-satellites in 1967. It was 30 years later, that the extragalactic nature of the gamma ray burst phenomenon was finally established by the Beppo–Sax satellite. Better telescopes were becoming available, by using spark chambers to record pair production at photon energies >30 MeV, and later by Compton telescopes

K. Pinkau (✉)
Max Planck Institut für Plasmaphysik, 85748 Garching, Germany
e-mail: Pinkau-Muenchen@t-online.de

for the 1–10 MeV range. In 1972, SAS-2 began to observe the Milky Way in high energy gamma rays, but, unfortunately, for a very brief observation time only due to a failure of tape recorders. COS-B from 1975 until 1982 with its wire spark chamber, and energy measurement by a total absorption counter, produced the first sky map, recording galactic continuum emission, mainly from interactions of cosmic rays with interstellar matter, and point sources (pulsars and unidentified objects). An integrated attempt at observing the gamma ray sky was launched with the Compton Observatory in 1991 which stayed in orbit for 9 years. This large shuttle-launched satellite carried a wire spark chamber "Energetic Gamma Ray Experiment Telescope" EGRET for energies >30 MeV which included a large Cesium Iodide crystal spectrometer, a "Compton Telescope" COMPTEL for the energy range 1–30 MeV, the gamma ray "Burst and Transient Source Experiment" BATSE , and the "Oriented Scintillation-Spectrometer Experiment" OSSE. The results from the "Compton Observatory" were further enlarged by the SIGMA mission, launched in 1989 with the aim to closely observe the galactic center in gamma rays, and INTEGRAL, launched in 2002. From these missions and their results, the major features of gamma ray astronomy are:

–Diffuse emission, i.e. interactions of cosmic rays with matter, and matter–antimatter annihilation; it is found, "...that a matter–antimatter symmetric universe is empirically excluded...."

–Nuclear lines, i.e. solar gamma rays, or lines from radioactive decay (nucleosynthesis), like the 1.809 MeV line of radioactive ^{26}Al;

–Localized sources, i.e. pulsars, active galactic nuclei, gamma ray burst sources (compact relativistic sources), and unidentified sources.

Keywords Gamma ray · Telescope · Astronomy

Gamma-rays originate in nuclear decay processes, nuclear interactions and interactions in strong electromagnetic fields. Therefore, Gamma-ray astronomy's aim is to investigate the nuclear and the high-energy universe. Originally, the nuclear universe had been presenting itself through the discovery of cosmic rays early in the twentieth century and the studies of their origin and propagation. After the discovery of the π-meson in 1947, and the π°-2γ decay, it was clear that γ-rays (as seen, for example, by extensive air showers) could be produced by charged cosmic rays. High-energy nuclear interactions and cascade showers were observed in nuclear emulsions [1]. Therefore, possible sources of cosmic rays, like supernovae and their remnants, and places of cosmic ray confinement, like the galactic plane were expected to become visible in gamma-ray astronomy [2]. In addition, sites of nucleosynthesis [3] and the possible existence of antimatter were expected to reveal themselves if the sky was investigated in the light of gamma-rays. Cosmic rays, nuclear physics and elementary particle physics are the ancestors of gamma-ray astronomy.

Cosmic gamma-rays are absorbed quickly in the upper atmosphere, and their signal is drowned in the local background produced by the interaction of cosmic rays with local matter. Thus, gamma-ray astronomy requires observations from space.

It is important to remember that nuclear physicists were used to construct and calibrate their own equipment and to translate their data into results of experimental observation. Working with one's hands has been an important ingredient in the process of "grasping" the meaning of the new discoveries. Therefore, the history of gamma-ray astronomy is the history of gamma-ray telescopes, constructed by the scientists themselves, flown on space missions. Some of these are very large observatories, others are more specialized experiments (Fig. 1).

It is a coincidence, though not completely fortuitous, that both, space observations and the study of high energy and elementary particle physics with large accelerators like CERN, became available in the late 1950s. At that time, elementary particle physics using cosmic rays came to an end, although most of the elementary particles had then been discovered using cosmic rays [4]. The cosmic ray community split up. Some went to the large accelerator laboratories, others into astrophysics, and gamma-ray astronomy became one of their new fields of interest.

The beginnings of gamma-ray astronomy were difficult. Philip Morrisons's paper [5] in 1958, in which he made an estimate of gamma-rays source strengths, raised hopes that balloon observations might permit gamma-ray

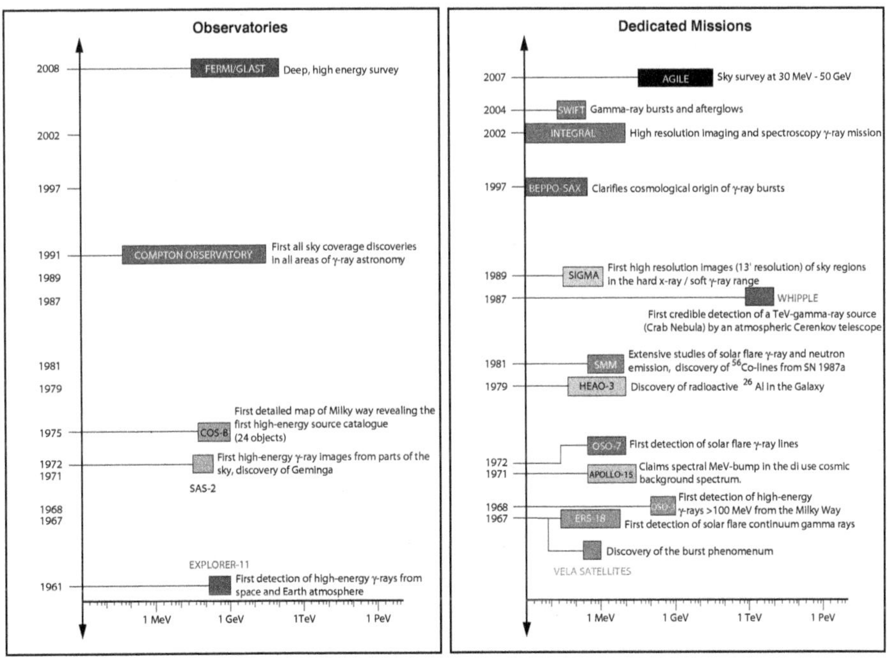

Fig. 1 Gamma-ray astronomy timeline

Fig. 2 Explorer XI

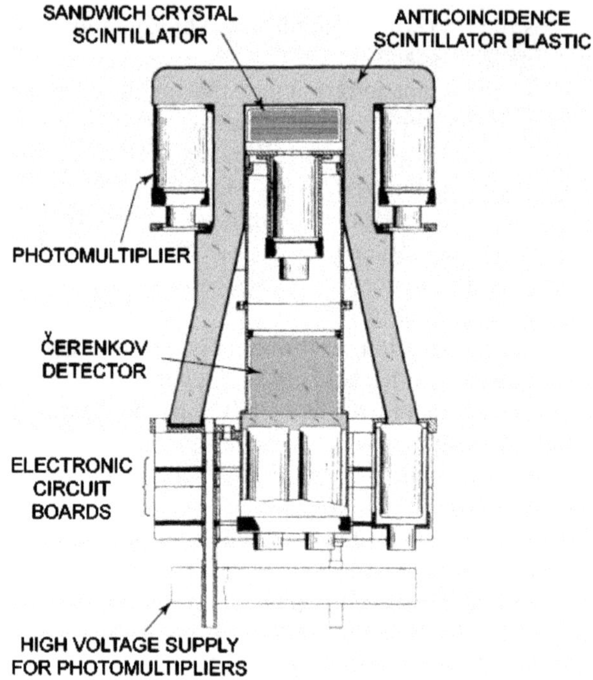

observations. But this was in vain. In reality, gamma-ray sources were so weak that observations with large equipment for extended periods from space were required. Explorer XI in 1962 [6], and its successor OSO-3 in 1968 [7] were the first successful gamma-ray astronomy experiments, indicating the prominence, in gamma-rays, of the galactic plane (Fig. 2).

The experiment on board of Explorer XI was the prototype of a gamma-ray telescope, consisting of an anticoincidence vessel, a converter of gamma-rays into electrons (by pair production, the Compton- or the photoeffect), de-

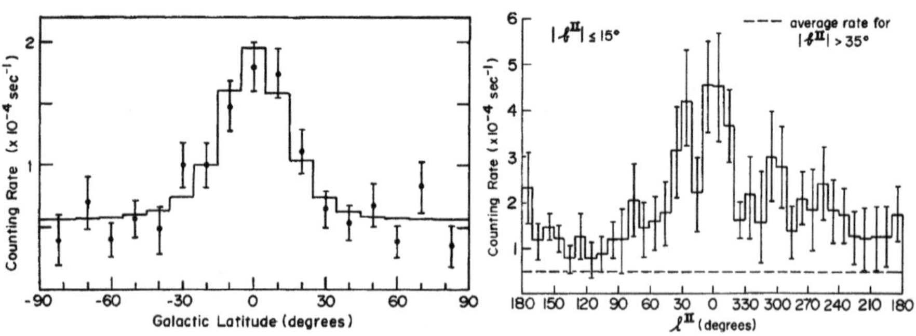

Fig. 3 Galactic plane observed by OSO 3

tecting them by a directional Cerenkov counter. Later instrumental improvements replaced the converter by a spark chamber, or by a Compton telescope, or by coded masks, in order to improve the directionality of the telescope (Fig. 3).

1 The galactic plane

The earliest signal of an astrophysical object in gamma-rays was the galactic plane, observed by OSO 3. From then on, one had a clear understanding of the sensitivities required. As a consequence, instruments in the energy range above 30 to 50 MeV were developed, where electrons and electron-positron-pairs were recorded and the gamma-ray direction determined from their visualisation in spark chambers. Their energies were determined either with scintillation calorimeters, or by multiple scattering measurements [8]. Gamma-rays resulting from the π°-decay have a mean energy of 70 MeV. SAS 2 (NASA) [9] was launched in 1972, COS – B (ESRO/ESA) in 1975. SAS 2, unfortunately, had a very short life only, but COS-B [10] observed the sky for 7 years. It was here in Noordwijk in 1970 that ESRO/ESA was finally convinced

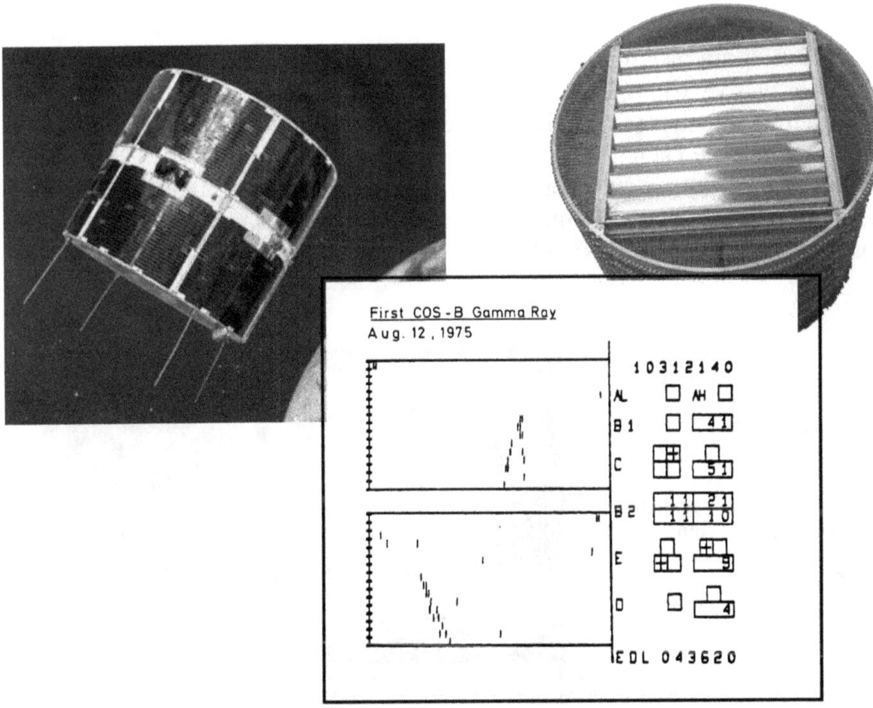

Fig. 4 The COS-B payload and the first COS-B gamma-ray

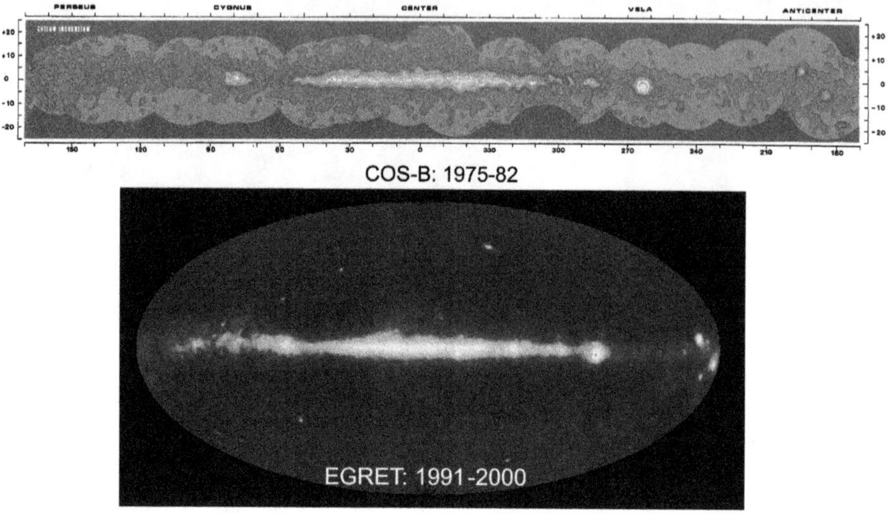

Fig. 5 Imaging with pair telescopes > 100 MeV. In the COS-B map, the Cygnus Region, Inner Galaxy, Vela Pulsar and Crab and Geminga Sources beyond the "galactic hole" between 210° and 250° galactic longitude are visible. In the EGRET map, extragalactic sources are added, like 3C279, PKS 0524 + 134, the LMC. PKS 0208-512, or 3C454.3

to commit itself to a single aim gamma-ray astronomy satellite, COS-B [11] (Fig. 4).

COS-B is an example of translating technology into science: Its spark chamber was very long lived due to metal-ceramic technology, thus providing a large time integral, so that the first detailed map of the galaxy in gamma-rays could be recorded (Fig. 5).

Beyond the diffuse emission from the galactic plane expected from the interaction of cosmic rays with interstellar matter, the Crab and Vela pulsars were discovered in gamma-rays and the periodicity of their light curves observed.

Fig. 6 The Compton Gamma-ray Observatory. The four instruments are: Oriented Scintillation-Spectrometer Experiment OSSE [14], Burst and Transient Source Expriment BATSE [15], Compton Telescope COMPTEL [16], and Energetic Gamma-ray Experiment Telescope EGRET [17]

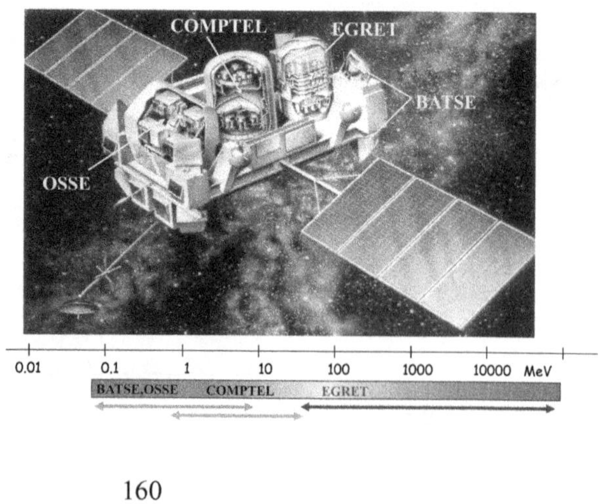

A first extragalactic source, 3 C 273 was seen by COS-B, although most of its observation time was directed towards the galactic plane.

In the late 1960s, NASA planned a number of High Energy Astrophysical Observatories (HEAO) which were to be free-flying very large satellites to be launched from the shuttle. At that time, Carl Fichtel and I agreed to collaborate and propose a very large Gamma-ray Experiment, which later (1978) became the proposal of C.E. Fichtel, R. Hofstadter and me of the Energetic Gamma-ray Experiment Telescope EGRET for HEAO. HEAO had a very complicated and involved history, but in the end it led to the Compton Gamma-ray Observatory (GRO) with EGRET part of its payload, launched in 1991. It is the status of gamma-ray astronomy after COS-B and before EGRET, which is well documented in C.E. Fichtel and J.I. Trombka's book [12], while the status in 2001 is well recorded in Schönfelder's compilation [13] (Fig. 6).

In the energy range from about 30 MeV to several GeV, impressive pictures of the gamma-ray sky have been produced by COS-B, EGRET, adding extra-galactic observations, and, very recently the Fermi gamma-ray space telescope (Fig. 7).

The diffuse emission from the plane of the galaxy, resulting from interactions of cosmic rays with interstellar matter is conspicuous. Also, the Large Magellanic Cloud is visible in gamma-rays, probably by the same process. In a way, the gamma-ray observations wrap up what we know of cosmic rays from the most recent observations of their composition and spectra: they are interstellar matter, accelerated by the turbulent motion of plasma and magnetic fields; they reside within the galaxy (galaxies), their intensity in intergalactic space is low.

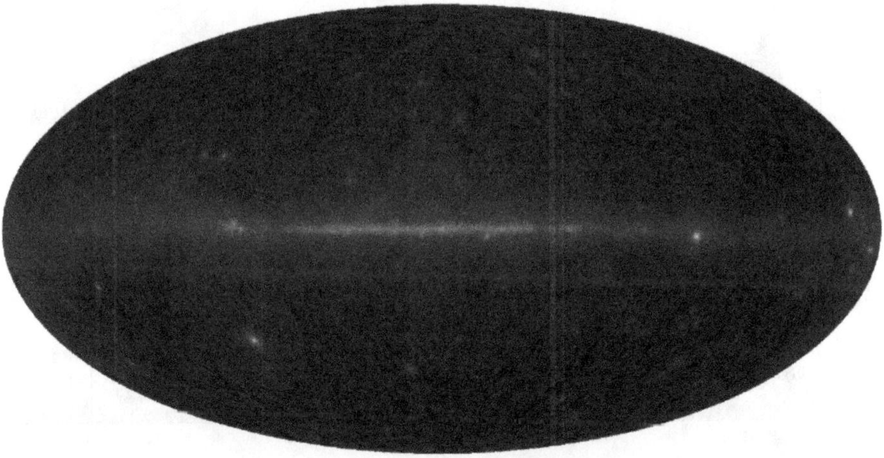

Fig. 7 Fermi gamma-ray space telescope 2008: 4 days exposure. Display of the enormous power of this most recent instrument (http://science.nasa.gov/headlines/y2008/26aug_firstlight.htm)

2 Localized sources

Embedded in the galactic plane, but also extending beyond the galaxy, gamma-ray point sources are visible. The Crab and Vela pulsars, spin-down neutron stars, are one class (Fig. 8). Other sources, known from x-ray astronomy, are accreting binaries. A sub-class of accreting binaries are superluminal jet sources—accreting, rapidly spinning black holes. And finally, supernovae and their remnants also are visible in gamma-rays.

A particular feature of gamma-ray point sources is their variability. Variability is a great help in the identification of gamma-ray point sources with astronomical objects which are known from other spectral regions. A complete understanding of the astrophysics involved requires these identifications, which are sometimes very difficult and uncertain [18].

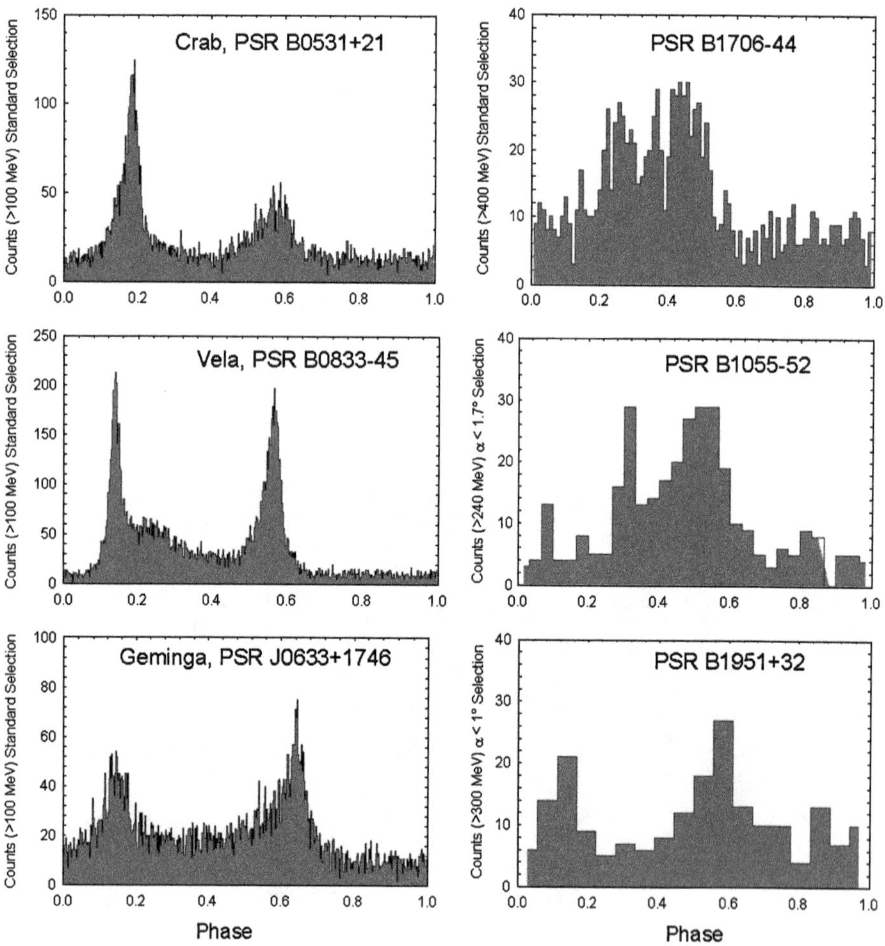

Fig. 8 High energy pulsar light curves

3 The sun

In the 1960s, interest in the sun as a gamma-ray source was large as part of the general interest in solar physics and the physics of the interplanetary medium, which received strong support from the aim to land a man on the moon, because the astronauts had to be protected from the impact of solar flares.

Solar flares were predicted to emit gamma-ray lines at 0.511 MeV due to positron annihilation, at 2.2 MeV due to neutron capture, at 4.4 and 6.1 MeV from the de-excitation of carbon and oxygen nuclei. Peterson and Winckler [19] discovered high-energy bremsstrahlung gamma-rays in a balloon experiment, and Chupp et al. [20] discovered nuclear lines in two large flares by the OSO-7 satellite. The sun can be the brightest source of gamma-rays in the sky, making it unnecessary to apply special pointing equipment.

4 Gamma-ray bursts

When the Vela-satellites in 1967 discovered the gamma-ray burst phenomenon, this was completely unexpected. The Vela satellites were operated in order to detect nuclear explosions and to enforce the nuclear test ban treaty. However, it soon became clear that the gamma-ray burst phenomenon observed had nothing to do with nuclear explosions on earth. The gamma-ray bursts are by far the brightest objects in the gamma-ray sky. The Burst and Transient Source Experiment BATSE (http://www.batse.msfc.nasa.gov/batse/) on board the Compton Gamma-ray Observatory detected more that 2,000 bursts until 1998, with an isotropic distribution. BATSE consisted of a number of flat scintillation counter discs, pointing in different directions. Because a gamma-ray burst is so luminous, the relative amplitude of the gamma-ray signal recorded in the various scintillators allows

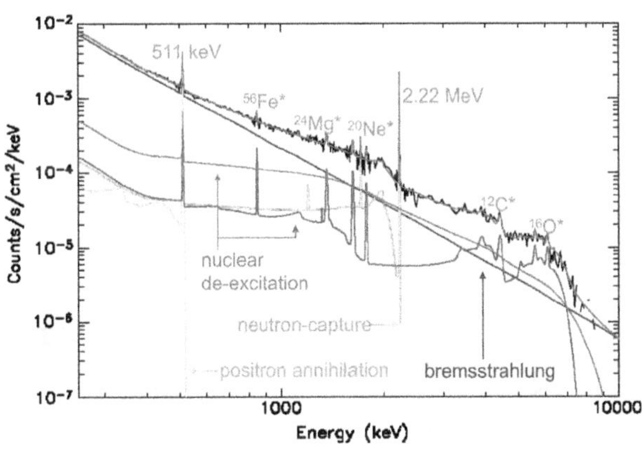

Fig. 9 Gamma-ray lines from solar flares. The lines from the de-excitation of nuclei are indicated, positron-electron annihilation results in the 511 keV, and neutron capture in the 2.2 MeV line. RHESSI stands for Ramaty High-Energy Solar Spectrometer Imager [21]

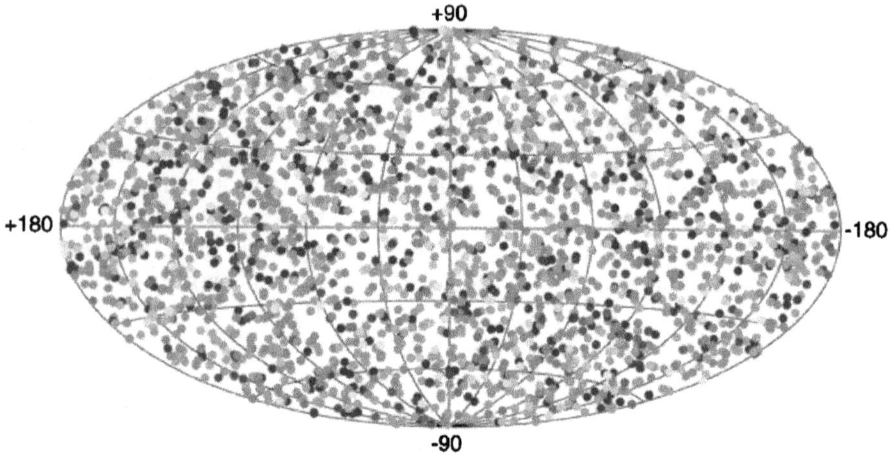

Fig. 10 BATSE catalogue of 2,704 gamma-ray bursts. The colour code of the data points refers to the fluence (i.e. integral energy flux) recorded: from red (10^{-4} erg/cm^2) to dark blue (10^{-7} erg/cm^2)

the direction of arrival of the gamma-rays to be derived to within about one degree.

In 1997, the Italian–Dutch satellite BeppoSax was able to identify several burst sources because X-ray and optical observations of the transient source al-

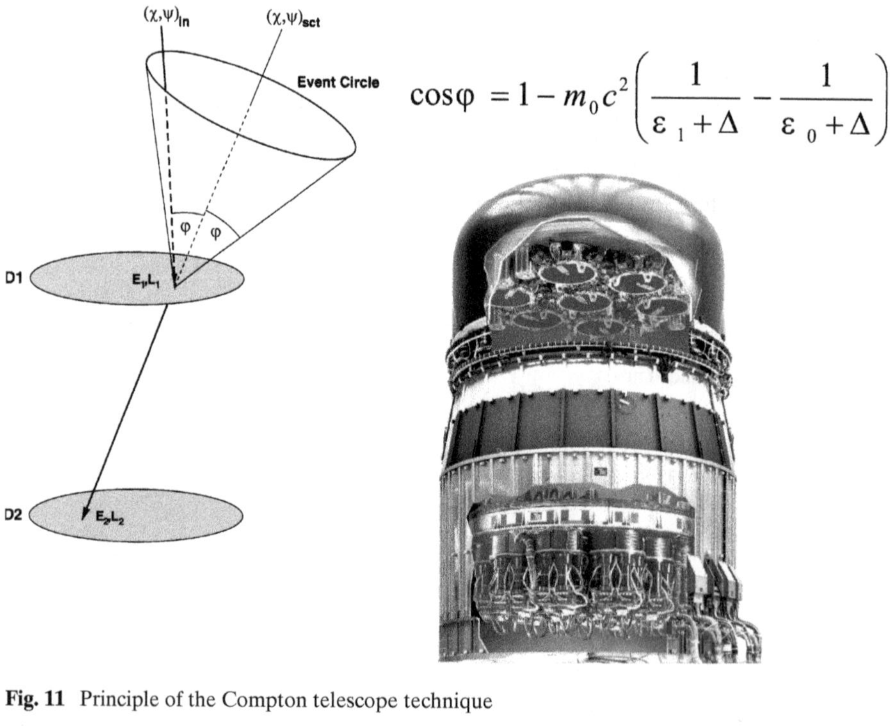

$$\cos\varphi = 1 - m_0 c^2 \left(\frac{1}{\varepsilon_1 + \Delta} - \frac{1}{\varepsilon_0 + \Delta} \right)$$

Fig. 11 Principle of the Compton telescope technique

lowed identification and observation of red shifts of 0.835 and 3.42, respectively [22]. Gamma-ray luminosities of 7×10^{51} erg and 3×10^{53} erg were derived, assuming isotropic emission. In 1999, Akerlof et al. [23] detected bright optical emission from a burst and derived a luminosity of 3.4×10^{54} erg assuming isotropic emission. For comparison, the rest mass energy of the sun is 1.6×10^{54} erg and therefore, very massive objects must be involved to produce these enormous energies.

5 Low-energy (1–30 MeV) gamma-rays

Another instrument on board of GRO was the Compton Telescope COMPTEL, observing the sky in the range from 1–30 MeV [11, 14].

The development of the Compton scattering technique in the 1960s at the Max Planck Institute in Garching followed the construction of a neutron telescope, in which two successive elastic recoil collisions of an incoming neutron with hydrogen nuclei were recorded in thin spark chambers. This successful experimental approach led to the development of a Compton Telescope, recording first, the recoil electron in a first Compton collision and secondly, the total remaining energy of the scattered photon. As shown in Fig. 9, this technique allows the gamma-ray's direction to be determined to lie on an "event circle". It would be an improvement if it were possible to determine the azimuthal angle of the recoil electron. For the time being, a statistical

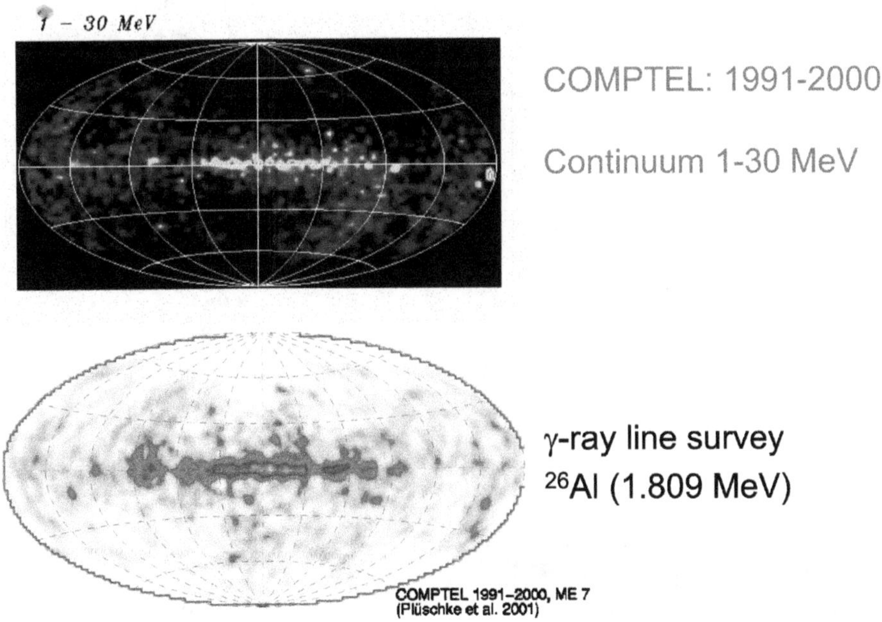

Fig. 12 Imaging with Comptel

analysis makes it possible to construct sky maps, and to observe localized sources (Figs. 10, 11, and 12).

The sky map of low-energy (1–30 MeV) gamma-rays [24] shows similar features to that at higher energies. The sky map of ^{26}Al [25] is taken as an example of the topic of nucleosynthesis and gamma-ray astronomy. Because its radioactive decay time is 1.04×10^6 years ^{26}Al accumulates in the interstellar medium from many different sources. Gamma-ray lines from radioactive decay can be well studied with the SPI instrument aboard INTEGRAL. This is a coded mask telescope with a solid state Germanium detector [26] (http://www.mpe-garching.mpg.de/gamma/instruments/integral/spi/www/) (Fig. 13).

Extragalactic gamma-ray sources have been studied by the SIGMA mission (covering the hard X-ray/low energy gamma-ray region) and GRO with its instruments OSSE, COMPTEL, and EGRET. They consist of active galactic nuclei (AGN), pulsars, supernovae and supernova remnants and unidentified objects, likely to involve massive objects. Variability is one of the characteristics of gamma-ray sources (Fig. 14).

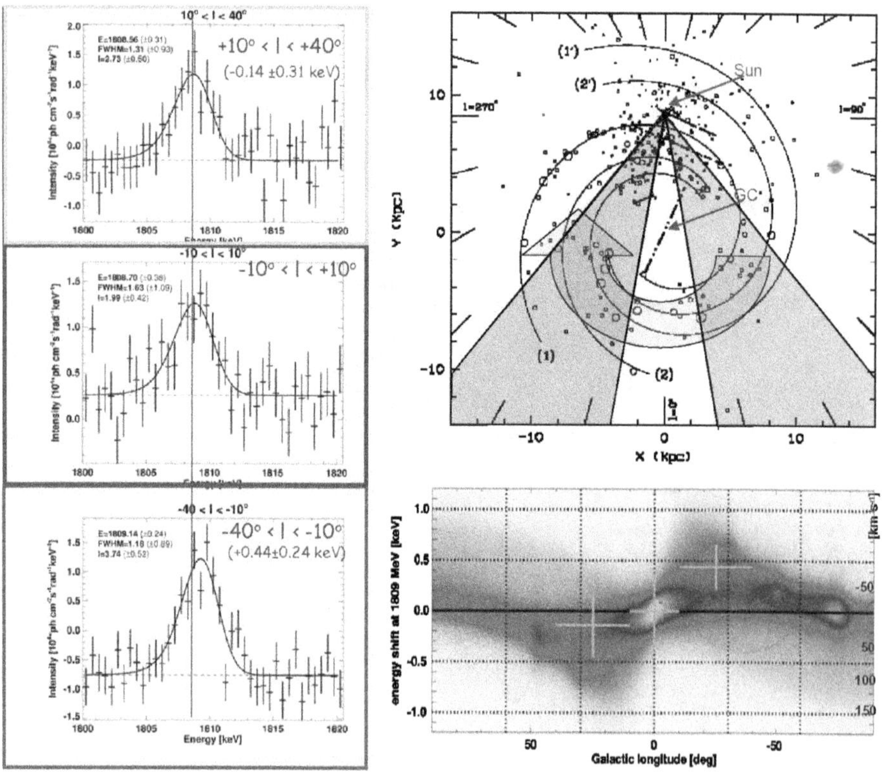

Fig. 13 ^{26}Al gamma-ray line as seen by INTEGRAL-SPI [27]

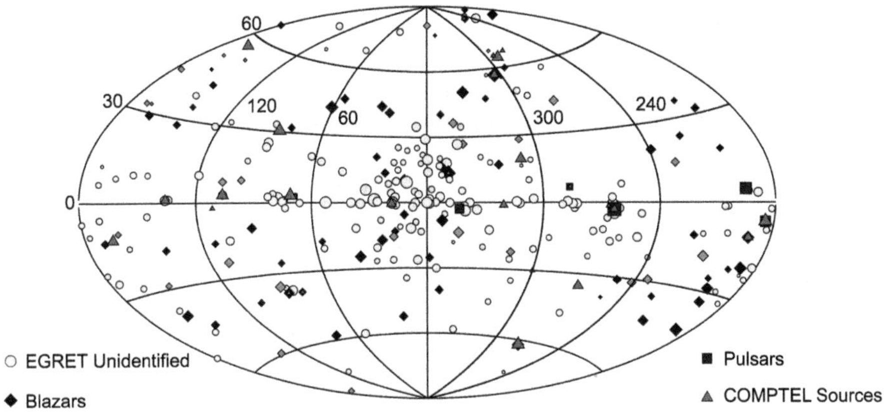

Fig. 14 Catalogue of gamma-ray sources. A compilation of the 3. EGRET Catalogue [28] and the 1. Comptel Catalogue [29]

In the study of individual sources, gamma-ray astronomy connects to astronomy at other energies (wavelengths). As mentioned before, gamma-ray sources in the TEV region have not been considered here. They will be the topic of the presentation by H. Völk.

However, there is particular interest in the nature of the gamma-ray background. At intermediate energies, careful analysis has led to the conclusion, that the gamma-ray background is produced by unresolved point sources. This leads to one result, rather exclusive to gamma-ray astronomy, and perhaps more fundamental than the other—albeit exciting—results: from their study of the 1 to 10 MeV gamma-ray background radiation, A.G. Cohen, A. de Rújula, and S.L. Glashow [30] "conclude that a matter–antimatter symmetric universe is empirically excluded."

6 Conclusions

If one asks what the particular contribution of gamma-ray astronomy is to our knowledge of the universe, then the following summary could be made:

- Gamma-rays are produced by cosmic ray interactions, also due to large-scale acceleration processes, using mechanical energy of the clouds of the interstellar medium; they reside inside the galaxies: the acceleration process is slow, using the mechanical energy of magnetized clouds;
- Other forms of acceleration processes also produce gamma-rays, be it directly or through interactions of the accelerated particles, like in solar flares, or massive accreting or objects. These sources are highly variable and belong to the most luminous objects in the sky: strong acceleration processes using the rotational energy of spinning or accreting massive and/or magnetized objects.

- Gamma-ray lines indicate the sites and the nature of processes of nucleosynthesis.
- A completely different aspect of high significance is given by the absence of nucleon-antinucleon annihilation gamma-rays, which should be seen in the 1 to 10 MeV region, red-shifted from the π°-decay. The diffuse gamma-ray flux from COMPTEL, lies several orders of magnitude below the annihilation flux expected for a universe, in which matter–antimatter symmetry is achieved by dividing it into regions populated exclusively by matter or antimatter.

Acknowledgement I thank Gottfried Kanbach for his help in the preparation of this presentation.

References

1. Edwards, B., Losty, J., Perkins, D.H., Pinkau, K., Reynolds, J.: Analysis of nuclear interactions of energies between 1000 and 100 000 BeV. Philos. Mag. **3**, 237 (1958). doi:10.1080/14786435808238218
2. Hayakawa, S.: Prog. Theor. Phys. **8**, 571 (1952). doi:10.1143/PTP.8.571
3. Burbidge, E.M., Burbidge, G., Fowler, W.A., Hoyle, F.: Rev. Mod. Phys. **29**, 547 (1957). doi:10.1103/RevModPhys.29.547
4. Powell, C.F., Fowler, P.H., Perkins, D.H.: The Study of Elementary Particles by the Photographic Method. Pergamon Press, London (1959)
5. Morrison, P.: Nuoco Cim. **7**, 858 (1958). doi:10.1007/BF02745590
6. Kraushaar, W.L., Clark, G.W.: Phys. Rev. Lett. **8**, 106 (1962). doi:10.1103/PhysRevLett.8.106
7. Kraushaar, W.L., Clark, G.W., Garmire, G.P., et al.: Astrophys. J. **177**, 341 (1972). doi:10.1086/151713
8. Pinkau, K.: Molière's theory of multiple scattering applied to the spark chamber. Zeitschr. f. Phys. **196**, 163 (1966). doi:10.1007/BF01326425
9. Kniffen, D.A., Fichtel, C.E.: Astrophys. J. **250**, 389 (1981). doi:10.1086/159387
10. Mayer-Hasselwander, H.A., Bennet, K., Bignami, G.F., Bucceri, R., Caraveo, P.A., Hermsen, W., Kanbach, G., Lebrun, F., Lichti, G.G., Masnou, J.L., Paul, J.A., Pinkau, K., Sacco, B., Scarsi, L., Swanenburg, B.N., Wills, R.D.: Large scale distribution of galactic gamma radiation observed by COS-B. Astron. Astrophys. **105**, 164–175 (1982)
11. Pinkau, K.: Objectives of future gamma-ray missions, open questions relating to observations of celestial gamma-rays. Proc. ESRO Coll. Noordwijk, ESRO SP-58, p. 33 (1970)
12. Fichtel, C.E., Trombka, J.I.: Gamma-ray astrophysics. New insight into the universe. NASA SP 453 (1981)
13. Schönfelder, V. (ed.): The Universe in Gamma-Rays. Springer, Berlin (2001)
14. Johnson, W.N., Kinzer, R.L., Kurfess, J.D., et al.: Ap. J. Suppl. **86**, 693 (1993). doi:10.1086/191795
15. Fishman, G.J., et al.: NASA Conf. Publ. **3137**, 26 (1992)
16. Schönfelder, V., Aarts, H., Bennett, K., et al.: Ap. J. Suppl. Ser. **86**, 657 (1993). doi:10.1086/191794
17. Kanbach, G., Bertsch, D.J., Favale, A., et al.: Space Sci. Rev. **49**, 69 (1988)
18. Carramiñana, A., Reimer, O., Thompson, D.J. (eds.): The nature of unidentified galactic high-energy gamma-ray sources. Astrophysics and Space Sciences Library, vol. 267. Kluwer, Boston (2001)
19. Peterson, I.E., Winckler, J.R.: Phys. Rev. Lett. **1**, 205 (1958). doi:10.1103/PhysRevLett.1.205

20. Chupp, E.L., Forrest, D.J., Higbie, P.R., et al.: Nature **241**, 333 (1973). doi:10.1038/241333a0
21. Lin, R.P., Hurford, G.J., Madden, N.W.: Proc. SPIE **3442**, 2 (1998). doi:10.1117/12.330245
22. Costa, E., et al.: Discovery of an X-ray afterglow associated with the γ-ray burst of 28 February 1997. Nature **387**, 783 (1997). doi:10.1038/42885
23. Akerlof, C., Balsano, R., Barthelmy, S., et al.: Nature **398**, 400 (1999). doi:10.1038/18837
24. Strong, et al.: Proc. of the 3rd INTEGRAL Workshop Taormina Italy. Astrophys. Letters and Communications, Vol. 39, p. 209 (1999)
25. Oberlack, U.P.D.: Thesis, TU München (1997)
26. Diehl, R., et al.: Nature **439**, 45 (2006). doi:10.1038/nature04364
27. Diehl, R., Hartmann, D.H.: Phys. Unserer Zeit **39**, 183 (2008). doi:10.1002/piuz.200801174
28. Hartmann, et al.: ApJS **123**, 79 (1999). doi:10.1086/313231
29. Schönfelder, et al.: A&AS **143**, 145 (2000)
30. Cohen, A.G., de Rújula, A., Glashow, S.L.: Ap. J. **495**, 539–549 (1998)

Exp Astron (2009) 25:173–191
DOI 10.1007/s10686-009-9151-z

REVIEW ARTICLE

Imaging very high energy gamma-ray telescopes

Heinrich J. Völk · Konrad Bernlöhr

Received: 28 November 2008 / Accepted: 4 February 2009 / Published online: 10 March 2009

Abstract The technique of γ-ray astronomy at very high energies (VHE:> 100 GeV) with ground-based imaging atmospheric Cherenkov telescopes is described, the H.E.S.S. array in Namibia serving as example. Mainly a discussion of the physical principles of the atmospheric Cherenkov technique is given, emphasizing its rapid development during the last decade. The present status is illustrated by two examples: the spectral and morphological characterization in VHE γ-rays of a shell-type supernova remnant together with its theoretical interpretation, and the results of a survey of the Galactic Plane that shows a large variety of non-thermal sources. The final part is devoted to an overview of the ongoing and future instrumental developments.

Keywords History and philosophy of astronomy · Instrumentation: telescopes · Gamma rays: observation · Techniques: image processing · Radiation mechanisms: non-thermal · ISM: supernova remnants

PACS 07.05.Tp · 95.55.Ka · 98.58.Mj · 98.70.Rz · 98.70.Sa

H. J. Völk (✉) · K. Bernlöhr
Max-Planck-Institut für Kernphysik, P.O. Box 103980, 69029 Heidelberg, Germany
e-mail: Heinrich.Voelk@mpi-hd.mpg.de

K. Bernlöhr
Institut für Physik, Humboldt-Universität zu Berlin, Berlin, Germany
e-mail: Konrad.Bernloehr@mpi-hd.mpg.de

1 Introduction

At γ-ray energies $E_\gamma \gtrsim 30$ MeV electron-positron pair production in the Coulomb field of an atomic nucleus dominates the interactions of photons with normal matter. Cosmic γ-rays of such energies can therefore be most effectively measured in either gas-filled or solid state satellite detectors above the atmosphere registering the tracks of the resulting $e^{\pm}$-pairs, or using the Earth's atmosphere itself as detector in observing secondary Cherenkov radiation with ground-based telescopes. Satellite detectors work from ~ 30 MeV up to some tens of GeV, where they become statistics-limited as a result of the limited size of detectors that can be transported by rockets into space. In the Very High Energy range (VHE; $E_\gamma > 100$ GeV) large optical telescopes are used, often with diameters of 10 m or more.

The history, development and results of satellite γ-ray detectors have been described in this volume by K. Pinkau. Addressing the complementary ground-based technique, we shall in the following summarize the physics of UV/Optical imaging atmospheric Cherenkov telescopes which register the Cherenkov radiation from the atmospheric secondary particles. Subsequently we shall discuss two important astrophysical results that demonstrate the power of this observation method. Given our personal involvement in the H.E.S.S. (High Energy Stereoscopic System) array in Namibia we shall in most instances use this experiment as example (see Fig. 1), although we shall also emphasize the other large arrays existing worldwide. The scientific results are astonishing, given the slow development of the field over the three decades before the mid-1990s. We conclude with a glimpse at the next generation European project that envisages a further sensitivity increase by an order of magnitude.

Fig. 1 The H.E.S.S. telescope array [1] in the Khomas Highland of Namibia at an elevation of 1800 m a.s.l. The four 13 m telescopes stand at the corners of a square with a side length of 120 m

2 Air showers, secondary atmospheric Cherenkov emission, imaging atmospheric Cherenkov telescopes

High-energy cosmic gamma rays produce $e^{\pm}$ pairs in the atmosphere over $\frac{9}{7}X_0$ (with the radiation length $X_0 \approx 36.5$ g/cm^2 in air). This is followed by Bremsstrahlung over the next radiation length which implies new γ-rays that generate new pairs, etc., until the energy of the final generation of electrons becomes so small that their fate is dominated by ionization losses which rapidly cool and thermalize them. The result of these processes is called an electromagnetic Air Shower that exists for about 10^{-4} s while traversing the atmosphere. The multiplication of the number of particles and their eventual removal by thermalization leads to a maximum number of shower particles at about 250 to 450 g/cm^2 for primary γ-rays of 20 GeV to 20 TeV, corresponding to an atmospheric height of about 7 to 12 km above the ground.

The most realistic and complete description of the physical processes and the corresponding results in the detector(s) is given by Monte Carlo (MC) simulations. Except when explicitly noted we use the simulations performed by one of us [2] which use the CORSIKA code [3]. It incorporates all the physics of the atmospheric interactions and emission processes and gives the possibility to statistically track their evolution in the atmosphere, down into the detector.

Not only γ-rays penetrate into the atmosphere, but also charged energetic nuclei (Cosmic Rays). They must be distinguished from the γ-rays. Energetic protons and nuclei undergo hadronic interactions and produce dominantly neutral (π^0) or charged ($\pi^{\pm}$) pions. Whereas the former decay into two gammas, the latter ultimately produce electrons, positrons and two neutrinos via the $\pi \rightarrow \mu \rightarrow e$ decay. This leads to background air showers which are of a mixed hadronic and electromagnetic nature. As a result of the large transverse momentum transfer in hadronic interactions the hadronic shower component is broad and irregular compared to the electromagnetic component (see Fig. 2).

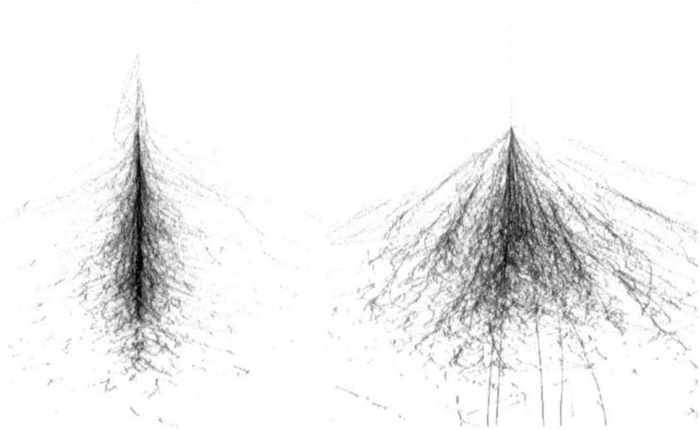

Fig. 2 The different character of gamma showers and hadronic showers. The gamma shower is slender and to lowest approximation axially symmetric about the direction of the primary. The hadronic shower is irregular and may contain electromagnetic subshowers as a result of the large transverse momenta generated in hadronic interactions (from K. Bernlöhr)

In general the flux of Cosmic Ray (CR) particles is much larger—by a factor of about 10^3—than that of γ-rays. This implies a large background for γ-ray astronomy from the ground. It must be separated from the γ-ray signal, because no anti-coincidence shield can be applied as in space detectors.

At primary γ-ray energies of about 100 GeV very few energetic photons or electrons reach the ground. But the shower electrons from the original γ-ray are still observable with optical telescopes through their Cherenkov radiation in the optical range, because this atmospheric Cherenkov emission reaches the ground without major absorption. Figure 4 below gives an impression of the overall configuration. The disadvantages of this very promising measurement technique are the weakness of the Cherenkov light and, to some extent, its optical character. They require large light collection devices and limit the observation time to clear and moonless nights. The observation efficiency is typically about 10%, depending critically on the astronomical quality of the site.

2.1 Cherenkov light pool

The atmospheric Cherenkov light emission from a single particle is characterized by a forward cone with an opening angle $\Theta \approx 1°$ that increases downwards. For a particle moving vertically downwards, the largest ring on the ground near sea level is from a height of 12 to 15 km (see Fig. 3).

The ensemble of shower electrons from an energetic primary γ-ray produces a rather uniformly illuminated "light pool" on the ground, centered on the shower core, with a radius of about 125 m, if the multiple scattering of the

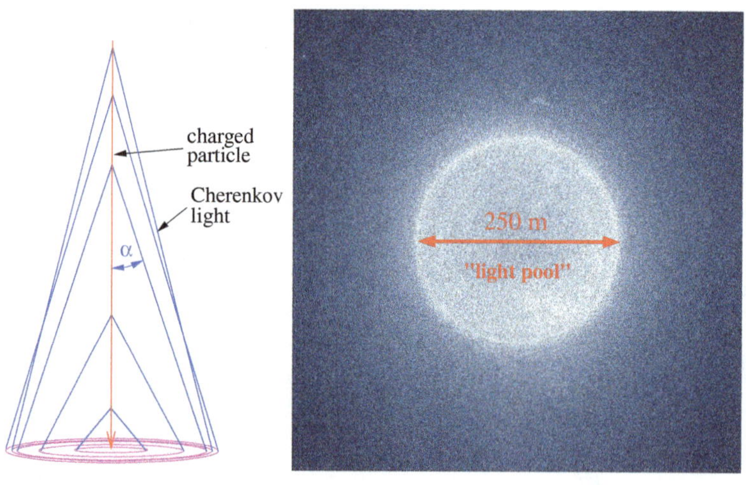

Fig. 3 *Left*: Atmospheric Cherenkov emission from a downward-moving single particle. *Right*: The "light pool" at an observation level at 1800 m above sea level from a γ-ray shower with a primary energy of 1 TeV (from K. Bernlöhr)

shower electrons is included (see Fig. 3). To first approximation it corresponds to the effective area of a telescope that images the shower.

A schematic picture of a shower from a cosmic γ-ray source, illuminating an array of telescopes on the ground, is given in Fig. 4. Since the atmospheric index of refraction is very close to 1, the Cherenkov light almost keeps pace with the radiating charged particles. Near the edge of the "light pool" most light from a γ-ray shower arrives within 2 ns (see Fig. 5). Thus a very short temporal window is possible, in order to suppress the dominant night sky background. This suggests the use of fast photomultiplier cameras for the telescopes.

2.2 Atmospheric Cherenkov telescopes

The lateral distribution of the Cherenkov photon density depends somewhat on the atmospheric profiles in its amplitude. However, the dependence on the distance from the shower core—the extrapolation of the direction of the primary γ-ray–is essentially independent of the atmospheric conditions. An example is shown in Fig. 5. At a γ-ray energy of 100 GeV about 1000 Cherenkov photons are produced in a 100 m^2 telescope. With a conversion efficiency of 10% this results in $\sim$ 100 photoelectrons in the image. Given that the total number of Cherenkov photons is about proportional to the primary γ-ray energy, this determines the threshold energy of the telescope.

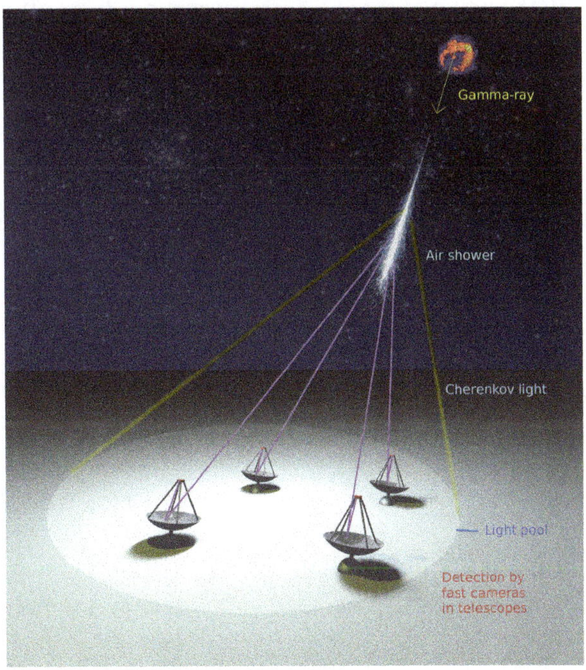

Fig. 4 Schematic of the Cherenkov light pool, originating from a primary γ-ray from within a cosmic-ray source (e.g. a supernova remnant) that illuminates an array of telescopes (from K. Bernlöhr)

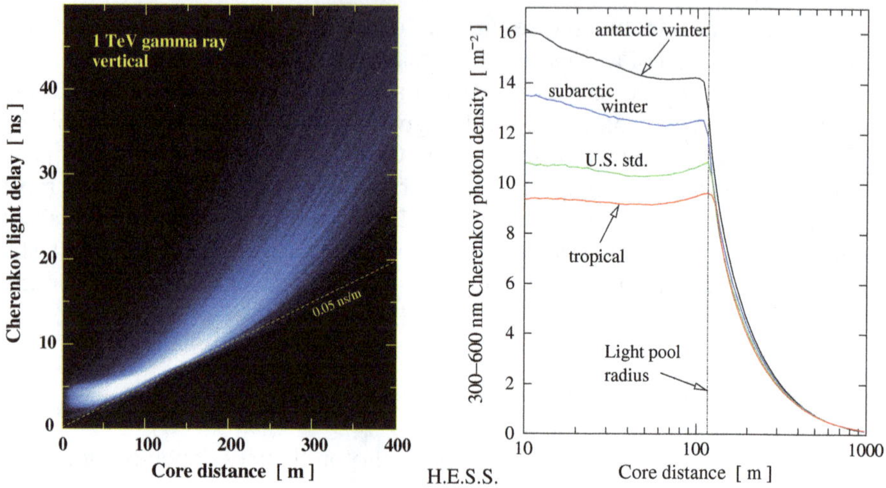

Fig. 5 *Left*: Distribution of arrival time and core distance of Cherenkov photons in a 1 TeV γ-ray-shower. Time is counted with respect to the time of arrival of the primary γ-ray on the ground assuming no interactions. *Right*: Lateral distribution per unit area of the optical Cherenkov emission from a shower with primary energy of 100 GeV, for various atmospheric profiles (from K. Bernlöhr)

After successful initial experiments concerning CR air showers—what we call today the CR background—in the 50ies in the UK [4] and in the USSR [5], the first dedicated γ-ray observations were attempted in the Crimea by the Lebedev group from Moscow, following the suggestion by [6]. At the time the results were basically negative [7]. Nevertheless they resulted in an upper limit to the VHE γ-ray flux from the Crab Nebula.[1,2]

In the USA the growing interest in VHE γ-ray astronomy led to the construction (in 1968) of a 10 m optical reflector with tessellated mirrors at the Whipple Observatory in southern Arizona (see Fig. 6).[3] The imaging of the shower's Cherenkov light is schematically shown in Fig. 6.

Despite intriguing indications the results remained controversial for about 20 years. With a multi-pixel photomultiplier camera in the focus since the mid-80ies and the introduction of image analysis finally the Crab Nebula could be

[1] A very interesting account of the early efforts in the USSR is given by A.S. Lidvansky in the talk "Air Cherenkov Methods in Cosmic Rays: A Review and Some History" presented at the centenary conference "P.A. Cherenkov and Modern Physics" (Moscow, 2004) [8].

[2] A condensed history of the field until 1994, very worthwhile reading, is to be found in [9].

[3] Of the scientists originally involved (G.G. Fazio, J.E. Grindlay, G.H. Rieke, T.C. Weekes, and others) several have later also become leaders in fields like Infrared Astronomy and X-ray Astronomy.

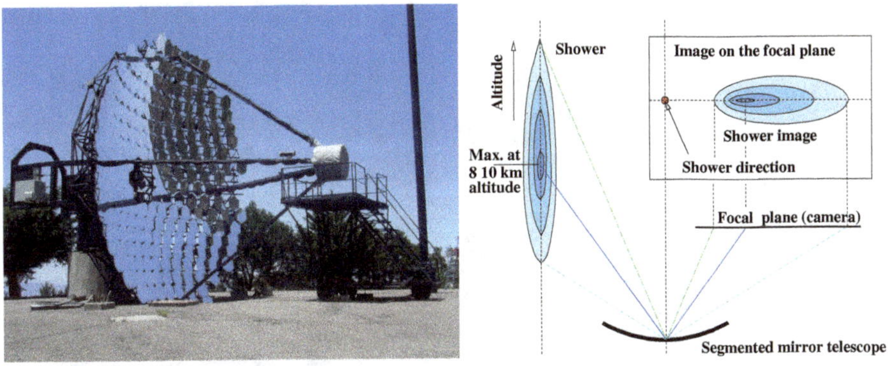

Fig. 6 *Left*: The 10 m Whipple telescope on Mt. Hopkins, USA [11]. *Right*: Shower imaging by a telescope. Despite some asymmetry, the shower image in the camera has the shape of an ellipse. The shower direction is a point somewhere on the extension of its major axis. For γ-ray primaries the image intensity gives the primary energy (from K. Bernlöhr)

detected with high (9σ) significance after 60 h of observation time [10]. Even after this breakthrough and a number of further significant detections, the field continued to evolve rather slowly until the mid-90s when a number of new telescopes started operation. They introduced the stereoscopic technique and very fine camera pixelation (see below), leading up to the present group of four major telescope systems worldwide.

The difficulties in the early γ-ray observations may be traced back to the massive background of nuclear CR events. In addition, the large spatial extension of air showers in the direction of the primary photon trajectory leads to a very extended image $\sim 1°$ in the camera plane (see Fig. 7), making the size of the field of view (FoV) of the camera a critically important parameter of the system. The lateral spread and irregularity of hadronic showers increases this extension into the second dimension and makes itself visible in the camera through an irregularly structured image.

This difference between γ-rayshowers and hadron showers can be used to distinguish the nature of the generating particles through *image analysis*, given a sufficiently dense array of photomultipliers in the camera [12].

The aim of Hillas' algorithm was to reject the dominant events from CR nuclei by suitable cuts on the images that are derived from Monte Carlo simulations. The application of this method has proven to be very successful. The remaining problem is that with a single telescope one obtains only one projection of the Cherenkov light "shower". In addition, a single telescope suffers from a second kind of background effect. It is produced by penetrating muons (from CR interactions) that reach the ground at the telescope mirror or its immediate neighborhood. These muon events are not easily distinguishable from γ-rays. The solution for these drawbacks consists in the stereoscopic technique. It was pioneered in the HEGRA telescope array on La Palma [13] that started full operation in the mid-90s (see Fig. 8).

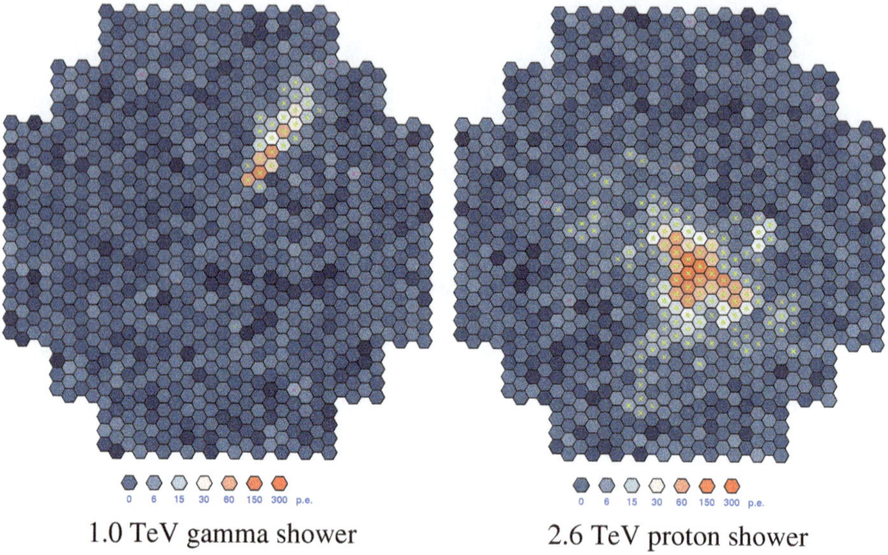

Fig. 7 Difference between the images of gamma-induced and hadron-induced showers in the camera (from K. Bernlöhr)

2.3 Stereoscopic method

The use of several telescopes observing the same shower in coincidence allows a unique determination of the shower direction by projecting the images in all triggered telescope cameras into one camera (see Fig. 9). Then the intersection point of the image major axes yields the shower direction. Compared to a single telescope the angular resolution, the energy resolution, the background rejection and the sensitivity are improved. In addition this method allows the 3-dimensional reconstruction of the shower, including the height of maximum

Fig. 8 The start of stereoscopy: HEGRA on La Palma (1995–2002) [14]. The five 3.5 m telescopes were situated in the center and at the 4 corners of a square of 100 m sidelength. The FoV was $\approx 5°$

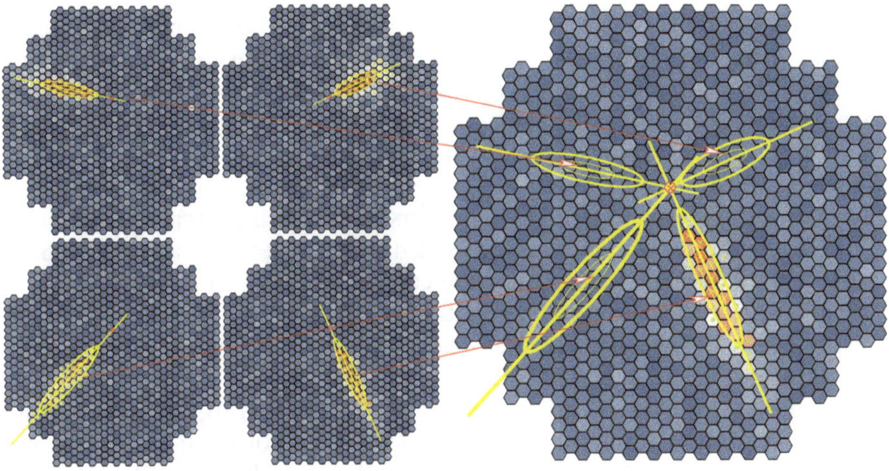

Fig. 9 Projection of several images from within the light pool of an event into one camera plane in a stereo system with four telescopes (from K. Bernlöhr)

particle number. The most advanced data analysis methods use 3-dimensional modeling of the shower, no longer confining themselves to the use of the 2-dimensional image parameters alone.

The second major step forwards in the stereoscopic observation mode is the suppression of the above-mentioned local muons with a stereo trigger: they leave an image only in the telescope concerned, but not in the other telescopes (unless the telescope light-gathering power is so enormous that even a single charged particle can trigger an event from an inter-telescope distance, a case which we will not discuss here). Therefore such events can be almost completely eliminated (see Fig. 10). This is most important near the energy threshold, where the shower images get weak and poorly defined. In other

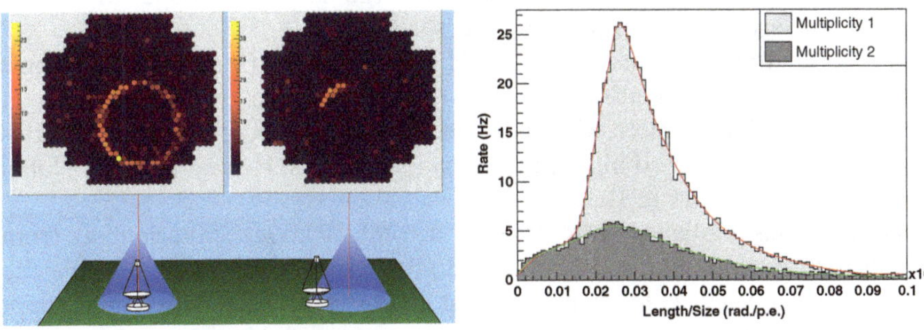

Fig. 10 Local muons hitting telescopes in the center and close to the periphery (*left*, courtesy G. Hermann and W. Hofmann). In peripheral encounters the resulting image is difficult if not impossible to distinguish from a low-energy γ-ray shower (*right*, based on H.E.S.S. data [15]). No such effect occurs in the other telescopes of a multiple system. The exclusion of muon events in stereoscopic systems reduces the muon background drastically

words, only stereoscopic systems can reach the theoretical energy threshold derived from the mirror size. Single telescopes are severely hampered by the muon background and must operate significantly above their theoretical threshold.

Despite the small mirror sizes of its components, the HEGRA array became the most sensitive VHE instrument of its kind. It could thus prove the promise of stereoscopy. Among others, several Blazar-type extragalactic sources, like the active galactic nuclei Mrk 421 and Mrk 501, discovered shortly before by the Whipple telescope, were confirmed or even measured with improved quality. The γ-ray emission of the young Galactic Supernova Remnant Cassiopeia A was discovered in a series of deep observations over a period of 232 h, and an unidentified VHE γ-ray source in the Cygnus region was detected; its astrophysical origin is subject of discussion to this day. In spite of these important results the HEGRA array was not large enough to see more than the "eight-thousanders" of the VHE range. Its scan of the Galactic scan revealed no new source and was in this sense not successful. The other existing instruments of the time—the Japanese–Australian CANGAROO I telescope at Woomera and the British Durham Mark 6 telescope in Narrabri, both in the southern hemisphere in Australia, the French CAT telescope in the French Pyrenees (championing very fine camera pixelation), and others, together with the Whipple telescope—all shared this problem (for an overview of these instruments see [16]).

2.4 Third generation instruments; H.E.S.S.

As a consequence, a third generation of imaging Cherenkov telescopes was envisioned and eventually realized. These were the VERITAS array [17], planned for the Mt. Hopkins area again, CANGAROO III [18] near Woomera, H.E.S.S. in Namibia [19]—also in the southern Hemisphere—and MAGIC [20] on La Palma at the HEGRA site. They all were to have high-definition cameras and, with the exception of MAGIC that was initially designed and operated until recently as a large single telescope, they were stereoscopic arrays, following HEGRA (see Fig. 11).

As an example let us consider H.E.S.S. (High Energy Stereoscopic System).[4] The experiment's name is at the same time a tribute to Victor F. Hess, the Austrian physicist who discovered the Cosmic Rays in 1912 in a series of carefully instrumented high-altitude balloon flights. He received the Nobel prize for this discovery in 1938.

The H.E.S.S. Collaboration involves a total of about 150 scientists from various European countries, Armenia, South Africa and Namibia. While the astrophysicists play a decisive role, the majority are physicists that adapt techniques from accelerator physics to this young field of astronomy. The

[4]The conceptual basis for stereoscopic systems of ~ 10 m telescopes is given in [21, 22].

Fig. 11 The third generation imaging Cherenkov telescopes: the Japanese-Australian array CANGAROO III [23] the two European initiatives H.E.S.S. [1] and MAGIC [24], and the US-observatory VERITAS [25]

pattern is reminiscent to that visible in the development of radio and X-ray astronomy some decades earlier.

The 4-telescope array (see also Fig. 1) is designed for maximum mechanical stability despite full steering and automatic remote alignment capability of the 382-piece mirror system that implies an area of 107 m^2 per telescope; the individual glass mirrors have a diameter of 60 cm. The camera in the primary focus weighs almost 1 ton (see Fig. 12).

As mentioned before, the large field of view of 5° of the telescope cameras is a key feature of H.E.S.S. Even though quite expensive in cost, operation and

Fig. 12 One of the H.E.S.S. cameras in its shelter [1]. It has 960 photomultiplier pixels of size 0.16°, resulting in a 5° field of view. The electronics with 1 GHz analog sampling rate is entirely contained in the camera body

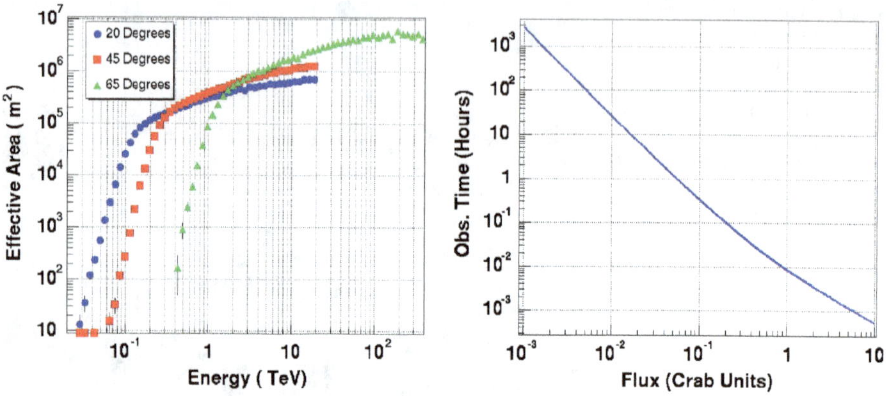

Fig. 13 Performance of the H.E.S.S. 4-telescope system. The effective area as a function of γ-ray energy is shown (*left*). The *right panel* gives the observation time required for a 5 σ detection as function of the γ-ray flux in units of the flux from the Crab Nebula ([19, 28])

maintenance, this requirement resulted from the realization that most nearby Galactic VHE sources should be extended by $\sim 1°$ degree or more, e.g. [26]. This should be also the case for the VHE emission from nearby rich clusters of galaxies, one of the key targets of extragalactic γ-ray astronomy [27].

Only a large field of view permits efficient observations of such extended objects. In addition it makes it possible to perform an efficient sky survey, especially in the Galactic Plane. Beyond that, the collection area of the telescopes increases with increasing γ-ray energy, since the Cherenkov light pool extends beyond 125 m for very energetic showers that produce a large amount of Cherenkov light. For the H.E.S.S. telescopes this implies a considerable increase of the collection area with energy up to γ-ray energies of about 100 TeV.

The resulting performance of the H.E.S.S. array is summarized in Fig. 13. After suitable data cuts on the expected direction of the source and on the shape of the images, derived from Monte Carlo simulations [28], the analysis accepts more than 50% of the gamma events and less than 0.1% of the events from CR nuclei; the angular resolution is better than 0.1° per event and the energy resolution is 10–15% per event. This compares favorably with satellite detectors. The energy threshold (at zenith) is 100 GeV. The collection area increases with γ-ray energy, reaching $> 1\ \mathrm{km}^2$ for observations at large zenith angles. The sensitivity in the overlapping energy range is a factor of ≈ 10 better than that of HEGRA and exceeds that of the single-telescope MAGIC by a factor ~ 3. This implies an observation time of 1 hr for a 5σ detection of an energy flux of 10^{-11} and 10^{-12} erg cm^{-2}s^{-1} at γ-ray energies of 100 GeV and 1 TeV, respectively. In other words, an instrument like H.E.S.S. can detect the (point-like) Crab Nebula in TeV γ-rays in about 30 s, compared with the $\sim$ 18 h required for the equivalent original detection 15 years earlier.[5]

[5]The Crab Nebula is generally treated as standard candle in VHE γ-ray astronomy.

3 Two major scientific results

In this section we shall illustrate the technique of ground-based γ-ray astronomy with two examples from the H.E.S.S. results. They share the ubiquitous feature of spatial extension, characteristic of so many of the non-thermal sources in the Universe. The first example is a Supernova Remnant (SNR), thus possibly belonging to the long-sought class of sources of the Galactic CRs. The other example concerns the results of the Galactic Plane scan in the VHE range that has been performed since 2004.

3.1 The supernova remnant RX J1713.7-3946

This SNR, located in the constellation Scorpius in the Galactic plane, was originally discovered with the X-ray satellite ROSAT [29]. In X-rays RX J1713.7-3946 has a diameter of 1°, twice the size of the full moon (see Fig. 14). The ASCA (1–3 keV) image shows an overall shell structure ([30]). Also in subsequent observations with Chandra and XMM the X-ray emission was "entirely" non-thermal: no thermal emission has yet been identified. The existence of a central X-ray point source, which is probably a neutron star, suggests the core collapse of a massive star. RX J1713.7-3946 was also detected with H.E.S.S. [31, 32] and there exists a close spatial correlation of the X-ray and the VHE γ-ray emission (see Fig. 14).

The γ-ray spectrum is very hard—harder than a spectrum containing equal energy per decade in the observed range—and extends up to γ-ray energies of about 100 TeV [33] (see Fig. 15). This means that the generating charged

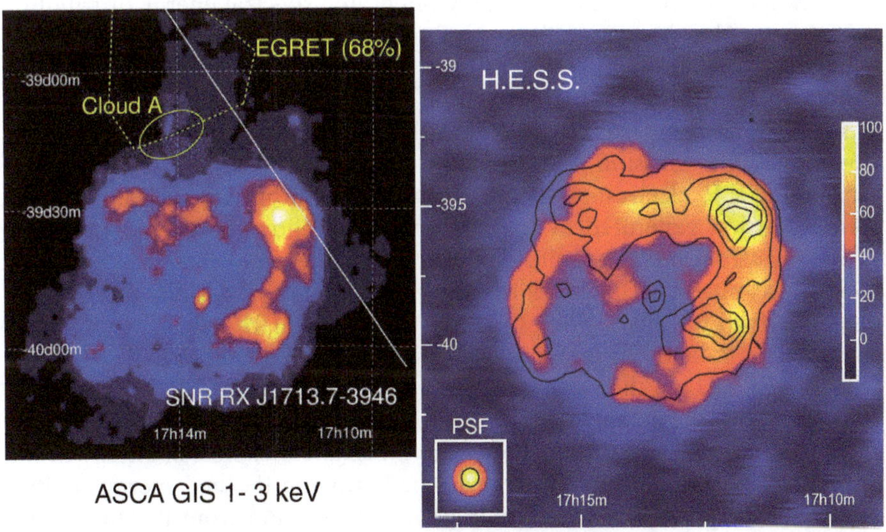

Fig. 14 *Left*: Image of the Supernova Remnant RX J1713.7-3946, as observed with the ASCA satellite in the 1–3 keV band [30]. *Right*: H.E.S.S. image of RX J1713.7-3946 at energies above 190 GeV (in false colors), with the ASCA contours superposed [33]

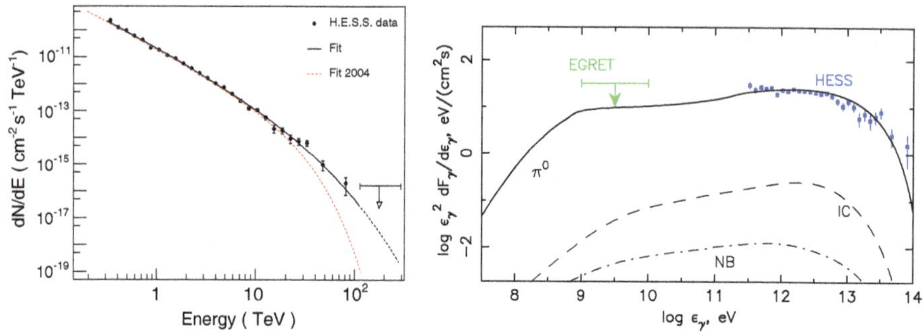

Fig. 15 *Left*: H.E.S.S. energy spectrum of RX J1713.7-3946 [33]. *Right*: Theoretical spectral energy distributions of RX J1713.7-3946 in the VHE range, assuming a magnetic field strength $B_{\mathrm{eff}} \approx 130\ \mu$G inside the remnant [35]

particles must have energies in excess of about 300 TeV and, in the case of nuclear particles, even of about 600 TeV. A reasonable phenomenological fit to the differential γ-ray energy spectrum is $dN/dE = N_0 E^{-\Gamma} \exp{(E/E_c)^{0.5}}$, with $\Gamma = 1.8 \pm 0.04$ and $E_c = 3.7$ TeV. Extrapolating this spectrum back to 1 GeV, using arguments from diffusive shock acceleration theory (see below), yields a total energy in energetic particles of about $10^{50}/\langle n \rangle$ erg. For an average gas density of $\langle n \rangle \approx 1\ \mathrm{cm}^{-3}$ this corresponds to roughly 10 percent of the expected total hydrodynamic explosion energy. Also spectral imaging and an energy-resolved morphological characterization was achieved.

Given the possible fundamental significance of such an empirical finding we show a comparison of the spatially-integrated spectral energy distribution (SED) from H.E.S.S. with a theoretical model spectrum [34, 35] that assumes a magnetic field strength $B_{\mathrm{eff}} \approx 130\ \mu$G inside the remnant and is able to fit the *amplitude* of the VHE spectrum (see Fig. 15) for a typical injection rate of supra-thermal nuclear particles into the acceleration process. The inverse Compton (IC) emission and Bremsstrahlung (NB) of the simultaneously accelerated CR electrons turn out to be negligible in comparison with the γ-ray emission generated by the π^0-production and subsequent decay into two γ-rays from inelastic collisions of the accelerated nuclear particles with thermal gas nuclei in the interior of the remnant. The upper limit in the GeV range, derived for this region of the sky from the EGRET detector, is shown as well. The IC and NB amplitudes in Fig. 15 are derived from the observed synchrotron SED in the radio and X-ray ranges below 100 keV (see Fig. 16), consistent with the magnetic field strength above.

That 10 or more percent of the entire mechanical explosion energy of the Supernova explosion should be transformed into non-thermal energy of ultra-relativistic particles is a rather outrageous proposition from the theoretical side. Yet nature might indeed comply with it, as the H.E.S.S. result shows. The predicted hadronic dominance of the γ-ray SED is expected to be confirmed by the recently launched Fermi Gamma-ray Space Telescope with an SED that

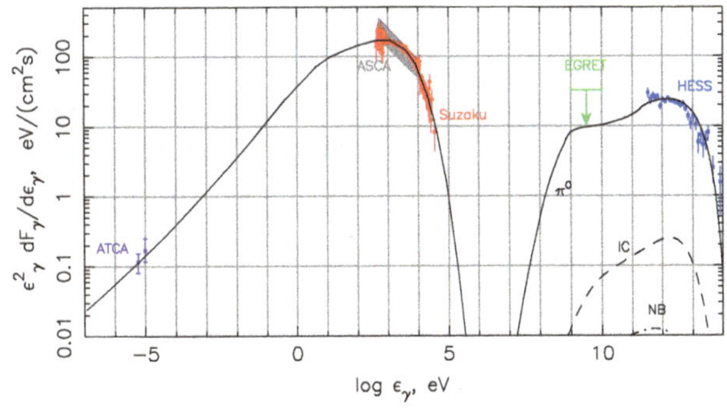

Fig. 16 Overall and spatially-integrated non-thermal spectral energy distribution of RX J1713.7-3946. The lower-energy synchrotron observations in the radio (ATCA) and X-ray (ASCA and Suzaku) ranges is compared with the theoretical model [35]. The γ-ray spectrum at higher energies is the same as in Fig. 15

should be only a factor ≈ 2.5 lower at 1 GeV than that measured at 1 TeV from Fig. 15.

3.2 Galactic Plane survey

H.E.S.S. has also performed a survey of the Galactic Plane within a band of $-3 < b < 3$ degrees in Galactic latitude b (see Fig. 17) with a 3 h exposure

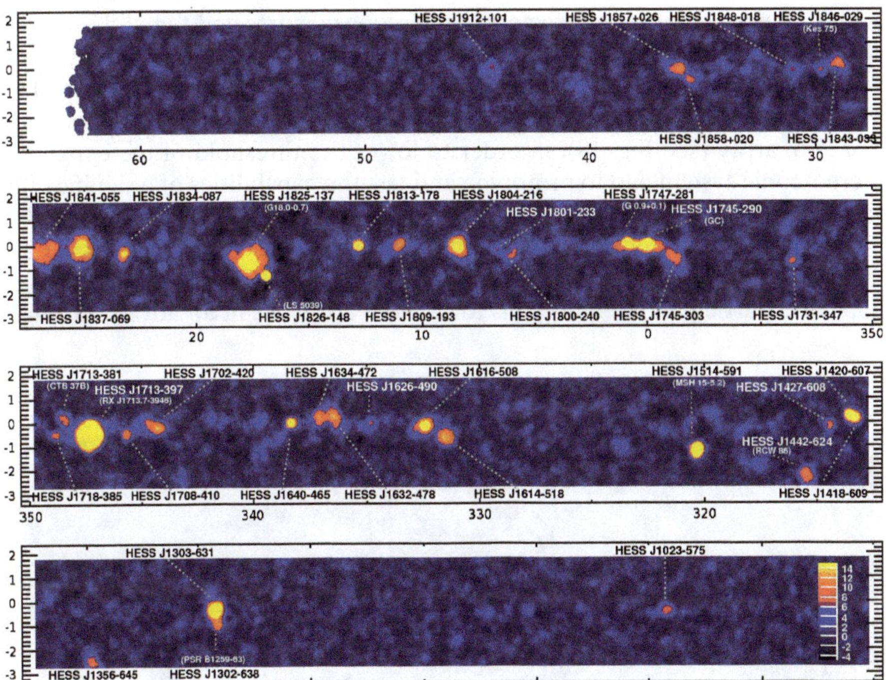

Fig. 17 Significance map of the Galactic Plane Survey by H.E.S.S., presented in Galactic coordinates: $280° < l < 60°$ and $-3° < b < 3°$ [36]. Most of the sources are spatially extended beyond the 0.1° resolution of the array. The Galactic Center (GC) region and the SNR RX J1713.7-3946 can be seen in the right half of the 2nd panel and the left half of the 3rd panel from the top, respectively

per field. The scan resulted in about 40 sources of at least 5σ significance (status 2007) in a band $280 < l < 60$ degrees of Galactic longitude l, most of them spatially extended [36]. About half of these sources are unidentified in other wavelength ranges. One gets the impression that the amazing number of sources and variety of source types almost fills the nearby part of the plane, as the significance map Fig. 17 suggests. The source types range from Pulsar Wind Nebulae, Pulsars in binary systems, X-ray Binaries, shell-type SNRs and OB associations to giant molecular clouds and the Galactic Center region. The survey is presently continued. The number of VHE sources, at least in this special part of the sky, appears mainly limited by the sensitivity of the instrument and possibly also by confusion, rather than by the availability of sources.

4 The future

The fact that the VHE sky is rich in sources—also in extragalactic space, as recent detections by several observatories have shown—has led to efforts to increase the power of the existing detectors. MAGIC started to experiment with advanced solid state detectors in telescope cameras and to build a second telescope, in order to be able to do stereoscopic observations. This second telescope is currently—fall 2008—in the commissioning phase (see Fig. 11). H.E.S.S. began in 2006 to build a 28 m telescope in the center of the original H.E.S.S. array (see Fig. 18), in order to lower the threshold of the expanded stereoscopic system and to prototype and test the capabilities of a significantly larger telescope (see Fig. 19).

Beyond these extensions of existing instruments, there are worldwide activities to build a fourth generation of arrays. In the European context, with Japanese participation, joint plans for a large array are in an advanced stage.

Fig. 18 Photomontage of H.E.S.S. Phase II (courtesy W. Hofmann). The 28 m telescope in the center of the original H.E.S.S. array should increase the sensitivity of the system by a factor ≈ 2, at a threshold energy of ~ 70 GeV in coincidence mode

Fig. 19 Build-up, in Namibia, of the dish and mirror support structure for the H.E.S.S. II telescope (courtesy M. Panter). First light is expected at the end of 2009

The project is called Cherenkov Telescope Array (CTA) [37] which should enhance the sensitivity at 1 TeV by a further order of magnitude, down to the level of 10^{-3} of the flux from the Crab Nebula (1 mCrab). In the US the project Advanced Gamma-ray Imaging System (AGIS) [38] is actively pursued, with a similar goal.

The CTA project involves different types and sizes of telescopes (see Fig. 20) designed to extend the present energy range in both directions: to lower energies of some tens of GeV and to higher energies in excess of 100 TeV. Apart from deep broadband investigations of Galactic objects, this should also allow the investigation of the non-thermal aspects of large-scale structure formation in the Universe by studying active galactic nuclei, clusters of galaxies, and the related phenomena of accretion, starbursts, and galaxy–galaxy mergers. These goals clearly require the coverage of a large energy range.

Fig. 20 Possible configuration of the CTA project, involving a mix of telescopes of various sizes and characteristics (courtesy W. Hofmann)

Acknowledgements We are indebted to our colleagues in the HEGRA and H.E.S.S. Collaborations, and especially in the Max-Planck-Institut für Kernphysik for many discussions over the years on the physics of γ-ray observations with imaging Cherenkov telescopes. We also thank W. Hofmann, G. Hermann and M. Panter for allowing us to use illustrative figures which they had designed for other purposes.

References

1. H.E.S.S. Collaboration: http://www.mpi-hd.mpg.de/HESS/ (2009)
2. Bernöhr, K.: Simulations of imaging atmospheric Cherenkov telescopes with CORSIKA and sim_telarray. Astropart. Phys. **30**, 149 (2008)
3. Heck, D. et al.: CORSIKA: a Monte Carlo code to simulate extensive air showers. Tech. report FZKA 6019. Forschungszentrum Karlsruhe (1998)
4. Galbraith, W., Jelly, J.V.: Light pulses from the night sky associated with Cosmic Rays. Nature **171**, 351 (1953)
5. Nesterova, N.M., Chudakov, A.E.: Observation of Cherenkov radiation accompanying extensive air showers of cosmic rays. Sov. Phys. JETP **28**, 385 (1955)
6. Zatsepin, V.I., Chudakov, A.E.: On the methods of searching for local sources of high energy photons. J. Exp. Theor. Phys. **41**, 655 (1961)
7. Chudakov, A.E., Dadykin, V.L., Zatsepin, V.I., Nesterova, N.M.: A search for photons with energy 10^{13} eV from local sources of cosmic radiation. Proc. P.N. Lebedev Phys. Inst. **26**, 118 (1964) (in Russian); English Translation: Consultants Bureau **26**, 99 (1965)
8. Lidvansky, A.S.: Air Cherenkov Methods in cosmic rays: a review and some history. Radiat. Phys. Chem. **75**, 891 (2006) (archive:astro-ph/0504269)
9. Weekes, T.C.: The atmospheric Cherenkov technique in very high energy gamma-ray astronomy. SSRv **75**, 1 (1996)
10. Weekes, T.C., et al. (Whipple Collaboration): Observation of TeV gamma rays from the Crab nebula using the atmospheric Cherenkov imaging technique. Aerosp. Power J. **342**, 379 (1989)
11. Humensky, B.: University of Chicago. http://astro.uchicago.edu/cosmos/projects/veritasstereo/ (2004)
12. Hillas, A.M.: Cherenkov light images of EAS produced by primary gamma rays and by nuclei. Proc. 19th Int. Cosmic Ray Conf. (La Jolla) **3**, 445 (1985)
13. Daum, A. (HEGRA Collaboration): First results on the performance of the HEGRA IACT array. Astropart. Phys. **8**, 1 (1997)
14. HEGRA Collaboration and Max Planck Institut für Kernphysik: http://www.mpi-hd.mpg.de/CT/ (1996)
15. Funk, S., et al.: The trigger system of the H.E.S.S. telescope array. Astropart. Phys. **22**, 285 (2004)
16. Ong, R.A.: Very high-energy gamma-ray astronomy. Phys. Rep. **305**, 93 (1998)
17. Weekes, T.C., et al.: VERITAS: the very energetic radiation imaging telescope array system. Astropart. Phys. **17**, 221 (2003)
18. Mori, M., and the CANGAROO-III Team: Recent results form CANGAROO-II&III. In: Aharonian, F.A., Völk, H.J., Horns, D. (eds.) High Energy Gamma-Ray Astronomy, AIP Proc., vol. 745, pp. 639ff. Melville, New York (2005)
19. Benbow, W. for the H.E.S.S. Collaboration: The status and performance of H.E.S.S.. In: Aharonian, F.A., Völk, H.J., Horns, D. (eds.) High Energy Gamma-Ray Astronomy, AIP Proc., vol. 745, pp.611 ff. Melville, New York (2005)
20. Lorenz, E. (for the MAGIC Collaboration): The MAGIC telescope, status and future plans. In: Aharonian, F.A., Völk, H.J., Horns, D. (eds.) High Energy Gamma-Ray Astronomy, AIP Proc., vol. 745, pp. 622ff. Melville, New York (2005)

21. Aharonian, F.A., Hofmann, W., Konopelko, A.K., Völk, H.J.: The potential of ground based arrays of imaging atmospheric Cherenkov telescopes; I. Determination of shower parameters. Astropart. Phys. **6**, 343 (1997)
22. Aharonian, F.A., Hofmann, W., Konopelko, A.K., Völk, H.J.: The potential of ground based arrays of imaging atmospheric Cherenkov telescopes; II. Gamma-ray flux sensitivities. Astropart. Phys. **6**, 369 (1997)
23. CANGAROO Collaboration: http://icrhp9.icrr.u-tokyo.ac.jp/gallery.html (2004)
24. MAGIC Collaboration: http://wwwmagic.mppmu.mpg.de/ (2008)
25. Criswell, S., VERITAS Collaboration: http://veritas.adlerplanetarium.org/ (2007)
26. Drury, L.O'C., Aharonian, F.A., Völk, H.J.: The gamma-ray isibility of supernova remnants. A test of cosmic ray origin. A&A **287**, 959 (1994)
27. Völk, H.J., Aharonian, F.A., Breitschwerdt, D.: The nonthermal energy content and gamma-ray emission of starburst galaxies and clusters of galaxies. SSRv **75**, 279 (1996)
28. Aharonian, F. et al. (H.E.S.S. Collaboration): Observations of the Crab Nebula with H.E.S.S. A&A **457**, 899 (2006)
29. Pfeffermann, E. & Aschenbach, B.: In: Zimmermann, H., Trümper, J. (eds.) Roentgenstrahlung from the Universe. MPE Report 263, Garching:MPE, 267 (1996)
30. Uchiyama, Y., Takahshi, T., Aharonian, F.A.: Flat spectrum X-ray emission from the direction of a molecular cloud associated with SNR RX J1713.7-3946. PASJ **54**, L73 (2002)
31. Aharonian, F. (H.E.S.S. Collaboration): Direct evidence for high-energy particle acceleration in the shell of a supernova remnant. Nature **432**, 75 (2004)
32. Aharonian, F. (H.E.S.S. Collaboration): A detailed spectral and morphological study of the gamma-ray supernova remnant RX J1713.7-3946 with H.E.S.S.. A&A **449**, 223 (2006)
33. Aharonian, F. (H.E.S.S. Collaboration): Primary particle acceleration above 100 TeV in the shell-type supernova remnant RX J1713.7-3946 with deep H.E.S.S.observations. A&A **464**, 235 (2007)
34. Berezhko, E.G., Völk, H.J.: Theory of cosmic ray production in the supernova remnant RX J1713.7-3946. A&A **451**, 981 (2006)
35. Berezhko, E.G., Völk, H.J.: Hadronic versus leptonic origin of the gamma-ray emission from Supernova Remnant RX J1713.6-3946. A&A arXiv:0810.0988 [astro-ph] (2008)
36. H.E.S.S. Collaboration: Source of the month: Dec. 2007. http://www.mpi-hd.mpg.de/hfm/HESS/pages/home/som/2007/12/ (2007)
37. Cherenkov Telescope Array: The Cherenkov Telescope Array (CTA) Project. http://www.cta-observatory.org/ (2009)
38. Buckley, J. et al.: The status and future of ground-based TeV gamma-ray astronomy. A white paper prepared for the Division of Astrophysics of the American Physical Society. arXiv:0810.0444 (2008)

Exp Astron (2009) 25:193–207
DOI 10.1007/s10686-009-9173-6

HISTORICAL REVIEW

History of solar telescopes

Oskar von der Lühe

Received: 17 April 2009 / Accepted: 4 June 2009 / Published online: 27 June 2009

Abstract Solar observations have been done with telescopes since their invention—already Galileo looked at the Sun. Despite the Sun's unusual brightness, telescopes which specialize in solar observations are fairly recent, dating from the late nineteenth century onwards. Today, many solar telescopes have rather little in common with nighttime telescopes. They are adapted to high light flux, a limited range of declination, and to the specifications of solar spectrographs and polarimeters. This paper presents the history of the modern optical solar telescope on the ground and in space, the accompanying evolution of scientific capabilities, and a brief outlook into the future.

Keywords Solar telescopes · Sunspots · Magnetic field · Coronagraph · Evacuated telescopes

1 Introduction

The Sun has been a target for observations by telescopes since their introduction into astronomy in the early seventeenth century. However, the history of the dedicated telescope which is built for observing the Sun is not nearly as long. Those telescopes were introduced when quantitative high-resolution spectroscopy required spectrographs with long focal lengths, making a laboratory setup a necessity. The history of the solar telescope therefore covers only the past 150 years.

O. von der Lühe (✉)
Kiepenheuer-Institut für Sonnenphysik, Schöneckstr. 6-7, 79104 Freiburg, Germany
e-mail: ovdluhe@kis.uni-freiburg.de
URL: www.kis.uni-freiburg.de

Many solar telescopes differ considerably from the typical night-time telescope. They deal with an extended, bright source and require attention to flux concentration and internal seeing. An ordinary refractor is always a good choice for a small solar telescope, when the Sun is viewed in projection on a screen or a special eyepiece is used. Apertures of half a meter and beyond are often installed in fixed telescopes, refractors or Schiefspiegler reflectors with a long focal length which are fed by coelostats and heliostats. The Gregorian telescope has recently become a popular configuration for large-aperture reflectors because of the real primary image, where a field stop rejects a large fraction of the solar image and reduces the flux through the remaining optics.

Today, about 100 telescopes world-wide, ranging from small synoptic to large-aperture facilities, observe the Sun regularly from the ground to collect scientific data. Many of them deviate significantly from conventional telescope designs. We will review the evolution of the dedicated solar telescope on the basis of several typical examples.

2 Early observations of the sun and sunspots

Establishing sunspots as a widely recognized solar phenomenon is undoubtedly one of the first accomplishments of astronomical research with telescopes. Sunspots were first observed by Fabricius, Galileo, Scheiner and Harriot in the first decade of the seventeenth century. Galileo definitely deserves the credit for establishing that sunspots are part of the sun, that they change with time, and that the sun rotates with a period of about a lunar month. Scheiner observed sunspots for a much longer time than Galileo. He confirmed Galileo's assertion of what now is known as activity belts, and his results play an important role in establishing the sunspot cycle in the early years of telescope observations. His drawings also show a clear distinction between the dark umbra in the centre of the spots and the peripheral, somewhat brighter penumbra. The first insight into the physics of sunspots was the discovery of an asymmetry in the shape of sunspot penumbrae when viewed at the extreme limb by Alexander Wilson in 1769. He interpreted this effect as a depression of the whole sunspot, which agrees with the modern interpretation of a depression in the continuum opacity above sunspots.

The phenomenon of the solar activity cycle remained undiscovered until in the nineteenth century Schwabe discovered a regular variation of the occurrence of sunspots, observing them for more than 20 years. In the second half of the nineteenth century Carrington established the heliographic coordinate system and the solar rotation nomenclature which is still in use today. He also discovered the average latitude decrease of sunspots from the beginning to the end of a cycle maximum. Wolf traced the history of sunspot observations back to Galileo's time, establishing that an 11-year cycle existed almost all of the time for more than 250 years. The most widely known solar activity index,

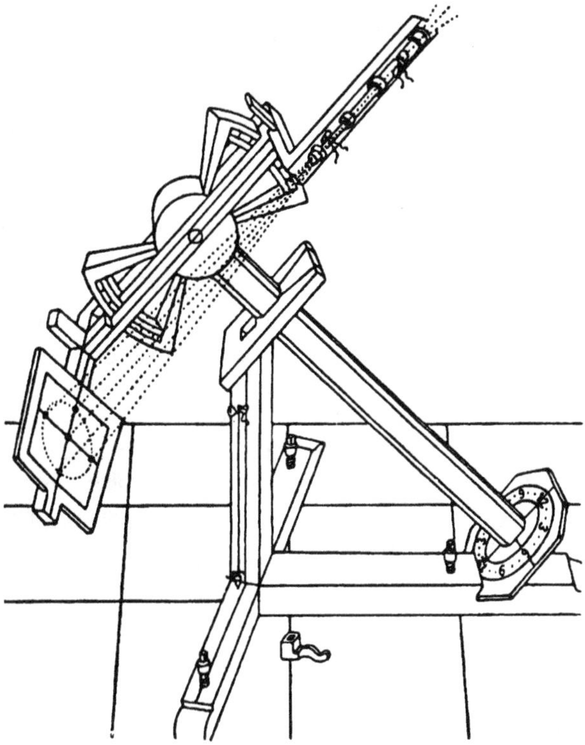

Fig. 1 Scheiners equatorially mounted telescope for observing sunspots

the Wolf sunspot number *R*, was established by him. A daily record of R, established by Wolf in Zürich, was maintained until the late twentieth century.

Observing sunspots requires little more than an ordinary refractor (Fig. 1). Because of its brightness, the solar image is projected conveniently onto a screen close to the eyepiece, which is moved some distance away from the focal plane of the objective to focus the image on the screen. Tracings of sunspot positions with respect to the solar limb on paper can be obtained easily and with fairly high precision. Sunspot recordings from various observatories dating back for decades still play a role in detailed studies of the solar cycle. Many small telescopes dedicated to specific, repeating synoptic observations today are refractors.

3 Photometry, spectroscopy and small scale structure—the Meudon Observatory

Solar observations played an important role when quantitative measurement by spectroscopy and photometry established the science of astrophysics. The solar spectrum, observed by Joseph von Fraunhofer in 1817, shows many absorption lines, the strongest of which were given letters from A to J. This nomenclature is still in use today (Na "D" and Ca II "H"). Gustav

Kirchhoff attributed the lines to the presence of chemical elements in the solar atmosphere, introducing spectroscopic analysis around 1860. The prism spectrograph made it possible to isolate features in the solar spectrum. Prominences, so far only visible during a total solar eclipse, could then be observed at any time by isolating the Balmer α line in emission above the solar limb. In 1868, P. J. Janssen in Paris and J. N. Lockyer in England were the first to observe prominences this way. Visual spectroscopic observations resulted in the discovery of the solar chromosphere and of spicules, with Father A. Secci being one of the most notable observers.

The second important invention was the photographic plate, which permitted an objective recording of solar structure. The first photoheliograph which permitted regular photographic recordings of the solar disk was constructed by Warren de la Rue in England, and was operated from 1861 onwards; it served mainly for recording sunspots. This instrument and its successors were later operated for a long time at the Royal Observatory at Greenwich. The pioneer of photographic recording of solar small scale structure is undoubtedly Pierre Jules Janssen in Meudon close to Paris. Figure 2 shows one of his photographs of a sunspot group, together with a modern granulation image representing one of the last photographic recordings of the Sun. While solar granulation with a scale of a few arcsec is clearly visible throughout most of the field, the

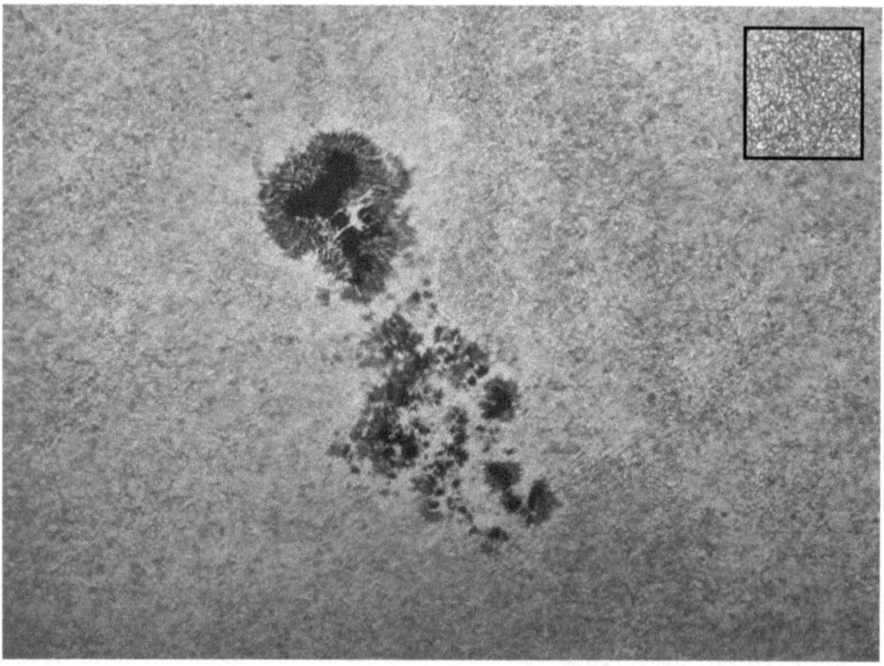

Fig. 2 Photograph of a sunspot by Janssen from the late nineteenth century (Meudon Observatory). The *insert* shows a quiet sun photograph from 1979 (Kiepenheuer-Institut)

structure of the penumbra is distorted by seeing and by the limited resolution of the photoheliograph.

It is evident that observers in the second half of the nineteenth century with a surprising fidelity recorded many features from visual observations which were nearly impossible to record photographically at that time. Figure 3 is an excellent example of a sunspot observed in 1873 showing small-scale features, in particular in the penumbra, which today require a large aperture telescope, solid-state detectors and high-resolution image-processing methods.

Janssen deserves the credit for being one of the most prolific solar observers of his time, and for establishing the first comprehensive solar observing facility on the grounds of the Meudon Observatory near Paris during the last quarter of the nineteenth century. Several buildings were devoted to solar observations, featuring large spectrographs and heliographs which made a fixed installation of the telescope necessary. Coelostats deliver sunlight to these installations, which is no observing constraint because of the Sun's limited range of declinations. The combination of a fixed, horizontal telescope which produces a solar image with a diameter of a dozen cm at the prime focus with feed optics for sunlight became widespread during the twentieth century. Meudon remains an eminent solar facility to this day, and some of the facilities introduced by Janssen are still in regular use.

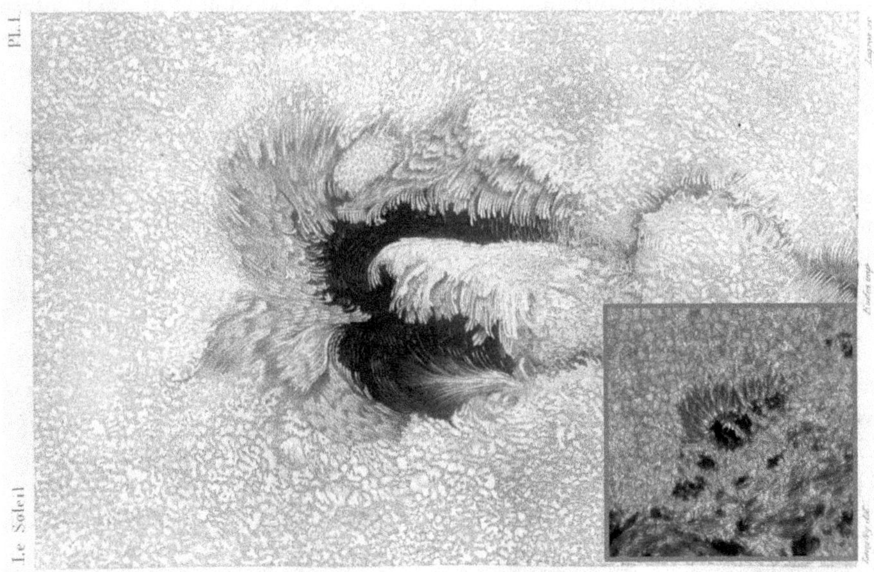

Fig. 3 Complex sunspot, observed by Langley on Christmas 1873 (A. Secci, Le Soleil Vol 2, 1877). The *insert* shows a modern high-resolution image of a part of a sunspot group at 430 nm with a comparable scale from May 2003 (Kiepenheuer-Institut)

4 The solar magnetic field and the Mt. Wilson Observatory

In the first decade of the twentieth century, George Ellery Hale founded the Mt. Wilson Observatory near Pasadena as a predominantly solar observing station. The first solar facility was a horizontal telescope with a spectroheliograph which was moved from the Yerkes Observatory. A vertical telescope on an 18-m (60 ft) high tower was soon added in order to overcome difficulties with the horizontal light path which suffered from internal turbulence. The telescope features a coelostat and an objective lens at the top of the tower, directing the sunlight vertically downwards to the primary focal plane near the ground. The spectrograph was again mounted vertically into a hole in the ground below the focal plane. It was found that the combination of a vertical telescope and a buried, vertical spectrograph was much better in terms of internal seeing. A few years later, another solar telescope on a 47-m (150 ft) tower and a 23-m vertical spectrograph was added to the ensemble to achieve even higher spatial and spectral resolution. Both vertical telescopes are still used today and play an important role in helioseismology.

The main achievement of Hale at that time was the discovery of magnetic fields in sunspots, stimulating an era of intense solar research at Mt. Wilson in the first half of the century. Hale was aware of the broadening and splitting of spectral lines in sunspots which had already been observed earlier. He assumed that vortex flows in sunspots, which he erroneously deduced from chromospheric observations, might cause a vertical magnetic field in the spots and concluded that the split lines should exhibit circular polarization. The longitudinal Zeeman effect in sunspots was discovered soon after polarization optics were added to the spectrographs, and although Hale's hypothesis has now been shown to be wrong, the presence of magnetism is one of the most important discoveries in solar physics.

The development of spectropolarimetry resulted also in an intense development of instrumentation at Mt. Wilson. The development of diffraction gratings permitted the construction of highly sensitive spectrographs and magnetographs. The art of spectroheliography, where the solar image and the recording photographic plate scanned entrance and exit slits of the spectrograph simultaneously, was perfected. The result is an image of the solar disk in a very narrow spectral region, marking the beginnings of field spectroscopy.

A prominent descendant from the Mt. Wilson solar telescopes is the Einsteinturm on the Telegrafenberg in Potsdam, Germany (Fig. 4). This telescope was constructed in the 1930s with the goal of measuring the solar gravitational red shift, one of the effects predicted by Einstein's theory of general relativity. The building features an inner wooden tower which supports the telescope objective and the coelostat (Fig. 5) on its top, while the mechanically detached, solid outer tower supports the dome and provides shielding from wind forces. A folding flat mirror directs the sunlight prior to the focal plane towards the horizontal spectrograph in the basement of the building. The location under ground of the spectrograph provides a stable environment for high-precision spectroscopy. This configuration, tower telescope and horizontal spectrograph,

Fig. 4 Einsteinturm and cross section (Astrophysikalisches Institut Potsdam)

is also commonly found. Intrinsic solar phenomena such as the correlation of intensity and velocity in the photosphere prevented the measurement of the gravitational red shift for a long time, so the Einsteinturm did not fulfil its original purpose. However, the telescope is known for its outstanding contributions to high-precision polarimetry, and it is undoubtedly one of the most beautiful solar telescopes.

Another unusual telescope was the Domeless Coudé Refractor which was built by K. O. Kiepenheuer in 1963 at the Capri Observatory of the then Fraunhofer-Institut in Freiburg (Fig. 6). This telescope consisted of a tube with a 35 cm diameter superachromat objective which could be pointed directly at the Sun. Two flat mirrors in the optical train provided a fixed solar image on an inclined optical bench with an—equally inclined—Littrow spectrograph. The

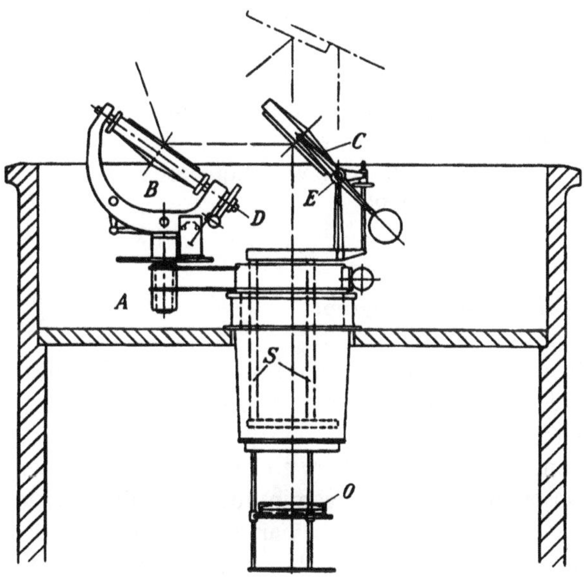

Fig. 5 Top part of the Einstein tower telescope showing the coelostat and the objective lens (*O*). *B* and *C* are the coelostat primary and secondary mirrors, the primary axis *D* is directed towards the Earth's poles and tracks the Sun with half the Earth's rotation rate. The Sun's declination determines the height of the secondary mirror with respect to the primary. The azimuth position of the primary mirror with respect to the vertical telescope axis can be adjusted to avoid obscuration by the secondary (Astrophysikalisches Institut Potsdam)

Fig. 6 Domeless coudé refractor of the Fraunhofer-Institut in Capri (Kiepenheuer-Institut)

intent behind the absence of a dome and the careful thermal control of the entire system was suppression of internal and near-by seeing.

One of the most prominent accomplishments of this telescope was the detection of solar oscillation Eigenmodes by F. L. Deubner in 1973. The site of the observatory, on a cliff only 150 m above the Mediterranean Sea level, was exceedingly pleasant and featured a beautiful telescope, but proved not to provide adequate weather and seeing conditions for high-resolution solar research. The telescope was therefore abandoned and operations were moved to Tenerife in 1988.

5 The Lyot coronagraph

The coronagraph invented by Bernard Lyot in the 1930s is one of the most cleverly devised telescopes. Like prominence observations before the spectrograph, the solar corona could only be observed during a solar eclipse, condemning researchers to endless campaigns to eclipse sites and limiting observations to the few minutes of totality. The high contrast between photosphere and corona—one part in 10^6—prevented regular observations even in coronal emission lines. The concept of using a coronagraphic mask in the focal

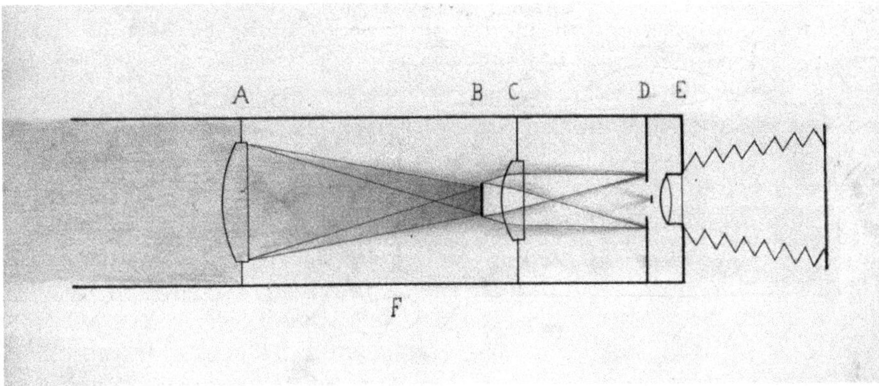

Fig. 7 Sketch of the solar coronagraph by B. Lyot (Meudon Observatory)

plane which covers the solar disk was known, but did not deliver the necessary dynamic range. It was clear to Lyot that stray light needs to be suppressed with all available means. He succeeded in observing actual coronal emission with a combination of (see Fig. 7)

1. a singlet lens with very smoothly polished surfaces as the objective,
2. a flat or cone-shaped field stop in the primary focus, rejecting the image of the solar disk into a light trap,
3. a field lens following the field stop, rendering an image of the objective in a subsequent conjugate,
4. an under-sized aperture stop ("Lyot stop") in that plane, collecting most of the light diffracted at the edge of the objective,
5. further stops as required, capturing spurious light reflected from edges or curved surfaces,
6. a collimator lens which produces the final image of the corona.

The combination permitted viewing at least the inner parts of the corona. The general principle of the coronagraph is used in many ground- and space based telescopes today. While space remains the prime location for coronal research in the spectrum from the visible to the extreme UV, ground-based coronagraphs are frequent. They require sites with very little aerosols which deliver "coronal conditions" often, and are found at 3,000 m altitude and more.

6 Evacuated telescopes

Internal seeing, i.e., turbulence in the air in the light path of telescopes and instruments has since long been identified as the cause of image deterioration. The concentration of flux in a solar telescope makes matters worse, which is why solar astronomers paid particular attention to thermal control. 2 arcsec seeing was considered "good", 1 arcsec or better was rare. An obvious remedy is putting the equipment outside the atmosphere, but space experiments as well as high-altitude balloon flights are limited, expensive and rare. Many ground-based telescopes are designed to obstruct natural air flow as little as possible; the Mt. Wilson towers are a prominent example. A radical solution is evacuating parts or the entire light path inside the facility. R. McMath was the first to evacuate a solar spectrograph with success at the University of Michigan in the 1950s, and he transferred the concept to the to date largest solar telescope, the 1.6 m McMath-Pierce facility of the National Solar Observatory at Kitt Peak in Arizona (Fig. 8). Although designed for vacuum, its spectrograph has never been evacuated for technical reasons.

Evacuating entire telescopes has been discussed intensely in the 1960s, and towards the end of that decade the first evacuated solar telescopes were commissioned. The first of these was the 61 cm Vacuum Solar Telescope and associated Vacuum Spectroheliograph which came into operation in the fall of 1969 at the San Fernando Solar Observatory in California. They were followed shortly after by the vacuum tower telescopes at Sacramento Peak, New Mexico (Fig. 9) and at Kitt Peak (Fig. 8, background). At about the same time, the 45 cm Gregory Coudé telescope of the Universitätssternwarte Göttingen at their Locarno observatory was transformed into an evacuated telescope and the Fraunhofer-Institut in Freiburg installed a 45 cm evacuated

Fig. 8 The McMath-Pierce Facility (*foreground*) and the vacuum tower with the SOLIS telescope (*background right*) of the National Solar Observatory at Kitt Peak (Kiepenheuer-Institut)

Fig. 9 Dunn Solar Telescope (DST) at Sacramento Peak. The cross section diagram to the right in the entrance hall of the DST shows that the major part of the telescope is underground (Kiepenheuer-Institut)

Newtonian solar telescope at the Teide Observatory of the Instituto Astrofisica de Canarias, Tenerife. All vacuum telescopes immediately demonstrated the viability of evacuation as a means to improve image quality.

The vacuum tower telescope at Sacramento Peak deserves further comment. It was constructed by R. B. Dunn and is now named after him. The telescope consists of a 76 cm vacuum window, followed by a turret with two flat mirrors which are illuminated at 45° and can be individually rotated, all on top of a 30 m tower. The light is directed vertically downwards towards the main mirror which is located some 60 m below ground. The 1.5 m main mirror is slightly tilted, so the reflected sunlight focuses to form a solar image near ground level at some distance from the incoming beam. The light can be directed towards several ports with exit windows of various sizes; there is a 51 cm unvignetted image of the entire solar disk at one of these ports. A large platform at ground level supports a multitude of post-focus instruments. The entire telescope including instrument platform, totaling 340 tons, floats in a mercury bearing near the top of the tower and rotates to maintain the field angle fixed. The telescope was an immediate success, frequently delivering sub-arcsec image quality due to its quality and the frequent good seeing at that site. It is now one of the most prolific solar telescopes and famous for its advanced solar adaptive optics program.

Essentially all major solar facilities which followed were designed as vacuum telescopes. Examples are the Domeless Solar Vacuum Telescope of the Kyoto University, which was commissioned at the Hida Observatory in 1979, and the Baikal Large Solar Vacuum Telescope of the Russian Institute of Solar-Terrestrial Physics, which is operated in Irkutsk since 1980.

7 The present and the future

The present generation of large—i.e., meter-class—solar telescopes was conceived and constructed in the last quarter of the twentieth century. Many of them in the 0.5 to 1 m range of apertures are evacuated. The activities of the European Joint Organisation for Solar Observations in the 1970s resulted in the construction of several major, nationally supported, solar telescopes in the Canary Islands of Tenerife and La Palma from 1980 to 2000. They represent a variety of unusual concepts. The Swedish Solar Telescope (SST) is now the world's largest solar refractor with a singlet objective close to 1 m in diameter. The objective also serves as the entrance window and has been designed accounting for the bending under vacuum load. As at the DST, a turret with two 1.5 m flat mirrors directs sunlight towards a focal plane in the basement of the building. Chromatic aberration of the objective can be compensated by inserting a Schupman lens in the light path. The SST is reputed for delivering the world's sharpest solar images. The Vacuum Tower Telescope (VTT) is a reflector with a coelostat on the top of a 40 m tower and a folded light path in vacuum, followed by a very stable Echelle spectrograph which extends 13 m below ground. The telescope is equipped with a comprehensive set of spectrographs for the visible, near IR and UV spectral ranges. A multi-conjugate adaptive optics (MCAO) system in the VTT system has closed the loop in 2004, for the first time with an astronomical telescope on sky. THEMIS is a Cassegrain telescope designed for low instrumental polarization and stands out for its multi-line spectroscopic capability. Unlike the other telescopes, THEMIS uses a thin entrance window for low stress birefringence and is filled with helium. Because of a low refractive index and high thermal conductivity, helium filling effectively quenches internal seeing.

A practical limit for the size of the aperture of an evacuated telescope is about 1 m. Containing the vacuum requires thick glass windows which pose problems of their own; uneven illumination may produce significant aberration which varies slowly with time (Fig. 10). Therefore, the concept of an evacuated telescope has been abandoned in recent, large aperture projects and replaced by open telescopes and domes which limit the obstruction of free air flow. The pioneer in this regard is the 45 cm Dutch Open Telescope (DOT) which has been in operation on La Palma since 1998 (Fig. 11). The telescope is designed for high stiffness and tracks easily in high wind speed conditions, effectively eliminating mirror seeing by free airflow. The dome consists of a stiff steel framework covered by weather-resistant cloth, and opens completely. The success of this telescope in terms of image quality has strongly influenced recent projects.

There is unanimous consent in the solar community that the observation of phenomena which are essential for understanding important physical processes, in particular in connection with the Sun's magnetic field, requires higher spatial resolution and more flux. Today, the diffraction limit of the meter-class telescope, corresponding to some 0.1 arcsec or 100 km on the solar surface, is frequently reached throughout the visible spectral range. At

Fig. 10 Wavefront tilt maps (Foucaultgrams) in the horizontal (*top row*) and vertical direction (*bottom row*) of the VTT, Tenerife. *Left*: original state indicating spherical aberration, *middle*: attempt to actively cool the entrance window, *right*: with baffle eliminating direct sunlight illuminating the entrance window (Kiepenheuer-Institut)

these scales, the Sun changes within fractions of a minute. Larger apertures are needed, not only to improve on resolution but also to collect enough photons to perform polarimetry with sufficient sensitivity. A new generation of large-aperture solar telescopes is therefore entering the scene these days. The 1.5 m Gregor telescope is—as the name implies—an open Gregorian telescope with three powered mirrors on a common axis, followed by coudé optics. It is being built by a German consortium led by the Kiepenheuer-

Fig. 11 Dutch Open Telescope on La Palma (Utrecht University)

Institut and is nearing completion at the Teide Observatory on Tenerife, next to the VTT. The lightweight Zerodur primary mirror is actively cooled from the back with a closed-circuit air system. Post-focus instruments are located below the telescope and include a multi-conjugate adaptive optics system with three deformable mirrors. The 1.6 m New Solar Telescope (NST) at Big Bear Lake, California is being built by the New Jersey Institute of Technology. Its off-axis Gregorian design maintains a completely unobstructed pupil which facilitates adaptive optics operation and improves access to the field stop at the primary focus. NST features an actively supported thin meniscus mirror which is also air-cooled. It will also be equipped with adaptive optics, following a coudé optical train.

The more distant future will see solar telescopes of the 4 m aperture class. The Advanced Technology Solar Telescope (ATST) of the National Solar Observatory (Fig. 12) is expected to receive construction funding as of this year (2009) and will see first light in the middle of the next decade. The configuration is an off-axis Gregorian where the main mirror is a 4 m diameter cut-out from a 12 m paraboloid. The reason for the off-axis design is the capability to observe the corona, which requires a clear pupil and very smoothly polished mirror surfaces. Also, disposing of many kW of optical power from the field stop at the primary focus is greatly facilitated. ATST will be erected with an "open dome" permitting as much air flow as possible while removing wind shake from the main structure. Most post-focus instruments will reside on a rotating platform below the telescope. Adaptive optics and MCAO are an integral part of the concept. The European solar community has recently started a comprehensive design study for its own 4 m class European Solar Telescope (EST).

Fig. 12 Rendering of the Advanced Technology Solar Telescope (National Solar Observatory)

Acknowledgements I am indebted to J. M. Beckers, D. Elmore, J. W. Harvey, W. C. Livingston, D. Neidig, C. Plymate, G. W. Simon, D. Vrabec, and H. Wöhl for discussion and information, in particular about the more recent history of solar telescopes.

Further Reading

1. Bray, R.J., Loughhead, R.E.: Sunspots, Int. Ast. Ser. Chapman & Hall, New York (1964)
2. Bray, R.J., Loughhead, R.E.: The Solar Chromosphere, Int. Ast. Ser. Chapman & Hall, New York (1974)
3. Bray, R.J., Loughhead, R.E., Durrant, C.J.: The Solar Granulation, 2nd Ed. Cambridge Ast. Ser. (1982) ISBN 0 521 247144
4. Danjon, A., Couder, A.: Lunettes et Télescopes. Rev. D'Optique Théor. Inst., Paris (1935)
5. Foukal, P.: Solar Astrophysics. Wiley, Hoboken (1990)
6. Hentschel, K., Zum Zusammenspiel von Instrument Experiment und Theorie. Kovač, Hamburg (1998)
7. König, A., Köhler, H.: Die Fernrohre und Entfernungsmesser, 3rd Edn. Springer, New York (1959)
8. Le Cocguen, R.: Les observations optiques du Soleil à Meudon. Thesis, Obs. de Paris (2008)
9. Manly, P.L.: Unusual Telescopes. Cambridge Univ. Press, Cambridge (1991)
10. Phillips, K.J.: Guide to the Sun. Cambridge Univ. Press, Cambridge (1992)
11. Shapley, H., Howarth, H.E.: A Source Book in Astronomy. McGraw-Hill, New York (1929)
12. Shapley, H. (ed.): Source Book in Astronomy 1900–1950. Harvard Univ. Press, Cambridge (1960)
13. Secci, A.: Le Soleil Pt. 1 and 2 (2nd Ed.). Gauthier-Villars, Paris (1877)

Exp Astron (2009) 25:209–224
DOI 10.1007/s10686-009-9162-9

REVIEW ARTICLE

History of neutrino telescope/astronomy

Atsuto Suzuki · Masatoshi Koshiba

Received: 25 March 2009 / Accepted: 26 March 2009 / Published online: 18 April 2009

Abstract Neutrinos represent a new window to the Universe. In this paper we discuss the attempts to detect neutrinos, starting with the Homestake experiment, which showed the deficit of solar neutrinos. The detection of neutrinos from SN 1987A gave a new impetus to neutrino research. By using successive generations of neutrino detectors it was possible to show that the solar neutrino deficit could be explained by a flavor change of massive neutrinos. With the latest detector, kamLAND, it is possible to investigate the interior of the Earth through the detection of geoneutrinos.

Keywords Neutrinos · Sun: interior · SN1987A · Earth: interior · Neutrino oscillations

1 Introduction

The first extraterrestrial neutrino observation was made with the motivation to look into the interior of the sun. In this way it was expected to test directly stellar evolution and nuclear energy generation in stars. In 1968, the Homestake experiment, led by R. Davis Jr., reported the first result, which showed the deficit of neutrino events relative to solar model predictions. This discrepancy is called the solar neutrino problem. In 1986 Kamiokande,

A. Suzuki (✉)
KEK, High Energy Accelerator Research Organization, Oho 1-1,
Tsukuba, Ibaraki, 305, Japan
e-mail: atsuto.suzuki@kek.jp

M. Koshiba
University of Tokyo, Hongo 7-3-1, Bunkyo-ku, Tokyo, Japan
e-mail: office@hfbs.or.jp

led by M. Koshiba was ready to detect solar neutrinos; it had the capability to measure the time, the direction and the energy spectrum of an incident neutrino. In 1989, Kamiokande confirmed the solar neutrino deficit first shown by the Homestake experiment. Within two months after the beginning of the Kamiokande operations, a bunch of 5×10^{10} extragalactic particles per cm^2 swept through the Earth. They were neutrinos from Supernova SN1987A. Starting at 16:35:35 (JST) on February 23, 1987, 11 neutrino events were observed in Kamiokande within 13 seconds. Based on the energies of each event and the number of observed events, it was concluded that the origin of a supernova is the gravitational collapse of the iron core of massive stars, and that 99 % of the released gravitational energy is carried out by neutrinos. The observed energy release by neutrinos in SN1987A was quite consistent with this scenario and it confirmed the basic idea of the core collapse model. In 1998, Super-Kamiokande presented for the first time decisive evidence that neutrinos have an intrinsic mass. Atmospheric neutrinos are produced by high-energy collisions of cosmic rays with the Earth's upper atmosphere and come in two types: the electron neutrino (ν_e) and the muon neutrino (ν_μ). Super-Kamiokande found that the numbers of ν_e were consistent with the theory, while the numbers of ν_μ were significantly lower. The only explanation given the data is that ν_μ was oscillating, or changing from one type of neutrino to another, which could occur only if they have non-zero masses. The three years from 2001 to 2003 were the golden years of solar neutrino research. In this period, the mystery of the solar neutrino deficit was solved. Combining the SNO charged- and neutral-current measurements revealed that the ν_e produced in the interior of the Sun had been changed into ν_μ or tau neutrinos (ν_τ) by the time they reach the Earth. Thus, the cause of the solar neutrino deficit was the solar neutrino oscillation. KamLAND observed a deficit in the detected number of anti-neutrinos emitted by nuclear power reactors. KamLAND verified independently the solar neutrino oscillation under the CPT invariance, and significantly improved our knowledge of the parameters that characterize neutrinos. In 2005, KamLAND measured for the first time neutrinos produced by radioactive decay deep inside the Earth, so-called geoneutrinos. The heat generation of the Earth is a basic factor in understanding the interior dynamics of plate tectonics, mantle convection and terrestrial magnetism. KamLAND opened up this new area of research.

2 What are neutrinos?

As we know, elementary particles comprise 6 different quarks and 6 different leptons. The τ, μ, electron and 3 neutrinos (ν_e, ν_μ, ν_τ) belong to the lepton family. These elementary particles posses their own antiparticles. An antiparticle has the opposite properties to its counterpart. For instance, the antiparticle of the electron is the positron, which has a positive charge. If two bodies that are made of particles and antiparticles respectively meet, they are annihilated instantaneously, emitting electromagnetic radiation in the process.

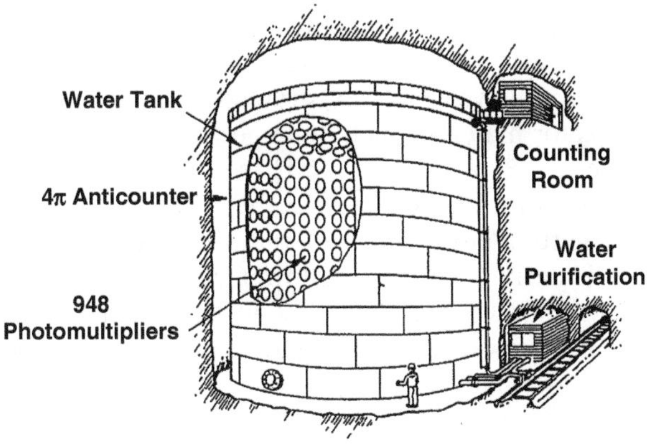

Fig. 1 First generation detector at Kamioka: Kamiokande

Neutrinos are weakly interacting particles, which can be shown from the following: the mean free paths in water for 10 MeV electrons and $\nu_e's$ are less than 0.1 cm and 10^{15} km (~10 lightyears), respectively. This is why neutrinos are called ghost particles, and it means that neutrinos can pass through the Earth and the Sun without leaving any footprintstraces. The weakness of the neutrino interaction means that they act as direct messengers from deep inside astronomical objects. For instance, the light produced at the center of the Sun takes roughly 2,000,000 years to reach the surface. It is obvious that information on the light production will have disappeared. On the other hand, neutrinos generated by nuclear fusion processes at the center of the Sun can reach the surface in 2 s, and travel to the Earth within 8 min. By detecting these neutrinos, we can investigate the stellar evolution mechanism inside the Sun. Neutrino astronomy is based on the weakly interacting properties of neutrinos.

The weakness of the neutrino interaction makes it difficult to detect neutrinos. There are three major requirements for neutrino astronomy detectors: Large detector volume, Quiet outside, and Quiet inside. A large detector volume increases neutrino interaction rates. On the ground, there are always

Fig. 2 Second generation detector: Super-Kamiokande

showers of cosmic-ray particles. These particles produce many background events inside detectors. To reduce these cosmic-ray-induced events, neutrino detectors have to be constructed at deep underground or underwater sites. Detector materials include large amounts of radioactive elements which produce α-, β- and γ- rays. These particles also generate events that mimic neutrino interactions inside the detector. A reduction of $(10^6 \sim 10^{10})$ from the natural/environmental radioactivity is required for neutrino detectors, particularly for detecting lower energy neutrinos. We have constructed the neutrino-telescope detectors in the Kamioka mine in Japan. Kamioka is $\sim$200 km west of Tokyo. The underground laboratory was constructed at a 1,000 m rockoverburden. The first-generation detector is Kamiokande which is a 3,000 ton imaging water Cherenkov detector (Fig. 1), and was in operation from 1983 to 1997. The second generation is Super-Kamiokande, which is also a water Cherenkov detector with 50,000 tonnes of water (Fig. 2) . Super-

Fig. 3 Third generation detector: KamLAND

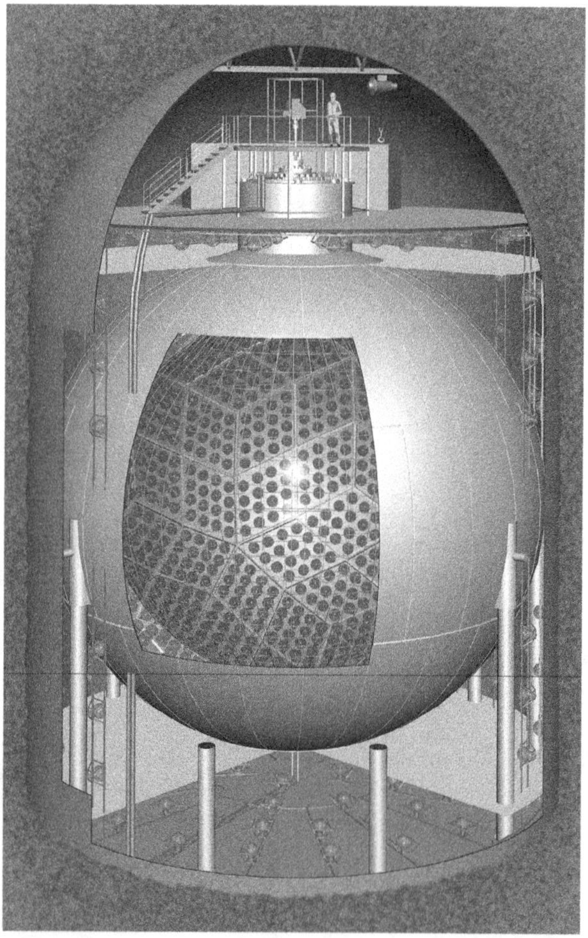

Kamiokande has been running since 1996. The third generation is KamLAND, which has 1,000 tonnes of liquid scintillator as the detector volume (Fig. 3). The experiment was launched in 2002.

3 Pioneers of neutrino astronomy

3.1 Solar neutrinos

Neutrino astronomy started with the detection of solar neutrinos. The sun shines, burning 4 hydrogen atoms to produce 1^4He(helium), $2e^+$ (positron), $2\nu_e$ and an energy of 3.8×10^{26}joule/s. This nuclear fusion reaction proceeds mainly through the proton-proton fusion chain. Neutrinos are produced in the following reactions: $p + p \rightarrow {}^2He + e^+ + \nu_e$ (pp-neutrino), $p + e^- + p \rightarrow {}^2He + \nu_e$ (pep-neutrino), ${}^7Be + e^- \rightarrow {}^7Li + \nu_e$ (^{7}Be-neutrino) and ${}^8B \rightarrow {}^8Be^* + e^+ + \nu_e$ (^{8}B-neutrino). Thus, detecting solar neutrinos brings direct information on the star evolution mechanism. Now, how many solar neutrinos do we receive on Earth. Solar neutrinos are produced isotropically, and some

Fig. 4 Homestake 520-ton radiochemical Cl-37 (C_2Cl_4) detector (Courtesy of Homestake Group)

of them can reach the Earth. The flux on the Earth is roughly $6 \times 10^{10}\,s^{-1}cm^{-2}$. This is what passes through us, day and night.

3.2 Birth of solar neutrino detection

The pioneer of detecting solar neutrinos is Ray Davis. He and his colleagues constructed a 520-ton chlorine-37 (C_2Cl_4) solar neutrino detector at a depth of 1,500 m at the Homestake Gold Mine (South Dakota) (see Fig. 4). When solar neutrinos enter the detector and interact with Cl, this Cl is transformed into argon (Ar). The number of Ar counts corresponds to the number of neutrinos. Due to the threshold energy of the Cl-to-Ar transformation, the detector is sensitive mainly to the ^{7}Be-neutrino and the ^{8}B-neutrino. The data-taking started in 1968. A summary of the Homestake data taken in the period of 1970 to 1995 is shown in Fig. 5. The average Ar counting rate is ~ 0.5 in units of Ar atoms/day, whereas the expectation from the standard solar model (SSM) is 1.8. A reduction by a factor of three from the SSM can be seen. This result was called “Missing Solar Neutrinos” or “Solar Neutrino Problem”. What was wanted was an experiment which could identify the solar neutrinos, because the Cl experiment only detected the neutrino event rate.

3.3 Confirmation of the solar neutrino problem

A new solar neutrino experiment started in 1986. Masatoshi Koshiba and his colleague constructed a 3,000 ton water Cherenkov detector, Kamiokande, at the Kamioka mine in 1983. The original goal of Kamiokande was to detect proton decay. After half a year of data taking, M. Koshiba proposed to upgrade the Kamiokande detector so as to observe the solar neutrinos. Kamiokande provides real-time measurements of the direction and the energy of neutrinos. When a solar neutrino enters the detector and hits an electron in the water, this electron is accelerated, emitting Cherenkov light at an angle of 42° relative to the direction of the electron velocity. The event-time measurement gives

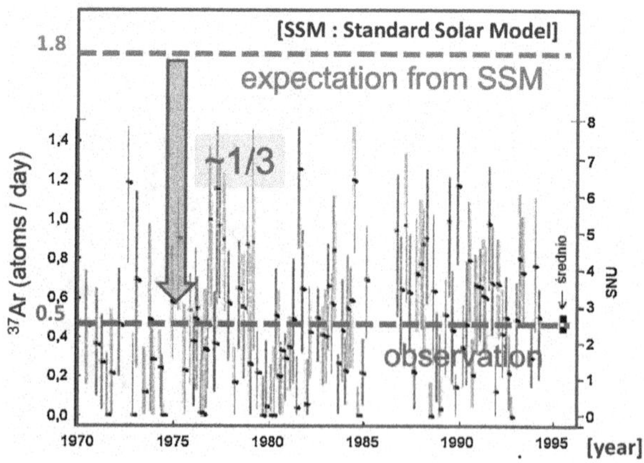

Fig. 5 Homestake data for 1970 to 1995 runs [1]

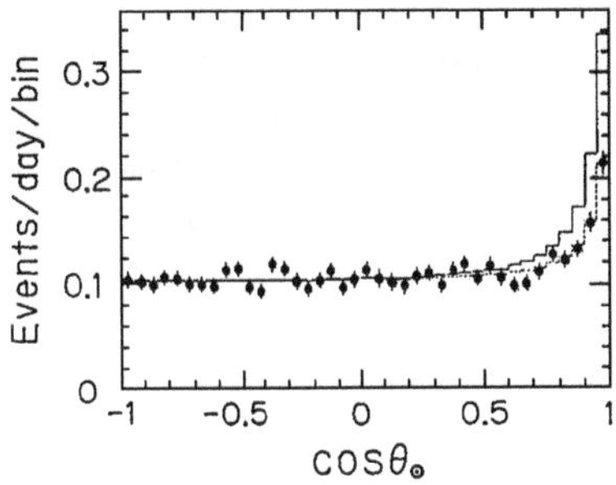

Fig. 6 Angular distribution with respect to the direction of the Sun. The histogram shows the expectation from the SSM

the position of the Sun. Combining the solar position and the reconstructed scattering position gives the direction from the Sun. The point is that this scattering angle is small. Therefore, the scattered electron keeps the direction of the incident neutrino. Kamiokande is sensitive only to the ^{8}B-neutrino. Figure 6 shows the angular distribution of electrons with respect to the direction of the Sun in Kamiokande [2]. A clear excess at $\cos\theta \sim 1$ is observed, indicating neutrinos from the Sun. However, the excess is lower than the SSM expectation shown by the histogram in Fig. 7 [2]. A reduction by about a factor 2 was observed. Thus Kamiokande confirmed that there was a "Solar Neutrino Problem" with a completely independent procedure from the Cl experiment. The observed energy spectrum of the scattered electrons is consistent with the energy spectrum calculated using that of ^{8}B-neutrinos. Kamiokande also showed that there was evidence of ^{8}B production inside the Sun.

Table 1 is the summary of solar neutrino observations up to 1995. The new experiments using the Ga target for neutrinos also showed a reduction by a factor of 2. Where have the solar neutrinos gone? Solutions to the solar neutrino problem were thought over both from the points of view of the SSM problem and the neutrino deficit. The solar neutrino problem was finally solved in the early 2000s, which is discussed below.

4 Supernova, SN1987A neutrinos

Within 2 months after the beginning of the solar neutrino data-taking in Kamiokande, extragalactic messengers swept through the Earth. On February

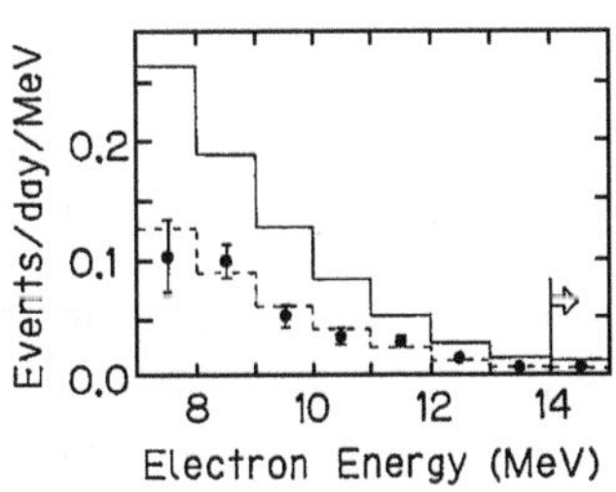

Fig. 7 Energy spectrum of scattered electrons. The histograms show the expectation (*solid*) and one half the expectation (*dashed*) from the SSM and ^{8}B-decay

Table 1 Summary of the ratio of observation to expectation for all solar neutrino experiments up to 1995

Experiment	Solar neutrinos	Data/theory
Homestake (Cl)	$^7Be + ^8B + \cdot$	0.29 ± 0.03
Kamiokande (H_2O)	^8B	0.48 ± 0.08
GALLEX (Ga)	$pp + ^7Be + ^8B + \cdot$	0.60 ± 0.09
SAGE (Ga)	$pp + ^7Be + ^8B + \cdot$	0.52 ± 0.09

23 of 1987, bursts of light and neutrinos reached the earth from the brightest supernova (SN1987A) in the 383 years since Kepler's supernova. Light from the explosion, ~170,000 light years away in the Large Magellanic Cloud (LMC), was visible only in the Southern Hemisphere. On February 25, the presence of hydrogen was observed in the spectrum of SN1987A. This feature indicates that SN1987A is a type-II supernova caused by the gravitational collapse of a massive star. By comparing the supernova's position with older photographs of the LMC, the progenitor star was identified to be a hot blue supergiant star, Sanduleak -69 202. According to the supernova theory, this signifies the birth of a neutron star accompanied by the emission of ~99 % of the released gravitational energy in the form of neutrinos within a few seconds. Thus, the neutrino burst signals the time of core collapse, which is earlier than the time of brightening by 1 to 10 hours, depending on the size of the progenitor.

Kamiokande could easily detect the event around 7:35 UT in the N_{hit} – T*ime* plot as shown in Fig. 8, where N_{hit} is proportional to the energy of the event and T*ime* stands for the time at which the event occurred. A neutrino burst was observed on 23 February 1987, 7:35:35 UT (±1min.) during a time interval of 13 s. The signal consisted of 11 electron events of energy 7 to 36 MeV, of which the first two point to the Large Magellanic Cloud with

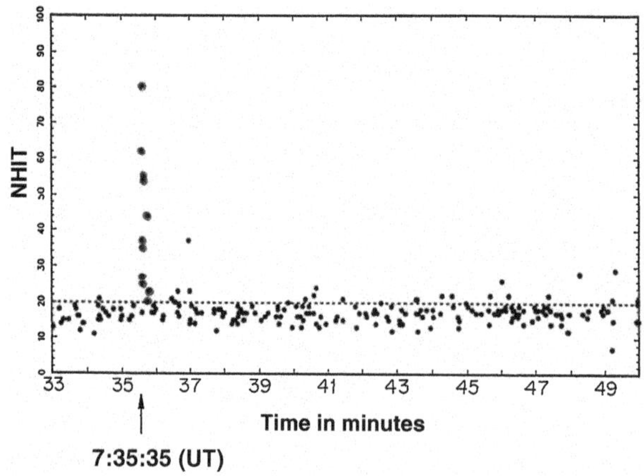

Fig. 8 Scatter plot of N_{hit} against time relative to 7:35:35 UT, February 23, 1987

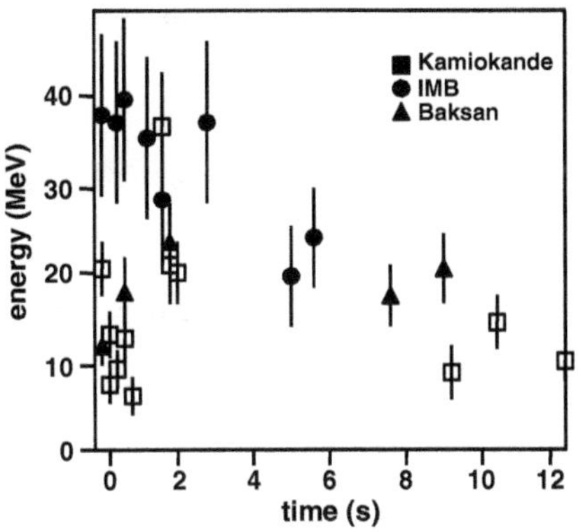

Fig. 9 SN1987A neutrino events observed by Kamiokande, IMB and Baksan showed that the neutrino burst lasted about 13 seconds [6]

angles $18° \pm 18°$ and $15° \pm 27°$. The data over a period of 2.7 days from 21 February to 24 February was searched to determine the statistical significance of the burst at 7:35:35. It is seen that the probability of accidental occurrence of a burst with 9 events per 10 s. is less than 5.7×10^{-8}. For events with $N_{hit} \geq 30$ the probability is less than 5.2×10^{-11}. The burst was therefore significant.

Other underground neutrino detectors (The Irvine-Michigan-Brookhaven collaboration IMB, the Baksan underground scintillator telescope, and the Liquid Scintillator Detector (LSD) in the Mont Blanc underground neutrino observatory also succeeded in observing the neutrino bursts. The time sequence of events detected by Kamiokande[3], IMB[4], and Baksan[5] at 7:35 UT is shown in Figs. 9[6] where the earliest events of the three groups are assigned $t = 0$.

Because of several incidents and misfortunes, Kamiokande had a narrow escape from missing the neutrino burst form SN1987A. We learned much about the difficulty of continuing to take data under good detector conditions which is requisite for observatories. The supernova burst arrived ~2 min after an interval in the data-taking on behalf of gain calibration of the electronics (Fig. 10). Gain calibration of the Kamiokande electronics, generating test pulses into all photomultiplier (PMT) channels, was repeated every 2 h. Each calibration took about 3 min. Based on this experience, the procedure of gain checks was modified by carrying them out channel by channel instead of for all channels simultaneously. Thus there has been no dead time for the gain check at a cost of one rest channel. IMB also narrowly escaped missing its supernova detection. On 23 February, 5:00:00.001 UT, about 2 h before the neutrino burst, high-voltage power supplies tripped off, shutting down a quarter of the detector's 2048 PMT's. This also caused the on-line data-filtering routines to stop functioning. Fortunately, the raw data could be recovered from the partially malfunctioning detector and suffered no serious biases due to the missing PMT's.From the above-mentioned stories, people might say that Kamiokande and IMB have been very lucky. However, according to the Japanese proverb, catching good luck is a matter of ability.

What could we learn from SN1987A ? The number of neutrino events gave a mass for the exploded star of $(8 \sim 15)$ M_{solar}. The burst duration time of

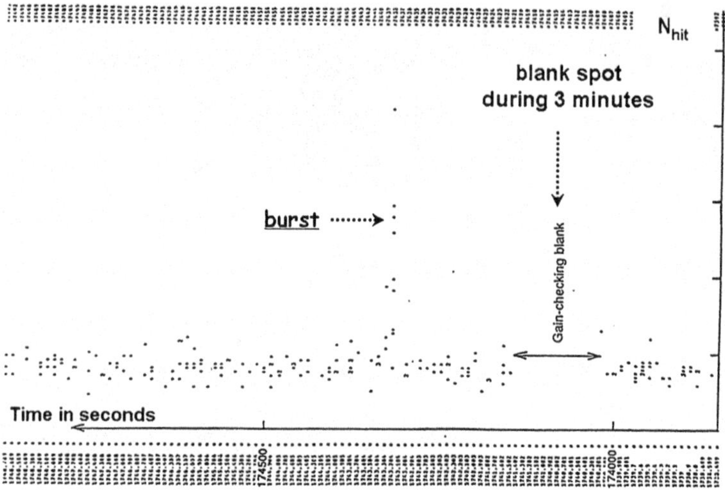

Fig. 10 Computer print-out: the scatter plot of N_{hit} vs. the event detection time [7]

13 s gave indirect evidence of neutron star formation. The total energy of the observed neutrinos predicts that the energy carried out by neutrinos is 10^{53} erg. This is 100 times larger than the energy brought by light. The SN1987A neutrinos verified the star evolution scenario. M. Koshiba and R. Davis received the Nobel Prize in Physics in 2002 for their pioneering contribution to astrophysics, particular for the detection of cosmic neutrinos.

5 Second-generation neutrino detector at Kamioka: Super-Kamiokande

The Kamiokande achievements in detecting neutrinos gave confidence for constructing the next-generation gigantic water Cherenkov detector, Super-Kamiokande (SK). SK is a 50,000 ton imaging water Cherenkov detector. The experiment started to take data on the 1st of April, 1996 after 5 years of construction. Two years later, the first announcement of the discovery of neutrino oscillation was made based on the observation of strong deficits of upward atmospheric ν_μ's [8]. Atmospheric neutrinos are produced when primary cosmic rays, mostly protons and ^{4}He, hit the top of the atmosphere and then produce pions, kaons, protons, neutrons etc. A produced pion decays into μ and ν through $\pi^\pm \rightarrow \mu^\pm + \nu_\mu(\bar{\nu}_\mu)$. A muon subsequently decays into an electron and neutrinos through $\mu^\pm \rightarrow \mathrm{e}^\pm + \nu_\mathrm{e}(\bar{\nu}_\mathrm{e}) + \bar{\nu}_\mu(\nu_\mu)$. These neutrinos, ν_e, $\bar{\nu}_\mathrm{e}$, ν_μ and $\bar{\nu}_\mu$ are called atmospheric neutrinos and can reach large depths. Atmospheric neutrinos are observed via their charged current interactions with nuclei in the detector water. The incoming direction reflects the flight path of neutrinos from the production source to the detector, which ranges from 10 km to 13,000 km. The atmospheric neutrinos have the right characteristics for studying neutrino oscillations which originate from the finite masses of neutrinos. Shown in Fig. 11 are the observed zenith angle distributions of $(\nu_e + \bar{\nu}_e)$ and $(\nu_\mu + \bar{\nu}_\mu)$ candidate events. Here, $\cos\theta = 1$ means neutrinos produced at the top of the atmosphere, $\cos\theta = 0$ at the horizon and $\cos\theta = -1$ in the atmosphere of the opposite hemisphere. The distribution of $(\nu_e + \bar{\nu}_e)$ events is consistent with the prediction (red histogram). However a deficit of $(\nu_\mu + \bar{\nu}_\mu)$ events can be seen in the $\cos\theta < 0$ bin, where the number of counts observed islees than predicted. This discrepancy between the data and the prediction is well explained by $\nu_\mu(\bar{\nu}_\mu) \rightarrow \nu_\tau(\bar{\nu}_\tau)$ oscillation (green histogram). SuperKamiokande announced the discovery of the neutrino oscillation in 1998 based on the observation of the strong deficit of the up-going muon neutrinos in the zenith angle distribution.

The discovery of neutrino oscillations gave the first evidence on th finite mass of neutrinos. The New York Times reported this result in June 1998, commenting "Mass Found in Elusive Particles; Universe May Never Be the Same". The SK project was led by Yoji Totsuka, who passed away on July 10, 2008 at the age of 66. The August 2008 issue of Physics World magazine put as a headline "Japanese particle-physics leader dies". The article stressed Totsuka's importance to neutrino physics, adding that Totsuka could well have won a Nobel Prize had he lived longer.

6 Solving the solar neutrino problem

6.1 Sudbury Neutrino Observatory: SNO

The solar neutrino problem was finally solved in 2002. The Kamiokande result ignited the solar neutrino research fever. Two new experiments using a gallium

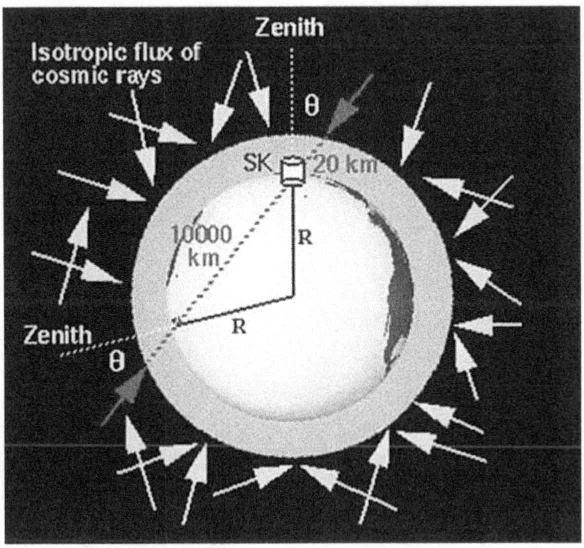

Fig. 11 *Left*: Illustration of the atmospheric neutrino oscillations. *Right*: The zenith angle distribution of $(\nu_\mu + \bar{\nu}_\mu)$(*top*) and $(\nu_e + \bar{\nu}_e)$ (*bottom*) candidate events. The *points* show the data, *red histograms* show the no-oscillation predictions and green histograms show the best-fit expectations for $\nu_\mu(\bar{\nu}_\mu) \rightarrow \nu_\tau(\bar{\nu}_\tau)$ oscillations

(^{71}Ga) target also observed the deficit. The importance of the Gd results is that the p-p neutrinos produced in the most fundamental fusion process are found to be missing. This extremely accurate result was given in 2001 by Super-Kamiokande which confirmed the Kamiokande result. The experiments mentioned above are chiefly sensitive to the ν_e.

The standard solar models provided very good agreement with other properties of the Sun. However, it was not known whether the solar neutrino flux discrepancy was due to limitations in the solar models or to unknown neutrino properties. Measurements that could distinguish between the two possibilities were required. The most favoured explanation, arising from new neutrino physics, was flavor change through the oscillation of massive neutrinos. We had to await a new experiment designed to make separate measurements of the flux of ν_e's from the Sun and of the flux of all active neutrino types (ν_e, ν_μ, ν_τ). The Sudbury Neutrino Observatory (SNO) was just the right detector to carry out experiment that made it possible to come to a definitive conclusion about whether solar neutrinos are changing their types before reaching the Earth by a direct comparison of these two fluxes. SNO uses 1,000 tonnes of heavy water (D_2O) in a Cherenkov detector (Fig. 12) situated 2 km underground in Ontario, Canada. The charged current (CC) reaction : $\nu_e + d \rightarrow p + p + e^-$, is sensitive only to ν_e's. The neutral current (NC) reaction : $\nu_x + d \rightarrow p + n + \nu_x$, is sensitive to all active neutrino flavors (x = e, μ, τ). The elastic scattering (ES) is sensitive to all flavors, but less sensitive to ν_μ and ν_τ.

In June 2002, SNO [9][10] reported a finite non-ν_e neutrino component (3.41 ± 0.45 (stat) $^{+0.48}_{-0.45}$ (syst)) at 5.3 sigma, which provided strong evidence for

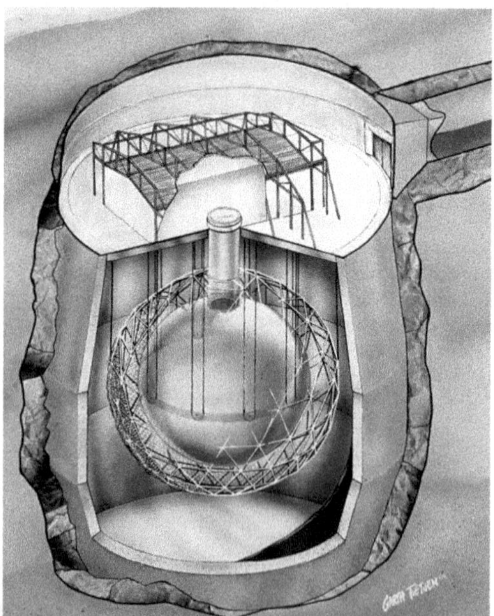

Fig. 12 1,000 ton heavy water Cherenkov detector at Sudbury Neutrino Observatory (SNO)

a flavor transformation of solar ν_e to ν_μ, ν_τ. The fluxes show clearly that about two-thirds of the ν_e's have changed their flavor to other neutrino types. The observed total flux summing up all neutrino types is in excellent agreement with the flux of ^{8}B-neutrinos obtained from the standard solar model. This provides a clear answer to the "Solar Neutrino Problem". The observed fluxes in all solar neutrino experiments are summarized in Table 2.

Although SNO demonstrated that solar neutrinos can change flavor, there were several alternative scenarios for these flavor changes to explain the solar neutrino observations [12]. It is necessary to pin down a unique solution to the solar neutrino problem from among these alternative scenarios. The third-generation neutrino detector at Kamioka, KamLAND was constructed to determine the correct scenario.

6.2 Third-generation neutrino detector at Kamioka: KamLAND

KamLAND is a monolithic liquid scintillator detector containing 1,000 tonnes of ultra pure liquid scintillator. The detector has been located at the old Kamiokande site after the Kamiokande detector was dismantled. In the context of the $\bar{\nu}_\mu \rightarrow \bar{\nu}_x$ oscillations with CPT invariance, electron neutrino (ν_e) oscillations are equivalent to antielectron neutrino ($\bar{\nu}_e$) oscillations. A nuclear reactor is a very intensive source of $\bar{\nu}_e$. There are many commercial nuclear power plants around Kamioka. These reactors, situated at distances ranging from 130–220 km from KamLAND generate 7% of the world total reactor power; 70 Gigawatt (GW)out of 1.1 Terawatt (TW). This situation provides an effective baseline of about 180 km for neutrino oscillation study and gives superior sensitivity to determine the neutrino masses. In addition, the baseline of 180 km and the average energy of reactor neutrinos make it possible to test one of the solar neutrino oscillation scenarios.

In December 2002, KamLAND reported the first result of reactor neutrino deficit with 99.95 % C.L. significance [13]. Improved statistics [14] strengthened the significance of neutrino disappearance to 99.9998 % in 2005 and observed the clear oscillation pattern for the first first time in the world (see Fig. 13). In the context of two-flavor neutrino oscillations with CPT invariance, all solutions to the solar neutrino problem except for the "Large Mixing Angle" solution are excluded. Thus, KamLAND succeeded in pinning down the solution to the solar neutrino problem. The solar neutrino problem

Table 2 Results of solar neutrino experiments

Experiment	Target	Data/theory
Homestake	^{37}Cl	0.30 ± 0.03
Kamiokande	e^- (H_2O)	0.48 ± 0.07
SAGE	^{71}Ga	0.51 ± 0.04
GALLEX+GNO	^{71}Ga	0.53 ± 0.04
Super-Kamiokande	e^- (H_2O)	0.41 ± 0.02
SNO pure D_2O CC	d (D_2O)	0.30 ± 0.03
SNO salt NC	d (D_2O)	0.85 ± 0.07

The observed neutrino flux is normalized by the SSM (Standard Solar Model) prediction

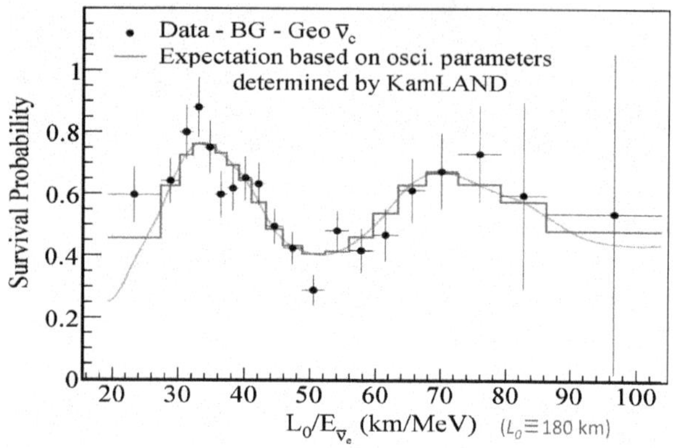

Fig. 13 The ratio of observed signal rate to non-oscillation expectation is plotted as a function of $L_0/E_{\bar{\nu}_e}$. Here, $L_0 \equiv 180$ km. The *curve* shows the expectation from the best-fit reactor neutrino oscillation

has been solved completely. This means the solar neutrino astronomy has just begun.

7 Geoneutrino observation

A basic factor in the interior dynamics of plate tectonics, mantle convection, terrestrial magnetism and the evolution of the present Earth is the radiogenic heat, ~90 % of which comes from the decay of ^{238}U and ^{232}Th. The radiogenic heat is supposed to contribute approximately half of the total measured heat dissipation rate from the Earth which is (44.2±1.0) trillion watts (TW) or (31±1) TW, depending on the analysis procedure. The other half is believed to come from the primordial energy of planetary accretion and latent heat of core solidification. It is well known that the radiogenic heat produced in the interior of the Earth by $\alpha - \beta$ radioactivity may be measurable by detecting

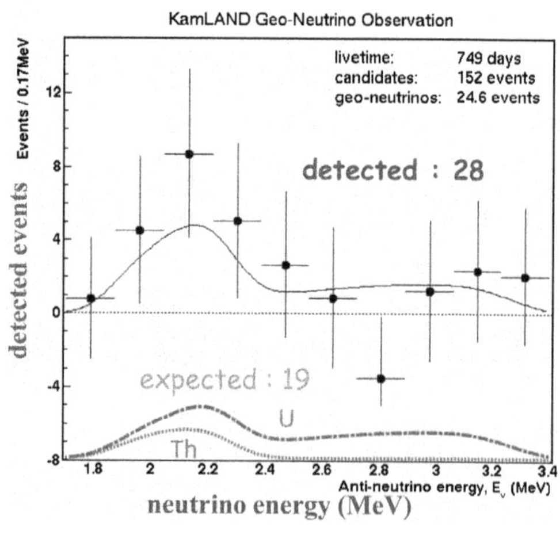

Fig. 14 $\bar{\nu}_e$ energy spectrum of the geoneutrino candidate events subtracted by the total backgrounds

antineutrinos $\bar{\nu}_e$ (geoneutrinos) from β-decay, since essentially all $\bar{\nu}_e$ reach the surface without interaction. ^{238}U and ^{232}Th decays via a series of well established α and β^- processes emitting 6 and 4 $\bar{\nu}_e$'s, respectively. It can be imagined that the Earth emits $\bar{\nu}_e$'s like an antineutrino star. Detecting neutrinos from radioactive elements in the Earth is expected to give direct insight into the deep Earth and is essential to study the above fundamental subjects. Particle physicists have long been aware of the possibility of measuring the heat produced inside the Earth by detecting geoneutrinos. George Gamow already mentioned this possibility to Fred Reines in 1953 when F. Reines just started the neutrino discovery experiment. However, it took the development of large-scale versions of liquid scintillators to make such measurements feasible. The first chance for searching geoneutrinos can be expected to be presented by the massive liquid scintillation detector, KamLAND.

In 2005, KamLAND succeeded in detecting geoneutrinos [15]. Figure 14 is the measured energy spectrum of geoneutrino candidate events. 28 events are detected, while from the Bulk Silicate Earth model 19 events are expected[16]. The observation is roughly consistent with the expectation within the present experimental accuracy. The 99% confidence upper limit of the total ^{238}U and ^{232}Th geoneutrino flux is $1.62 \times 10^7 \text{cm}^{-2}\text{s}^{-1}$ which corresponds to an upper limit on radiogenic power of 60 TW.

Geoneutrinos made their debut in 2005. About 50 years had passed since G. Gamow's suggestion. KamLAND has constantly been accumulating geoneutrino events. More accurate result should appear soon. While the results from KamLAND do not strongly constrain the models of the Earth's interior as yet, this first measurement marks the beginning of the era of being able to see the Earth's interior with neutrinos, much as is done with medical scans of humans (see Fig. 15). An upgraded KamLAND, and future detectors on land and at sea making multiple measurements, will be able to sort out the contributions from the various Earth layers.

Fig. 15 *Left*: Geoneutrino generated points observed by KamLAND (Monte Carlo simulation). The current geophysics assumes that no ^{238}U and ^{232}Th inside the core. *Right*: The Earth structure

References

1. Cleveland, B.T., et al.: Measurement of the solar electron neutrino flux with the homestake chlorine detector. Astrophys. J. **496**, 505 (1998)
2. Hirata, K., et al.: Results from one thousand days of real-time, directional solar-neutrino data. Phys. Rev. Lett. **65**, 1297 (1990)
3. Hirata, K., et al.: Observation of a neutrino burst from the supernova SN1987A. Phys. Rev. Lett. **58**, 1490 (1987)
4. Bionta, R.M., et al.: Observation of a neutrino burst in coincidence with supernova 1987A in the large magellanic cloud. Phys. Rev. Lett. **58**,1494 (1987)
5. Alexeyev, E.N., et al.: On possible detection of a neutrino burst on February 23, 1987 at the BAKSAN underground scintillation telescope. In: Proc. of the Leptonic Session of the 22nd Recontre De Moriond, p. 739 (1987)
6. Nakahata, M.: SN1987A heralds the start of neutrino astronomy. CERN COURIER **47**, 23 (2007)
7. Koshiba, M.: Observational neutrino astrophysics. Phys. Rep. **220**, 229 (1992)
8. Fukuda, Y., et al.: Evidence for oscillation of atmospheric neutrinos by super-kamiokande. Phys. Rev. Lett. **81**, 1562 (1998)
9. Ahmad, Q.R., et al.: Direct evidence for neutrino flavor transformation from neutral-current interactions in the sudbury neutrino observatory. Phys. Rev. Lett. **89**, 011301 (2002)
10. Ahmad, Q.R., et al.: Measurement of day and night neutrino energy spectra at SNO and constraints on neutrino mixing parameters. Phys. Rev. Lett. **89**, 011302 (2002)
11. Smy, M.: Super-kamiokande solar neutrino results. In: Proc. of Int. Conference on Neutrino Physics and Astrophysics, Neutrino (2002)
12. Smirnov, A.: Solar neutrinos: interpretation of results. In: Proc. of Int. Conference on Neutrino Physics and Astrophysics, Neutrino (2002)
13. Eguchi, K., et al.: First results from KamLAND: evidence for reactor antineutrino disappearance. Phys. Rev. Lett. **90**, 021802 (2003)
14. Araki, T., et al.: Measurement of neutrino oscillation with KamLAND: evidence of spectral distortion. Phys. Rev. Lett. **94**, 081801 (2005)
15. Araki, T., et al.: Experimental investigation of geologically produced antineutrinos with KamLAND. Nature **436**, 499 (2005)
16. McDonough, W.F., Sun, S.-s.: The composition of the earth. Chem. Geol. **120** 223 (1995)

Exp Astron (2009) 25:225–240
DOI 10.1007/s10686-009-9156-7

REVIEW ARTICLE

400 years of astrometry: from Tycho Brahe to Hipparcos

Erik Høg

Received: 30 October 2008 / Accepted: 25 February 2009 / Published online: 24 March 2009

Abstract Galileo Galilei's use of the newly invented telescope for astronomical observation resulted immediately in epochal discoveries about the physical nature of celestial bodies, but the advantage for astrometry came much later. The quadrant and sextant were pre-telescopic instruments for measurement of large angles between stars, improved by Tycho Brahe in the years 1570–1590. Fitted with telescopic sights after 1660, such instruments were quite successful, especially in the hands of John Flamsteed. The meridian circle was a new type of astrometric instrument, already invented and used by Ole Rømer in about 1705, but it took a hundred years before it could fully take over. The centuries-long evolution of techniques is reviewed, including the use of photoelectric astrometry and space technology in the first astrometry satellite, Hipparcos, launched by ESA in 1989. Hipparcos made accurate measurement of large angles a million times more efficiently than could be done in about 1950 from the ground, and it will soon be followed by Gaia which is expected to be another one million times more efficient for optical astrometry.

Keyword History of astrometry

1 Introduction

The prospects for astrometry looked bleak at the middle of the twentieth century. If an astrometrist retired the vacancy was usually filled with an astrophysicist, and astrophysics was moving towards the exciting new extragalactic

E. Høg (✉)
Niels Bohr Institute, Juliane Maries Vej 30, 2100 Copenhagen Ø, Denmark
e-mail: erik.hoeg@get2net.dk

astronomy. But I did not feel any pressure from this trend when I studied in Copenhagen (1950–1956) where both my teachers at the observatory, Bengt Strömgren and Peter Naur, were very familiar with astrometry, and it was natural to follow their advice. As a boy, I had read about Tycho Brahe and Ole Rømer, the two Danish heroes in astronomy, who both in fact worked on what is now called astrometry. Astrometric catalogues on the library shelves like the Albany General Catalogue (GC) [2], the AGK2, and the Jenkins catalogue of parallaxes attracted me, though I did not of course know that 40 years later I should lead the construction of the Tycho-2 Catalogue [14]. This catalogue has replaced all previous reference catalogues with its positions and proper motions derived from observations with the Hipparcos satellite and 100 years observations with ground-based telescopes. This presentation is focused on optical astrometry over the past 400 years. More details about the recent history of astrometry are given in [15–17].

The term astrometry does not apply to astronomical measurement in general as the word suggests, but only to the measurement of positions on the sky of stars and other celestial objects. The position of a star changes with time due to its proper motion, to the parallactic motion created by the motion of the Earth around the Sun, and to the orbital motion in the case of a binary star. The radial velocity of a star along the line of sight as measured by the shift of lines in the spectrum is the third component of the space velocity, which is needed for many applications, but it does not belong to astrometry. It should be noted that the radial velocity can affect the proper motion of nearby stars to such extent that accurate measurement of the radial velocity can be obtained from observations of positions over long intervals of time, according to e.g. Lindegren et al. [21]. The term astrometry came into use to distinguish it from astrophysics, especially after the introduction of stellar spectroscopy 150 years ago and of atomic theory later on, which were used to analyse the spectra. For the two millennia prior to that, astrometry had in fact been the main task of astronomy. Astrometric observational data have been the basis for a deep astronomical understanding of stars, star systems, planetary motions, and the underlying physical laws.

I will follow the history of astrometric instruments from the introduction of telescopic sighting and wire micrometers in the seventeenth century, via the transit instrument and the meridian circle in the eighteenth to photographic astrometry in the late nineteenth and photoelectric astrometry in the twentieth century, including satellite astrometry, and finally CCD detectors. Some of the main astronomical results obtained by optical astrometry during the centuries will be outlined.

Today, positions and proper motions are given as celestial coordinates in the main astrometric reference system which is the International Celestial Reference System (ICRS). This system, adopted by the International Astronomical Union, is defined by the radio positions of 212 quasars, supposed to have negligible proper motions. It is represented in the optical by positions and proper motions in the Hipparcos Catalogue, officially called the Hipparcos Celestial Reference Frame (HCRF). ICRS is a coordinate system in right ascension and declination (RA and Dec) very close to the previous adopted system which was called J2000.0. Tycho Brahe used a system of ecliptic longitude and latitude. Later on RA-Dec systems tied to the Earth's rotation axis were used although the change with precession and nutation had to be taken into account. With the adoption of the ICRS for all catalogues the astronomical application of catalogues is simplified since precession and nutation only need be taken into account in connection with the pointing of ground-based telescopes as a computed coordinate transformation. The transformations are computed by means of vectors, no longer by spherical trigonometry as was usual some 30 years ago.

The history of other aspects of astrometry deserves equal appreciation. Such aspects are mathematical methods of data reduction, computing techniques, electronic control of instruments, electronic data acquisition, accurate clocks, and machines for measuring photographic plates, but these could not be included here. Briefly, however, on computing: Tycho Brahe had used a method called *prosthaphaeresis*, which had been invented by the Arab mathematician Ibn Jounis in the eleventh century. It replaced multiplication with the addition of trigonometric functions. Logarithmic tables came into use after they had been introduced by John Napier in 1614 and had been enthusiastically supported by Kepler. In the subsequent period of over 300 years astronomical formulae were developed in logarithmic form to facilitate calculations, and books appeared with logarithms of seven and more decimals of trigonometric and many other functions. The time of logarithms had run out about 1950. We had used logarithms in school, but we were using mechanical and electrical calculators in astronomy in the 1950s. The first electronic computer, an IBM 650 with punched card in- and output, came to Copenhagen in 1954 and I took a programming course. Already 2 years earlier Peter Naur had told of his experience with the electronic computer EDSAC in Cambridge at a lecture in the university. The room was overfull, I was sitting with others on the floor, and even Niels Bohr had come to listen and to wonder at the fantastic punched tapes with rows of holes for numbers, which Naur rolled out on the floor and then gave us freely. I was soon engaged to interpolate in Naur's ephemerides of the minor planet 51 Nemausa computed with EDSAC, using Leslie Comrie's interpolation tables and a mechanical calculator.

2 Tycho Brahe's legacy: instruments and Newton's theory

Tycho Brahe had set a new standard for astronomical measurements by his 20 years of work on Hven. He improved the then classical instruments, the sextant and the quadrant, to give positions of stars with errors some five times smaller than those of his contemporary colleague Wilhelm of Hesse, i.e. about 1 min of arc. Tycho made many improvements and had the means to carry them out thanks to a lavish support by the Danish king, Frederik II. The king wanted to promote science and saw that Tycho Brahe complied with this ambition. Tycho received the island of Hven close to Copenhagen in 1572 and then enjoyed a support equivalent to one or two per cent of the king's annual income, altogether the value of "a barrel of gold". With this basis he could make instrumental improvements and obtain observations in amounts and qualities never seen before.

Tycho had learned from his experience with the large quadrant, which he built in Augsburg 1570, that size alone was not a safe way to better accuracy. The quadrant was not stable, exposed to rain and wind as it stood, and it tipped over in a storm after a few years. Moderate size would lead to more accurate manufacture which was another of Tycho's achievements. For instance, on Hven he developed the sighting device slits-and-plate (Fig. 1) which became the preferred tool for his followers until the telescopic sight could take over in about 1660.

In Danzig, Johannes Hevelius (1611–1687) continued to use the slits-and-plate sighting device until his observatory was destroyed by fire in 1679. His observations, accurate to about 15–20″, could compete easily with any of the early telescopic observations, and were not seriously rivalled until Flamsteed and Rømer had developed their own observing techniques after Hevelius' death [5].

Tycho left a catalogue with positions of 1,000 stars, the same number of stars as in the famous catalogue of Ptolemy. It remained unsurpassed for a century, until Hevelius' catalogue with 1564 stars appeared posthumously in 1690. Tycho's observations of the planets, especially of Mars, came to make the strongest impact on science and civilization. Johannes Kepler derived the three famous laws of planetary motion: about the elliptical form of the orbits, the speed in the orbit, and the size and period of all orbits in the system of planets. The three laws, completed in 1619, found an explanation in Isaac Newton's *Principia* of 1687 in terms of the universal gravitational attraction between masses and the laws of mechanics concerning force, velocity, and acceleration. These laws became the basis for the subsequent theories of celestial mechanics for planets and satellites.

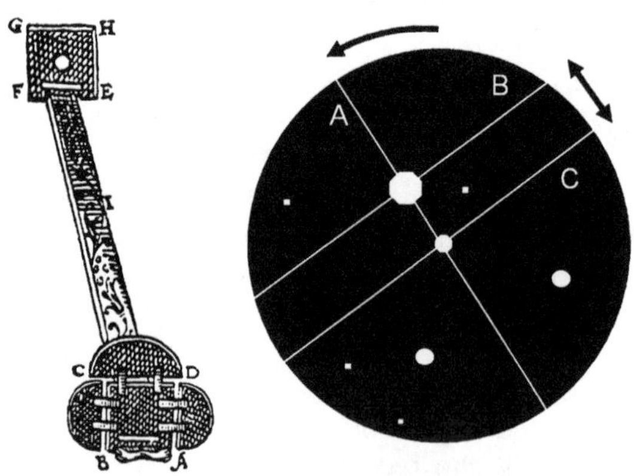

Fig. 1 *Left* Tycho Brahe's sighting device, after 1570, slits-and-plate. *Right* wire micrometer, after 1660. The observer centres one of the stars on the *A–C* cross of fixed wires by moving the telescope. He turns the micrometer to place wire A on both stars, a scale shows the position angle. He moves wire B to the star and obtains the separation

3 Seventeenth century: telescopic sight

The Galilean telescope magnified the object: it increased the angular resolution beyond the several minutes of arc of the unaided eye. This enabled Galilei in 1609–1610 to see mountains on the Moon, satellites around Jupiter, the changing form of Saturn, spots on the Sun, phases of Venus, and to resolve the Milky Way into individual stars. The angular resolution opened up a new view of the physical nature of the heavens, simply by watching what you could see inside the quite small field of view. But this type of telescope was unsuited as a sighting device; it could therefore not be used in classical astrometry concerned with the measurement of large angles between stars.

This problem could be solved with the telescope invented by Kepler in 1611 in which a convex lens is used at the eye, instead of the concave lens in Galilei's telescope. The convex eye-lens is placed with its focus where the star images are formed by the front lens. This allows a cross hair to be placed at the common focus of the objective and the eyepiece lens, thus defining a line and then one has a sighting device. This invention was described in a letter by the English amateur astronomer William Gascoigne in 1641, in which he also tells how the cross hair can be made visible at night by illumination [5]. Gascoigne developed the filar micrometer on the principle shown in Fig. 1, and he made a few observations in 1640 of the angular diameters of Jupiter, Mars, and Venus and of the angular separation of stars in the Pleiades cluster. But Gascoigne's work on the telescopic sight (and on the filar micrometer) fell into neglect after the inventor's death in the Civil War in 1644.

Christiaan Huygens independently invented the filar micrometer in about 1660 and made accurate measurements of the diameters of the Moon and all the planets. The invention was made independently in Italy and by 1675 the telescopic sight with a cross wire was in common use. The micrometer became a standard tool for measuring angles within the field of view during more than 300 years, manufactured in ever more accurate versions. This was the birth of *small-angle astrometry* which soon achieved accuracies of 1 s of arc and even better, while the accuracy of *large-angle astrometry* with quadrants and later on with meridian circles improved more slowly.

The regular swing of a pendulum was known to Galilei. To use it to control a mechanical clock was not simple, but Huygens succeeded in 1658. The pendulum clock in ever better versions was used by astronomers until the quartz clock could take over by 1950. An accurate clock was required in connection with observation of the transit time over the north–south meridian. In practice, the time was recorded when the star crossed one or several wires parallel to the meridian in the telescope field of view. Thanks to the regular rotation of the Earth this measurement corresponds to the right ascension coordinate of the star.

The other coordinate, the declination, could be measured when the telescope was precisely set to let the star follow a wire in the field. Reading the altitude angle of the telescope (cf. Fig. 2) on a finely divided circle was then required. The manufacturing of ever more accurate divided circles was a high

Fig. 2 *Left* the John Bird quadrant of 1773, Greenwich. *Right* the Copenhagen meridian circle from 1859 and a grid used for experiments in 1925 with photoelectric recording of star transits

art [5] and crucial for astrometry up to the 1990s when the reference systems of stars provided in the Hipparcos and Tycho-2 catalogues made the accurate circles virtually obsolete.

Flamsteed in Greenwich used mainly a sextant with two telescopic sights from 1676 to 1690. He then built a mural "quadrant", but of 140° arc so that he could directly observe the Pole Star. This instrument served him for the Great Catalogue and the last recorded observation was made in 1719, shortly before his death.

4 Eighteenth century: quadrants prevail

Astronomers of the new century learned that the Earth's motion around the Sun makes all stars write an ellipse on the sky in 1 year and that the Earth's axis wobbles. They realized that the fixed stars are moving, that the Sun moves among the stars, and that some stars orbit each other. They learned that comets return, and a new planet, Uranus, was found. These discoveries were results of measuring and mapping the stars ever more accurately and in ever larger number, and of recording the motion of the objects in the solar system. The latter observations were studied in celestial mechanics, and the scientists were challenged by the planets, and especially by the intricate motion of the Moon.

In 1718, Edmund Halley had discovered that fixed stars are moving since he found that three bright stars had changed their position since the ancient Greeks. On the same occasion Halley stated that the stars were at least 20,000 or 30,000 times as distant as the Sun. That could be concluded from the lack of positive evidence of a parallax in the most accurate position observations of the time. Already in 1748 James Bradley could extend this lower limit to at least 400,000 times since the failure to measure an annual parallax with his precision instrumentation showed that it must be less than half a second of arc.

Astronomers had been attempting to measure the annual parallax of a star since Copernicus in 1543 had proposed that the Earth orbits the Sun in 1 year.

Bradley was one of those and had a telescope built, a zenith sector, which could measure the position of stars near the zenith. The small angles to be measured and the stable mounting gave a high accuracy about 1 s of arc, though only in the north–south direction. His measurements in 1725 showed a shift, but not in a direction that could be explained by the shift of the Earth relative to the Sun. Bradley gave the true explanation which was that the velocity of the Earth in its orbit causes the direction to a star to be shifted forward in the direction of the Earth velocity. This was a new effect due to the very large, but finite velocity of light which had been discovered in 1675. The effect is called aberration and amounts to 20 s of arc for a given star at some times of the year. This is a large amount, and aberration was taken into account in all subsequent astrometry, resulting in much better accuracy of star positions.

With the higher accuracy Bradley was able to discover a smaller effect; the stars wobble with a period of 18 years and an amplitude of 9 s of arc. In 1748 he explained the effect as a wobble of the Earth rotation axis due to a variation of the Moon's orbit.

The quadrants were still dominating astrometric observations in the eighteenth century, but a different type of astrometric instrument was being developed. Already in 1675 Ole Rømer had introduced the transit instrument, which he set up in his Copenhagen residence in 1691. It consists of a telescope mounted perpendicularly on an axis, and the ends of the axis are placed on pivots in east and west. The telescope is pointed at a star before it crosses the meridian, and the transit time is measured, just as at a quadrant with telescopic sight. In 1704, 30 years after he had the first idea [8], Rømer set up a similar instrument, but now with a divided circle also mounted on the axis. Rømer introduced fixed microscopes to read the circle, thus obtaining the declination coordinate. This was *Rota Meridiana*, the meridian or transit circle (cf. Fig. 2), and it became the most accurate instrument for measuring large angles on the sky up until 1990. But it took a century before the meridian circles took over, for reasons explained in the following section. Rømer was motivated by the search for parallaxes and believed for a while that he had succeeded in finding one.

All Rømer's observations were destroyed in the Copenhagen city fire in 1728, except those from three nights, 20–23 October 1706, called *Triduum*, of which several copies had been made and placed at different locations. The external standard errors of the mean positions were 3.4″ in RA and 4.5″ in declination [23]. Tobias Mayer in Göttingen chose 80 of the stars from *Triduum* for re-observation and derived proper motions by comparing them with Rømer's observations in 1706, with his own and with Lacaille's observations in 1750 and 1756. His aim was to see if there was a systematic motion indicating that the Sun was moving towards any specific part of the sky. In 1760 he concluded that this was not the case. In 1783, however, William Herschel concluded from the motions of only 13 stars that the Sun was moving towards Hercules. Innumerable studies of this phenomenon, the Solar Apex, and of other systematic motions of stars were to follow whenever a new set of proper motions became available. Incidentally, William Herschel used the motions from Mayer, without mentioning Mayer or Rømer, but only that he had the motions from a book by Jerome Lalande (F. Mignard 2008, private communication).

Halley computed parabolic orbits of 24 comets from the preceding three centuries and in 1705 drew attention to the fact that three of them had nearly identical orbits in space and that they were therefore successive appearances of the same comet. He predicted its next return in 1758. In the popular mind comets had always been portents of disaster, while even to astronomers the nature of comets and their role in the cosmic order were still shrouded in mystery. The year came and Halley's Comet reappeared, spectacularly for everybody to see, and a triumph for astronomy, showing that the events of the world are predictable, as many in fact believed.

Bradley left the raw data of a large number of observations. They had been made from 1750 until his death in 1762 with a transit instrument for the right ascensions and a mural quadrant for the declinations. Since Bradley had carefully recorded temperatures and various calibrations the observations were considered to be very valuable and they were printed in full between 1798 and 1805. Then Friedrich Bessel took the reduction in hand and produced a good catalogue. Later in the nineteenth century A. Auwers improved the reduction further, resulting in a catalogue of over 3,000 stars. Brosche and Schwan [3] have shown by direct comparison with the Hipparcos catalogue that the uncertainty in both coordinates for a subset of 2,450 entries is only 1.1″. This catalogue was used to derive proper motions even for the General Catalogue published in 1937 and for FK5 [7]. Much larger star catalogues were published from observations later in the eighteenth century, especially the one of 50,000 stars by Jerome Lalande in *Histoire Céleste Francaise*.

William Herschel wanted to measure parallaxes and thought he could do it by measuring pairs of a bright and a faint star with his wire micrometer. Their separation would change with the time of year since the brightest star was closer to Earth than the faint one, so he believed to begin with. He knew that to measure the small separation in the field of view would be much more accurate than to measure large angles with a quadrant as others were doing. Hence he started to examine all the bright stars attentively with his 7-ft (focal length) telescope, to see whether they had faint companions nearby. On a night in 1781 he noticed a bright star that appeared larger than the others, about 4″ diameter, which he could see because of the high magnification and the excellent quality of his self-made mirror telescope. He suspected it to be a comet, but it was a new planet, later called Uranus. A new planet beyond Saturn was a world sensation. It was followed by astrometric position measurements in the following years, and it was found that its position had already been measured in 1756 and 1690, but it had been taken for a star.

Herschel's survey of companions to the bright stars begun in 1779 turned out to be important in itself. The large number of cases in which bright stars had a close companion, faint or bright, far surpassed what could be expected through chance distribution of stars. This had already been pointed out in 1767 by the Rev. John Michell. Herschel soon thought of these stars as real binary systems. Twenty years later he again measured the relative position of some of his stars and found 50 pairs where the position angle had changed by between 5° and 51°. Thus began double star astrometry by which the masses of stars were to be determined in great number as time allowed the stars to revolve so much that an orbit could be calculated. The first orbit was obtained for the binary *Xi* Ursae Majoris by Felix Savary and published in 1827.

5 Nineteenth century: finally parallaxes and meridian circles

Three men in the previous century, Rømer, Bradley, and Herschel, have been mentioned as motivated by the parallax question, but without the final success. Their efforts had however far-reaching consequences: The transit instrument and the meridian circle were invented; first-epoch positions were obtained in 1706 from which proper motions were derived in such number that the solar apex motion could be discovered in 1783; and double-star astrometry was begun in 1781 leading to the discovery of physical binaries and then orbits.

By 1840 parallaxes had been measured by three men, simultaneously and independently: Friedrich Bessel, Wilhelm Struve and Thomas Henderson, who had all been observing stars with large proper motions since that was now considered to be an indication of nearness. They published credible parallaxes for 61 Cygni, Vega, and α Centauri, respectively. John Herschel said: "It is the greatest and most glorious triumph which practical astronomy has ever witnessed", speaking as President of the Royal Astronomical Society when he awarded the gold medal to Bessel. Bessel received the medal because he had shown the reality of the parallax most convincingly by his analysis.

The instruments used were respectively a heliometer, a wire micrometer, and a mural circle which had reached the necessary perfection through decades of technological developments, not least through a co-operation with the astronomer, who by his demands drove the technician to persevere with improving the instruments. The heliometer was originally developed in 1753 by London's John Dollond, the same man who had marketed the world's first achromatic refractors. Its purpose was to measure the diameter of the Sun, hence its name. It was also called a "divided-lens micrometer" since the objective of the telescope was cut along a diameter into two semi-circles which could be positioned very accurately when they were shifted along the diameter. The observer would see two superposed images of the star field, one from each half, and he could shift the images, thus measuring separations of stars along the diameter. Bessel had used an outstanding example of a heliometer built by Joseph Fraunhofer in Munich. Struve used the largest refractor in the world, also from Fraunhofer. Henderson had only an ordinary mural circle, far less accurate; but his observations afforded a large parallax for α Cen of 0.91″ which has in our time been reduced to 0.742″ from Hipparcos observations.

This initial success spurred further activity, and results for the first dozens of parallaxes allowed some general conclusions to be drawn, e.g. that there is such a large diversity of luminosity among the stars that one can be millions of time brighter than another. But further progress was hampered by systematic errors, as could be seen when results from observations with different instruments by different observers were compared. The introduction of photography towards the end of the century did not bring a significant improvement to begin with, also because of systematic errors.

Measuring the distance to the Sun, needed to convert the stellar parallax measures into kilometres, renders an equally fascinating story as for the stars. Kepler wrote in 1620 that the distance must be at least three times larger than the antique value [29]. Ptolemy's value was 1,160 times the Earth's radius. Kepler's value corresponds to a solar parallax of 1 min of arc, i.e. the "horizontal parallax", the angle subtended by the Earth radius as seen from the Sun. Publication by Flamsteed and Cassini in 1673 gave a value between 9.5″ and 10″, based on observations from a favourable opposition of Mars in the year before. This is approximately correct as we now know, and implies a distance to the Sun 20 times larger than the antique value; what a widening of the cosmos within a century! The following centuries brought a "struggle for the next decimal" by expensive expeditions to observe transits of Venus across the solar disc and by equally expensive observations of the minor planet Eros at opposition, as reported by Pannekoek [24]. The value was given in 1942 as 8.790″ with a claimed accuracy of 0.001, and 30 years later 8.794,18″ with an uncertainty of 0.000,05.

The new century began with a remarkable discovery by Guiseppe Piazzi of Palermo, who was using his alt-azimuth circle to assemble a star catalogue of greater accuracy than any of his predecessors. On New Year's Day 1801 he had measured a star that appeared to have changed its position when he measured again on a subsequent night. It turned out to be an object moving in an orbit between Mars and Jupiter, where a vast empty interval of distances from the Sun often had made astronomers wonder. By 1807 three more objects, Pallas, Juno, and Vesta, had been found in the belt of asteroids, as they were later called. By 1891, more than 300 asteroids had been found, and the pace of discovery then greatly increased with the application of photography.

Uranus and Ceres had been found quite unexpectedly during the study of stars. But in 1846 the large planet Neptune was discovered in an orbit beyond Uranus as a result of calculations based on many observed positions of Uranus. These positions deviated from the ones predicted by the mathematicians from Newton's law of gravity. J.C. Adams and U.J.J. Le Verrier were able to predict that an unknown planet could be responsible, and Neptune was found by J.G. Galle and H.L. d'Arrest in the predicted area of the sky on the first night. The discovery of Neptune was a spectacular and widely publicized victory of mathematical astronomy (e.g. [10]).

Rømer's meridian circle of 1704 remained unique for a century during which mural circles were preferred for the measurement of star positions. But systematic differences between positions from different instruments of more than 10″ were an increasing problem. The history of the meridian circle and how it became the preferred instrument by about 1820 has been studied by Herbst [8] and Chapman [5]. The potential advantage of the full divided circle over the quarter circle had been realised by the astronomers, and the technical ability to manufacture and accurately divide the circle had come with the technical evolution. Achromatic objective lenses could now be made, but many technical issues played a role, and a rethinking of the astronomers took time. A transition period from 1780 to 1820 can be recognised; Piazzi's alt-azimuth of 1789, built in England, was the first instrument with a full circle, but it was in Germany that superlative meridian circles were first manufactured. It began with Repsold in 1802, then Reichenbach and Ertel, and now with full mechanical symmetry about the meridian plane which Rømer's instrument lacked. By 1820 continuous production of meridian circles was going on, and from 1850 the meridian circle had become the main instrument of an astronomical observatory.

Between 1859 and 1863, F.W. Argelander of Bonn published a three-volume catalogue of 325,000 stars, known as the *Bonner Durchmusterung* (BD), and a 40-plate atlas. It was the work of a tiny group of Bonn observers using a small refractor, and this survey was to prove immensely useful for observers for more than a century, but the positions were necessarily inaccurate. In 1867, therefore, Argelander proposed to the Astronomische Gesellschaft that a project be organized to measure, this time with great accuracy, the positions of the BD stars down to the ninth magnitude, and the work should be shared among observatories, each observing a zone of about 5° in declination. The same idea had been expressed 50 years earlier by Bessel (1822), whose assistant Argelander had been at the time. The work started almost immediately, but proceeded slowly in some places, the last results appearing in 1910 for the northern declinations and in 1954 for the southern.

The slow progress with the meridian circles and the new photographic technique (the dry plate was invented in 1871) in 1885 led the director of the Paris Observatory, Admiral E.B. Mouchez, to suggest the possibility of a great photographic star chart, which became the *Astrographic Catalogue* (AC). The long history of this great project includes the happy ending with a catalogue containing over 4.5 million star positions, published as AC 2000.2 by Urban et al. [27] which was used as the major source of old positions to derive the proper motions of the 2.5 million stars in the Tycho-2 Catalogue.

6 Twentieth century: photography, radio astrometry, and a satellite

Photography was a very powerful astrometric technique during most of the twentieth century, ending observationally about 1995 when photographic plates of the required quality were no longer manufactured. But the accumulated plates remain a valuable resource which is being exploited in catalogues with up to 1,000 million stars. Combined also with CCD astrometry at new epochs, accurate proper motions are obtained. The idea of deriving absolute proper motions by differential, small-angle measures with respect to galaxies was initiated by W.H. Wright in 1934 [28] which lead to the NPM and SPM programs [18]. An account of the history up to 2008 of the observations and the resulting catalogues of positions and proper motions is given in [16], including tables of selected catalogues. Present-day catalogues of astronomical and especially astrometric data are listed in [30].

Meridian circles are able to measure large arcs on the sky and were therefore used to provide ever more accurate fundamental catalogues with positions and proper motions for about 1,000 stars. Hertzsprung [9] used proper motions from Auwers' Fundamental-Catalog (1879) [1] as a measure of stellar distances, so-called secular parallaxes, when he discovered the dichotomy of red stars into giants and dwarfs; distances from trigonometric parallaxes were not accurate enough at the time.

Further reference stars were tied to the given fundamental catalogue by meridian circle observations, and by means of these stars the photographic plates could be reduced to a proper celestial reference system. Visual observation of the stars dominated meridian circle work during most of the century. It appears from Table 1 in [16] that the error of a position in an observation catalogue improved from 0.9″ in 1856 to 0.25″ about 1910, and to 0.15″ with the photoelectric observations in the Perth70 Catalogue [12]. The weight of this catalogue of 1976 was ten times higher than that of Küstner's large catalogue of 1908, partly because Perth70 was observed in the many clear nights of Western Australia.

The CMC1–11 [4] catalogues of 1999 with the Carlsberg Meridian Circle on La Palma were observed with a photoelectric slit micrometer similar to the one used for Perth70, but with automatic control of micrometer and telescope, giving a much higher efficiency. Observed in the better seeing on La Palma and during 14 years instead of 5 years for Perth70 the weight of the catalogue is larger by a factor 30.

The accuracy of the CMC1–11 catalogues comes close to the limit set by atmospheric image motion for the measurement of large angles between stars, according to a formula by Høg [11]. Small angles can be measured much more accurately from the ground than large angles, according both to experience and to a formula by Lindegren [20]. The CCD astrometry, explained in [16], gives better accuracy because the stars are referred to the dense set of reference stars in Tycho-2 within the small field of the CCD. The CMC14 catalogue of 2005 was observed with CCDs, resulting in 0.034″ accuracy from only two images of stars brighter than 13th mag. Altogether, the catalogue weight is 500 times higher with CCDs on the same meridian circle on La Palma than with the slit micrometer, in spite of the shorter observing period.

Photoelectric techniques came to revolutionize astrometry with meridian circles, and even more with the Hipparcos satellite by ESA, observing in the years 1989–1993. The evolution of photoelectric astrometry from the experiments by Bengt Strömgren in 1925 (cf. Fig. 2) to the Hipparcos satellite has recently been presented by Høg [15]. The satellite is described in [6], the optical system is shown in Fig. 3, and results are summarized in [16] and in the following section.

Radio astrometry with high precision was introduced in about 1970 by Ryle [26] and others. It has since played an important role through observation of

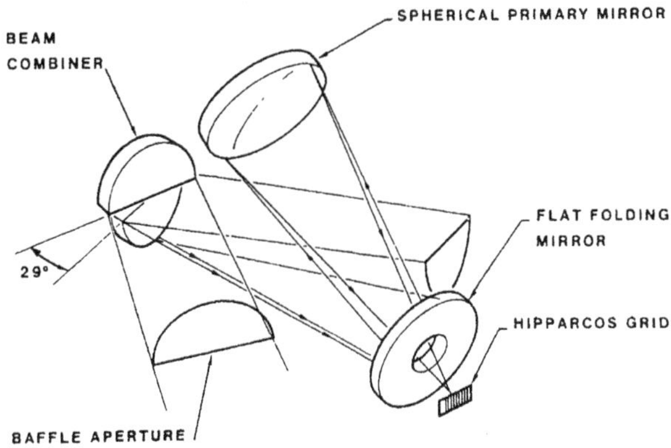

Fig. 3 Hipparcos: Schmidt system with 29 cm diameter aperture, 1.4 m focal length and two viewing directions, all mirrors silver coated for maximum reflectivity. The satellite rotation makes the stars cross the modulating grid and the Tycho star mapper slits

quasars because they are extragalactic and therefore have very small proper motions, if any, i.e. the quasars represent a non-rotating celestial coordinate frame. A non-rotating frame has traditionally, as in the fundamental catalogue FK5, been established dynamically by the requirement that position observations of the Sun and planets must obey Newton's laws. Radio astrometry of the static frame of quasars is ideal for following the complicated rotational movements of the planet Earth. A selected set of 212 quasars defines the International Celestial Reference System (ICRS) to which the stars in the Hipparcos catalogue have been tied [19].

7 Astrometry during 400 years

Errors of star positions and parallaxes in accurate catalogues are shown in Fig. 4. This means the *median external standard error* per star and coordinate in a catalogue, *if available*. In most catalogues the positions of bright stars are more accurate than those of faint ones. The representative median error, dominated by faint stars, is given for most catalogues.

It appears that the Landgrave of Hesse was able to measure positions with errors about 6 min of arc, ten times better than Hipparchus/Ptolemy in Antiquity. A few years after the Landgrave, and thanks to generous support from the king of Denmark, Frederik II, Tycho Brahe reduced the errors by a further factor of 6. The Landgrave and Tycho both wanted to equal Hipparchus by reaching the same number of 1,000 stars. A period of 400 years followed with gradual improvement of the accuracy as astronomers always made use of the best technical possibilities of their time, especially with better time-keeping equipment and accurate manufacturing of mechanics, optics, and with electronics. The accuracy was improved by a factor of about 250 in 400 years, i.e. a factor 4 per century, and the number of stars was greatly increased.

The introduction of space techniques, however, with the Hipparcos mission [6, 22, 25] gave a veritable jump in accuracy by a factor of 100 with respect

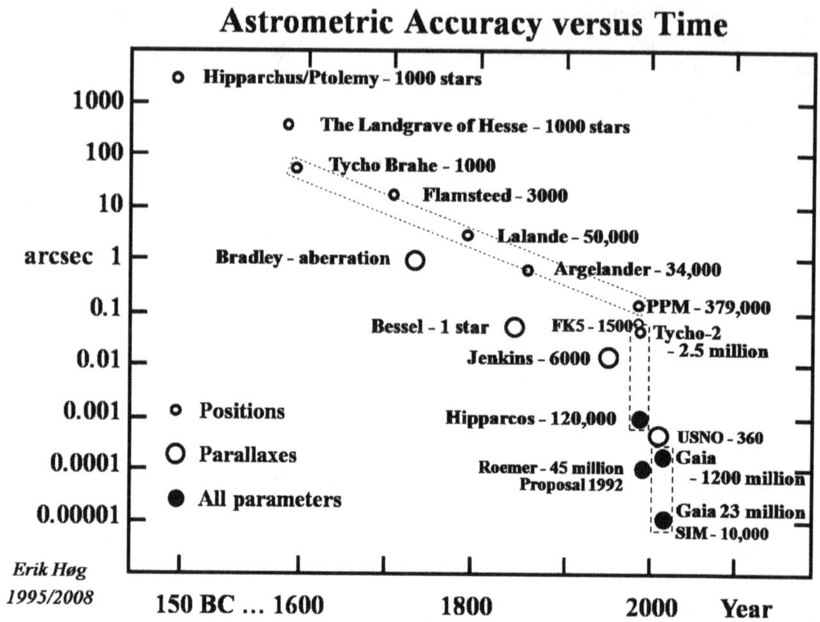

Fig. 4 Astrometric accuracy during the past 2,000 years. The accuracy was greatly improved shortly before 1600 by Tycho Brahe. The following 400 years brought even larger but much more gradual improvement before space techniques with the Hipparcos satellite started a new era of astrometry

to FK5, the most accurate ground-based catalogue ever. Hipparcos obtained a median accuracy of 0.001″ for positions, annual proper motions and parallaxes of 120,000 stars. The positions even in the Tycho-2 Catalogue with 2.5 million stars are as accurate as the positions in FK5 containing only 1,500 bright stars. Tycho-2 includes proper motions, derived from Tycho-2 positions and more than 140 ground-based position catalogues, but no parallaxes. The median standard error for positions of all stars in Tycho-2 is 60 mas, and it is 7 mas for stars brighter than 9 mag. The median error of all proper motions is 2.5 mas/year.

The points marked "parallaxes" might be labelled "small-angle astrometry" or "relative astrometry", and all ground-based measurements of parallaxes are of that kind. This is about ten times more accurate than large-angle astrometry which was required to measure the positions shown in the diagram. The first such point is "Bradley-aberration" shown at 1.0″, the accuracy which Bradley obtained for the constant of aberration with his zenith sector. The accuracy of ground-based parallaxes begins with Bessel's single star in 1838, followed by a factor 100 improvement in accuracy at the U.S. Naval Observatory in Flagstaff since about 1990 for faint stars.

"All parameters" means that about the same accuracy is obtained for annual proper motions, positions and parallaxes, as was in fact achieved with Hipparcos, for the first time in the history of astronomy. The Roemer proposal of 1992 [13] introduced CCDs in integrating scanning mode in a space mission, instead of photoelectric detectors as in Hipparcos. Roemer promised a factor 10 better accuracy than Hipparcos for many more stars, and a development began which led to the Gaia mission due for launch in 2011. For Gaia an improvement by a factor of 100 over Hipparcos is predicted for the 23 million stars brighter than 14 mag, i.e. 10 microarcsec median error. The median accuracy is expected to be 180 microarcsec for the 1,200 million stars in the Gaia catalogue brighter than 20 mag, much better than the accuracy of Hipparcos. The two dots for Gaia thus show the expected accuracy for bright and faint stars. Finally, in view of the expected Gaia results, studies are due about the scientific goals for ground-based optical astrometry after Gaia.

Acknowledgements I am indebted to Adriaan Blaauw for the kind invitation to contribute to the symposium on this vast subject. Without the invitation I would never have engaged myself in this quite large undertaking. Comments to previous versions of the paper from F. Arenou, P. Brosche, A. Chapman, T. Corbin, D.W. Evans, C. Fabricius, F. Mignard, H. Pedersen, P.K. Seidelmann, C. Turon, S.E. Urban, W.F. van Altena, and N. Zacharias are gratefully acknowledged.

References

1. Auwers, A.: Fundamental-Catalog für die Zonen-Beobactungen am nördlichen Himmel. Herausgegeben von A. Auwers. Publication der Astronomischen Gesellschaft XIV (1879)
2. Boss, B.: Albany General Catalogue (GC) of 33342 stars (1937)
3. Brosche, P., Schwan, H.: Der Zodiakalstern-Katalog von Zach und Barry aus den Jahre 1806, Mitteilungen Nr. 44 der Gauss-Gesellschaft e.V. Göttingen (2007)
4. CMC1–11: Carlsberg Meridian Catalogues La Palma 1–11, Copenhagen Univ. Obs., Royal Greenwich Obs., Real Inst. y Obs. de la Armada en San Fernando (1999)
5. Chapman, A.: Dividing the circle: the development of critical angular measurement in astronomy 1500–1850. Contains extensive lists of references (1990)
6. ESA: The Hipparcos and Tycho Catalogues. ESA-SP-1200 (1997)
7. FK5: Fifth fundamental catalogue. In: Fricke, W., Schwan, H., Lederle, T. (eds.) Veröff. Astron. Rechen-Institut, Heidelberg, no. 32 (1988)
8. Herbst, K.-D.: Die Entwicklung des Meridiankreises 1700–1850. GNT Verlag (1996)
9. Hertzsprung, E.: Zur Strahlung der Sterne (I), Zeitsch. für wiss. Photographie, S. 429–442. Reprinted in Ostwalds Klassiker der exakten Naturwissenschaften, Band 255 (1905)
10. Hoskin, M. (ed.): The Cambridge Concise History of Astronomy (1999)
11. Høg, E.: Refraction anomalies: the mean power spectrum of star image motion. Z. Astrophys. **69**, 313 (1968)
12. Høg, E., von der Heide J.: Perth 70—a catalogue of positions of 24,900 stars. In: Abhandlungen aus der Hamburger Sternwarte, vol. IX (1976)
13. Høg, E.: Astrometry and Photometry of 400 Million Stars Brighter than 18 Mag. In: IAU Symp., no. 156, p. 37. Shanghai, 15–19 Sept. 1992 (1993)
14. Høg, E., Fabricius, C., Makarov, V.V., Urban, S.E., Corbin, T.E., Wycoff, G.L., Bastian, U., Schwekendiek, P., Wicenec, A.: The Tycho-2 catalogue of the 2.5 million brightest stars. Astron. Astrophys. **355**, L27–L30 (2000)
15. Høg, E.: Development of photoelectric astrometry including the Hipparcos mission. In: Andersen J., Bland-Hawthorn J., Nordström B. (eds.) Proceedings of IAU Symposium, no. 254, The Galaxy Disk in Cosmological Context. http://www.astro.ku.dk/~erik/Accuracy.pdf (2009)
16. Høg, E.: Selected Astrometric Catalogues. http://www.astro.ku.dk/~erik/AstrometricCats.pdf (2008)

17. Høg, E.: Astrometric accuracy during the past 2000 years. http://www.astro.ku.dk/~erik/Accuracy.pdf (2008)
18. Klemola, A.R., Jones, B.F., Hanson, R.B.: Lick northern proper motion program. I - Goals, organization, and methods. Astron. J. **94**, 501 (1987)
19. Kovalevsky, J., et al.: The HIPPARCOS catalogue as a realisation of the extragalactic reference system. Astron. Astrophys. **323**, 620 (1997)
20. Lindegren, L.: Atmospheric limitations of narrow-field optical astrometry. Astron. Astrophys. **89**, 41–47 (1980)
21. Lindegren, L., Madsen, S., Dravins, D.: Astrometric radial velocities. II. Maximum-likelihood estimation of radial velocities in moving clusters. Astron. Astrophys. **356**, 1119–1135 (2000)
22. Mignard, F., Froeschlé, M.: Global and local bias in the FK5 from the Hipparcos data. Astron. Astrophys. **354**, 732 (2000)
23. Nielsen, A.V.: Ole Rømer and his meridian circle. In: Vistas Astr., vol. 10, p. 105. Oxford (1968)
24. Pannekoek, A.: A History of Astronomy. Translated from a Dutch book of 1951. London (1961)
25. Perryman, M.A.C., et al.: The HIPPARCOS catalogue. Astron. Astrophys. **323**, L49–L52 (1997)
26. Ryle, M.: The 5-km radio telescope at Cambridge. Nature **239**, 435 (1972)
27. Urban, S.E., Corbin, T.E., Wycoff, G.L., Makarov, V.V., Høg, E., Fabricius, C.: The AC 2000.2 catalogue. Bull. Am. Astron. Soc. **33**(4), 1494 (2001)
28. Wright, W.H.: On a proposal to use the extragalactic nebulae in measuring the proper motions of stars, and in evaluating the precessional constant. Proc. Am. Philos. Soc. **94**, 1 (1950)
29. Van Helden, A.: Measuring the Universe: Cosmic Dimensions from Aristarchus to Halley. Chicago & London (1985)
30. Zacharias, N., Gaume, R., Dorland, B., Urban, S.E.: Catalog Information and Recommendations. http://ad.usno.navy.mil/star/starcats_rec.shtml (2004)

Exp Astron (2009) 25:241–259
DOI 10.1007/s10686-009-9145-x

REVIEW ARTICLE

History of astronomical discoveries

Malcolm Longair

Received: 4 November 2008 / Accepted: 20 January 2009 / Published online: 5 February 2009

Abstract The wide diversity of routes to astronomical, astrophysical and cosmological discovery is discussed through a number of historical case studies. Prime ingredients for success include new technology, precision observation, extensive databases, capitalising upon discoveries in cognate disciplines, imagination and luck. Being in the right place at the right time is a huge advantage. The changing perspectives on the essential tools for tackling frontier problems and astronomical advance are discussed.

Keywords Astronomy · Astrophysics · Cosmology · Discovery · History

1 Introduction

The route to astronomical, astrophysical and cosmological discovery is often tortuous, but the essential ingredients are well known.

- New technology, including instrumentation,
- New ways of doing astronomy,
- Precision observation,
- Extensive databases,
- Capitalising upon discoveries in other disciplines,
- Relevant theory,
- Imagination,
- Luck.

M. Longair (✉)
Cavendish Laboratory, University of Cambridge, Cambridge, UK
e-mail: msl1000@cam.ac.uk

It is also an enormous advantage to be in the right place at the right time and to know that one is in that enviable position.

In my teaching of physics and astronomy to students fresh to University, I have presented an idealised overview of what can crudely be described as our 'scientific method'—how we validate our theories by experiments and their interpretation. My favorite example is the route to the discovery of Newton's laws of gravity and motion and the subsequent development of general relativity [15]. The sequence goes as follows:

1. *Experiment or Observation.* Tycho Brahe's observations of the motions of the planets (1575–1595).
2. *Interpretation of the data.* Kepler's laws of Planetary Motion (1605–1619)
3. *Development of Theory.* Newton's law of gravity (1665)—circular planetary orbits.
4. *Comparison of Experiment/Observation with Theory.* Success, but the orbits are ellipses.
5. *Refinement of Theory.* Newton's laws of gravity and motion (1687)—elliptical planetary orbits.
6. *Refinement of Experiment/Observation.* The advance of perihelion of Mercury discovered by Le Verrier (1855).
7. *Return to step 3.* Einstein's General Theory of Relativity (1915).

If only it were that simple!

Once students reach the third or fourth year, one can adopt a more realistic model for the process of discovery. It cannot be better expressed than in the words of Douglas Gough [7]:

> 'I believe that one should never approach a new scientific problem with an unbiased mind. Without prior knowledge of the answer, how is one to know whether one has obtained the right result? But with prior knowledge, on the other hand, one can usually correct one's observations or one's theory until the outcome is correct. However, there are rare occasions on which, no matter how hard one tries, one cannot arrive at the correct result. Once one has exhausted all possibilities for error, one is finally forced to abandon a prejudice, and redefine what one means by 'correct'. So painful is this experience that one does not forget it. The subsequent replacement of an old prejudice by a new one is what constitutes a gain in real knowledge. And that is what we, as scientists, continually pursue.'

In fact, the process of astronomical discovery is much more eclectic than either of these models suggest. The really great discoveries which advance our real knowledge and change our perspectives are rarely without controversy and conflict. Once the breakthrough is achieved, however, the long task of assimilating the new discovery into its rightful place in the astronomical landscape begins. This is just as important as the initial breakthrough and is the route to further discoveries.

At this conference, it is fitting that we should begin with the observations of Galileo, as described in his *Sidereus Nuncius* (The Sidereal Messenger), surely one of the most amazing publications in astronomy—it appeared in March 1610, within weeks of his astronomical observations.[1] By the time of its publication, Kepler had established the first two of his laws of planetary motion, the first describing the elliptical orbits of the planets with the Sun in one focus and the second the areal law, that the radius vector from the Sun to the planet sweeps out equal areas in equal times. These laws were discovered as a result of his brilliant analysis of the observed motions of the planets detailed in the magnificent datasets bequeathed to him by Tycho Brahe. These reinforced his strongly pro-Copernican position.

Galileo's *Sidereus Nuncius* included day-to-day observations of the four bright satellites of Jupiter, Io, Europa, Ganymede and Callisto. Here was a scale model of the Copernican picture of the Solar System. Kepler published his own enthusiastic commentary on Galileo's achievement in his booklet *Conversation with Galileo's Sidereal Messenger*, including his interpretation of the significance of Jupiter's moons:

> 'The conclusion is quite clear. Our moon exists for us on Earth, not for the other globes. Those four little moons exist for Jupiter, not for us. Each planet in turn, together with its occupants, is served by its own satellites. From this line of reasoning we deduce with the highest degree of probability that Jupiter is inhabited.'

What is one to say? Kepler was simply allowing his imagination free rein in the interpretation of startling new observations. I invite readers to come up with contemporary examples.

Kepler was the most distinguished mathematician in Europe and in 1619 published his great *Harmonices Mundi* (The Harmony of the World). This was the Grand Unified Theory of its day, as can be seen from the contents page—geometry, architectonics, harmony, metaphysics, psychology, astrology, astronomy, and so on (Fig. 1). By this time, he had much more accurate values for the radii and periods of the orbits of the planets. Suddenly, when he had more or less completed the writing of his treatise, the crucial third law appeared from nowhere. These astronomical data were crucial in Newton's derivation of the law of gravity. Kepler's third law states that $v \propto r^{-1/2}$ and centripetal balance requires $F \propto v^2/r$. It follows that $F \propto r^{-2}$, Newton's law of gravity in its primitive form. Newton's genius was to show that the same force which caused apples to fall was the same as that which held the planets in their orbits.

The subsequent extraordinary success of Newtonian mechanics and dynamics had a profound effect upon the sciences and all branches of intellectual

[1] I have given many more details of the achievements of Tycho Brahe, Kepler, Galileo and Newton in my book *Theoretical Concepts in Physics* [15].

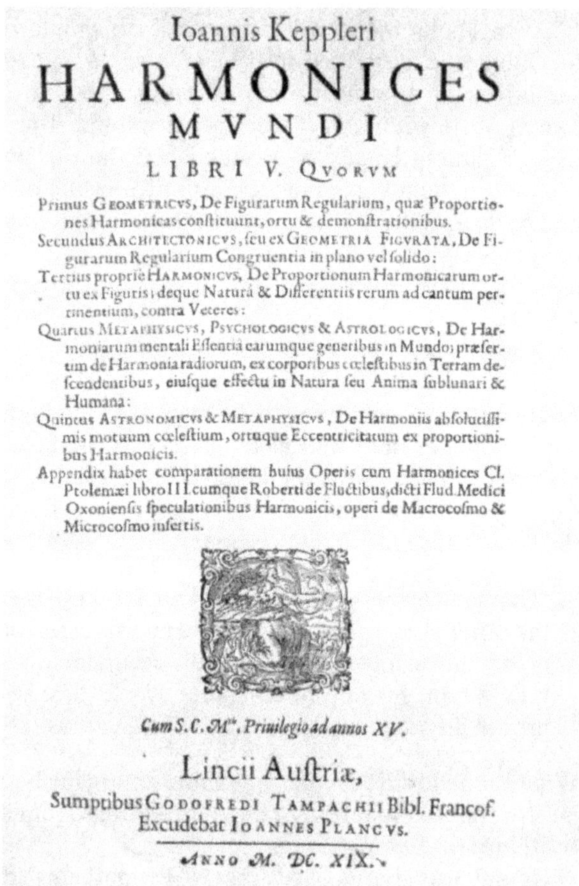

Ioannis Keppleri

HARMONICES
MVNDI
LIBRI V. QVORVM

Primus GEOMETRICVS, De Figurarum Regularium, quæ Proportiones Harmonicas constituunt, ortu & demonstrationibus.
Secundus ARCHITECTONICVS, seu ex GEOMETRIA FIGVRATA, De Figurarum Regularium Congruentia in plano vel solido:
Tertius propriè HARMONICVS, De Proportionum Harmonicarum ortu ex Figuris; deque Naturâ & Differentiis rerum ad cantum pertinentium, contra Veteres:
Quartus METAPHYSICVS, PSYCHOLOGICVS & ASTROLOGICVS, De Harmoniarum mentali Essentiâ earumque generibus in Mundo; præsertim de Harmonia radiorum, ex corporibus cœlestibus in Terram descendentibus, eiusque effectu in Natura seu Anima sublunari & Humana:
Quintus ASTRONOMICVS & METAPHYSICVS, De Harmoniis absolutissimis motuum cœlestium, ortuque Eccentricitatum ex proportionibus Harmonicis.
Appendix habet comparationem huius Operis cum Harmonices Cl. Ptolemæi libro III. cumque Roberti de Fluctibus, dicti Flud. Medici Oxoniensis speculationibus Harmonicis, operi de Macrocosmo & Microcosmo insertis.

Cum S. C. M^tis. Priuilegio ad annos XV.

Lincii Austriæ,
Sumptibus GODOFREDI TAMPACHII Bibl. Francof.
Excudebat IOANNES PLANCVS.
ANNO M. DC. XIX.

Fig. 1 The frontispiece and contents page of Kepler's *Harmonices Mundi*, the Harmony of the World, of 1619

endeavour. It is salutary, however, to note the words of Hermann Bondi on Newton's achievement in deriving the laws of gravity and motion [2].

> 'Newton's solution of the problem of motion in the Solar System was so complete, so total, so stunning, that it was taken as the model of what any decent theory should be like, not just in physics, but in all fields of human endeavour…I regard this as profoundly misleading. In my view, most of science is not like the Solar System but much more like weather-forecasting.'

2 Stellar structure and evolution

The Hertzsprung-Russell diagram is a splendid example of discovery resulting from precision observation.[2] Russell's approach to stellar astrophysics

---

[2]Many more details of the discoveries discussed in the remainder of this essay are included in my book *The Cosmic Century: A History of Astrophysics and Cosmology* [16].

involved the difficult task of obtaining accurate distances for nearby stars. In collaboration with Arthur Hinks of the Cambridge Observatories, he perfected the techniques of determining photographic parallaxes. Accurate photometry and spectral types were provided by Pickering for the 300 stars for which parallaxes were available. These were the data which he used in the famous 'Russell diagram' of 1914 (Fig. 2). Hertzsprung's and Russell's achievements set the scene for the development of the mathematical theory of stellar structure and evolution.

It is intriguing that Russell's diagram included the very faint A star, o-Eridani, with absolute magnitude about +11 (see Fig. 2). The story of its discovery is delightfully told in Russell's memoirs [28]. As recounted above, Russell had suggested to Edward Pickering in 1910 that it would be useful to obtain the spectra of stars for which parallaxes had been measured. According to Russell:

> 'Pickering said 'Well, name one of these stars'. 'Well', said I, 'for example, the faint component of Omicron Eridani'. So Pickering said, 'Well, we make rather a specialty of being able to answer questions like that'. And

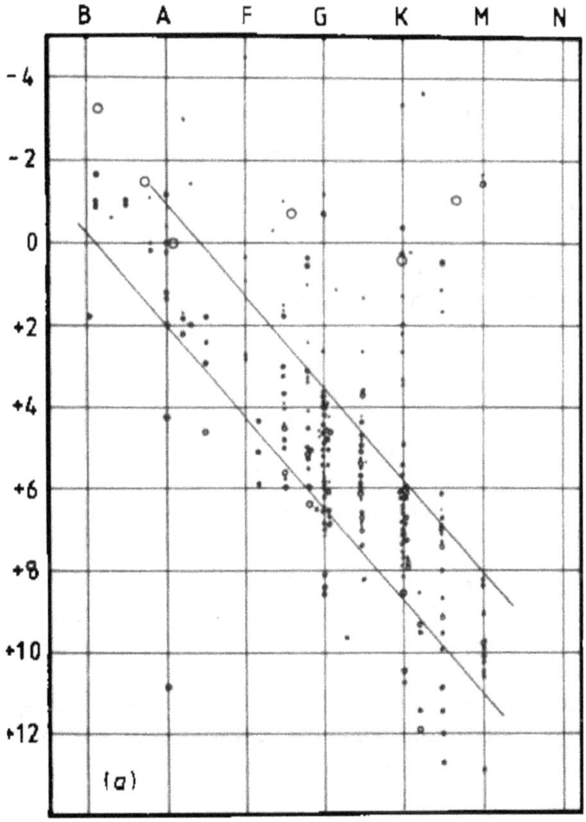

Fig. 2 Russell's original version of the Hertzsprung-Russell diagram of 1914 in which spectral type is plotted along the abscissa and absolute magnitude along the ordinate [25–27]

> so we telephoned down to the office of Mrs. Fleming and Mrs. Fleming said, yes, she'd look it up. In half an hour she came up and said, 'I've got it here, unquestionably spectral type A.' I knew enough, even then, to know what that meant. I was flabbergasted. I was really baffled trying to make out what it meant. Then Pickering thought for a moment and then said with a kindly smile, 'I wouldn't worry. It's just these things which we can't explain that lead to advances in our knowledge.' Well, at that moment, Pickering, Mrs. Fleming and I were the only people in the world who knew of the existence of white dwarfs.'

Even at that time, it was the general view that the sequence of stellar types was in some sense a temperature sequence. Hertzsprung had already applied Planck's law to understand the colours and sizes of the stars and so this 'white dwarf' was a truly remarkable object. The next white dwarf to be discovered was Sirius B by Adams in 1914. In 1926, Eddington used his theory of stellar structure to predict that the gravitational redshift of Sirius B should be 20 km s^{-1}. In the same year, the gravitational redshift of Sirius B was measured by Adams to be 19 km s^{-1} [4]. Eddington was jubilant:

> 'Prof. Adams has killed two birds with one stone; he has carried out a new test of Einstein's theory of general relativity and he has confirmed our suspicion that matter 2000 times denser than platinum is not only possible, but is actually present in the universe.' Douglas [3]

The story reads like a classic example of the combination of the genius of observation and theory. The only problem is that the best estimate of the gravitational redshift of Sirius B from recent HST observations is 80.42 ± 4.83 km s^{-1}. I invite the reader to reflect on this story. With hindsight, we can appreciate that the problem of scattered light from Sirius A may have confused the observations, and the bolometric corrections for a star like Sirius B was not well known. This story gives some substance to what in the 1960s we in Cambridge used to call *Bondi's Dictum* and *Redman's Theorem*:

> *Bondi's dictum* 'All observations are wrong.'
> *Redman's theorem* 'Any competent theorist can reconcile any given set of observations with any given theory.'

I would emphasise that this is the low point of this essay.

Eddington's great insights into the theory of stellar structure and evolution were not gained without considerable controversy. The physical data needed to construct stellar models were not then available. For example, the density and temperature dependence of the energy generation rate and the opacity of the stellar material were not known. Some sweeping approximations were needed to make progress and Eddington had the right qualities of fearless imagination and technical skill to make the problems tractable. For example, in order to simplify the mathematics of his standard stellar model, Eddington made what Leon Mestel refers to as the 'hair-raising approximation' that the radiation pressure is a constant fraction of the total pressure throughout the star. It is no surprise that the subject of the internal constitution of the stars provoked heated debate. In fact, Eddington was fortunate in that the form of the mass-luminosity relation for stars is remarkably independent of the precise processes of energy production and the opacity law. But, in the process of gaining this understanding, there were some fiery exchanges between Jeans, his principal adversary, and Eddington. The tone of the debates is faithfully recorded in the proceedings of the meetings of the Royal Astronomical Society in *The Observatory*. For example, in a letter to *The Observatory*, Jeans writes in 1926

> 'May I conclude by assuring Prof. Eddington it would give me great pleasure if he could remove a long-standing source of friction between us by abstaining in future from making wild attacks on my work which he cannot substantiate, and by making the usual acknowledgements whenever he finds that my previous work is of use to him? I attach all the more importance to the second part of the request because I find that some of the most fruitful ideas which I have introduced into astronomical physics—for example, the annihilation of matter as the source of stellar energy, and highly dissociated atoms and free electrons as the substance of the stars—are by now fairly generally attributed to Prof. Eddington.' Jeans [12]

3 The expanding universe

The velocity-distance relation for galaxies was the first of the great cosmological discoveries which were to provide the observational foundation of the present standard cosmological models. Expanding solutions of Einstein's field equations were implicit in Friedman's models of 1922 and 1924, although this did not attract any significant attention at the time [5, 6]. Friedman's solutions for uniformly expanding Universes were independently discovered by Lemaître in 1927 who derived the 'apparent Doppler effect' according to which the recession velocities of extragalactic nebulae were a cosmical effect of the expansion of the Universe. In 1928, Robertson found $v = cl/R$, where l is distance, and from the recession velocities of nearby galaxies he found the equivalent of a Hubble constant of 500 km s^{-1} Mpc^{-1}. These studies predated Hubble's paper on the velocity–distance relation for galaxies of 1929 [10].

The fact that most galaxies have recession rather than random velocities with respect to our own Galaxy was already known from the pioneering efforts of Vesto Slipher. He realised that it was not the size of the telescope which was important in measuring the spectra of faint diffuse objects such as galaxies, but the f-ratio of the spectrograph.

In 1921, Carl Wilhelm Wirtz searched for correlations between the velocities of spiral galaxies and other observable properties and concluded that, when the data were averaged in a suitable way,

> '... an approximate linear dependence of velocity upon apparent magnitude is visible. This dependence is in the sense that the nearby nebulae tend to approach our galaxy whereas the distant ones move away ... The dependence of the magnitudes indicates that the spiral nebulae nearest to us have a lower outward velocity than the distant ones.' Wirtz [32]

It seems undeniable that this is the first enunciation of what became known as Hubble's law.

After Hubble had established beyond any doubt the extragalactic nature of the spiral nebulae in 1925, he published in the following year a wonderful paper in which he set out the basic types of galaxies, their masses and the contribution they made to the mass density of the Universe [9]. He was fully aware of the importance of his estimate of the mean mass density of the Universe since, according to Einstein's static model of the Universe of 1917, once this was prescribed, all the other properties of the Universe followed immediately.

By 1929, he had assembled data on the distances of galaxies for which velocities had been measured (Fig. 3). The distances of the nearest seven objects, all within 500 kpc, were the best determined and used the Cepheid variable technique; the distances of the next 13 objects were found by assuming that the most luminous stars in the galaxies had an upper limit of absolute magnitude $M = -6.3$; the last four objects, believed to be members of the Virgo cluster, had distances assigned on the basis of the mean luminosities of the nebulae in the cluster. As Hubble acknowledged, most of the velocities used in his 1929 paper were due to Slipher—of the 44 galaxy redshifts known in 1925, 39 of them had been measured by Slipher. From these meagre data, the velocity-distance relation shown in Fig. 3 was derived. It is remarkable that he found the relation from such a nearby sample of galaxies, given that we now know that the random velocities of galaxies are of the order of 300 km s^{-1}. He had, however, other evidence even at that time. In his brief paper, Hubble

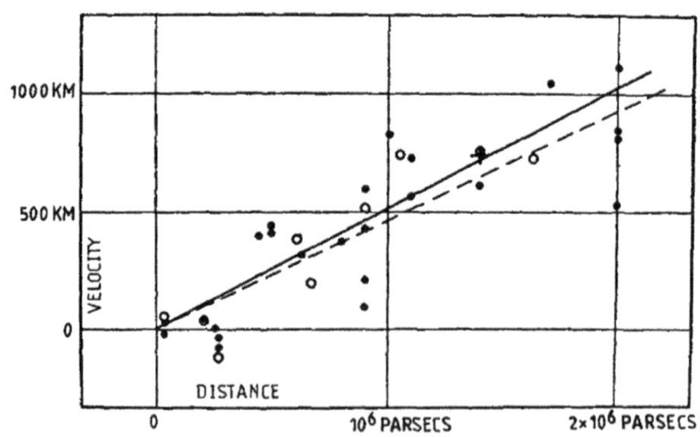

Fig. 3 Hubble's original version of the velocity–distance relation of 1929 [10]

noted that Milton Humason had measured a velocity of 3,910 km s^{-1} for the galaxy NGC 7619, consistent with a extrapolation of the velocity-distance relation to greater distances.

4 The impact of the second world war

The Second World War had an enormous impact upon the whole of science. For astronomy, the most important legacy was the opening up of the electromagnetic spectrum for astronomical observation. Major advances in radio techniques had been driven by the need to enhance radar location, and the V2 rockets developed by the German scientists and engineers opened up the possibility of observations above the Earth's atmosphere. Scientific computing was on the horizon. Just as important were the major investments made in basic science, largely sponsored by the US military, recognising the enormous contributions which physics had played during the Second World War.

Many of the great astronomical discoveries of the subsequent decades can be traced to one or more of these wartime developments. But there was more to the new perspective than simply advanced technology. The scientists themselves had become used to carrying out huge programmes under enormous pressure on a scale far greater than in the pre-war years. As Bernard Lovell expressed it in 1987, the astronomers adopted an approach ...

> '...utterly different from that deriving from the pre-war environment. The involvement with massive operations had conditioned them to think and behave in ways which would have shocked the pre-war university administrators. All these facts were critical in the large-scale development of astronomy.' Lovell [17]

Astronomy was about to embark upon a course which would inevitably lead to it becoming one of the Big Sciences.

4.1 Radio astronomy and the origin of high energy astrophysics

Immediately after the War, a number of University Groups began to investigate the nature of the cosmic radio emission which had been discovered by Jansky in 1934. This radiation had been the subject of study by the UK Telecommunications Research Establishment during the War because the extraterrestrial noise could confuse radar signals. In the immediate post-War era, the principal groups involved in these studies were at Cambridge, Manchester and Sydney. The observations, which were carried out at long radio wavelengths, soon established the existence of populations of galactic and extragalactic radio sources.

By the early 1950s, the Australian radio astronomers had established that some of the extragalactic radio sources were associated with rather strange massive galaxies. For example, the source Virgo A was associated with the massive elliptical galaxy M87 which possessed a remarkable optical jet and

Centaurus A with the nearby massive 'colliding galaxy' NGC 5128. Thanks to an accurate radio interferometric position obtained by Ryle and Graham Smith, Baade and Minkowski were able to associate the brightest extragalactic radio source in the northern sky, Cygnus A, with a very distant galaxy [1]. In 1953, Jennison and Das Gupta at Jodrell Bank showed that Cygnus A had a double radio structure, what are now termed radio lobes, extending far beyond the confines of the associated massive galaxy [13]. The potential for new approaches to astrophysics and cosmology was apparent. By the mid 1950s, the nature of the radio emission was established as synchrotron radiation and so, to account for the properties of the radio emission, two major new components of the Universe were required—huge fluxes of ultrarelativistic electrons and large-scale magnetic fields. Both had to be ejected well beyond the confines of the parent galaxy.

During the 1950s, as the positional accuracy of the radio sources improved, several of the brightest radio sources were identified with bright giant elliptical or cD galaxies. These studies led to the discovery of the quasi-stellar radio sources 3C 48, 196, 286 by Matthews and Sandage in the early 1960s. The optical images of these objects were star-like but had spectra unlike those of any known type of star. The breakthrough came with the discovery of the quasar 3C 273. In 1962, Hazard's radio lunar occultation of 3C 273 led to an accurate determination of its position and radio structure, which in turn led to its identification as a quasar. The optical spectrum was obtained by Schmidt in 1963. The spectrum was again unrecognisable until he realised that it contained the characteristic lines of the Balmer series of hydrogen in emission, but with a redshift of 0.17 [29]. Once it was established that the quasars had large redshifts, the floodgates opened and many quasars were identified in the subsequent years, including the radio quiet variety in 1965. By the same year, the redshifts of the quasars had extended to values greater than 2.

In many ways, the quasars were discovered too early. Although Carl Seyfert had make pioneering spectroscopic observations of the compact, blue nuclei of a number of spiral galaxies, what we would now call Seyfert galaxies, these observations, published in 1943, were largely ignored [30]. The extreme luminosities and variability of the quasars were unprecedented in extragalactic astrophysics. It was quickly recognised that relativistic astrophysics must be involved in a central way in understanding their properties. At the closing dinner of the 1963 Texas Symposium, Thomas Gold summarised these new opportunities as follows:

> 'Everyone is pleased: the relativists who feel they are being appreciated, who are suddenly experts in a field which they hardly knew existed; the astrophysicists for having enlarged their domain, their empire by the annexation of another subject—general relativity.'

The remarkable upsurge of interest in general relativity in subsequent years was undoubtedly stimulated by the discovery of radio galaxies and quasars.

4.2 The discovery of pulsars

In March 1965, Antony Hewish was awarded a grant of £17,286 by the Department of Scientific and Industrial Research to construct a large radio array to study the 'twinkling' of radio sources with the joint objectives of discovering quasars, which were strongly scintillating sources, and studying the properties of the distorting screen, the interplanetary medium. To obtain adequate sensitivity at the low observing frequency of 81.5 MHz, the array had to be very large, $4\frac{1}{2}$ acres (1.8 ha) in area, in order to record the rapidly fluctuating intensities of bright radio sources on a time-scale of one tenth of a second. This was the key technological development which led to the discovery of radio pulsars.

A pulsating radio source was discovered during the commissioning observations in August 1967 by Hewish's graduate student Jocelyn Bell. It disappeared, but was observed with greater time resolution in November 1967—the pulses, once every 1.33 s, were individually resolved [8]. This was a totally unexpected discovery, although Franco Pacini had made the prescient suggestion that neutron stars might be observable as oblique rotators in 1967 [19]. Within months of their discovery, the parent bodies of the pulsars were identified with magnetised, rotating neutron stars, their masses being typically about 1.4 $M_{\odot}$, their radii 10 km and rotation periods 1 s.

The existence of neutron stars had been predicted by Baade and Zwicky in 1934, just two years after the discovery of the neutron. In the preface to his remarkable volume, *Catalogue of Selected Compact Galaxies and of Post-eruptive Galaxies* of 1968, Zwicky described how these ideas were received:

> 'In the Los Angeles Times of January 19 1934, there appeared an insert in one of the comic strips, entitled *Be Scientific with Ol'Doc Dabble*, quoting me as having stated
>
> 'Cosmic rays are caused by exploding stars which burn with a fire equal to 100 million stars and then shrivel from 1/2 million miles diameter to little spheres 14 miles thick',
>
> says Prof. Fritz Zwicky, Swiss Physicist. This, in all modesty, I claim to be one of the most concise triple predictions ever made in science. More than thirty years were to pass before the statement was proved to be true in every respect.' Zwicky [34]

The neutron stars were one of the key discoveries of modern astronomy. In these stars, general relativistic corrections are no longer small, but are central to understanding their stability. Relativistic stars really existed and their discovery had implications for understanding all types of explosive events, including the violent events occurring in active galaxies. At the Brighton General Assembly of the IAU in 1970, Iosef Shklovsky asked me to introduce him to Jocelyn Bell. He said, 'Miss Bell, you have made the greatest astronomical discovery of the 20th century'.

Once the initial discovery was made, the long process of discovering large samples of pulsars began and neutron stars rapidly became part of the furniture of modern astrophysics. As a result of the Arecibo surveys, the binary pulsar PSR 1913+16 was discovered by Hulse and Taylor in 1974 [11]. The existence of such systems initially came as a surprise to the astrophysical community, but their value as probes of relativistic theories of gravity was immediately appreciated. The more recent discovery of the short-period double pulsar PSR J0737-3039A and B promises even tighter constraints on relativistic theories of gravity.

4.3 X-ray astronomy

The early history of X-ray astronomy has been splendidly presented at this conference by Riccardo Giacconi. As in the case of radio astronomy, these unexpected discoveries revitalised many astrophysical disciplines and inaugurated others. From a global perspective, I would highlight the astrophysical significance of establishing accretion as an energy source for active systems, of providing a means for discovering solar mass black holes and of using hot gas as a tracer of the mass distribution in the Universe.

This bald statement disguises the problems of interpretation which the pioneers of X-ray astronomy faced. From Giacconi's famous rocket flight of 1962 to the launch of the UHURU satellite in December 1970, the picture was confused, largely because rocket flights only allowed about 5 min observation above the Earth's atmosphere. With the launch of the UHURU satellite, many of the strange results fell into place. For example, some of the most intense Galactic X-ray sources were associated with compact objects in binary stellar systems. The pulsating X-ray sources Centaurus X-3 and Hercules X-1 were of special importance, because, not only was the X-ray emission pulsed with pulse periods similar to those of the radio pulsars, but the orbital motions of the X-ray sources and their occultations by the primary star were observed. Combined with optical observations of the primary star, the masses of the X-ray sources could be determined. The pulsating X-ray sources were convincingly associated with systems in which matter is accreted onto neutron stars. In addition, a few of the binary X-ray sources did not display pulsations but rather extreme short time-scale variability. In a number of these cases, the masses of the X-ray emitters were found to be greater than three times the mass of the Sun, indicating that the parent bodies of the X-ray sources were solar mass black holes. A recent compilation of the masses of X-ray sources in binary systems is shown in Fig. 4. These sources provide laboratories for the study of the behaviour of matter in strong gravitational fields.

The first evidence for the presence of dark matter in clusters of galaxies was discovered in 1933 by Zwicky who made the first dynamical estimates of the masses of clusters of galaxies [33]. He found a mass-to-light ratio of 500 for the cluster as a whole, very much greater than the values of about 3 found in the visible parts of galaxies such as our own. All subsequent studies have confirmed Zwicky's key result. As discussed by Giacconi, the discovery of the diffuse

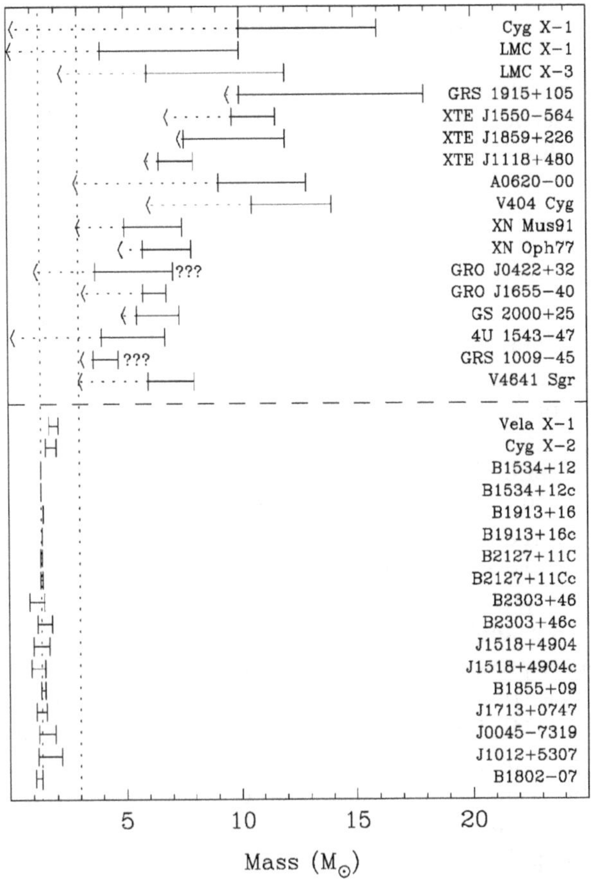

Fig. 4 A recent compilation of the masses of the parent bodies of X-ray sources in binary systems by Jerome Orosz (see http://mintaka.sdsu.edu/faculty/orosz/web/). The objects above the horizontal dashed line are black hole candidates, while those below the dashed line are neutron stars

X-ray emission of clusters of galaxies led to the realisation that the hot gas acts as a tracer of the underlying gravitational potential in which the galaxies move under the influence of the dominant dark matter. This in turn provides important constraints on the origin and evolution of large scale structures in the Universe.

The rapid variability of the X-ray emission from active galaxies indicated that the radiation must originate on very small scales. Just how small this region must be has been indicated by the observation of the line shapes of the 6.4 keV iron fluorescence line in the X-ray spectrum of the Seyfert galaxy MCG-6-30-15 which was first observed by the ASCA X-ray observatory [31] and subsequently by the XMM-Newton X-ray observatory. The line is highly asymmetric and broadened, the broadening being so great that relativistic phenomena must be involved. The most promising interpretation is that the X-ray iron fluorescence line is associated with matter in the accretion disc about a supermassive black hole, the largest redshifts being associated with matter close to the last stable orbit about the black hole.

5 Black holes—the story continued

No better example exists of the importance of new technology for astronomical discovery than the continued efforts to understand the astrophysics of black holes. As described by Reinhard Genzel in this volume, the combination of infrared array technology and adaptive optics enabled him and his colleagues to map out the motions of infrared stars orbiting the black hole in the nucleus of our own Galaxy with the Very Large Telescope in Chile. As he explained, these superb observations have enabled many key questions of black hole physics to be addressed observationally. It is unambiguous that there is a black hole in the centre of our Galaxy with mass $\approx 2.6 \times 10^6 M_\odot$. One of the infrared stars passes within 17 light-hours of the black hole at its distance of closest approach.

Another unexpected surprise dates from the early days of VLBI when superluminal motions were observed in the radio structure of the core of the bright radio quasar 3C273—a radio component was observed to move out from the radio core at about ten times the speed of light [20]. The conventional picture is that this phenomenon is an optical illusion associated with the fact that jets of radio emitting material travel out from the core at speeds $v \sim c$ at an angle close to the line of sight. This preferred model was developed in a prescient series of papers by Martin Rees in the 1960s in the context of accounting for the extreme optical variability of the emission of quasars [23, 24]. Unexpectedly, Mirabel and Rodrigues discovered galactic counterparts of superluminal sources among X-ray sources containing stellar mass black holes [18]. These microquasars appear to have all the properties of the extragalactic variety but scaled down by a factor of about 1,000,000. This has the great advantage that, since many of the properties of active nuclei scale proportionally to the mass of the black hole, their evolution can be studied over a periods of weeks and months, rather than the time-scales of millions of years which would be required for the extragalactic black holes.

6 γ-ray astronomy

The history of discovery in γ-ray astronomy has been splendidly told by Gehrels in this volume and need not be repeated here. Undoubtedly, the most surprising discovery was that of γ-ray bursts, first noted in 1967 by the Vela surveillance satellites but only released publicly in 1973, by which time many of their properties were well characterised [14]. It proved to be a real challenge to determine their nature because the bursts only lasted from less than a second to several minutes. Observations of the γ-ray bursts over the lifetime of the Compton γ-ray observatory showed that they are uniformly distributed over the sky. The key observations which enabled their nature to be established were the afterglows at longer wavelengths which gave the ground-based observers time to turn their telescopes to the location of the burst. As a result, their extragalactic nature was convincingly established. To account for their extreme luminosities and the short timescales of their variability, both collimation of the γ-ray emission and relativistic beaming are required. These sources also have cosmological importance since they have been observed at redshifts greater than 6 and so can act as luminous cosmological probes of the reionisation and, potentially, pre-reionisation eras.

The spectacular discoveries of ground-based ultra-high energy γ-ray astronomy using the atmospheric Cerenkov technique are described by Heinz Völk in this volume. These have their origin in the development of array detectors combined with the stereoscopic capabilities of telescope arrays such as the High Energy Stereoscopic System (HESS) in Namibia. The spectacular maps of ultra high-energy γ-rays from supernova remnants constitute a wonderful discovery, providing direct evidence for the acceleration of cosmic ray protons in the shells of supernova remnants. Another key discovery has been the detection of the extreme variable sources Markarian 421 and 501 as ultra-high energy γ-ray sources. The understanding of the astrophysics of these sources involves a huge range of high energy astrophysical processes, including relativistic beaming of the high energy γ-ray emission.

The upshot of these many unexpected discoveries is that extreme high energy events must necessarily involve relativistically beamed emissions. The relativistic effects are not small corrections but are large and involve the motion of relativistic material in bulk moving at extreme relativistic speeds.

7 Optical and infrared astronomy

Optical and infrared astronomy have been revolutionised over the last thirty years in the approaches both to the construction of the telescopes and to their instrumentation. The history of the telescope has been dealt with extensively by many of the speakers at this meeting and the prospects for even larger telescopes in the medium-term future look distinctly promising, provided they are affordable. In parallel with the revolution in telescope design, the great innovation for optical and infrared astronomy has been the replacement of the photographic plate by digital detectors such as Charge-Coupled Devices (CCDs) and Infrared Arrays. Credit for the discovery of CCDs goes to Willard Boyle and George Smith of the Bell Telephone Laboratories who demonstrated the imaging capabilities of their patented CCD camera in 1974 (Fig. 5). Their objective was to develop a 'Picturephone' which would enable telephone callers to see each other. In my view, one of the most daring decisions of the Hubble Space Telescope programme in 1977 was to use CCD detectors in the Wide Field–Planetary Camera. At that stage, the prototype detectors were only beginning to be used on large ground-based telescopes. The decision more than paid off in the superb quality of the images obtained once the telescope optics had been corrected and full operations began.

It is salutary to look at what we said the HST would achieve in 1977 with what has actually been achieved. There is no question but that the HST has far exceeded our most optimistic expectations, both in terms of the quality and

Fig. 5 Willard S. Boyle (*left*) and George E. Smith, the inventors of the charge-coupled device (CCD), demonstrating the imaging capabilities of their patented CCD camera in 1974. (Courtesy of the AT&T Laboratories)

quantity of superb data which has been obtained. One of my favorite examples is the case of the Orion Nebula, the region of formation of massive stars closest to the Earth. This was observed by Robert O'Dell and resulted in the detection of protoplanetary discs about low-mass stars through the silhouettes which they cast on the bright background emission of the nebula. No-one had thought about this way of observing such discs until the observations themselves showed that this indeed occurs in nature. The fact that the discs typically have angular sizes of about 1 arcsecond meant that they could not be reliably observed with the ground-based technology available at that time.

The other major lesson of the Hubble Space Telescope programme is the enormous impact the images have had upon the public imagination. In the long term, I am certain this is as important as any scientific breakthrough for the long-term health of observational astrophysics. Ultimately, it is the tax-payer who will be supporting our ability to continue the process of astronomical discovery.

8 The discovery of the cosmic microwave background radiation

No discussion of discoveries in astronomy is complete without mention of the Cosmic Microwave Background Radiation by Penzias and Wilson in 1965 [22]. This salutary tale of prediction, neglect and discovery in modern astronomy is too well-known to require detailed discussion here, except to note the forthcoming book edited by Peebles, Page and Partridge entitled *Finding the Big Bang*, in which many of those involved describe their versions of that history [21].

I simply add one footnote to that story. The discovery was made with the Bell Laboratories 20-foot horn antennae which was designed for satellite communication at centimetre wavelengths. Penzias and Wilson had access to a 7.3 cm cooled maser receiver, with which they planned to undertake radio astronomical observations. The persistent background radiation which they

Table 1 Penzias and Wilson's detection of the cosmic microwave background radiation [22]

Signal	Noise signal T/K
Total zenith noise temperature	6.7 ± 0.3
Atmospheric emission	2.3 ± 0.3
Ohmic losses	0.8 ± 0.4
Backlobe response	≤ 0.1
Cosmic background radiation	3.5 ± 1.0

discovered wherever they pointed their telescope on the sky was a weak signal, as illustrated by the data presented in their discovery paper (Table 1). It will be noticed that their detection of the background amounted to a 3.5σ results. How often do we tell our students to beware of 3σ result. It is often remarked that Kepler's crucial discovery of the elliptical orbits of the planets about the Sun was based upon a 4-sigma result, the minimum deviation he could find from purely circular orbits. Penzias and Wilson did not do any better, but confirmation of their result flooded in through independent measurements over the next few years. Not all 3σ results are spurious!

The results of the Wilkinson Microwave Anisotropy Probe (WMAP) have been spectacular and we can expect a further roughly order of magnitude improvement in our knowledge of the properties of the Cosmic Microwave Background Radiation with the ESA Planck experiment. There is the possibility of the detection of primordial gravitational waves through the detection of the tensor component of the fluctuations on large angular scales, which would undoubtedly be one of the greatest achievements of observational cosmology.

9 The future

A good impression of the future of astronomy is provided by the document entitled *ASTRONET: A Science Vision for European Astronomy* which provides a comprehensive survey of the challenges and opportunities for future facilities for astronomy, astrophysics and cosmology (see http://www.astronet-eu.org/). The report highlights a number of 'essential' projects to build on the present momentum of European astronomy. Here is a list of some of these projects with a bias towards extragalactic astrophysics and cosmology, with my own interpretation of the principal areas of science which will benefit from them.

- Large X-ray Telescope: Tests of General Relativity in the strong field limit.
- Laser Interferometer Space Array (LISA): Search for low frequency gravitational waves. Tests of General Relativity.
- Extremely Large Telescope: Large redshift type 1a supernovae, γ-ray bursts, dark matter/dark energy probes, astrophysics of galaxies at very large redshifts, growth of perturbations during and after the reionization era, variations of fundamental constants with cosmic epoch.

- Square Kilometre Array: Neutral hydrogen redshifts out to $z = 2$, baryon acoustic oscillations, weak lensing, deviations from general relativity, huge samples of pulsars and tests of general relativity, astrophysics of normal galaxies and radio sources at early epochs.
- X-ray Survey Satellite: Large complete samples of X-ray clusters of galaxies, tests of the concordance model.
- Cherenkov telescope array.
- Very Long baseline interferometry at submillimetre wavelengths.
- Satellites for γ-ray bursts at very large redshifts.
- 4–8 m ultraviolet space telescope.
- 4–8 m infrared cooled space telescope.
- Euclid: Baryon acoustic oscillations, weak gravitational lensing, deviations from general relativity, huge samples of galaxies with colours and spectra.

This is undoubtedly a very exciting and ambitious programme, but it is seriously expensive. My parting shot is

'Don't forget the operating costs!'

In the steady state, operating costs are dominant. If the available budget for operations is x € per year and y% the percentage of the capital costs per year needed for operations, the total capital costs of operating facilities at any time cannot exceed $100x/y$ €. This is a very important equation for the effective construction, operation and continued exploitation of large scale facilities for science. For example, if $x = 100$ M€, $y = 10$, the cost of the capital facilities being operated at any time cannot exceed 1 B€. The ultimate scientific success of large facilities is crucially dependent upon the ability to operate them on behalf of the user community in an efficient and effective way, with adequate resources to provide new instrumentation and upgrades of the facilities in response to new scientific opportunities and technologies.

References

1. Baade, W., Minkowski, R.: Identification of the radio sources in cassiopeia, cygnus a, and puppis a. Astrophys. J. **119**, 206–214 (1954)
2. Bondi, H.: Newton and the twentieth century—a personal view. In: Fauvel, J., Flood, R., Shortland, M., Wilson, R. (eds.) Let Newton Be: a New Perspective on His Life and Works, pp. 242–248. Oxford University Press, Oxford (1988)
3. Douglas, A.V.: The Life of Arthur Stanley Eddington. Thomas Nelson, London (1956)
4. Eddington, A.S.: Diffuse matter in interstellar space. Proc. R. Soc. **A111**, 424–456 (1926)
5. Friedman, A.A.: On the curvature of space. Z. Phys. **10**, 377–386 (1922)
6. Friedman, A.A.: On the possibility of a world with constant negative curvature. Z. Phys. **12**, 326–332 (1924)
7. Gough, D.O.: Impact of observations on prejudice and input physics. In: Weiss, W.W., Baglin, A. (eds.) Inside the Stars. Astronomical Society of the Pacific Conference Series, No. 40, pp. 767–775. Astronomical Society of the Pacific, San Francisco (1993)
8. Hewish, A., Bell, S.J., Pilkington, J.D.H., Scott, P.F., Collins, R.A.: Observations of a rapidly pulsating radio source. Nature **217**, 709–713 (1968)
9. Hubble, E.P.: Extra-galactic nebulae. Astrophys. J. **64**, 321–369 (1926)

10. Hubble, E.P.: A relation between distance and radial velocity among extra-galactic nebulae. Proc. Natl. Acad. Sci. U. S. A. **15**, 168–173 (1929)
11. Hulse, R.A., Taylor, J.H.: Discovery of a pulsar in a binary system. Astrophys. J. Lett. **195**, L51–L53 (1975)
12. Jeans, J.H.: Diffuse matter in interstellar space: letter to the editor. Observatory **49**, 333–335 (1926) (See the final paragraph)
13. Jennison, R.C., Das Gupta, M.K.: Fine structure of the extra-terrestrial radio source cygnus 1. Nature **172**, 996–997 (1953)
14. Klebesadel, R.W., Strong, I.B., Olson, R.A.: Observations of gamma-ray bursts of cosmic origin. Astrophys. J. Lett. **182**, L85–L88 (1973)
15. Longair, M.S.: Theoretical Concepts in Physics: an Alternative View of Theoretical Reasoning in Physics. Cambridge University Press, Cambridge (2003)
16. Longair, M.S.: The Cosmic Century: a History of Astrophysics and Cosmology. Cambridge University Press, Cambridge (2006)
17. Lovell, A.C.B.: The emergence of radio astronomy in the uk after world war ii. Q. J. R. Astron. Soc. **28**, 1–9 (1987)
18. Mirabel, I.F., Rodrigues, L.F.: Microquasars in our galaxy. Nature **392**, 673–676 (1998)
19. Pacini, F.: Energy emission from a neutron star. Nature **216**, 567–568 (1967)
20. Pearson, T.J., Unwin, S.C., Cohen, M.H., Linfield, R.P., Readhead, A.C.S., Seielstad, G.A., Simon, R.S., Walker, R.C.: Superluminal expansion of quasar 3c273. Nature **290**, 365–368 (1981)
21. Peebles, P.J.E., Page, L.A., Partridge, R.B.: Finding the Big Bang. Cambridge University Press, Cambridge (2009)
22. Penzias, A.A., Wilson, R.W.: A measurement of excess antenna temperature at 4080 mhz. Astrophys. J. **142**, 419–421 (1965)
23. Rees, M.J.: Appearance of relativistically expanding radio sources. Nature **211**, 468–470 (1966)
24. Rees, M.J.: Studies in radio source structure—i. a relativistically expanding model for variable quasi-stellar radio sources. Mon. Not. R. Astron. Soc. **135**, 345–360 (1967)
25. Russell, H.N.: Relations between the spectra and other characteristics of the stars, i. Historical. Nature **93**, 227–230 (1914)
26. Russell, H.N.: Relations between the spectra and other characteristics of the stars, ii. Brightness and spectral class. Nature **93**, 252–258 (1914)
27. Russell, H.N.: Relations between the spectra and other characteristics of the stars, iii. Nature **93**, 281–286 (1914)
28. Russell, H.N.: In: Davis Philip, A.G., DeVorkin, D.H. (eds.) In Memory of Henry Norris Russell, pp. 90–107. Dudley Observatory Reports No. 13 (1977)
29. Schmidt, M.: 3c 273: a star-like object with large red-shift. Nature **197**, 1040–1040 (1963)
30. Seyfert, C.K.: Nuclear emission in spiral nebulae. Astrophys. J. **97**, 28–40 (1943)
31. Tanaka, Y., Nandra, K., Fabian, A.C., Inoue, H., Otani, C., Dotani, T., Hayashida, K., Iwasawa, K., Kii, T., Kunieda, H., Makino, F., Matsuoka, M.: Gravitationally redshifted emission implying an accretion disk and massive black-hole in the active galaxy mcg:-6-30-15. Nature **375**, 659–661 (1995)
32. Wirtz, C.W.: Einiges zur statistik der radialgeschwindigkeiten von spiralnebeln und kugelsternhaufen (some remarks on the statistics of the radial velocities of spiral nebulae and star clusters). Astron. Nachr. **215**, 349–354 (1922)
33. Zwicky, F.: Rotverschiebung von extragalaktischen nebeln, (the redshift of extragalactic nebulae). Helv. Phys. Acta **6**, 110–118 (1933)
34. Zwicky, F.: Catalogue of Selected Compact Galaxies and of Post-Eruptive Galaxies. F. Zwicky, Guemlingen (1968)

Capabilities of amateur telescopes - and their development over the last 35 years

D. Koschny

Abstract Amateur astronomers have been using optical telescopes since many years. In the last ~10 years, the introduction of cameras with Charged-Coupled Devices (CCDs) and the advent of automated telescope mounts and reasonably-priced high-quality optics has dramatically improved the capabilities of amateur astronomers. This paper will sketch the development of both equipment and techniques of amateurs over the last 35 years.

Keywords Telescopes · Amateur astronomers

1 Introduction

This paper discusses the development of the equipment and techniques of amateur astronomers over the last 35 years. From a very limited choice of telescopes – either very cheap and of low quality, or of reasonable quality but highly priced – the equipment developed to a broad selection of different types and qualities of telescopes. While 35 years ago most amateur work was restricted to visual observations or imaging on film, the CCD camera has revolutionized the work of amateurs and allows them to contribute data of professional quality to science.

This brings up the question of what an amateur is. Searching the internet one finds the following definition [6]: "'Amateur' comes from Latin amator "lover", or from amare "to love". An *amateur* is generally considered a person attached to a particular pursuit, study, or science, without formal training or pay. Conversely, an expert is generally considered a person with extensive

D. Koschny (✉)
European Space Agency, ESA/SRE-SM, Keplerlaan 1, 2200 Noordwijk, The Netherlands
e-mail: Detlef.Koschny@esa.int

knowledge, ability, and/or training in a particular area of study, while a professional is someone who also makes a living from it.

However, the word 'amateur' can have a quite different meaning: Someone who is unqualified or insufficiently skilful. '*The entire thing was built by some* ***amateurs*** *with screwdrivers and plywood*"'.

Typically, amateur astronomers do *not* belong to the second kind. Rather, they are people that pursue the observations of the heavens because they enjoy doing it. Normally they get plenty of informal training by reading books or using standardized software, and they develop excellent skills in operating their equipment by many years of training.

Some amateur astronomers contribute significantly to science by producing excellent data.

2 Equipment in the 70s

The author became interested in astronomy as a hobby in 1972. In those days, at least in his home country Germany, there were basically two options for obtaining a telescope. One could buy either reasonably cheap but low-quality telescopes of Japanese production from large department stores, or expensive equipment from the company 'Kosmos', which was the only one in the country that sold good-quality astronomical telescopes (Figs. 1 and 2).

The catalogue of the company Kosmos demonstrates what additional equipment was available for the amateur astronomer:

- simple motor drives for the mounts to track the stars;
- eyepieces of different focal length, most of them of type Kellner or Huygens;
- a plate camera for photography;
- a projection screen to observe the sun.

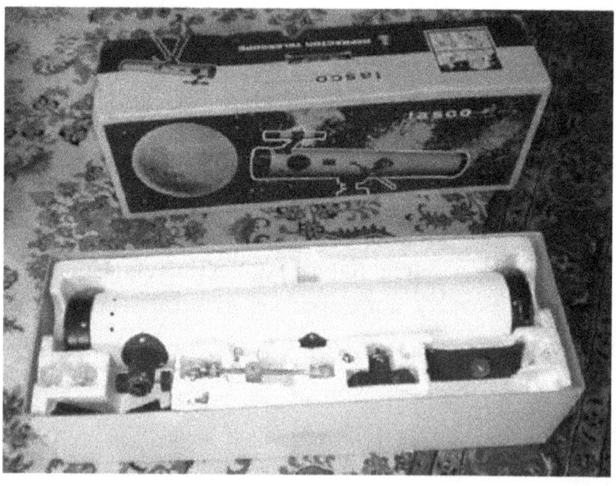

Fig. 1 The first telescope of the author. It cost about 250 DEM in 1973 and was available via large department stores. It was produced by the now defunct company Tasco from Japan

Fig. 2 A high-quality telescope available in 1972 in Germany from the company 'Kosmos'. The price for this 90-mm refractor was 1020 DEM including the parallactic mount and pier. Tracking was done by hand

3 Techniques in the 70s

Most observations in the 70s were done visually, *i.e.* using the eye. Amateurs used their telescopes to look at planets and produced drawings of what they saw, thus recording the changing appearance of *e.g.* the clouds of Jupiter, the albedo structures on Mars, or the numbers and shapes of sunspots. Estimates of the brightness of variable stars were done by visually comparing them to the magnitude of nearby stars of known magnitude.

Photographic film was used to permanently record the appearance of celestial objects. Figure 3 shows Jupiter as recorded on photographic film in 1977 [1]. Photography of planets was a difficult undertaking. To obtain reasonable resolution, the image of the planet had to be enlarged such that the brightness of the image was not very high, and exposure times of several seconds were needed. Due to the atmospheric turbulence during the exposure, the resolution of the imagc was limited to several arcseconds. Only the main cloud bands of Jupiter are visible in the image. The so-called Great Read Spot is visible as a dark oval patch in the upper cloud bands, but shows no structure. Compare this to Fig. 4, which is explained in the Section 5.

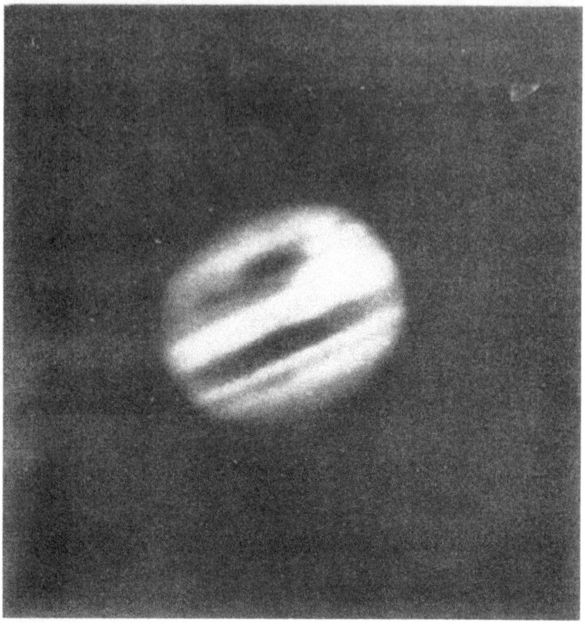

Fig. 3 Jupiter in 1977 as imaged by H. Bernard with the 12" Cassegrain of the Public Observatory in Munich, Germany [1]

More details on the planets were visible by combining drawings of visual observers. These would take advantage of short periods of good seeing (*i.e.* little air turbulence). However, it was difficult to reproduce these observations and many misconceptions followed. For example, the visual observation of structures in the atmosphere of Venus, or the attribution of so-called Transient

Fig. 4 Jupiter on 06 Mar 2003, 21h17m UT, imaged with a 12" Meade Schmidt-Cassegrain by Thierry Legault. He used a webcam Philips Vesta Pro and stacked the best of 960 image frames [4]

Light Phenomena (TLPs) on the Moon to impact events, were and still are disputed. Still, many amateurs enjoyed performing visual observations – and remember, that is what an amateur is supposed to do, enjoying his or her observations! The scientific usefulness of these observations, however, was obviously limited.

Deep-sky observers were either producing drawings, or imaging their objects with photographic film using long exposures. These were hampered by the so-called Schwarzschild Effect – photographic emulsion would lose sensitivity during exposures of minutes or longer. To double the intensity of an image, four times the exposure time was needed... This made deep-space imaging a challenge for the specialist. Also it was better not to leave developing the film to to commercial photo studios, as they often did not know how to develop film which looked essentially black. The small number of amateurs specializing in photography thus mainly opted for developing film themselves, a cumbersome process which took lots of practice.

4 Equipment today

When opening an amateur astronomer magazine today, one finds that the commercial advertisements sometimes span two pages. While the cheap, low-cost equipment of the 70s is still available, one can get decent equipment for only slightly more money. The main difference is that the choice of telescopes is much larger - one can buy low-quality but cheap Refractors and Newtonians, but also cheap Newtonians based on the Dobson design which are of high optical quality but where production costs have been saved by having only a very simple alt-azimuth mount built in wood. For those with more money, high-quality apochromatic refractors or large-aperture Ritchey-Chrétien telescopes are available.

Additional equipment can be obtained:

- mounts in all sizes;
- Roll-off roofs, domes – both for manual operation and for automation;
- CCD still cameras;
- Light-sensitive video cameras;
- Filter wheels and standard filters, both for red/green/blue (RGB) color imaging or the astronomical standard UBVRI;
- Automatic guiding systems and means of correcting the period error of worm wheels;
- Telescope localisation via the Global Positioning System (GPS)
- Heaters for removing dew from objective lenses or eyepieces:
- Software for telescope control or data analysis;
- And more.

Amateurs and commercial vendors have even agreed on a 'COM' (Common Object Model) standard for opening up the interface between different software tools, called the ASCOM standard (ASCOM = Astronomy Common

Object Model). This allows *e.g.* planetarium software which supports ASCOM to drive a telescope mount whose software is ASCOM-compatible. In this area, the amateur community is even more advanced than the professional astronomers.

5 Techniques today

The advent of electronic cameras, so-called CCD cameras (CCD = Charged Coupled Device) revolutionised the field of amateur astronomy. There are three different flavours of this advance:

(a) Long-exposure, cooled CCD cameras have become readily available to support deep-sky imaging;
(b) Light-sensitive video cameras have become available to support planetary imaging and applications where precise timing is important (meteor observations, asteroid occultations);
(c) Guiding has been automated by tracking a star either via an off-axis guiding camera or one mounted to a separate guidescope and interfacing directly to the telescope mount.

To see the advances of techniques, compare Fig. 4 to Fig. 3. Figure 4 shows Jupiter as recorded with a webcam in the year 2003 [4], with the same size telescope as in Fig. 3, namely 12" or 300 mm diameter. The photography is limited by the turbulence of the air during the exposure time of over one second. The image shown here was obtained using the 'lucky imaging' technique [3]. A simple webcam is used to record Jupiter for several minutes. Freeware software is used to select the sharpest images out of the recorded sequence and stack them together to obtain the image. The achieved resolution is close to the diffraction limit of the telescope, in this case around 0.3".

Long-exposure deep-sky images obtained with 10–16" amateur telescopes now rival those obtained 30 years ago with the venerable 5-m telescope on Mount Palomar using photographic film. Cooled CCD cameras with scientific grade sensors are available for prices starting at 2000 Euros. While this is a lot of money, many amateurs are willing to spend such an amount or higher to achieve top-quality imaging results.

Amateurs now use these capabilities to *e.g.* perform high-quality measurements of the position of asteroids or monitoring the brightness of variable stars. For asteroids, amateurs have obtained good astrometric measurements for asteroids as faint as 22 magnitudes.

6 Why the change?

The amateur astronomer's equipment market changed significantly over the last 30 years or so. In the 70s and 80s, a very small number of telescope and equipment vendors were available. Telescopes were either Newtonians,

Cassegrains, or Refractors. Schmidt-Cassegrain telescopes were available in Europe but comparatively expensive due to the exchange rate with the dollar. Optical parts for the do-it-yourself hobbyist were available, but not in a very large variety.

Nowadays, the market is much broader. Telescopes come in all flavours and price ranges. The same money may buy a small (say 80-mm aperture) but highest-quality apochromatic refractor, or a Dobsonian telescope (a Newton mirror telescope on a very simple but sturdy wooden mount) with 10 to 12" aperture. Telescopes of very high quality up to 20" (50 cm) aperture are available to the amateur. A wide choice of ancillary equipment is available, not only eyepieces and Barlow lenses, but lens heaters, GPS-synchronised clocks, flip-mirror systems, polar alignment scopes, color filters for laptop screens, and much more. Also, the market for the telescope maker has expanded significantly, not only focussing on optical elements but also including mechanical parts like special low-friction material used in altitude bearings, fasteners for trusses in truss-tube telescopes, and much more.

Here are a few thoughts about what caused these change in the equipment market and in techniques.

(a) Competition in the market. In particular in the US, several companies produced very compact Schmidt-Cassegrain telescopes and offered each other fierce competition. Some of them, like Criterion, lost the competition and had to leave the market. Others, like Questar, focussed on a niche, in this case very high-quality but expensive telescopes. Today the main competitors are Meade and Celestron. To stay in the market, they try to improve their production facilities to reduce the manufacturing cost. This can be negative for the customer, as it may reduce the quality of the telescopes, and indeed for the mechanical parts there is a tendency for lower and lower quality. However, competition also leads to creativity, and both companies have been competing with new products like standard Newtonian and Refractor telescopes, but also with new ideas like electronic alignment methods, usage of the GPS (Global Positioning System) signal to let the telescope know where it is located on Earth and what the actual time is, focal reducers which allow imaging with a larger field of view, and other innovative ideas.

(b) Halley's comet made astronomy understandable to the broad public. At least in the home country of the author there was a clear increase in the telescope market in the years 1980 to 1985, just before the appearance of Halley's comet in 1986. Young physics students decided to start their own business selling telescopes, not only from Meade and Celestron, but also importing cheap optics imported from Russia and China. Others realized that it is fairly straight-forward to build your own telescope, but building a solid telescope mount needs mechanical skills. A few people who were trained in this field - and were also amateur astronomers - decided to start their own business by producing good-quality telescope mounts and selling them (*e.g.* [5]).

(c) The acceptance of the Dobsonian telescope. The so-called Dobsonian design was implemented and pushed by John Dobson, an American amateur telescope maker who perfected the design of a simple, alt-azimuth wooden mount for a Newtonian mirror telescope. While in the 70s this design was considered by many amateurs as 'primitive', it became more and more accepted as a way of realizing large-aperture telescopes for visual observations for a reasonable budget. By now, companies like 'Obsession' sell telescopes with mirror diameters of 25" for less than 10000 Euros.

(d) The availability of personal computers allowed two important developments. It became possible to control telescopes and their mounts by software; and amateurs wrote software for data analysis, *e.g.* the measurement of features in Jupiter's clouds, which made the data analysis accessible to people who did not have the theoretical background to understand the details of the problems. The use of software tools ensures that people follow certain procedures when analysing their data, making it much easier to achieve good results and guaranteeing reproducibility.

(e) The availability of reasonably-priced CCD cameras and cheap web-cams. 35 years ago it was difficult to permanently record heavenly objects. Photography was available, but the sensitivity of film was low and long exposures were additionally hampered by the previously mentioned Schwarzschild effect. Nowadays, digital photography makes it comparatively easy to record the sky; even with normal consumer cameras excellent images are possible. The advantages of high-speed imaging for obtaining high-resolution planetary images have been discussed already in the section on 'techniques'.

(f) A GPS receiver not only makes it possible to determine one's position on the planet to high accuracy, it also allows precise timing with 10s of milliseconds accuracy. Using so-called time inserters which display GPS-synchronized time in an analogue video stream allows amateurs by simple means to perform highly accurate measurements of *e.g.* an asteroid occulting a star.

(g) A more social reason for the change in the field of amateur astronomy is the fact that people have more time. Even though our lives become more and more hectic, the average number of working hours has been reduced over the last 30 years, at least for the average worker. For most people, both Saturday and Sunday are free; many people save time by going to work with their car rather than with a bicycle or walking. Quite a number of people want to use their free time for something rewarding and something that has to do with nature. One possibility is to begin observing the stars, and a small but significant percentage of these people sooner or later start to contribute real scientific data.

(h) Last but not least, the internet has contributed to popularizing amateur astronomy a lot. While 35 years ago the night sky was considered something static, the fast turn-around time of the internet and email communication allows amateurs to be alerted to supernova outbursts, incoming comets, or asteroid occultations at very short notice. Additionally, observations can be shared widely via the internet, motivating other people to get involved too.

7 What can amateurs contribute to science?

Three centuries ago, amateur astronomers collected large amounts of data like magnitude estimates of variable stars; however, most measurements were done visually and thus of limited usefulness for scientific purposes. Today, mainly using CCD cameras, amateurs contribute significantly to certain areas of science. Some examples from the area of planetary science are given here.

(a) Asteroid follow-up observations. While most discoveries of asteroids are currently done by automated survey programs, these programs normally do not perform follow-up observations, *i.e.* observations of the newly discovered objects a day or several days after discovery. This, however, is very important to extend the observed arc of the path of the object and to thus ensure that the object's trajectory can be more accurately determined. Amateurs are regularly performing these observations and performing the required measurements. They are using shareware software that ensures that the data is processed properly and that different observers follow the same methodology. The data is then submitted to the central clearing house for minor planet observations, the Minor Planet Center (MPC).

(b) Asteroid occultations. When an asteroid's path crosses the position of a star, the star will be occulted for a short period of time (typically seconds). These occultations can be observed with small telescopes, equipped with a video camera and a time inserter synchronised via GPS. If several observers perform these observations, the size and possibly the shape of the object can be reconstructed. Also, very precise position information can be obtained. Amateurs have organised themselves to combine their measurements and submit precise positional information to the Minor Planet Center. Shape reconstructions and detailed observations are available on the internet (*e.g.* the Asteroid Occultation Reports Database AORDB by M. Kretlow, http://mkvh.de/ast/?Occultation_Reports_DB).

(c) Global and regular coverage of planetary atmospheres and surfaces via high-resolution planetary imaging. Amateurs use a combination of video cameras and special software available as freeware to apply the 'lucky-imaging technique' to planets [3]. A planet is recorded in video rate for several minutes, and thus hundreds to thousands of short-exposure images of the planet are obtained. Software tools (*e.g.* Giotto – see Dittie 2008, http://www.videoastronomy.org/, or Registax – see http://www.astronomie.be/registax/) are then used to identify the sharpest images which are then stacked together to obtain high-contrast images with a resolution which can reach the diffraction limit of the telescope (0.3" for a 40-cm telescope). This technique essentially freezes the turbulence of the atmosphere and yields a sharpness previously not achievable from ground. See Fig. 4 for an example.

(d) For many years, meteor observations have been the domain of amateur astronomers. This is one of the few areas where still today visual observations are producing relevant data. The 'Zenithal Hourly Rate' curves denoting the activity of meteor showers produced by the International Meteor Organisation are *the* standard also for professionals. These curves are obtained from visual meteor counts done by numerous amateurs across the globe. The main change which occurred in this field over the last 30 years was that in the 80s a big effort was made to standardize observations, introduce calibration factors (*e.g.* to record the 'limiting magnitude' during the observations to see whether a larger number of meteors is a result of a real increase or just clearer sky) and to set up one single data collecting entity, the International Meteor Organisation.
Recently, more and more video camera networks recording the night sky became operational. Software written by amateur astronomers (but very often by professional programmers) analyses the live video signal and detects meteors in real time, storing the images and additional data. Due to the enthusiasm and commitment of the amateurs these networks are regularly maintained and expand more and more. The main product of these networks is orbital data of the observed meteoroids, which makes it possible to draw conclusions on the ejection process of the dust particles on the meteoroid parent bodies.

The above points are only a few examples, there are many more possibilities nowadays for amateur astronomers to contribute to 'real science'.

8 Limitations of amateur equipment and techniques

Obviously, mass-produced telescopes cannot have the same quality as an expensive, one-off telescope construction. Thus, most amateur telescopes suffer from problems in pointing and tracking accuracy. However, with the use of modern software tools, these mechanical weaknesses can be corrected by large amounts. For example, a program called TPoint (http://www.bisque.com/products/TPoint/) can measure the flexure of a telescope mount and correct for it electronically.

However, one point that still needs more attention is the fact that many amateurs do not analyse the properties of their equipment in great detail, or properly calibrate and/or archive their data. For those amateurs who only want to obtain 'pretty pictures' this is not an issue. However, for the data to be scientifically useful, a full understanding of the equipment, proper calibration (*i.e.* dark image subtraction and flatfielding for CCD images), and proper archiving with all necessary meta-information is essential. Koschny [2] mentions the following keywords to improve the quality of amateur data:

- Understand your equipment
- Calibrate
- Be goal-oriented
- Use procedures
- Document your work
- Publish and archive properly.

For more details on these points, see [2].

9 Conclusions

Both the equipment and the techniques used by amateur astronomers have undergone dramatic changes in the last 30 to 35 years. Whereas in the 70s most observations were done visually, today's amateurs use electronic CCD cameras and video cameras, and high-precision clocks based on the GPS system, and they can take advantage of free software or shareware to do complicated mathematical processing. Amateurs have organised themselves to be able to contribute to professional databases and have become relevant and respected players in several areas of planetary science and astronomy. Still, improvements in the way of working of amateurs are still possible, mainly by focusing increasingly on calibrating and properly archiving their data.

Those amateurs who do not want to contribute to 'real science' are also in a much better situation than 35 years ago. The choice of equipment is much wider, and telescopes with apertures of up to 0.5 m in diameter have become affordable.

Most importantly, people nowadays have more time and – luckily – many of them want to do something useful with their time. What better way than taking out a telescope during a clear night and enjoying the wonders of our universe! Whether these observations are then used for scientific analysis, or just done for their own pleasure, is not relevant - do not forget that 'amateur' comes from 'to love'.

References

1. Bernard, H.: In: Jahresbericht der Volkssternwarte München (1977)
2. Koschny, D.: Making amateur data scientifically useful. European Planetary Science Congress, pp. 21–26. Münster, Germany. EPSC2008-A-00021 (2008)
3. Law, N.M., Mackay, C.D., Baldwin, J.E.: Lucky imaging: high angular resolution imaging in the visible from the ground. A & A **446**, 739–745 (2006)
4. Legault, T.: http://legault.club.fr/jupiter.html (2003)
5. Mauz, M.: http://www.astronomiegeraetebau.de (2008)
6. Wikipedia: http://en.wikipedia.org/wiki/Amateur (2008)

Exp Astron (2009) 25:261–272
DOI 10.1007/s10686-008-9130-9

ORIGINAL ARTICLE

Creation of the Hubble Space Telescope

C. R. O'Dell

Received: 9 October 2008 / Accepted: 25 November 2008 / Published online: 16 January 2009

Abstract The Hubble Space Telescope has been the most successful space astronomy project to date, producing images that put the public in awe and images and spectra that have produced many scientific discoveries. It is the natural culmination of a dream envisioned when rocket flight into space was first projected and a goal set for the US space program soon after NASA was created. The design and construction period lasted almost two decades and its operations have already lasted almost as long. The capabilities of the observatory have evolved and expanded with periodic upgrading of its instrumentation, thus realizing the advantages of its unique design. The success of this long-lived observatory is closely tied to the availability of the Space Shuttle and the end of the Shuttle program means that the end of the Hubble program will follow before long.

Keywords Hubble Space Telescope · NASA · ESA · Space astronomy

1 Background and the beginnings

The creation of an observatory in space, free from the limitations of the Earth's atmosphere, was a natural goal for astronomers. However, doing this meant having the ability to put the telescope there and it is not surprising that the first mention was in the small book "Die Rakete zu den Planetenraumen" published in 1923 by Hermann Oberth. This was to be his doctoral thesis on the

400 Years of Astronomical Telescopes.

C. R. O'Dell (✉)
Department of Physics and Astronomy, Vanderbilt University, Nashville, TN, USA
e-mail: cr.odell@vanderbilt.edu
URL: http://www.vanderbilt.edu/AnS/physics/cv/odell.html

potential for travel into space by means of rockets, but in the end was considered too speculative and was only published privately. It was clear that a space observatory could avoid the limitations of the Earth's atmosphere, allowing it to look at all wavelengths of radiation and to also achieve angular resolutions of about the ratio of the wavelength being studied to the aperture of the telescope. The improved angular resolution would allow such a telescope to observe fainter objects that a telescope of much greater size. Although the American Robert Goddard had developed early liquid fuel rockets, realistic consideration of spaceflight only began when German V-2 rockets and their developers were brought to the USA in 1946 and a number of the rockets were used to make brief observations from above the Earth's atmosphere. At the same time the RAND corporation was charged with identifying the potential uses of space and they charged the young Yale professor Lyman Spitzer with identifying applications within astronomy. Spitzer identified the operation of a diffraction limited large telescope from earth-orbit as a natural culmination of a program of space astronomy. Originally this study was classified, but the idea naturally resurfaced when the National Aeronautics and Space Administration (NASA) was created in 1958 and was identified in 1962 as a goal of NASA's science program, an audacious aspiration for an organization that had placed into orbit only a few small and short-lived satellites.

1.1 The earliest steps

Many preliminary steps were required before something like today's HST could be considered. In the USA there were a series of Orbiting Astronomical Observatory spacecraft dedicated to astronomy (two of the four being operated successfully) and a host of suborbital rocket observations. In Europe the European Space Agency (ESA) successfully launched the TD-1A and Astronomical Netherlands Satellite. NASA and ESA then combined skills to create the highly successful synchronous orbit International Ultraviolet Observatory, which was launched in 1978.

Serious consideration of the Hubble Space Telescope (HST, then called the Large Space Telescope) by NASA began in 1971, when competitive conceptual designs were considered. These covered a wide range of possibilities since it was not clear how a long lifetime was to be achieved (orbital servicing or frequent launch of upgraded observatories as the predecessor failed) and what type of detector was to be used (film, which fogs in space due to the radiation environment and requires frequent replacement, or electronic detectors, which had limited imaging capability). Once NASA began development of the Shuttle Transportation System (STS), with the expectation of frequent and low cost access to space, this became the designated mode of transportation into earth-orbit. The observatory design was then optimized around the expectation of being able to periodically (the identified interval was three years) visit the spacecraft in order to replace failed components, install improved components, and upgrade the scientific capability. The advantages of on-orbit servicing by astronauts were accompanied by the need to do something very new, to design a scientific spacecraft that was astronaut-friendly. This required testing design concepts in the zero-gravity simulating Neutral Buoyancy Simulator at NASA's Marshall Space Flight Center (MSFC).

1.2 Creating support for the HST

It is remarkable today to recall that in the early 1970's there was not broad support for the HST. The technology was perceived as being well into the future and the common opinion was that it would be better to spend the several hundred million dollars in building multiple 5 m telescopes. There was a geographic division in support for the HST between the clear-sky and large-telescope rich western astronomers and those from the midwest and northeast. I joined the project as Project Scientist in September, 1972 and my first major task was to break down this division. This was done by causing NASA headquarters to issue a call for participants in the preliminary design of the observatory and its scientific instruments (SI). This was followed up in-depth presentations in Boston, Chicago, and Los Angeles. After this time community support started to become grow, both because of the scientific potential of the observatory, but also because institutions and individuals saw opportunities in this for them. However, the single water-shed event was a session on the HST as part of an American Institute of Aeronautics and Astronautics meeting, in January, 1974 where senior big-telescope west-coast astronomers came out supporting the HST. Those who argued that the same high quality images would be available in a few years through compensation for atmospheric turbulence were not silenced, even though that capability has now, three decades later, only been realized in the much more benign infrared region.

1.3 The preliminary design period

The preliminary design period was also the period when funding for the final design and construction of the observatory was sought since there was no assurance that we would funded for this much more expensive period. From the beginning there was a conflict between what NASA headquarters foresaw as available and what it would actually cost to build this expensive observatory. In December, 1974 the NASA administrator dictated a funding maximum of \$300 million and the project was forced to look at alternatives to the 3 m aperture instrument that had been the design goal until then. This resulted in the selection of a 2.4 m design. At that diameter the heavy support equipment could be packaged in a toroidal ring around the primary mirror and remain within the clearance envelope inside the STS payload bay. This also allowed the project to draw on technology that had been developed for use in reconnaissance satellites and reduced the moment of inertia since the support equipment had been packaged at the after end of the telescope in the three meter design. This reduction in the moment of inertia allowed the adoption of a simpler and lower cost pointing system and at this point the design solidified around the 2.4 m aperture. This drive towards cost reduction also resulted in the opening of negotiations with ESA about their participation. This lead to full participation of ESA in the preliminary design of the SI's and eventually (1976) to an inter-agency agreement to provide one of the SI's and the solar arrays that would power the observatory.

Operational flexibility was achieved by making many of the most vulnerable components replaceable while in orbit and by compartmentalizing the SI's. The central field of view was divided into five portions. The central portion was diverted perpendicular to the optical axis and fed to a "radial" SI, while the nearby portions went directly into one of four "axial" SI's. Around this central field of view there were three fine guidance sensors (FGS), which employed star images to measure small motions of the spacecraft pointing, combine that signal with information from highly precise on-board gyroscopes and through variations in large spinning disks, provide the necessary guiding stability of 0.007″. Since only two FGS were necessary to operate the pointing system, the third FGS could be used form astrometry science. The five SI's and the three FGS's were all individually compartmentalized so that they could be withdrawn and replaced with a better unit during a servicing mission.

2 The design of the HST

The final design and construction of the HST was approved as part of the NASA budget for the fiscal year beginning in October, 1977. What followed was the refinement of the design produced during the Preliminary Design phase and the development of a set of SI's, some of which were quite different from those considered during Preliminary Design. The resulting overall design is shown in cartoon form in Fig. 1.

The heart of any optical telescope is the imaging system. In the case of the HST the Ritchey–Chretien variation of the Cassegrain design was selected, which provided a coma-free image over a wide concave field well behind the primary mirror (Fig. 2). In order to save weight, the primary mirror is a mostly open "egg-crate" construction, with thin front and back plates fused to a mesh of very thin reinforcing pieces. Structurally the foundation of the observatory is a titanium main ring encircling the primary mirror and eventually all loads connect to it. The optical elements are kept in alignment with an open truss structure of graphite–epoxy material, which is both very strong per unit weight and also has a low thermal coefficient of expansion. This material was also used to construct the framework behind the primary mirror where the modularized SI's and the FGS's were attached.

The power for the system is provided by solar arrays that are folded-up during launch. The energy from these arrays is stored in six NiH batteries to power the observatory during the half of the orbit that the HST is in the Earth's shadow.

The batteries, the computers that control the SI's and the telescope, the reaction wheels that effect the pointing and stability, and the six reference

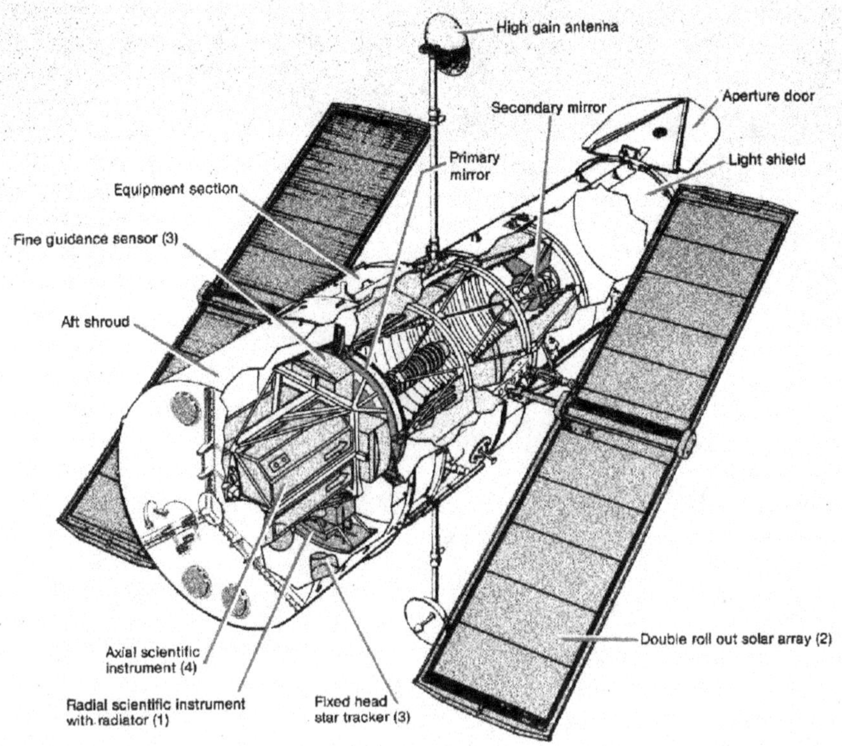

Fig. 1 This cutaway drawing of the HST as launched shows the major elements of the observatory. For scale consider that the primary mirror is 2.4 m in diameter and the length of the body of the telescope, with the aperture door closed is 12.9 m

gyros were all located in the "equipment section" that straddles the main ring. Small telescopes looking nearly perpendicular to the optical axis are located in the aft portion of the observatory and are there to help with the crudest level of pointing of the telescope. Although there are low omni directional antennas for use when the observatory is not precisely under control, the primary source of data flow is through a pair of steerable high gain antenna, that are directed outward towards one of several geosynchronous relay satellites. This allows the HST, in its low Earth orbit about 570 km above the Earth, to be continuously seen and controlled from a ground station at GSFC on the north side of Washington, DC.

The exterior of the observatory is covered by highly reflective aluminum coated layers of mylar, whose primary purpose is to stabilize the interior temperature of the HST during the wide range of conditions of illumination on the day and night side of the orbit. The underlying structure is light tight and there is an aperture door that is closed only during launch and servicing missions. The "jaunty" angle of this aperture door was selected to allow

Fig. 2 The HST is shown in orbit immediately after SM3B

observing Venus at maximum elongation without sunlight entering the main tube of the telescope.

3 The construction phase

The HST program was greatly affected by the rather arbitrary cost ceiling imposed by the NASA Administrator. The true costs to build the HST only surfaced with the actual construction. This meant that cost was always a constraint, causing NASA to deviate from its usual policy of building a series of versions starting with prototypes of increasing sophistication and to build a single (protoflight) spacecraft. This meant that the development path was very unforgiving of developmental problems and highly vulnerable to cost growth and schedule delay. The situation was exacerbated by constraints on the manpower level that could be put on the project at the lead NASA center (MSFC). This was because of the HST used some technology that had been developed for classified programs. A final complication was a cumbersome management structure, with MSFC responsible for the observatory, GSFC for the SI's and eventually the operation, the Lockheed Corporation responsible for the overall hardware but with the Perkin-Elmer Corporation responsible for the telescope, science operations being planned for a non-NASA facility, and finally, ESA providing an SI and the solar arrays.

Actual construction of HST hardware began with the primary mirror even before the official start of the construction phase of the program. This was because it was considered the pacing item in terms of development and assembly of the observatory. A backup primary mirror was also constructed in parallel as insurance against a catastrophic failure of the first mirror. In the end it was the FGS that proved to be the greatest challenge and a redesign of this component garnered most of the management attention as the primary mirror was being finished.

Cost growth and schedule slips became the norm, usually be held in check by trade-offs in testing, performance, and on-orbit serviceability. When the situation rose to crisis proportions a new total budget would be reluctantly allocated and a new launch date established. This happened twice, with the launch finally scheduled for September, 1986 but in January of that year the Challenger STS accident occurred and there were a series of launch dates established before launch finally occurred in April, 1990.

4 The method of operation

Until the growth in investment in American science after World War II, the largest observatories were usually operated by single institutions. Almost simultaneous with the creation of NASA this pattern begin to break down with the creation of national facilities for optical astronomy (Kitt Peak National Observatory and Cerro Tololo Observatory) and radio astronomy (National Radio Astronomy Observatory) funded by National Science Foundation funds and operated under contract by consortia of universities and open to use by all, although the scientific data were treated as proprietary. Although the dollar investment in the early short lived space observatories were comparable to that in a large ground based facility, the responsibility and scientific reward was limited to the institution developing each one. Data was considered the proprietary property of the scientists who had developed the instruments and released only with publication of the results. This system was not changed until the International Ultraviolet Explorer was launched in 1978.

To the non-NASA people involved in the Preliminary Design phase of the HST, which started in 1972, it was clear that the traditional way of doing things was not appropriate and this group begin to agitate for the creation of a special facility dedicated to conducting the science of the HST, funded by NASA, but operated by a non-NASA corporation. After much hesitation and the recommendation of a Blue Ribbon panel in 1976, NASA finally adopted this approach and contracted for creation of the Space Telescope Science Institute adjacent to the Johns Hopkins University in Baltimore, Maryland, about an hour's drive north of GSFC. The name Space Telescope Science Institute predates the final naming of the HST in 1983.

This highly successful organization conducts an annual solicitation of observing programs and leads the activity of reviewing these proposals by qualified outsiders. Although there is a guideline in the inter-agency agreement that ESA member nations should get 15% utilization fore ESA's 15% financial contribution, this is not considered in the reviews and it is only in the most recent reviews that Europeans have received less. The winning scientists are given software packages that simulate the high level functions of the individual SI's and the HST and then use these to develop observing plans for their programs. These programs are then merged at the Space Telescope Science Institute (STScI) and submitted (often with iterations) to the Goddard Space Flight Center (GSFC), who then takes the lead in their execution. The resulting scientific data is pre-processed at GSFC and pipe-line processed at the STScI. They are usually available to the scientists within 48 h. The proprietary period for the data (with only a few exceptions) is never more than 1 year from the time of execution of the observation and is carefully archived. This has created a growing body of publicly available data that has lead to many archival studies as reflected in the continued growth in the number of refereed papers published annually.

5 Problems uncovered after launch

Rude shocks awaited the HST project when the observatory was first used for obtaining images. Although star images had a sharp core, they were surrounded by a halo of light that contained about 80% of the total signal. The problem was further compounded by the fact that vibrations in the observatory were induced by the solar arrays during each transit into and out of the sunlit side of the orbit, preventing the taking of any images for several minutes. The latter problem was ameliorated by changing software in the pointing system, but the image quality problem remained. Fortunately, it had always been planned to be able to enhance the images by data processing and this enabled much scientifically useful data to be obtained, but work in crowded fields of view and on faint objects was excluded.

Scientists and engineers at the STScI took the lead in characterizing the nature of the problem with the images and concluded that the problem could be identified as "spherical aberration", just as if the primary mirror had been manufactured to the wrong shape. A subsequent investigation revealed that an error had been made in the spacing of optical components in the testing devise used to monitor the figure of the mirror as it was being polished into its final surface. This meant that the mirror had been manufactured with great precision, but with poor accuracy. The mirror looked perfect, but with a flawed measuring method. Because of the severe cost constraints on the program there had been no plans for redundant testing and since the mirror appeared to be satisfactory there was no reason to consider using the backup mirror.

Like most accidents there were multiple contributing factors. The immediate problem was a small error in making a simple blackened cap fit over the end of a very precise measuring rod. A chip of black paint produced a reflective surface whose position was measured instead of the end of the rod. There was some hint of a problem recorded by the technicians. The record of this was not examined critically by the manufacturer and the report was never sent to the NASA personnel monitoring the work. An underlying cause of this error not being detected was the manpower restraints under which MSFC supervised the program for reasons of security and a certain laxness in monitoring features that were considered mature technology.

Fortunately, the investigation board found (then a decade later) the test device remained as it had been used and it was a simple matter to calculate the shape of the primary mirror. This meant that any future SI's could compensate for this in the optical train that was always present for transferring the image form by the HST onto the detector for that SI. A compartment simulating the size and thermal environment of one of the SI's had already been built and STScI personnel designed a novel procedure of adding deployable optics to this unit. Once that unit, then renamed COSTAR (Corrective Optics Space Telescope Axial Replacement unit), was installed in the place of an axial SI, it could deploy optics in front of the entrance apertures of the other axial SI's and they could operate normally.

6 The scientific instruments

The packaging of the individual SI's into standardized axial (four) or radial (one) compartments allowed their development to be pursued separately and eventually for them to be replaced in orbit by superior SI's. Because of the long lead time for developing each of the original SI's and the many schedule slips of the overall observatory, the first set of SI's was the simplest. As new technology and approaches became available, these instruments could and have been replaced by superior devises.

The original set of SI's covered the normal full range of capabilities of a large ground based observatory in having two cameras, two spectrographs, and a photometer. The WF/PC was the wider field of view camera and was mounted axially, thus allowing direct radiant cooling of the instrument. There were two separate optical trains, both employing an array of four Charge Coupled Device detectors, one giving a wider field but lower resolution (the WF mode) and the other a narrower field and higher resolution (the PC mode). The Faint Object Camera was developed by ESA had was a highly versatile camera with the highest resolution of all the cameras (thus able to detect the faintest objects) and several low resolution spectroscopic modes. The Goddard High Resolution Spectrograph (GHRS) offered high spectral resolutions with a linear array of detectors and the Faint Object Spectrograph (FOS) providing low resolutions with a linear array. The High Speed Photometer (HSP) was the simplest of the instruments but was capable of highly accurate, fast time resolution photometry of compact sources.

The evolution of the complement of SI's on the HST has been a major element in the HST remaining an observatory that is unique. Each of the instruments was seriously affected by the initial spherical aberration of the imaging they were given. The HSP was sacrificed during the first servicing mission to make room for installation of the COSTAR and the WF/PC was replaced by a simpler version designated as WFPC2 that employed only four CCD's in a single array, one of which was of the original higher resolution focal length, thus producing an irregular outline field of view of the mosaic combined images. The GHRS spectrograph was replaced by a higher spectral resolution, two dimensional detector instrument (STIS, Space Telescope Imaging Spectrograph) during the second servicing mission, which also saw the replacement of the FOS with NICMOS (Near Infrared Camera and Multi Object Spectrograph), which added near infrared capabilities to the optical and ultraviolet wavelength instruments already operating. This instrument exhausted its on-board cryogens much more rapidly than expected. A developed refrigerator was installed during SM3B that restored NICMOS operation. This continued until the refrigerator was turned off as preparations began for SM4 and the unit failed to back into operation. The fate of this unit is uncertain at the time of preparing this article. The FOC was replaced during servicing mission 3B with the Advanced Camera for Surveys (ACS). ACS offers about the same resolution as the PC portion of the original WF/PC camera but does this over an area 25 times larger. This gain is largely the result of the improvement in the ability to manufacture space qualified CCD's. During the forthcoming final servicing mission the WFPC2 will be replaced with the Wide Field Camera 3 (WFC3) which offers about a ten times gain in imaging capability over the WFPC2 and with a wide selection of diagnostic filters that is similar to WFPC2 and superior to those in ACS. This same mission will see the COSTAR replaced by a moderate resolution spectrograph (COS, Cosmic Origins Spectrograph) capable of reaching very faint sources. Astrometry continues to be conducted with the FGS's, even though they must work with the aberration distorted images formed by the telescope.

7 Servicing missions

Servicing missions were always an integral part of the planning for the HST, in fact, the observatory was built around this capability. However, the original plan for visits every 3 years has been overcome by the realities of events. The STS has proven to be much more expensive to operate that initially presented, with the result that fewer flights have been available, a situation compounded by hiatus in flights following both of the STS accidents (Challenger in January, 1986 and Columbia in February, 2003). In addition, the needs of the HST have varied from the plan, as instruments and spacecraft components failed at unpredicted times.

The first servicing mission (SM1) occurred in December, 1993 and saw installation of the WFPC2 and COSTAR. The other major event was the removal of the solar arrays and replacement by similar ones that had been redesigned to be free of the vibrations induced at sunlight terminator crossings.

The second servicing mission (SM2) occurred in February, 1997 and featured the installation of the STIS and NICMOS, in addition to trading out one of the FGS, which was returned to Earth and refurbished for reinstallation later.

The third servicing mission was divided into two flights, the first (SM3A) occurring in December, 1999 which was earlier than originally planned and was caused by premature failure of most of the FGS units necessary for providing fine guiding. This mission also saw the tradeout of another FGS and the installation of a much more capable computer. The second component (SM3B) occurred in March, 2002 and saw the installation of the ACS and the tradeout of the solar arrays for those of a new design with higher performance. There was also the installation of a cooling device on the outside of the HST that would cool the nitrogen used to chill the NICMOS. The NICMOS coolant had thawed faster than anticipated due to an unexpected thermal leak and this cooling device restored NICMOS to full operation, a development never directly anticipated in the design of the observatory.

The final servicing mission (SM4) is now planned for spring in 2009. This ambitious mission will see the installation of the WFC3 and the COS, the complement of batteries, the Data Handling box that handles all data interfaces with the Sis (the primary unit for this failed in late September, 2008), and again all of the gyros (four of the six having failed). Moreover, servicing the STIS and ACS will also be done. These servicing actions had never been anticipated in the design of the instruments and the fact that they will be attempted is a tribute to the flexibility of the system.

In addition to the listed actions that were conducted on each servicing mission there has been a myriad of less visible but always important changes and repairs to the spacecraft systems.

The HST was designed with a fifteen year lifetime as the goal. However, the expectation was always that it would be operated as long as it was considered worth the cost to continue its service and operation. As this article is submitted only one SI (WFPC2) and two of the six gyros are operating. With the possible exception of the NICMOS all of the SIs will be operating after SM4 and the observatory will then be at the height of its capabilities. The long term problem lies with accessibility to the STS. Once the USA became part of the Russian Mir space station program and then construction of the International Space Station (ISS), an increasing and by now an almost exclusive use of the STS has been for visits to the ISS. SM4 to the HST is the only scheduled non-ISS flight of the STS following the Columbia accident and only seven flights of the STS are planned in 2009 and 2010, after which the STS fleet is to be retired as a new transportation system is developed. Should the operational lifetime of the STS be extended then it may be that additional servicing missions will be considered.

8 Epilogue and recommended reading

As the rockets envisioned by Hermann Oberth allowed astronomers to envision the HST, the loss of a particular rocket powered aircraft forecasts an end to the HST program. However, the science of the HST will not end even on that indefinite date in the future when the HST stops functioning. The data from the observations have been carefully archived and will continue to be of scientific use far into the future, much as some astronomers today benefit from data obtained a century or more ago.

8.1 Recommended reading

There have been a large number of publications about the HST and presenting its images. Not all of the descriptions of the history of the project are accurate and I list below what I consider to be the best.

The outstanding historical book was prepared by someone who quietly observed history in the making: Smith, R.W.: 1993, "The Space Telescope", second edition (Cambridge University Press: Cambridge).

A detailed history of the Marshall Space Flight Center involvement is in: Dunar, A. J., & Waring, S. P.: 1999 "Power to Explore—History of the Marshall Space Flight Center 1960–1990" (U.S. Government Printing Office: Washington) Chapter 12.

A concise attempt to tell the story of the HST from one scientist's point of view as an insider is: O'Dell, C. R.: 2003 "The Orion Nebula: Where Stars are Born" (Harvard University Press: Cambridge), Chapter 10.

A wonderfully humorous and prophetic book about the scientific results of the HST, written before it was launched is: Longair, M.: 1989 "Alice and the Space Telescope" (Johns Hopkins University Press: Baltimore).

Acknowledgements The HST would not have been created without the efforts of scientists, managers, and engineers working in universities, government facilities, and corporations too numerous to list here. Lyman Spitzer was the original scientist-advocate who was joined by John Bahcall at a critical stage in convincing the legislative decision-makers of the USA that the HST was worth the price. Jean R. Olivier was NASA's chief engineer for the HST throughout its design and development. Arthur D. Code enabled the author to enter the field of astronomy in one of the first departments to consider space astronomy to be real astronomy. He also led the efforts to establish the highly successful way in which this observatory is used to do science. It is well stated "Success has many fathers, but failure dies an orphan." These people are particularly worth acknowledging as they were among that core group who retained faith in the project when the HST project went through its darkest days.

Exp Astron (2009) 26:1–18
DOI 10.1007/s10686-009-9161-x

REVIEW ARTICLE

History of mirror casting, figuring, segmentation and active optics

Lothar Noethe

Received: 10 November 2008 / Accepted: 6 March 2009 / Published online: 14 April 2009

Abstract Since the invention of the telescope the mirrors in reflecting telescopes have evolved from small pieces of polished speculum to sophisticated, computer-controlled systems. This review describes the major problems with the casting, figuring and support of these mirrors, and how fundamental inventions like depositing a silver layer on glass or electronic devices like computers opened the path to new solutions for fabricating larger mirrors with improved performance.

Keywords Telescope mirrors · Casting · Figuring · Segmentation · Active optics

1 Introduction

Mirrors are probably the most important and delicate devices in a reflecting telescope. Over 400 years the primary mirrors, in particular, have evolved from small pieces of polished metal into diversified complex, computer-controlled systems. Common to all of them is that they have to deal with a number of very different tasks and requirements:

- The light has to be reflected efficiently over a broad range of wavelengths
- Reflections require a smooth surface, called a *true surface* by Lord Rosse
- In addition, focussing the light requires a well-defined overall shape of the mirror surface, called an *accurate general figure* by Lord Rosse

L. Noethe (✉)
European Southern Observatory
Karl-Schwarzschild-Str.2, 85748 Garching, Germany
e-mail: lnoethe@eso.org

- The reflecting area should be as large as possible
- The mirror figure has to be stable irrespective of the inclination of the mirror
- The mirror figure has to be stable under varying environmental conditions, in particular temperature variations

In Fig. 1 these tasks are represented by the horizontal bars. The subsystems and features which are potentially capable of fulfilling the requirements are listed in the column on the left. They are the blank, a support system, a reflecting layer, segmentation, and a control system also called active optics. The figure shows the evolution from a situation where a single subsystem, namely a polished piece of metal, had to cope with all the requirements to one where the requirements are fulfilled by several subsystems, the most important of them being the control system.

Broadly, one may divide the history of the mirrors into three eras. During the first and by far longest one, the speculum era, the only solution to fulfill all the requirements as well as possible was to polish a piece of speculum, a bronze alloy consisting of roughly 70% of copper and 30% of tin. Shortly before the end of this era a first specialization took place when the invention of two types of support systems relieved the mirror from the task of supplying all of the figure stability. However, the fundamental invention which marks the beginning of the second era, the glass era, lasting for about 100 years, was the development of depositing a silver layer on glass, relieving the mirror substrate also from supplying the reflectivity. The switch to glass as a carrier for the thin reflective layer initially made the manufacture of blanks easier until the glass mirrors reached dimensions of a few meters. The low thermal conductivity of the types of glass used in the early times of this era generated thermal inhomogeneities and therefore stresses and deformations in the mirror

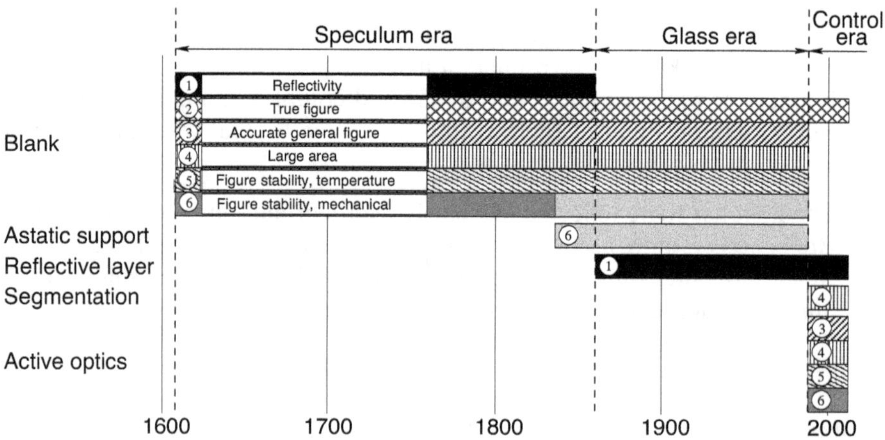

Fig. 1 Fulfillment of requirements, represented by *horizontal bars*, by subsystems, shown in the column on the left

during casting, polishing and operation. The major advances marking this era, also justifying the name glass era, came from the development of glasses first with low and in the end effectively zero expansion coefficients, which solved these problems to a great extent. What then limited the quality of the mirrors was their figure stability under changes of their inclination.

The solution to this problem, marking the beginning of the third era, the segmentation and active optics era or maybe more briefly the control era, came with the modern development of electronic devices, in particular computers and CCD-cameras. The fulfillment of the stability requirements and the quest for large area could now be accomplished by control systems, which also offered the possibility of building up large mirrors of medium-sized segments.

2 The speculum period

2.1 Required general figure

The accurate general figures for the mirrors with which one could achieve the optimum image quality over small fields of view were discovered during the first 70 years after the invention of the reflecting telescope with remarkable progress in the optical theory of reflecting telescopes. Further advances were only made at the beginning of the twentieth century by Karl Schwarzschild. A first design for a reflecting telescope was suggested in 1616 by Zucchius, with a dispersing lens as an eyepiece and a tilted mirror to avoid the obstruction by the eyepiece and the observer.

In 1636 Mersenne invented an afocal telescope consisting of two parabolic mirrors with the convenient feature of viewing the image from behind. Focal telescopes giving a real image were designed in 1633 by Gregory, consisting of two aspheric mirrors and having an intermediate focus, in 1668 by Newton, consisting of a parabolic primary and a flat secondary mirror, reflecting the light to the side, and, finally, around 1670 by Cassegrain, who proposed a hyperbolic convex mirror instead of the elliptical concave one in the Gregory design. All these designs gave excellent images for objects on the optical axis only. However, errors for off-axis objects were no issue with the large focal lengths used at that time and the small fields given by the eye as the detector.

2.2 Achieving a true and accurate general figure

Initially, the mirrors were small and the only problems were to obtain a true and accurate general figure. Polishing mirrors is more critical than polishing lenses, since the image quality is four times more sensitive to surface errors on a mirror than to surface errors on a lens. The basic polishing technique had been developed early on. Newton used a tool with a pitch surface and putty of tinoxide as the polishing agent. Both pitch and metal oxides are still the standard polishing substrate and agents nowadays. Unfortunately, true and accurate general figures could only be accomplished with conflicting characteristics of the pitch. A soft pitch layer would produce a better true surface whereas a hard pitch would produce a better accurate general figure. Compromises were found with the cutting of grooves into the pitch layer by John Edwards in 1790 and, later, by Lord Rosse, using square pieces of pitch consisting of two layers, a soft one on top of a hard one.

The mirrors were initially polished by hand. William Herschel attached handles to the rims of his largest mirrors which were then used by ten men to move the mirror during grinding and polishing. The invention of steam engines enabled Lord Rosse to construct polishing machines and thereby to achieve repeatable results and improve systematically the polishing procedure. To generate the random movements which are required to obtain a spherical surface, his mirrors rotated on a platform with two excenters.

Usually, a spherical surface was produced first before the mirror figure was aspherized to a parabolic or near-parabolic shape. Various techniques were applied for the aspherization. Edwards used elliptical tools, thereby modifying the distribution of the pitch, and Lord Rosse used variations of the length of the radial strokes. An additional randomization was achieved by Lassell who introduced circular instead of pure radial strokes by the polishing machine.

Sensitive and reliable test methods were essential for accurate polishing. The quality of the mirror could be estimated to some extent from the images but is was difficult to relate errors in the image to errors on the surface of the mirror. For spherical mirrors, in 1721 Hadley invented the autocollimation pinhole test at the center of curvature.

Aspherical surfaces could be tested by measuring the focal lengths related to circular zones on the mirror generated by masking the mirror, a technique first used by John Mudge at around 1780. Another early method to test aspheric mirrors was invented by James Short who thereby succeeded in fabricating a Gregorian telescope with two aspheric mirrors which was as good as the refracting ones at that time. He defocused the image and polished the mirrors until the image was uniformly illuminated. However, no details are known about his manufacturing methods and the polishing of aspheres, especially for the secondary mirrors, was so difficult that Short seems to have been the first and for many years also the only telescope maker who could master these problems. Therefore, until the middle of the nineteenth century, the Newton design was chosen for all large telescopes. It gave reasonably good images with the mirrors having large focal ratios at that time, even if the primary mirrors were not well aspherized or even simply spherical. The difference between a spherical and a parabolic primary mirror was quite small, for example only 2.6 μm for the 72 in. mirror produced by Lord Rosse. The biggest problem in those days was therefore to produce a good spherical surface in the first place.

Despite the lack of accurate test methods, the reflecting telescopes in the 18th century achieved the same resolution as the refracting ones. William Herschel achieved a resolution of 0.484 arcseconds with a 10 in. aperture, which was tested in vertical position, and he could even see the central core produced by diffraction. Lord Rosse could resolve a 0.5 arcsecond double star with his 72 in. telescope. Some of the telescope mirrors could therefore possibly be classified as diffraction-limited, that is their performance was close to the limits set by physical optics, but they were certainly seeing-limited, that is they did not further reduce the limitations for the resolution imposed by the frequently strong atmospheric effects prevailing in the British Isles.

2.3 Casting of mirrors

With the requirement of large reflecting surfaces, the casting of the speculum mirrors also became a problem. Figure 2 shows for a few selected mirrors the evolution of the diameters of the speculum mirrors, starting with 2 in. by Newton in 1671, 6 in. by John Hadley in 1723, 24 in. by James Short in 1736, 19 and 48 in. by William Herschel in 1784 and 1789, finally culminating in the largest speculum mirror with a diameter of 72 in. by William Parsons, the Earl of Rosse. His successors went back to more manageable mirrors with diameters of 48 in., for example Lassell in 1861 and the telescope makers building the Melbourne reflector in 1869.

The major problem for the casting was that the metal blanks had to fulfill several potentially conflicting requirements:

- Fractures had to be avoided, both overall fractures of the blank as well as small fractures especially in the front surface. This required ductile materials and low stresses in the substrate.
- The surface had to be polishable, which required surfaces which were both hard and homogeneous.

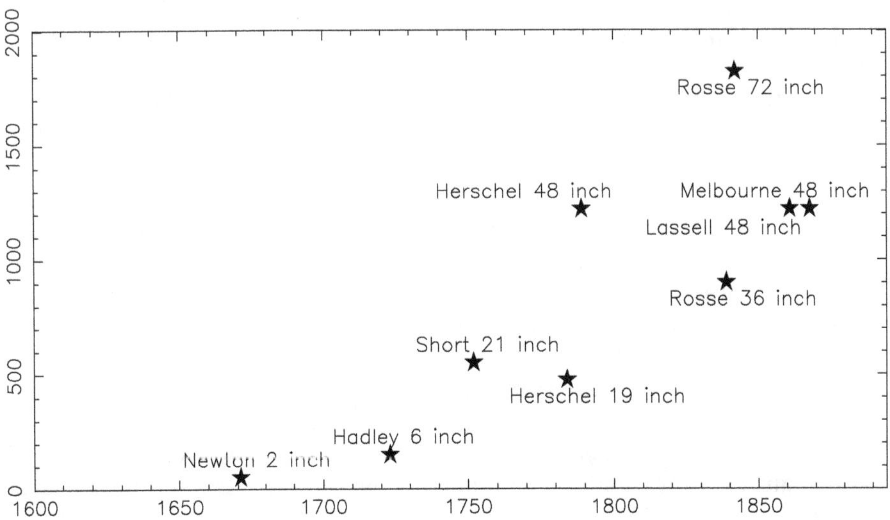

Fig. 2 Diameters of speculum mirrors

- The reflectivity had to be as high as possible. The optimum reflectivity between 65% and 75% for visible light is achieved with a fraction of 45% of tin.
- The material had to be resistant to tarnishing.

With only one solution at hand, namely that the blank had to deal with all these requirements, the number of parameters which could be tuned to fulfill the requirements as well as possible was naturally limited. The two most important parameters were the fraction of the tin and the cooling speed. Unfortunately, a particular choice for one of these parameters could be beneficial for some requirements, but detrimental for others. The following table shows in a qualitative way the impacts of such choices on the requirements listed above.

low	Fraction of tin	high	slow	Cooling speed	fast
+	Ductility	–	+	Stresses	–
–	Reflectivity	+	+	Surface hardness	–
–	Tarnishing	+	–	Homogeneity	+

For example, with a low fraction of tin the alloy would be ductile and therefore be less prone to developing fractures, but the reflectivity would be low and it would tarnish rapidly. Similarly, cooling down the blank slowly after casting would avoid large stresses and yield a hard surface, but could lead to inhomogeneities due to recrystallization.

To minimise the stresses during the casting, large temperature differences had to be avoided. The blank was therefore cast in one go as fast as possible. For example, the four tons of metal required for Lord Rosse's 72 in. mirror were cast from three containers in a mere three seconds. The metal was poured into an iron mould, a technique Rosse had taken over from the London optician Cuthbert. The thickness of the form was tuned to solidify the polishable surface homogeneously and at a speed which would give a dense, hard surface without disturbing crystallisation. In addition, hoop-iron was used to let excess air escape and avoid bubbles. The substrate was solid after 20 min and cooling down took 6 weeks. Despite all the precautions taken, two blanks fractured completely, one was porous because residues of tallow had evaporated, one had small fractures at the surface, and only the fifth casting was a full success. However, also the one with the small fractures was polished. This way the telescope, like others in the speculum period, had two mirrors. One was installed in the telescope while the other one was repolished, after tarnishing had reduced the reflectivity below acceptable limits. To improve the reflectivity the effects of admixtures of other metals like arsenic in the alloy were investigated, but with little success. So, in the end, the casting procedure was always a compromise between conflicting requirements, and, despite tremendous efforts by the telescope makers, there was not much room for significant improvements. Figure 3 shows a 48 in. mirror cast in 1789 by William Herschel, and a 72 in. mirror, the largest speculum mirror ever produced, cast in 1842 by Lord Rosse. Herschel had to reduce the tin content for his 19 in. mirror to 28% and to even less for his 48 in. mirror, resulting

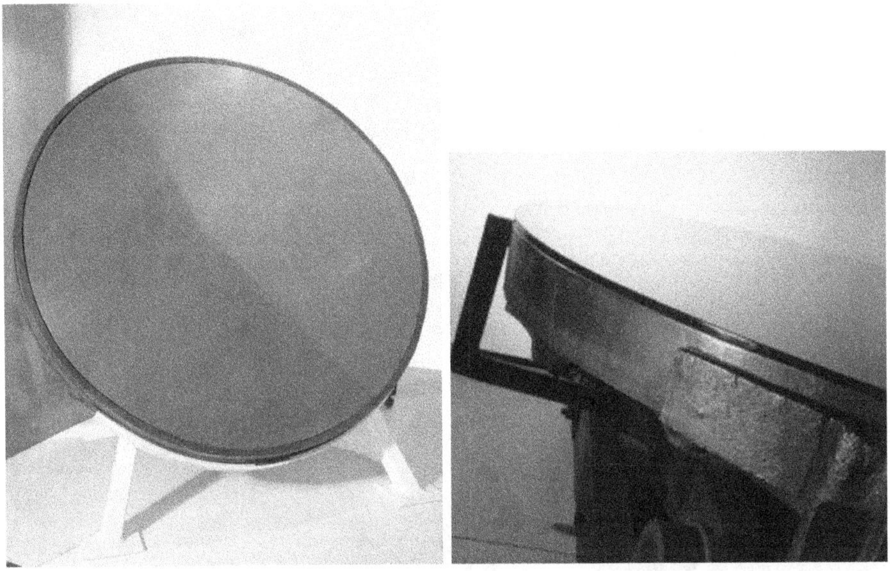

Fig. 3 *Left*: 48 in. mirror fabricated by William Herschel, *Right*: 72 in. mirror fabricated by Lord Rosse (Courtesy British Museum of Science)

in a lower reflectivity and guaranteeing rapid tarnishing. 45 years later Lord Rosse had succeeded in raising the fraction of tin to 31.8% for his even larger mirror, the precise percentage showing the care with which the fraction had been selected after several tests. Figure 4 shows the casting of the mirror for the Melbourne reflector, demonstrating the great public interest in this project.

Another disadvantage of speculum was its high specific weight. The total mass of the 72 in. mirror was a staggering 3 tons. Therefore, Lord Rosse also experimented with structured mirrors. Soldering together flat pieces of metal, he manufactured a 36 in. mirror, having the same diameter as one of his solid mirrors. The results he achieved with the two mirrors in his telescopes were nearly identical. However, he did not pursue the development of structured mirrors any further.

2.4 Support of the mirrors

As the mirrors became larger they also became more flexible. The figure of a mirror is strongly influenced by the forces acting on it. There are therefore two possibilities to maintain the proper figure at all angles of inclination. One is to build a mirror with sufficient intrinsic stiffness, such that the external forces do not have a strong impact on its shape. The second one is to design a support system which fulfills two requirements, first, that the sets of forces at different inclinations do not change the shape of the mirror, and, second, that the applied forces be independent of deformations of the mirror cell.

Herschel's mirrors were not supported from the back but only by a ring around their outer rim. Such a support did not fulfill the two requirements

Fig. 4 Casting of the 48 in. mirror for the Melbourne reflector in Dublin in 1866 (Reproduced from R.N. Wilson, Reflective Telescope Optics I)

mentioned above at all and the mirrors therefore had to rely completely on their intrinsic stiffness for figure stability. However, the casting process put a limit on the thickness of the blanks. Herschel's first 48 in. mirror was only 50 mm thick, too thin as it turned out during polishing. A second one had a thickness of 90 mm. Fortunately, even with the simple ring support, the difference of the figure in the form of spherical aberration between the mirror being in horizontal and vertical position was of the same order as the difference between a parabola and the closest sphere, and therefore to a large extent tolerable.

Nevertheless, two important inventions offered solutions for the two requirements. The first was the whiffle tree support, designed by Thomas Grubb and used for the first time in 1835 for a 15 in. mirror, and the second the astatic lever support, introduced by William Lassell and used for the first time in 1842 for a 9 in. mirror.

The principle of the whiffle tree support is shown in Fig. 5 on the left. For such a hypothetical two-dimensional system the mirror would be coupled to the mirror cell only at two points, for a real three-dimensional space it is coupled at three points. The locations of these points define the position of the mirror but do not affect its shape, thereby complying with the requirement of the independence of the mirror shape from its cell. The geometry of the whiffle tree defines the fraction of the total weight of the mirror which is carried at each support point. The support forces depend on the inclination of the mirror.

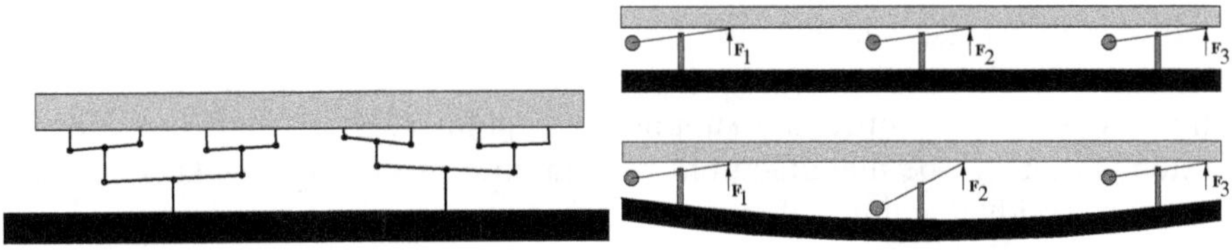

Fig. 5 *Left*: Principle of the whiffle tree support, *Right*: Principle of the astatic lever support

With a proper choice of this geometry, usually one where the support points are evenly distributed and apply approximately the same support force, one can indeed fulfill also the second requirement, namely the independence of the figure from the inclination of the mirror. This type of support was also used by Lord Rosse for his mirrors and is still nowadays the standard support for mirrors with diameters up to 2 m.

The principle of the astatic lever support is shown in Fig. 6 on the right. The applied forces are independent of the inclination of the lever as long as the support point, the axis and the center of gravity of the counterweight are located on one line. A deformation of the cell would change the inclination of the lever but not the applied forces and would therefore not modify the figure of the mirror. Again, with a proper distribution of supports and choice of the counterweights, similar to the one of the whiffle tree support, the figure of the mirror is independent of the inclination. This support is still the standard support for mirrors with diameters of about 4 m.

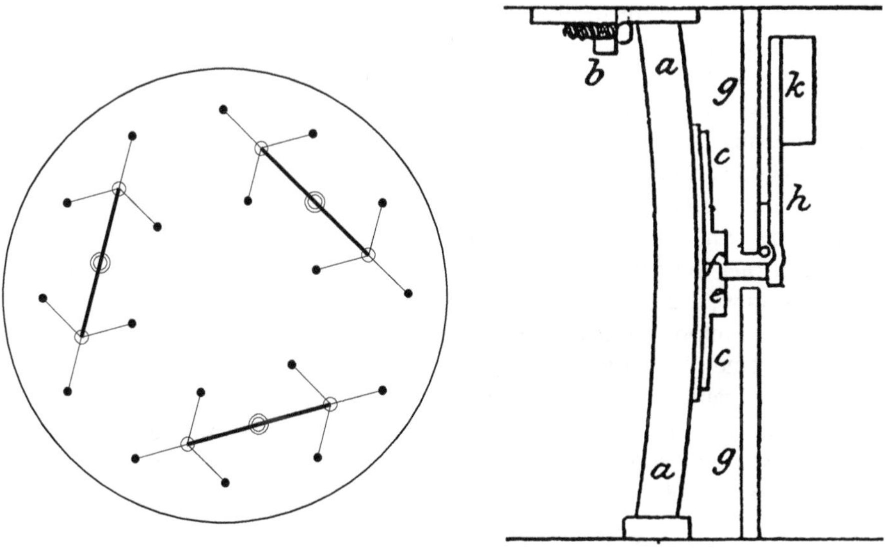

Fig. 6 *Left*: Sketch of a two-stage whiffle tree, with double circles representing the 3 pivots of the first stage, single circles the 6 pivots of the second stage, and filled circles the 18 support points, *Right*: Contemporary drawing of an astatic lever (Picture on the right reproduced from R.N. Wilson, Reflective Telescope Optics I)

In general, these support systems guarantee a substantial fraction of the figure stability and relieve the blank from the task of supplying a large intrinsic stiffness. However, since the supports are point supports and the support system can never be manufactured with infinite precision, the blank still has to have a residual amount of intrinsic stiffness.

2.5 Situation at the end of the speculum era

The only task of which the blank material had been relieved was to supply a very large intrinsic stiffness. However, regardless of all these efforts and developments, the problems related to the rather poor reflectivity and the rapid tarnishing remained unsolved until the end of the speculum era. A serious disadvantage was that the removal of the tarnished layer effectively required a new time-consuming re-polishing operation. This was the major reason why all the three telescopes with the largest speculum mirrors had a short lifetime and were never routinely used.

In addition, the mirrors were rather heavy. Nevertheless, in the middle of the nineteenth century towards the end of the speculum era, 50-in.-class telescopes had been developed which, under favourable conditions, delivered sub-arcsecond images. Furthermore, the 6 ft mirror of Lord Rosse remained the largest mirror for 75 years.

3 The era of the glass mirrors in passive telescopes

However, in the middle of the nineteenth century, two inventions opened up a new path. One, made around 1840 by Varnish and Mellish and, independently, by Liebig was the silver coating. The blank material was therefore also relieved of supplying the reflectivity and glass could now be used as a carrier for a thin reflecting layer. This offered several advantages. First, a fresh perfectly silver-coated surface reflected approximately 95% of the visible light. Second, glass was three times lighter than speculum and therefore required less massive support systems and telescope structures. Third, the surface was generally homogeneous, very hard and therefore much easier to polish. Early on, the first mirrors were coated with a silver layer by Steinheil in 1856.

The other important invention was the knife-edge test by Foucault in 1856. It was extremely sensitive to surface errors and showed many more details. It enabled Foucault to proceed, during the polishing process, from a spherical surface through an elliptical to a parabolic surface, testing the figure at all intermediate steps. Foucault was then also the first telescope maker to exploit both inventions. He built several telescopes, the largest one in 1862 with a silver-coated glass mirror with a diameter of 800 mm.

However, glass also had a serious drawback, its low thermal conductivity. In combination with a non-negligible coefficient of thermal expansion of the plate glasses used at the beginning of the glass era, it gave rise to large stresses, which made the casting of large blanks as difficult as casting large metal blanks. Long cooling times were required to avoid fractures, which, however, could result in partial devitrification. For the same reasons the mirrors were sensitive to temperature variations during operation. Another problem during the casting arose from air bubbles in the blank. The two largest mirrors which were produced under these constraints were a 60-in. in 1901 and the 100-in. mirror for the Hooker telescope in 1917, with thicknesses of about 8 and 13 in., respectively.

The polishing technology used by Ritchey for these glass mirrors was essentially the same as the one used for the speculum glass mirrors. Careful attention was, however, given to a number of details. For example, the content of dust in the air was minimized, grinding material was added through bores in the grinding shell, the polishing tool was made of pine wood with a last layer of cherry wood, pitch was replaced by sap with an additional layer of beeswax on the sap, testing was done in the morning and only during the warmer parts of the year because of the sensitivity of the blanks to temperature variations. All these details show that polishing was still very much a kind of art. The parabolization was achieved by a variation of the size of the squares on the polishing tool. Although carborundum was used as an abrasive during grinding, the grinding and polishing of the 100-in. mirror took 5 years.

The step to larger diameters could only be taken nearly 40 years later after the invention of low-expansion glasses, namely borosilicates, fused quartz, and, finally, glass ceramics with effectively zero expansion coefficients. The first mirrors made of low-expansion glass with diameters up to 120 in. were completed between 1933 and 1939, followed by a leap forward in size by the 200 in. mirror of the Palomar Telescope in 1949, shown in Fig. 7. Although a larger mirror with a diameter of 6 m was fabricated in 1976 for the Russian

Fig. 7 Honeycomb structure of the 200 in. mirror of the Hale Telescope on Mount Palomar (reproduced from R.N. Wilson, Reflective Telescope Optics I)

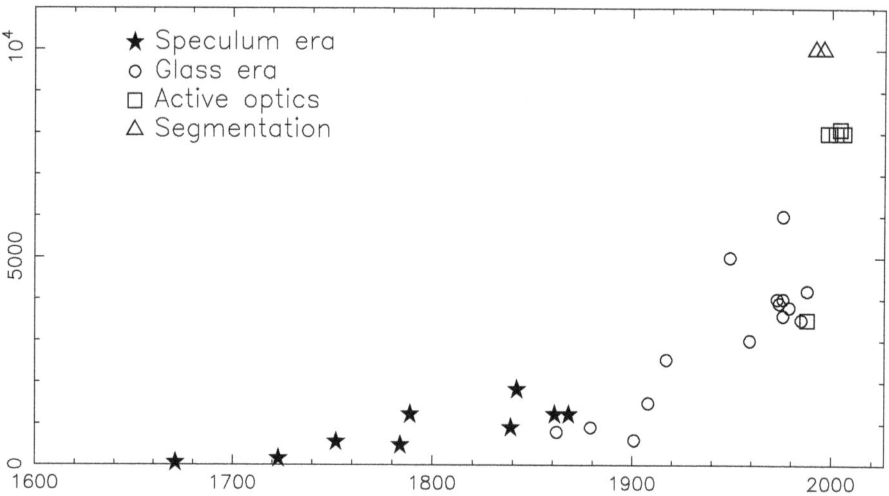

Fig. 8 Diameters of mirrors. *Stars*: speculum mirrors, *Circles*: glass mirrors, *Squares*: glass mirrors with active supports, *Triangles*: segmented mirrors

Zelenchuk Telescope, the Palomar mirror marked the onset of another consolidation phase which lasted approximately 40 years, as shown in Fig. 8.

The reason for the consolidation phase was the rapid increase of the weight of the mirrors with their diameter. To preserve the stiffness of a mirror and thereby avoid strong sensitivities of its figure to external disturbances, an increase in the diameter had to be accompanied by an even stronger increase in the thickness, leading to bulky and unacceptably heavy mirrors. The heaviest one ever cast was the solid 6 m mirror of the Russian Zelenchuk telescope with a weight of 45 tons.

One way to reduce the mass was to fabricate mirrors with a honeycomb structure, a solution adopted for the 200-in. Hale telescope. 5 tons of glass had to be removed during the grinding of the flat front surface to a spherical shape. The mirror was finally finished in 1947. It was a remarkable achievement, but also a monumental effort and it became obvious that an extrapolation of this technology to larger diameters was very difficult. Therefore, all subsequent mirrors until 1994, when a honeycomb mirror was cast at the University of Arizona for the WIYN Telescope, were solid. To limit the weight to approximately ten tons, the largest mirrors, except for the Russian 6 m mirror, had diameters of the order of 4 m and thicknesses of approximately 500 mm.

3.1 Situation at the end of the glass era

At the end of the glass era the blank was relieved of supplying the reflectivity and the bulk of the figure stability. The standard diameter of the mirrors of large telescopes was approximately 4 m. During the glass period substantial progress had been made in the field of glass technology, especially the development of low- and zero-expansion glasses. This enabled the casting of blanks with weights of up to 45 tons and made the figure of the mirror fairly insensitive to temperature variations. In addition, technologies had been developed to deposit thin layers of other metals like aluminium on the glass. However, little progress had been made in the field of supporting the mirrors. Therefore, at the end of the glass era, large mirrors, having insufficient intrinsic stiffness, were suffering from deformations introduced by imperfect support systems. As a consequence, the optical quality was often barely sufficient to exploit good seeing conditions.

4 The era of segmentation and active optics

Another fundamental development opened the door for the solution of the stability problem, namely the rapid development of electronic devices, in particular the computer and the CCD camera. The next big step forward was therefore the enhancement or replacement of passive devices by control systems for the fulfillment of some of the requirements, in particular the figure stability. This control became known as active optics. The first implementations of this technique were extensions of the passive astatic supports. The passive support still supplied most of the figure stability and the effects of the inaccuracies of the force settings were corrected by control systems which could modify the forces within a limited range. For example, the forces applied by an astatic support could be changed with a movable counterweight. The required signals were obtained with a wavefront analyser, recording the wavefront error and calculating the appropriate differential forces to be applied to the mirror and, in the whole telescope, also the differential displacements of the secondary mirror, as shown in Fig. 9. A first paper investigating such a scheme with the title 'Control of the optical surface of a thin, deformable primary mirror with application to an orbiting astronomical observatory' was published in 1973 by J.F. Creedon and A.G. Lindgren [3]. An experiment with a mirror with a diameter of 762 mm, a thickness of 12.7 mm and 61 actuators proved in the laboratory that a strongly deformed diffraction pattern of the uncontrolled mirror could be transformed into a diffraction pattern close to the diffraction limit. Spectacular as these results were, they did not convince a broad community. Understandably, such a technique was not tested first in space as suggested by the title of the publication. However, initially no one was prepared to implement such a technique even in a ground-based telescope. Furthermore, since the paper appeared in *Automatica*, a journal for control engineering, it remained largely unknown in the astronomical community.

Independently, in 1977, R.N. Wilson made systematic measurements of the telescope quality at the 3.6 m telescope of ESO, using the newly invented Shack-Hartmann method which gave easily interpretable information on the errors in the wavefront [9]. To separate the errors generated by the telescope from the ones due to the atmosphere, the latter were largely averaged out by long exposure times of the order of one minute. The most important remaining

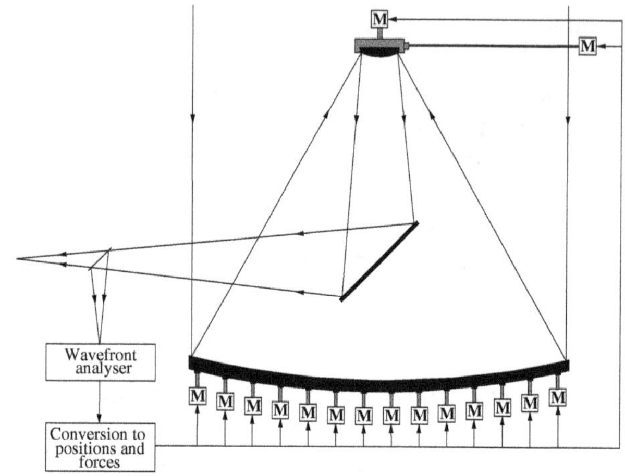

Fig. 9 Principle of active optics

errors were smooth aberrations like defocus, decentering coma mainly generated by a misalignment of the mirrors, spherical aberration mainly due to incorrect figuring, and some aberrations generated by mirror deformations, especially saddle-shaped astigmatism. The idea behind this scheme is that the deformations of a rather rigid elastic body like a mirror generated by disturbing external forces can only be rather smooth. This can be seen in Fig. 10, where the diameter of the circle containing 80% of the point spread function generated by the telescope is used as the criterion for the telescope quality. Apparently, after subtracting the most important errors, the initial telescope quality, ranging from 0.4 to 0.6 arcseconds, was reduced to less than 0.3 arcseconds, largely independently of the sky position. The conclusion

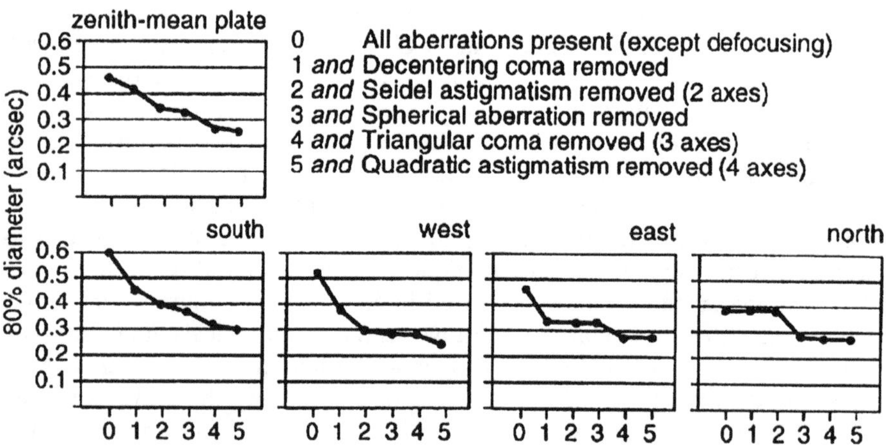

Fig. 10 Image quality of the ESO 3.6 m Telescope for different sky positions, given in units of 80% of the encircled energy, without and after mathematical removal of polynomial terms (Reproduced from R.N. Wilson, Reflective Telescope Optics II)

was therefore that one could recover the *intrinsic quality* of the telescope by measuring and correcting only these dominant errors, which were produced by misalignments and by deformations of the large primary mirror.

An experiment similar to the one by Creedon and Lindgren was done with a mirror with a diameter of 1 m, a thickness of 19 mm, and 78 supports on four rings. The geometry was identical to the support arrangement of the proposed 3.5 m New Technology Telescope, which, quite remarkably, was initially conceived by ESO as a testbed for new technologies, in particular active optics. Later, this was modified in the sense that the NTT had also to be capable of doing astronomical observations even if the active optics failed, a decision which, in the end, may have helped to convince the astronomical community of the feasibility and the advantages of active optics. However, this added the requirement that, even without active optics, the telescope still had to deliver adequate images. Therefore, the mirror thickness could not be reduced as substantially as one might have wished, but was only reduced to 241 mm, approximately half of the thickness of conventional mirrors with that diameter. The active optics design was the one shown in Fig. 9, with the force changes applied by astatic levers equipped with movable counterweights. Figure 11 shows interferograms of the NTT mirror in the workshop of the optical manufacturer without and with active corrections. The immediate success of the NTT on the mountain proved the feasibility of this active optics concept, and all of the large 8 m class telescopes have made use of this or similar techniques. The fundamental advantage was that the residual intrinsic stiffness, which was required in passive telescopes to cope with the inaccuracies of the forces applied by the astatic supports, could now be reduced even further, which made the fabrication of telescopes with mirror diameters of 8 m feasible and affordable. With closed-loop active optics control and improved polishing, greatly facilitated by computerized wavefront measurements

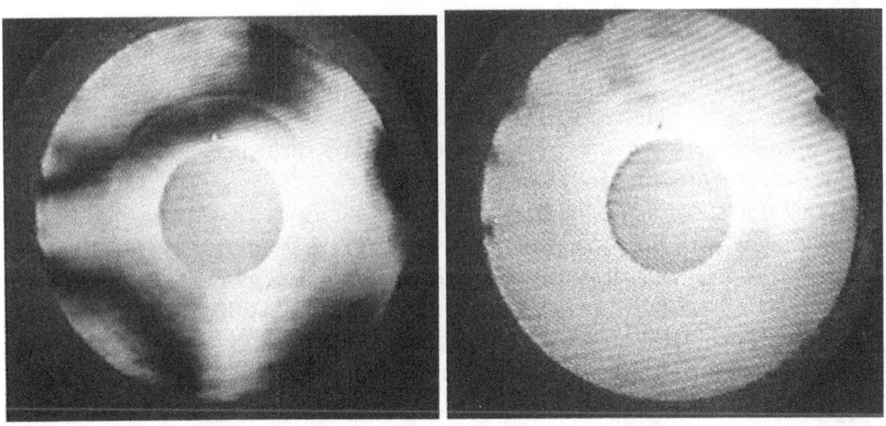

Fig. 11 Interferograms of the NTT primary mirror in the optical workshop without control (on the *left*), showing large amounts of astigmatism, and after active optics corrections (on the *right*), showing only the uncorrectable high-spatial frequency errors

and polishing procedures, the mirrors are effectively diffraction-limited, with deviations from the required surface of the order of only 30 nm.

One way to maximize the intrinsic stiffness and to minimize the control efforts for large mirrors is to build structured mirrors like that of the Hale telescope. This technique has probably reached its pinnacle with the casting of the 8.4 m honeycomb blanks by a team of the University of Arizona led by Roger Angel [1]. The mirror, cast in 1997 of borosilicate glass by spinning the mould to achieve a surface figure close the required parabola, has 1662 cells and an effective lightweighting of roughly 80%. Despite its 40% lower mass compared with a solid 8 m mirror, it is approximately ten times stiffer. Another motivation for this solution was that, due to the lower mass and the delicate structure with thin ribs, the mirror, supported by active ventilation, adapts much faster to the ambient temperature. However, even a mirror with these characteristics requires active corrections.

The other 8 m blanks are extensions of the NTT concept described above. In principle, this technique could be applied to mirrors of any size. The fabrication of monolitholic blanks significantly larger than the current maximum size should be feasible with fused quartz. However, a strong increase in cost and problems of transport seem to put a limit on the diameter of monolithic mirrors.

Currently, the most promising way of reaching larger diameters is the segmentation of the mirror. This approach was pioneered by a team under J. Nelson for the 10 m Keck Telescopes [5]. The individual segments can be rather thin, for example 75 mm for a segment diameter of 1.8 m from corner to corner in the Keck Telescopes. The individual segments are coupled to the mirror cell by three position actuators and are supported by passive whiffle trees with additional devices which allow the manual correction of the segment figures. Since the segments would follow the deformations of the mirror cell, neither the support system nor any other passive devices supply any figure stability for the complete segmented mirror. The intrinsic stiffness of the segmented mirror now required and therefore its figure stability is accomplished completely by another type of active optics, a closed-loop control based on sensors at the borders between segments which are capable of measuring the relative height and, in the Keck Telescope design also the relative angle between adjacent segments, and on three position actuators per segment correcting their rigid-body positions.

Another way of achieving the full figure stability by control is to support a thin mirror with position actuators only. Such a solution has been realized in the Hexapod Telescope of the University of Bochum. The mirror with a diameter of 1.5 m and a thickness of 55 mm is coupled to a mirror cell, made of a fairly rigid structure of carbon-fiber trusses, by 36 piezo actuators (see [10]). The support system is therefore much simpler than a combined passive/active force-based system. However, since the mirror would follow directly the deformations of the mirror cell, corrections have to be made much more frequently.

Whereas all systems discussed so far would not need a control system if the mirror cell was infinitely rigid or the passive support forces were infinitely accurate, the Chinese LAMOST Schmidt telescope [7], inaugurated in 2008, requires an active optics system even under such unrealistic assumptions. Conventional Schmidt telescopes were limited in size by the limitations on the diameter of the aspheric refracting plate which corrects the spherical aberration generated by the spherical mirror. Replacing this refracting plate by a flat aspheric mirror, in combination with the segmentation of the mirrors, gives the possibility of building much larger Schmidt telescopes. In the LAMOST all-reflective Schmidt design, the light is reflected towards a spherical mirror, which is fixed in space, by a flat siderostat mirror located at the center of curvature of the spherical mirror. Both mirrors are segmented. During observation the inclination of the flat mirror and therefore also the angle between the flat mirror and the spherical mirror change continuously. Since the required figure of the flat aspheric mirror depends on this angle the rigid-body positions as well as the figures of the individual segments have to be adjusted continuously by active optics control.

Finally, there is one type of mirror which needs very little control, namely a liquid mirror with a surface of parabolic shape which is naturally assumed by a liquid in a rotating container [2]. If mercury is used, the blank will supply the true figure, the accurate general figure and the reflectivity. The figure stability is achieved by a careful control of the speed of rotation. A liquid mirror can only operate vertically. The largest mirror built so far has a diameter of 6 m.

5 Current status

Apart from the liquid mirrors, in all other modern mirrors the substrate is only responsible for the true figure. A thin, in general, aluminium layer provides the reflectivity. Most of the other requirements are fulfilled by control systems. With the surface quality of today's large mirrors one gets close to the diffraction limit even for visible wavelengths. Glass is still the dominant substrate, but may be replaced in the future by aluminium or lighter and more rigid materials like silicon carbide. The maximum diameter of the mirrors is currently 10 m, but telescopes with segmented mirrors with diameters of 30 to 40 m are in planning.

Acknowledgements The author would like to thank Ray Wilson for a critical assessment of the draft version leading to many improvements on the historical details. Most of the information has been derived from the books by King [4] and Riekher [6] and from the historical sections in the book on reflecting telescope optics by Wilson [8].

References

1. Angle, J.R.P., Hill, J.M.: The manufacture of large glass honeycomb mirrors. Proc. SPIE **332**, 298–306 (1982)

2. Borra, E.F.: Liquid mirrors. Can. J. Phys. **73**, 109–125 (1995)
3. Creedon, J.F., Lindgren, A.G.: Control of the optical surface of a thin, deformable primary mirror with application to an orbiting astronomical observatory. Automatica **6**(5), 643–660 (1970)
4. King, H.C.: The History of the Telescope. Griffin, London (1955)
5. Nelson, J.: Segmented mirror telescopes. In: Foy, R. and Foy, F.C. (eds.) Optics in Astrophysics, NATO Science Series, II. Mathematics, Physics and Chemistry, vol. 198, pp. 61–72. Springer, Heidelberg (2005)
6. Riekher, R.: Fernrohre und ihre Meister, 2. Auflage, Verlag Technik GmbH, Berlin (1990)
7. Wang, S.G., Su, D.Q., Cui, X.Q., Chu, Y.Q., Wang, Y.N.: Special configuration of a very large Schmidt telescope for extensive astronomical spectroscopic observations. Appl. Opt. **32**, 5155–5161 (1996)
8. Wilson, R.N.: Reflecting Telescope Optics I, 2nd edn. Springer, Heidelberg (2004)
9. Wilson, R.N.: Reflecting Telescope Optics II, pp. 274–298. Springer, Heidelberg (1999)
10. Wilson, R.N.: Reflecting Telescope Optics II, pp. 214–216. Springer, Heidelberg (1999)

Exp Astron (2009) 26:19–34
DOI 10.1007/s10686-009-9157-6

REVIEW ARTICLE

Telescope mounts and dome structures

Jason Spyromilio

Received: 26 November 2008 / Accepted: 25 February 2009 / Published online: 23 April 2009

Abstract The evolution of telescope mounts and domes is presented as the astronomical use of the telescope is about to reach its 400th anniversary. The changes in the mechanical designs are traced to requirements generated from the evolution of detectors and instrumentation.

Keywords Telescope mounts · Domes

1 Introduction

The astronomical telescope enters its fifth century this year, or maybe next. The date and process of the invention of the telescope is an interesting problem to address and it is extensively covered in these proceedings and in many publications and books.[1] Within the context of writing about mounts and domes it is clearly self-serving to claim that while plenty of opticians may have put two lenses of appropriate curvature in a straight line and looked through them, the inventor of the telescope was the one who mounted them. Today the definition of a telescope[2] is not the optical system alone but rather the complete device or as we would nowadays refer to it as the system. So for the purposes of this paper, everything that is not a mirror (for which the reader is encouraged to see the paper by Noethe in these proceedings) is a mount.

[1] *The history of the telescope* by Henry King is an excellent reference on the subject.

[2] An optical instrument designed to make distant objects appear nearer, containing an arrangement of lenses, or of curved mirrors and lenses, by which rays of light are collected and focused and the resulting image magnified. (OED)

J. Spyromilio (✉)
European Southern Observatory, Karl-Schwarzschild-Str.2, 85748 Garching, Germany
e-mail: Jason.Spyromilio@eso.org

Instrumentation will be discussed only to the extent that it influences the design of the mount.

Considerable evolution in the mounts of telescopes has taken place since the days of Lippershey & Galileo. We shall be tracing this evolution in the context of the detectors used to sense the light collected. It shall be argued that the technology needed for the mounts to perform has been driven by the evolution of the requirements placed upon the systems rather than, as is often said, driving that evolution. Similarly, housing telescopes inside buildings is not particularly new. The evolution of the dome from being a protector of the telescope from inclement weather to an integral part of the control of the environment during observations, as well as during the day, also merits discussion.

This paper will only address the mounts and domes of optical telescopes. As coatings have allowed telescopes that work in the visible to obtain data in the ultraviolet and near infrared these same telescopes now extend their reach. In fact they are now oft optimized for near-infrared work but fundamentally the drivers for their designs remain the same. Moreover, given that these proceedings cover the Hubble space telescope and other space borne observatories extensively with dedicated presentations, this paper will keep itself firmly earthbound.

The author cautions the reader on the subjective nature of this article and recommends that, for learned treatises on the subject of telescopes, the reader consult Ray Wilson's 'Reflective Telescope Optics' volumes I and II and 'The Design and Construction of Large Optical Telescopes' edited by Pierre Bely.

2 The early days

The mounting of lenses such that one could see farther marks the creation of the telescope and even Galileo stipulated that this was to be credited to the opticians in the Low Countries. However, the astronomical telescope must be considered to be the one used for looking at the heavens rather than at ships at sea or enemies on land. At the time of its invention there was no difference to be found in the mechanics of the device depending on where it would be pointed. However, the requirements for the astronomical telescope differ and soon emerge as drivers for its evolution. Why did Galileo look at Jupiter and Venus? Except for the sun and the moon, these are the brightest things out there and it is hardly surprising that one, at least an inquisitive one, would point a device with a certain degree of magnification at these objects. Telescopes not only made distant objects appear nearer, they revealed invisible distant objects and they still do. What has any of this to do with the mount? The original mount had a relatively straightforward task to achieve, namely, keep the lenses pointing at each other and limit the view of the observer to the good parts of the lenses. Described elsewhere in these proceedings the reader will find descriptions of the brilliance of the opticians at the time of the invention. It was not only their ability to obtain, or make, two lenses with appropriate curvatures nor to line them up, but also their understanding of the problems leading them to stop them down to the size of the human pupil, avoiding the poor polishing and negating the poor optical quality of the glasses, that made the telescope work. Throughout its evolution telescope design has been a process of matching the optics to the detector.

As anyone who has tried to find Jupiter with a small hand-held refractor can certify, this is a difficult task. Not only was Galileo an exceptional physicist, theorist and experimentalist, he also must have been an incredibly skilled observational astronomer. The Galilean mount was good enough should one to be looking for ships at sea or at battlefield encampments from an unsteady horse. It was light and compact. However, for looking for a source a few arcseconds in size in the sky, paper tubes holding lenses were not the way to go. A device to hold the tube in place would be necessary if lesser mortals were to use the telescope.

The detector used in all early observations was well suited to the quality of the optics and the designs of the telescopes. The human eye is able to focus, has lousy resolution and cannot integrate for any meaningful length of time. At any given place on earth the sky goes by pretty slowly at 15 s of arc per second of time and unless the magnification is significant, the astronomical telescope simply acts as a flux bucket. With increasing magnification and the light gathering power of a many times the human eye, objects can be distinguished from their neighbours. Unsurprisingly, the science firmly focused on what the device and its detectors were good at: binary stars. Orbits began the era of modern astronomy and as long as technology did not improve, remained the principal object of study for quite some time.

The telescope went down a technological cul-de-sac as, in spite of Newton's caution,[3] the method of increasing magnification was to make ever longer telescopes. For binary stars you need a relatively small field of view and a mount that points reasonably well.[4] The mount followed the requirements and innovation was ever present. Hevelius' telescopes should be held as objects of admiration rather than astonishment. The use of pulleys in combination with load spreaders (multiple connections to the tube) for the accurate control are both concepts that have continued to be adopted more than two centuries later in telescopes. Wind shake would have plagued these structures and indeed they could only be used on the calmest of nights. Interestingly the concept of a dome to house these devices appears to have been a challenge too far.[5]

[3]Newton Opticks *For very long Tubes are cumbersome, and scarce to be readily managed, and by reason of their length are very apt to bend and shake...* although he uses this to explain why Huygens seems to have chosen to fix the object glass on a pole (see below).

[4]Hevelius in 1673 wrote to Flamsteed in a discussion whether naked eye observations or the use of sights would be better: *For it is not only a question of seeing the stars somewhat more distinctly (..) but whether the instruments point correctly in every direction, whether the telescopic sights of the instrument can be accurately directed many times to any observations, and can be reliably maintained.*

[5]However, with a goniometer in between them, Hevelius and his wife Elizabeth, the earliest of night assistants, are shown observing from the comfort of their house.

Huygens separated the two lenses, mounting them on separate structures, and hereby demonstrating in practice his command of aberrations. Separating the pointing of the optics from the pointing of the telescope gives an additional degree of freedom to the observer. Aligning for best image within the field of view rather than relying on the telescope to point to the sky and keep the optics pointing at each other is a technique only recently rediscovered with the invention of active optics.[6] Huygens' split mount was limited by the sensing technology employed the human eye.

The breakthroughs of the achromatic doublets are well described in Watson's very readable book [1] as well as in these proceedings and do not particularly affect the mounts or domes. The demands for better lenses did however bring to the fore instrumentalists that produced the most elegant and beautiful of mounts. Fraunhofer not only made the very best glasses, his preeminence in this field can best be indicated by the lengths that the British authorities went to in order to try, and in the end fail, to obtain similar quality glasses. In the history of telescope mounts Fraunhofer stands without peer in his time for the delivery of the Dorpat reflector.[7] The German equatorial mount, providing a perfectly balanced system, has remained since then a most popular solution to mounting a telescope. The Fraunhofer telescope mounts provided for sidereal tracking using pulleys and weights.

The earliest reflector by Newton should have been invaluable to astronomers. Not so much for the optical design which after all was surpassed by Gregory, Cassegrain and others, but rather for the fact that it clearly showed to Newton that the limitations of the optics were to be overshadowed by the limitations of the atmosphere.[8] The first seeing-limited telescope clearly needed a better-seeing site. Given the likely quality of the optics one can only suspect that being seeing-limited required a pretty dreadful site, something England was likely to be able to achieve without effort. The eventual move of British telescopes to better sites than could be found in the British isles has been brilliantly described by Hoyle[9] within the context of bureaucratic administration of science.

Mirrors rather than lenses made telescopes much bigger. Noethe (these proceedings) explains the evolution of ever better mirrors. The telescope mounts had to follow. Large structures were necessary to hold the mirrors in place and suitable locations for observing found. However, the detector was

[6]On the 8-m UTs of the VLT the focal plane is held 'under stress' in position suitable for the instrumentation but not a priori best for the mount. The differential loading on the mount is offloaded via a series of nested loops.

[7]*'the Dorpat glass in common with nearly all Fraunhoferes other objectives, surpassed the best English specimens and compares well with modern achromats.'* Henry C. King, The History of the telescope, Dover.

[8]Newton Opticks: 'Long telescopes may cause Objects to appear brighter and larger than short ones can do, but they cannot be so formed as to take away the confusion of the rays which arises from the Tremors of the Atmosphere.'

[9]Hoyle, Home is where the wind blows, University Science books, quoting Baade '[*siting a telescope at Herstmonceux*] *is like building a submarine in the Arizona desert'*.

still the human eye. Lord Rosse's telescope had limited tracking time. This was not a particular drawback. Staring for longer did not make the object any clearer. In fact one can surmise that, if the atmospheric conditions did not get you, the field aberrations would certainly only allow a good glimpse at the object in particular locations. However, the massive increase in flux collection combined with a realisation that magnification had its limits, resulted in the detection of spiral nebulae. Phenomenological qualitative investigations could proceed. Similarly the beautiful altitude over azimuth mountings by the Herschels served them well. The Herschels set out to map the heavens and therefore required sky coverage and reasonable astrometry. For that you need a steerable mount. The basic mount technologies, however, remained those of Hevelius: wood & pulleys. It is hardly surprising as wood is light and pulleys, thanks to millennia of engineering evolution, well understood. The azimuth rotation of the Herschel telescopes relied upon an early version of the bogie systems that are currently in use for radio telescopes and optical domes, and were also considered for the massive OWL concept.

Innovation in the support of the mirrors came in the form of the wiffle tree invented by Grubb and the astatic lever by Lassell but these will be dealt with in the contribution by Noethe. The Lassell mount appears to have inspired, or to have been re-invented, for the ESO 3.6-m telescope.

3 Spectrographs

Experimental astronomy can be summarized by the statement 'it's over there, it's rather faint and rather blue'.[10] With sufficient accuracy, astrometry, photometry and spectroscopy have given us a profound understanding of our universe and its contents. Spectroscopic techniques have revolutionized astronomy (see Watson's Stargazer). For the design of telescopes and domes initially spectroscopy was not a driver but became such after the introduction of sensitive detectors. Spectrographs became increasingly large with time and, with ever increasing telescope powers, increasing resolutions were demanded by astronomers. Quantitative stellar absorption line spectroscopy naturally requires resolutions around 30,000, requiring a rather massive spectrograph. Mounting the high-resolution spectrographs on the telescope tube itself became impossible and it was never really tried until the telescopes themselves grew much heavier. Multiple evolutionary paths for telescopes and domes converge towards coudé rooms, coudé relays and coudé spectrographs. However, while spectra were obtained before photography, in so far as domes and mounts are concerned, the challenges came from the change of detector. We shall concentrate on the remarkable demands that integration posed upon the telescope.

[10] Making the use of interferometry to determine the 'it's so big' a true parameter opener.

4 The era of photography

The photographic plate did not take long to revolutionize everything about telescopes. The optical design evolved rapidly to decrease the f/ratio of the beam and make the telescopes faster, much as Newton had suggested. Naturally this selected for reflectors. Indeed since the invention of silvering on glass and the spectacular Foucault telescope in France reflectors were overcoming the challenges of speculum. The Foucault, which can still be seen in Marseille, is a beautiful demonstration of the wedge mount[11] which allows it to be moved from one latitude to another. Others in these proceedings have the pleasure of describing this 'perfect telescope'. With a two-dimensional detector available field of view became an issue. Field aberrations and tracking mattered. With integration all manner of ills surfaced. With photographic plates the record of the telescope performance became evident for all to see. Tracking errors and jumps in the positions of optics must have been beautifully recorded and promptly thrown away as poor observations.

From the perspective of the telescope mount the biggest challenge was to be able to track accurately. With integration times of minutes and hours, jumps or drifts in the position of the telescope could not be tolerated. Already in the late 1800s the high-level specifications required tracking errors not to exceed 0.1 arc sec on the sky. Surprisingly, or maybe not so much so,[12] it is the specification that almost all modern telescopes also aim to meet. Simple pulleys or even motors with traditional gears and weights would not achieve the required accuracy. Much attention was diverted to the gears connected to the driving mechanisms. An example of the complexity in the drives can be found in Grubbs phase-locked clock drive [2]. Backlash and periodic errors were to be banished from telescope mounts. For the 100-in. Hooker the teeth of gear were cut and run in after it was mounted on the telescope. The operation was of such significance for the final performance that the name of the Italian instrument maker who performed it remains with us to this day: C. Jacomini. Periodic errors in the telescope tracking remained present and have plagued telescope builders to this day.[13]

Telescopes were rapidly becoming bigger and heavier. Already the 60-in. at Mount Wilson had a moving mass of 22 tons. The heavy thick glass required a suitably sturdy cell to support it. In addition, stiffness was necessary to retain collimation and it was achieved at the expense of adding weight to the telescope tube. To support the telescope and ensure a smooth operation a conventional bearing with the technology of the 1900s would not have sufficed. The 60- and the 100-in. telescopes were supported by a mercury bath that took

[11] An inclined alt-az mount.

[12] A tracking error of 0.1 arc sec allows the telescope to remain seeing-limited in all but the very best of nights.

[13] As an example, at the AAT the periodic error surfaced many years after commissioning when the encoder coupling rather than the drives was found to be mildly elliptical. The telescope could track exceptionally well but was instructed by the control system to jump.

most of the weight. The bearings then are used to retain the alignment of the mount. The hydrostatic bearing introduced at the 200-in. and later adopted in almost all major telescopes could be considered an evolutionary step from the mercury bath. Still pintle bearings are often considered in telescope designs to maintain the alignment while the hydrostatic bearings carry the load.

We briefly discussed the spectrographs before, when their first use on telescopes was mentioned. Their influence on the telescopes and domes has been profound. Large equatorial telescopes provide two natural foci, prime and Cassegrain, both of which have variable gravity vectors. As soon as long exposures with spectrographs became possible (thanks to the photographic plate) provision of a gravity-stable focus was much sought after by the astronomers. The coudé solution is natural for equatorial telescopes necessitating a bending of the beam (therefore coudé) out of the telescope tube and down through the elevation axis into the back of the telescope. The popularity of the asymmetrical yoke mount (as for e.g. the 82-in. Otto Struve at McDonald observatory) arises from the relative ease (small number of reflections) with which the beam is extracted from the tube and sent to a gravity-invariant focus. Coudé spectrographs have been the workhorse instruments for stellar astronomy for almost a century. The immense increase in quantum efficiency arising from the change from photographic plates to electronic detectors opened the extragalactic horizons for these high-resolution spectrographs.[14]

Solutions for the telescope mount arising from the combination of these requirements have been found by a number of engineers. However, within the context of the telescope mount, Marc Serrurier stands out without doubt as having a most profound influence. The Serrurier truss passively maintains the collimation of the primary and secondary mirrors of the telescope by allowing them to 'fall' by the same amount and parallel to each other. This was one of many innovations that the 200-in. brought to the telescope world. It is safe to say that the soul of the 200-in. can be found in almost all telescopes to this day.

With the postwar increase in availability of funds for astronomy, the ambitions of most astronomers could be met. The astronomers outside California also wanted access to a 200-in. class telescope. Budgets and schedules tempered the ambitions to the 4-m class. The mounts varied, some keeping close to the horseshoe variant 'English' equatorial that the 200-in. employed, as is the case at the Calar Alto Max Planck 3.5-m and CFHT. At the KPNO, Blanco and AAT the tube was moved to sit in the horseshoe. At the 3.6-m on La Silla the tube was moved forwards of the horseshoe, in effect reproducing the work of Lassell a century earlier. Thanks to improved mechanical techniques, the inconvenience of blocking the telescope from observing the pole was removed.[15]

[14]The high stability attainable in spectrographs is providing on of the most exciting science cases for the EELT namely the direct detection of the expansion of the universe using CODEX.

[15]Lassell's telescope mount required bracing between the fork presumably to provide sufficient stiffness.

All designs work well and none of these telescopes were limited by their mounts. Interestingly they seem to have been limited by the quality of their optics and their domes.[16] To compensate for their smaller apertures with respect to the 200-in., these 4-m telescopes were placed at 'better' sites. Seeing-limited remained the name of the game but with natural seeing now understood to be better than the best images the telescopes delivered, much effort was devoted to accurate guiding, cooling of the telescope and ventilation inside the dome. For example the AAT employed enormous fans to ensure that the dome was appropriately ventilated, although rumour has it that it did not make much difference whether the fans sucked or blew air, as long as they did one or the other.

The use of computers to control the positions of the telescopes combined with feedback from electronic detectors made 'blind' tracking a moot point. The skill of the observer in being able to 'see' fainter was no longer critical. The mounts required tolerable pointing and tracking until a guide star could be acquired. Excellent pointing such as achieved at the AAT made acquisition faster and observing more efficient but it did not change the scientific capabilities of the telescope.

5 The return to altitude–azimuth telescopes

The altitude over azimuth mount was popular with the Herschels, Nasmyth and others. As tracking requirements became tougher the equatorial was clearly advantageous. The BTA 6-m telescope is widely acknowledged to be the modern-era leader in the use of altitude-azimuth mounts. Although the 6-m design started in 1959, in the western world the generation of telescopes designed or built in the 60s remained equatorials.[17] Neither the dimensions of the telescopes nor their performance was limited by this choice. While alt-az mounts are widely credited with making telescopes cheaper this is only partially due to the mount. The dimensions of the domes were not dictated by the mounts. Rather it was the speed of the optics that required an evolution in polishing and support of the mirrors to change from the f/3.3 of the 200-in. to the f/1.8 of the VLT. Almost the entire 4-m class of telescopes of the 60s had f-ratios of order 3. The BTA 6-m has a primary f-ratio of 4 and subsequently a long tube and an enormous dome.

The MMT, WHT and NTT finally broke the back of the equatorials. With faster f/ratios on the primary (e.g. NTT at f/2.2) the telescopes became stubbier and subsequently also stiffer. This evolution was as much due to the improved polishing techniques as to the confidence that with active optics any residual

[16]The limitations of expectations may be as much to blame with 1 arc sec images invariably deemed to be better than expected and therefore good enough.

[17]For an educational view of the choice of mount and the conservatism of big science see Hoyle, The Anglo-Australian Telescope, University of Cardiff 1981.

errors left in the mirrors after polishing could be corrected at the observatory.[18] Stiffer telescopes allowed for smaller domes and less wind protection of the structures. The NTT worked spectacularly well and although the author risks the wrath of those involved, success cannot be attributed to any single technological advance. It was not the active optics that made it good, for it was many years that the system reliability was poor and there was limited use of closed loop operations. The exceptional images the telescope delivered had much to do with the ability of the ESO staff to properly align the telescope using the tools of active optics and the exceptional conditions of the site. The venerable 3.6-m telescope on La Silla after it was properly aligned also delivered exceptional images.

The breakthrough of the NTT was profound, as much psychological as technical. The active optics technology gave telescope builders the confidence to deploy thinner mirrors. Conservatively, the NTT primary was made thick enough to operate without active optics and was almost identical in thickness to the larger 3.8-m mirror installed at UKIRT[19] that predates the NTT by almost decade. The success of the early closed loop operations gave the confidence necessary for the next generation of monolithic primary mirror telescopes.[20] That telescopes would be operated in an active fashion became the norm, albeit with some cases of exaggeration where it is not strictly necessary resulting in unnecessary complications.However, of all the large telescopes that followed only the ESO telescopes operate in strict closed loop fashion without the use of calibrated lookup tables.

One limitation of the alt-az mount is the lack of true gravity-invariant foci. Of the major alt-az telescopes in operation today only the VLT has a provision for a coudé focus. It should be noted however that this requirement was generated from the need to combine the four 8.2-m telescopes either coherently or incoherently rather than by 8-m instrumentation. Although equipped to be able to host a gravity invariant instrument, the VLT coudé focus is used only as a relay for the combination of the beams (albeit equipped with sophisticated adaptive optics systems, beam separators for interferometry and other such facilities). For a limited field of view beam de-rotators can provide the equivalent of a gravity invariant focus at the Nasmyth focus of an alt-az mount. Furthermore, the use of optical fibres to transport the beams has overcome the challenges of incorporating a coudé mirror relay into most alt-az telescopes. However, with telescopes of ever increasing dimensions the distances over which fibres need to transport the light can be prohibitive for broad wavelength applications.

[18] On-site polishing was the case for the 200-in., while the NTT was saved from an error in the primary by changing the forces on the primary and refocusing the telescope.

[19] The UKIRT primary dimensions were selected for low cost rather than performance.

[20] A prima facie case of trust in engineering, as the telescope did not operate in closed loop for astronomy until 1997 by which time the VLT was well into its construction phase.

6 The 8 to 10-m telescopes

The move from 4-m to telescopes twice that size required new concepts. Much in the spirit of the 200-in., the design and construction of the Keck telescopes was a truly audacious enterprise. Segmentation was not a new idea but certainly making it work, and on the biggest telescope, was a spectacular achievement. It is not difficult to see why it is so profound and why enough credit cannot be given to the team that achieved this. The 6-m BTA telescope has a primary mirror that weighs 42 tons. Each of the 36 segments making up the Keck 10-m primary mirror weighs only 400 kilograms for a total of a bit over 14 tons. The brilliance of being able to command the technology to polish the off-axis segments cannot be overestimated especially when the final primary f-ratio of 1.75 is considered. For the mount, the release from having to carry the enormous weight of the mirror plus its cell transformed into providing a stiff platform to support the individual segments without them moving too far from their original position.[21] The use of an altitude over azimuth mount for the telescope appears natural. From a construction point of view the alt-az mount makes it significantly easier to provide improved stiffness while maintaining the orthogonality of the axes that makes control relatively straightforward.

Keeping the primary dimensions close to 8 m, glass manufacturers, both commercial and academic, were able to cast, anneal and polish monolithic mirrors. Telescopes with thin meniscus mirrors were favoured by some while others opted for light-weighted thicker mirrors. Seven 8-m telescopes are in operation today with thin meniscus mirrors while three 6.5-m and one twin 8.4-m telescope are in routine operation using light-weighted thicker mirrors. All of these telescopes rely upon active optics to maintain their shapes. Actuation took the place of the Serrurier truss while the astatic levers or pneumatic supports already deployed in telescopes from the 200-in. onwards became an integral part of the control strategy. Significant variations in the designs can be found between the various 8-m class telescopes: The UTs of the VLT appearing most classical maintaining a Serrurier-like structure; Gemini has a very light upper structure and lacks the Nasmyth platforms; LBT lacks Nasmyth platforms and uses a novel rocking chair mount supporting a number of secondary configurations on retractable arms.

A variety of driving mechanisms have been deployed at the 8 to 10-m class of telescopes from roller friction drives to gear drives to direct drives. Again, much as was the case with the designs of the 4-m class mounts, they all work. There is no unique solution that has been universally adopted.

With increasing diameter and higher exposure to the wind, some of the mounts were no longer able to maintain the telescope pointing/tracking. At the VLT the tip-tilt secondary mirror is permanently used in field stabilization

[21]The relative positioning of the segments is critical as errors in position appear as aberrations of individual segments.

mode to correct for the effects of wind shake on the telescope mount. This is not the case at all 8-m telescopes where better wind protection may be afforded to the mount. However, the exceptional tracking performance of the VLT is largely to be credited to the integrated approach of control of the focal plane, including field stabilisation and all aberrations.

7 Domes of the past

Early engravings of telescopes and descriptions of observing do not put much emphasis on domes or structures to protect the telescope. Indeed leaving the telescope out in the open did not appear to do it much harm. There exists some evidence that for astrometric work some observers preferred the comfort of a room with a big window but as telescopes grew this became less of an option. Almost everywhere a telescope can be deployed the weather can safely be assumed to occasionally turn bad.[22]

Protecting the telescope from the elements is a multi-faceted job. Even today the optical elements do not appreciate large changes in temperature. In early reflectors the speculum could, and would, crack, while prior to the use of low thermal expansion glass the mirror shape would change sufficiently to spoil observations. Much effort was put into limiting the excursions in temperature of the 60- and 100-in. mirrors. Including insulation and conditioning of domes is as much a necessity today as it was almost a century ago. Today the requirements on keeping the temperature of the mirror close or below ambient arises from the need to eliminate mirror seeing rather than maintaining shape. The end result is the same. Massive air-conditioning units keep domes cooled and ready for the beginning of the coming night.[23]

With the telescope tube lengths defining the dimensions of the system, hemispherical domes were obvious solutions. However, the inclusion of cranes and other handling devices increased the volume of the domes and increased costs. The requirement for multiple focal stations required the exchange of the secondary mirrors. For different solutions to this challenge one can consider the AAT automatic crane operation exchanging the entire top ring and the 3.6-m on La Silla that loads different secondary barrels into the telescope. Theoretically these are automatic operations. Although the AAT can exchange a secondary and resume operations in under 1 hour, it is not considered to be an automatic operation.

To escape the turbulence associated with the ground (surface layer), telescopes were elevated well above the ground. This had both positive and negative side effects. On the negative side the enormous buildings also functioned as laboratories and storage and operations centers thereby generating

[22]Even at Paranal the author has experienced rain, albeit once in a decade.

[23]At Paranal almost 1 MW of power is consumed during the day by the air-conditioning systems on the domes.

unnecessary heat loading on the base of the telescope. Pioneering work was undertaken at the CFHT, cooling the floors of the telescope and isolating the observing space from the warmth that humans seek. The large domes presented ventilation problems for the telescope and created dome seeing. On the positive side, the domes protected the telescopes from the wind. The relatively weak structures of the equatorials of the 60s and the not so stiff control mechanisms would not have been able to deliver the tracking performance under high wind loading. The large domes appear to have been a necessity rather than unfortunate consequences of mount design or handling.

Domes however were and remain very expensive. One method to drive innovation is to limit the budget. At the MMT and later at the NTT the dome was made to co-rotate with the telescope at all times, resulting in a reduction in at least one dimension of the dome. With the smaller dome sizes came less dome seeing as adequate ventilation was easier to achieve. The fast primary mirrors of the telescopes made them stubby and stiff. Control of the altitude against a stiff azimuth fork made for good wind rejection. On the NTT walls were built into the dome to separate the instrument rooms from the telescope and reduce heating. The entire building is air-conditioned to ensure that heat does not enter the telescope chamber. However, the drawback of the instrument rooms is that there is poorer ventilation of the telescope chamber when the wind is coming from the side. Most large telescopes have now cut ventilation doors into the cladding of the domes and thereby improve the airflow and remove trapped air from their domes. Without much ado, the community pretty much ignored the co-rotating dome. The reasons are relatively straightforward and are mostly related to cost. With the exception of the somewhat simpler control, co-rotating domes were not smaller. The size of the Nasmyth instrumentation installed at most telescopes negated any real volume advantage that a co-rotating dome would provide.

The 8-m domes have generally followed the same principles although, to save volume and costs, some have very limited crane capabilities. On the VLT the dome crane has a load limit of 5 tons barely enough to lift the instruments to the Nasmyth platform. The exceptions are the Subaru dome with a top-end changing capability necessary for the unique prime focus of that telescope and the LBT dome that appears to have doubled as a construction crane used for lifting and handling large pieces of the telescope from the ground floor into the telescope chamber.[24]

The 8-m domes were constructed following detailed analysis of airflow using wind tunnels and sophisticated computer simulations. During the design phase of the telescopes concerns arose about the stability of the thin meniscus mirrors under wind loading. Although the effect of mirror deformation under wind loading is present, in routine operations it is a negligible source of error.

[24]The environmental constraints at Mount Graham necessitated a minimal footprint for the telescope.

The smaller domes however no longer protect the telescopes as well from the wind. On the VLT the wind speeds and turbulence at the level of the secondary mirror can exceed the external wind speed. The Unit Telescopes' altitude axes (all four of them) cannot totally reject these disturbances. This effect was predicted and. as mentioned above, the telescopes employ field-stabilizing secondary mirrors that are used in closed-loop operations to reject the effects of wind shake.

Not only are the telescope mounts no longer responsible for collimation, they are also allowed to dynamically misspoint. The combination of the active optics and the other systems of the telescope maintain the focal plane in position and with negligible aberrations. A corollary of this control strategy is that the pointing of the mount is no longer critical,[25] relaxing dramatically the requirements on the mount. The mount no longer strives to point but rather to maintain alignment to the extent that the closure of the active loops can be fast.

8 The telescopes and domes of the future

At the time of writing three major new extremely large telescopes are in a detailed design phase: the TMT,[26] GMT[27] and EELT.[28] The 30-m TMT and 42-m EELT are finely segmented telescopes with primary mirrors assembled from a large number of 1.45-m point-to-point pseudo-hexagonal segments. Both telescopes have very fast, f/1, primary mirrors. The 21-m Gregorian GMT is also segmented combining seven 8.4-m circular segments. The mounts are alt-az rocking-chair mounts. All pretence of a Serrurier truss is gone. The GMT exploiting the extreme speed of its primary mirror segments (f/0.7) and the natural gaps between segments has a tripod support of the seven adaptive secondaries. The TMT, after some consideration of a Gregorian design, has reverted to a Ritchey-Chrétien baseline. Both telescopes have exceptional credentials in their design and execution teams with TMT largely following in the footsteps of Keck while benefiting from the Gemini project and the GMT having on board the mirror-making skills of the university of Arizona and being led by the team that successfully delivered the twin Magellan telescopes, regarded to be peerless in image quality and cost-effectiveness.[29]

[25]With the exception of the case where the field rotation is unguided and relies upon the position of the telescope being the 'true' pointing.

[26]www.tmt.org

[27]www.gmto.org

[28]www.eso.org

[29]This is a clear bias of the author.

The EELT has adopted a novel optical design in response to its top-level requirements.[30] These requirements are that it is to have a primary mirror of 42-m, have adaptive optics built into the telescope and provide a gravity-invariant full-field focus for instrumentation. The later requirement necessitates a Nasmyth or Nasmyth-like platform where the beam can be folded downwards towards a focal station. Building adaptive optics into the telescope rather than as a postfocal system creates a significant complication in the design. While in a futuristic telescope an adaptive primary could be considered, even its advocates appreciated the challenge of doing so within constraints of a project planned for operation in the next decade. A Ritchey-Chrétien design for the EELT would necessitate a very large secondary mirror to be made adaptive. In the case of a Gregorian design for the EELT the secondary would be even larger. Preliminary estimates of the schedule and cost of the adaptive secondary mirror in the case of the Gregorian design indicated that not only would it be the single most expensive item in the telescope, it would also be the most technologically challenging. Even with the most optimistic schedules, the secondary mirror would dictate the critical path for the project. The combination of all of these factors together with other considerations such as the dimensions of the dome, the reduced performance of a longer mount and the poor suitability of the Gregorian design for the foreseen instrumentation necessitated a different solution for the EELT. The optical design finally selected is a three-mirror anastigmat used on-axis with two folding flats that are used to bring the beam out of the mount. The f/1 primary mirror reflects the beam to a convex 6-m secondary mirror. The f/4 beam is focused at the hole of the quaternary mirror and goes through to the mildly aspheric tertiary. The return beam comes off the slightly inclined quaternary and a field-stabilizing fifth mirror to reach the Nasmyth platform.

The EELT selected mount design is a rocking-chair design that is naturally balanced while retaining the altitude rotation axis above the primary mirror. A driving requirement for the design is the need to ensure the steel structure supporting the primary mirror units retains the primary mirror shape to the maximum extent possible. A 2,500-ton structure, such as the altitude 'tube' of the EELT, is unlikely to maintain its shape as the orientation of the gravity vector changes. For the purposes of guaranteeing the safety of the mirror and in seeking to minimize the cost of the actuators supporting the mirror segments, the stiffness of the mirror substructure is critical. Therefore the design effort is strongly focused on minimizing the deflections of the primary mirror of the telescope. Although the current passive design meets the specifications, provision for actively changing the lateral loads on the cradles supporting the telescope is being considered. The structure itself becoming active is a natural extension of active optics.

[30]Following on the Huygens proposition *Quanquam lentes non frustra sint multiplicandae, quod et vitri crassitudine et iteratis reflexionibus non parum lucis depereat; hic tamen utiliter id fieri experientia docuit* loosely translated as *much light is lost due to extra reflections so avoid them unless it is useful not to do so.*

The first eigenmode of the EELT telescope mount is above 2 Hz and similar performances are found in the design of the TMT mount. This places it well below the 8 Hz of the VLT structures but within a very respectable range for such an enormous structure. The deflections of the secondary mirror with respect to the location of the primary (itself not a well-defined entity) are below 5 mm. Although for a fast optical system these numbers are large they are within range of active collimation and, with help from the adaptive optics and the field stabilization units, it is expected that the residual focal plane aberrations will be controlled. All mirrors in the EELT, with the exception of the tip-tilt M5, are active. Each primary mirror segment is equipped with warping harnesses that redistribute the load amongst the wiffle tree supports. The secondary and tertiary mirrors are classical and, as in the NTT/VLT primaries, active. The deformable quaternary is 2.5 m in diameter with thousands of actuators while the fifth mirror is expected to correct for as much as 2 arc sec of tilt error from the mount.

The domes of the future telescopes are also differ much amongst each other. GMT has chosen a carousel type dome, as for the VLT and LBT. A very large number of openings is envisaged to the extent that the telescope can be almost totally exposed to the external air. The GMT has a very low centre of gravity (below the primary mirror) and is very stiff with minimal support necessary for the secondary mirrors and therefore can be expected to perform better than its competitors in the presence of wind. TMT has selected a calotte style of dome rarely used before in large telescopes.[31] The dome is compact, economical and according to the detailed CFD and wind tunnel analysis provides exceptional protection to the telescope without restricting ventilation. The EELT dome design is more traditional, pushing the hemispherical dome to be as close as possible to the allowable free volume of the telescope mount. The sizes of these domes are comparable to football stadia, giving a useful metric for the challenges of construction such a structure on top of 10,000-ft or even higher mountains.

9 Concluding remarks

Ground-based telescopes continue to increase their flux collecting and resolving powers through ever increasing diameters of mirrors and adaptive optics systems built into the telescopes or post-focal systems. It is unclear if the current generation of giants being designed constitutes another stage in the evolution of telescopes or a cul-de-sac. Only time will tell.

It is clear that the community for which the telescopes are being built is rapidly evolving. The large community of users[32] is often detached from the

[31] The Pic-du-Midi Bernard Lyot 2-m is the only telescope of significant size employing such a dome with mixed opinions as to its success.

[32] 4,500 registered users populate the VLT database.

apparatus and nowadays much emphasis is placed on 'science-ready' data. Much as the automation of the 4-m class telescopes negated the advantage of skilled observers, the automation of data acquisition and reduction is negating the advantages of the current, and somewhat aging, generation of observers. Assuming we succeed in the design and construction of the forthcoming generation of telescopes, we will be making the species of experimental astronomer extinct and the new species of virtual astronomer will rule the day. Evolution and progress unfortunately are not necessarily the same and while the former is unavoidable, the latter requires effort.

Acknowledgements The author would like to thank the ESO EELT project team for educating him and notes that all errors and opinions in the text are his alone. Lothar Noethe not only has had the patience to educate the author but also to read early and late drafts of this text and much improve them.

References

1. Watson, F.: Stargazer. Da Capo, Cambridge (2005)
2. Glass, I.S.: Victorian Telescope Makers. IoP, Bristol (1997)

Exp Astron (2009) 26:35–48
DOI 10.1007/s10686-009-9155-8

REVIEW ARTICLE

Adaptive optics: a breakthrough in astronomy

Pierre Léna

Received: 5 December 2008 / Accepted: 24 February 2009 / Published online: 24 March 2009

Abstract Until the 1970s, atmospheric seeing was considered as an absolute limitation for angular resolution of ground-based optical telescopes, exactly at the time of the conception of the new generation of giant optical telescopes, as the VLT and the Keck. Emerging in the context of the cold war with many constraints due to the research being classified, but with the new possibilities of digital control, astronomical adaptive optics was shown to be feasible in 1989 and gradually convinced an initially skeptical astronomical community of its potential. Twenty years later, it is a mandatory ingredient for the planning of Extremely Large Telescopes on the surface of the Earth, and has allowed many discoveries concerning galactic and extragalactic objects. Some directions for new developments are discussed.

Keywords Adaptive optics · High angular resolution · Astronomical imaging

1 Introduction

Adaptive optics (AO) is a technique which emerged in the 1980s to correct the effects of atmospheric seeing on images produced by optical telescopes, observing from the surface of the Earth at visible or near-infrared wavelengths. It basically consists of a scheme for improving the performances in angular resolution of an optical imaging system, through image analysis and feedback to one or several optical elements, by using digital computing. Both active and adaptive optics operate on this principle, the difference being that the former

P. Léna (✉)
Université Paris Diderot & Observatoire de Paris - LESIA, Paris, France
e-mail: pierre.lena@obspm.fr

uses low-frequency feedback loops (usually $f \leq 1$ Hz), while the latter works at the higher frequencies which are characteristic of atmospheric seeing ($f \leq$ several kHz). Figure 1 illustrates the three basic components of such a system: one (or several) deformable mirror(s), a wavefront sensor and a control loop.

The advent of AO, made possible by the performances of modern computing tools, has produced a profound change in optical astronomy from ground-based observatories, where AO-equipped large telescopes can now reach their diffraction limit, at least beyond $\lambda \approx 1$ μm. Not only has this increased the angular resolving power of large optical telescopes by a factor close to 100, but it had also made possible the use of these large telescopes for optical interferometry on long baselines. Finally, the feasibility of AO opens the way to a new generation of optical Extremely Large Telescopes, with a further gain in angular resolution made possible. In addition, many subtle developments will considerably increase the capabilities of current AO systems.

This article presents the early history of adaptive optics in Section 1. After recalling in Section 2 the main properties of atmospheric seeing, and characteristics of AO systems, Section 3 presents the first and hesitant AO steps, and Section 4 the blossoming of astronomical AO in the last two decades. Finally, Section 5 examines briefly the future of AO in various astronomical contexts, such as the observation of exoplanets, the imaging of remote galaxies and the concept of Extremely Large Telescopes.

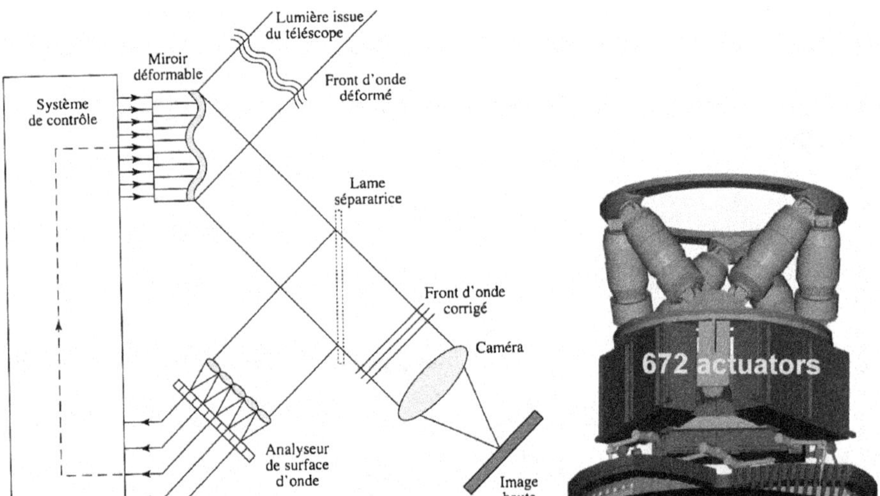

Fig. 1 *Left:* All adaptive optics system use the same basic scheme. A deformable mirror corrects the phase deformations of the incoming wavefront. They are measured at a wavelength λ_{ref} by a wavefront sensor, and the corrected image is formed on a camera operating at λ_{obs}. A computer analyses the WFS signal and activates the mirror. *Right*: A recent development of adaptive optics, where the secondary mirror of the telescope is the deformable element: the Large Binocular Telescope (Arizona) is equipped with such a mirror at the end of the years 2000

The reader will find detailed reviews, established during the 1990s, on the fundamentals of astronomical AO in [1–4].

2 Atmospheric seeing and adaptive optics

Atmospheric seeing has long been considered as an inevitable plague to optical astronomy. Newton, in his *Opticks*, properly understood its origin and advised to place the telescopes at high altitude in quiet air. Solar (e.g. Jean Rösch at Pic du Midi in the 1950s) and double-star observers realized later that the random evolution of images may allow the high spatial frequencies to be sometimes observable in an image. In 1969, Antoine Labeyrie systematized this intuition and introduced speckle interferometry, as an a posteriori treatment of snap-shot images, making it possible to recover in part the information up to the diffraction limit of the telescope. Speckle interferometry was developed in the 1970s at optical wavelengths both in the visible by Gerd Weigelt [38] and the infrared by Pierre Léna [39]. Beside its scientific results, it brought a great deal of detailed understanding of the atmospheric properties, which were outlined by François Roddier in his 1981 paper: *The effects of atmospheric turbulence on optical astronomy* [13]. This understanding would prove very useful for the early development of adaptive optics.

The turbulent structure of the atmosphere was becoming well described, with a ground layer, a mid-altitude layer (3–5 km high) due to orographic influences, and a layer at the tropopause (10–13 km). Depending on the site, these three main layers may account for the final image perturbation by comparable amounts or, on the contrary, one of them may strongly dominate the others.

The four classical parameters describing the seeing-perturbed images are: the Fried parameter $r_o(\lambda)$, leading to an image of width λ/r_o instead of the diffraction value λ/D (D telescope diameter); the atmospheric coherence time $\tau_c(\lambda)$, during which the image remains approximately unchanged; the isoplanatic angle $\alpha_o(\lambda)$, defining the angular field over which the image perturbation remains approximately constant; the phase excursion variance over an aperture of diameter D, given by $\sigma_\phi^2 = 1.03(D/r_o)^{5/3}$. These four fundamental parameters all vary with λ, the first ones as $\lambda^{1.2}$, the last one as λ^2, hence providing the above quoted infrared advantage. In a good site, at $\lambda = 1\,\mu$m, typical values are $r_o \approx 20$ cm, $\tau_o \approx 20$ ms, $\alpha_o \approx 10$ arcsec and $\sigma_\phi = 4$ rd (for $D = 10$ m) but these values may vary extensively with local conditions.

The seeing parameters translate into the AO specifying quantities, in the following approximative way: $N_\lambda \approx [D/r_o(\lambda)]^2$ is the number of degrees of freedom the deformable mirror must have to correct at the wavelength λ, and in consequence the number of sub-pupils to be analyzed by the WFS; $f_{\mathrm{loop}} \approx 10 \times (\tau_c)^{-1}$ is the required closed loop operation frequency; the rms amplitude of the correcting mirror is the achromatic quantity $\lambda\sigma_\phi/2\pi$.

An AO system corrects for the seeing effects and aims at recovering a perfect, stable diffraction-limited image. In practice this is never fully achieved, and compromises have to be made, depending on the scientific objectives. Performances of AO systems can be evaluated on the basis of a few quantities. For the image of a point source at a given wavelength, the Strehl ratio is defined as $S(\lambda) = I_{\mathrm{max}}/I_{\mathrm{diff}}$, namely the ratio of the corrected image maximum intensity to the intensity that pure diffraction would lead to. For a wavefront sensor operating at λ_{wfs}, the limiting magnitude m_{wfs} is the faintest one at which the loop can be operated on the reference source, giving then a minimal correction; one can also define a more useful limiting magnitude, namely the faintest one where a given and reasonably large value of S (e.g. 50%) is achieved at the image wavelength λ.

3 A long history: the first AO steps

Nature invented active/adaptive optics (AO) long before the scientists did. The lens in the eye changes its curvature, hence its focal length, in order to sharpen the image, through a muscular mechanism which implies a feed-back signal from the visual system to the muscle. One often quotes Archimedes, who used bronze shields of soldiers to concentrate solar light on the sails of enemy vessels in Syracuse (212 b.C.) and set them on fire.[1] Direct testimonies communicated to the author indicate that C.V. Raman used hand-held mirrors, constantly adjusted by servants, to track the Sun and illuminate the spectrograph which allowed him to discover the Raman effect (1932). But the seminal paper on the possible use of AO for astronomy appeared in 1953 when Horace Babcock described the principle and considered it for correcting solar images [5]. Soon after and probably independently, the idea was published by V.P. Linnik for stellar observation [6]. In the 1960s, interest increased in the propagation of light in turbulent media, such as the Earth atmosphere, with the essential contributions of Tatarski [7] and Fried [8]. Still, no practical application of adaptive optics was envisaged in astronomy. For example, at the 1977 Geneva conference exploring the future of optical telescopes [9], Babcock himself stressed the importance of good seeing in site selection, but did not mention AO! Only one paper in this Conference, by G. Bourdet, indicates that *... a very efficient technique for improving the quality of atmospheric turbulence-limited image-forming instruments is the real-time phase control system...first suggested by Babcock in 1953.*

During the 1970s, military work was carried out in the United States and in the USSR to use AO for improving the images of satellites. A system was built at the Itek company by Hardy and coworkers [10], and a larger version was installed in 1982 at the US Air Force Maui Optical Site on Haleakeala (Hawaii). The major part of this work remained classified until the early 1990s,

[1] An attempt to repeat this in order to examine its real feasibility is reported on kwc.org/mythbusters/2006/01/episode_46_archimedes_death_ra.html.

when after the fall of the Berlin wall the National Science Foundation in the United States asked for a declassification for the benefit of the astronomers. This was done in 1992 and completed in 1994 [11]. Before this, a very limited amount of information or equipment was openly available. Yet some attempts were made at Lick Observatory and a limited correction obtained on the star Sirius was published [12]. The possibilities of astronomical AO were clearly mentioned by Roddier in his fundamental paper of 1981 [13].

In Europe, plans for a new large telescope in the Southern hemisphere were discussed in 1978 [9] and further at a conference called by the European Southern Observatory in 1981, focusing on high angular resolution [14]. Indeed, it had been realized, both theoretically and experimentally by astronomical speckle interferometry [15–17], that the infrared domain was especially favorable for extracting high-resolution information in distorted images given by large telescopes, since several of the dominant seeing factors were improving with $\lambda^{1.2}$ (see above).

Once the concept of the European Very Large Telescope (VLT) was defined as a array of four 8-m telescopes, the question of a possible use of this array as a coherent optical interferometer, already brought up by Labeyrie at the 1978 Conference [9], became more acute. It soon was clear that this combination

Fig. 2 The Oracle meeting in Arizona. The small community of astronomers in Europe and United States dealing with optical interferometry and adaptive optics discussed in 1987 plans for high angular resolution with the VLT and other telescopes [20]

would not be efficient if individual apertures were not phased coherently: adaptive optics appeared the only way to reach the ultimate interferometric sensitivity the 8-m pupils could provide [18, 19]. As ONERA and the company CGE in France had some on-going studies which could be made accessible to astronomers, the project of an astronomical AO demonstrator emerged in 1984, jointly between France and ESO. At the Canada-France-Hawaii Telescope, the work of René Racine on seeing led to a workshop dedicated to AO in 1984, while in 1986 at NOAO in Tucson, Jacques Beckers and François Roddier began also to plan for AO: a joint US-European meeting was held in Oracle in 1987 (Fig. 2), closely tying AO and the future of optical interferometry [20]. In this decade two fundamental ideas emerged: the potential use of artificial guide stars for AO, hinted at by Linnik [6], made possible with lasers [21], was finally proposed in detail by Foy and Labeyrie in 1984 [22]; the concept of *multi-conjugate* AO, already suggested in 1975 by Robert Dicke [23], was developed by Jacques Beckers [24]. Both ideas would only come to fruition a decade later.

The ESO-France AO instrument COME-ON successfully achieved diffraction-limited images at $\lambda = 2.2$ μm on the double star γ_2 And in october 1989 at the Observatoire de Haute-Provence, with the 1.52-m telescope (Fig. 3) [25].

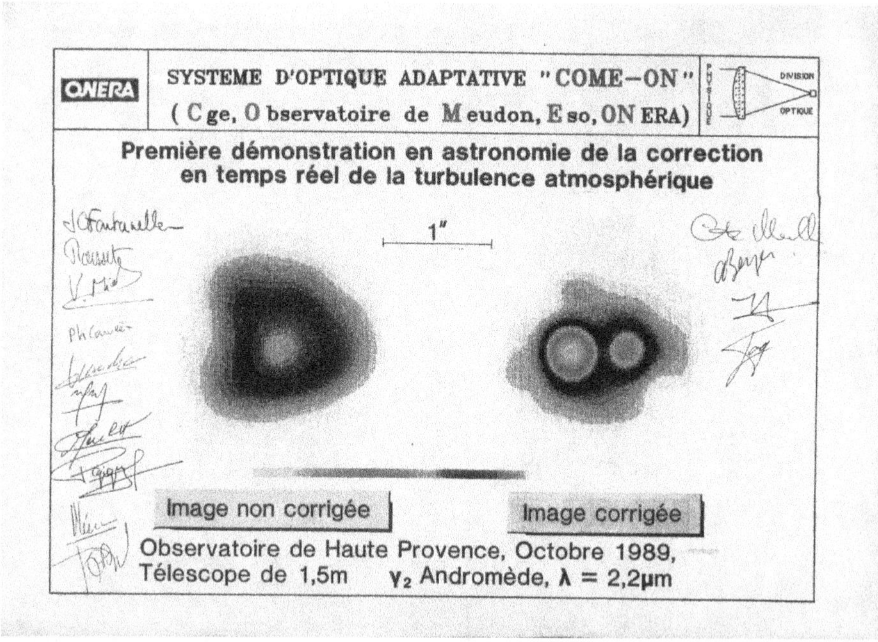

Fig. 3 The first astronomical adaptive-optics observation (1989). A nearly-diffraction limited image of the double star γ_2 And is resolved at the 1.52-m telescope of the Observatoire de Haute-Provence in October 1989, with the COME-ON instrument. Signatures from team members J.-C. Fontanella, P. Gigan, P. Léna, F. Merkle, G. Rousset and a few others celebrate the event

The AO system, with its 19-elements deformable mirror, was immediately transferred to the 3.6-m ESO telescope at La Silla (Chile), where it began a set of successful observations, opening the way to a practical use of AO in astronomy. At the same time, in December 1989, the VLT project was definitively approved by the ESO Council; it included adaptive optics as well as an interferometric mode, despite the skepticism of some astronomers: given the limited experience existing at the time in both techniques, one can easily understand their point of view, viz. that such a forward leap was excessive.

4 The blossoming of adaptive optics

The two decades 1990–2010 see the confirmation and expansion of adaptive optics in astronomy, to an unexpected scale and with important scientific results, mostly in the near infrared (1–10 μm). Figure 4 shows the progress of AO equipped telescopes. The progress in performances is related primarily to the number N of actuators of the active mirrors, which has increased from 10 to 10^3, but also to the loop correction frequency and the sensitivity of the wavefront sensors. This evolution allows a better image correction (high Strehl ratio), a better sky coverage with natural guide stars and a shorter wavelength

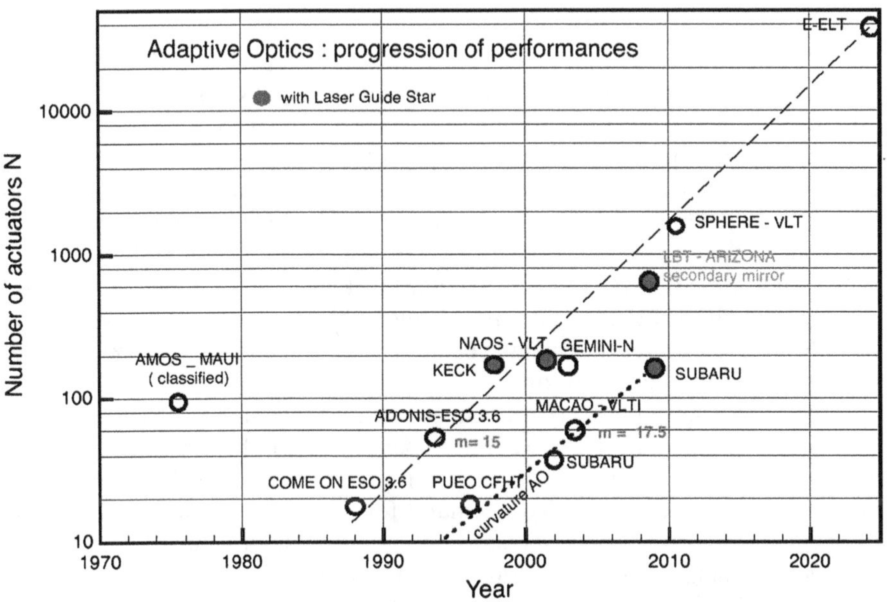

Fig. 4 The emergence and prospects of adaptive optics for the years 1970–2020. The progress is here measured in terms of N, the number of degrees of freedom of the deformable mirror surface. The limiting magnitude, usually for a WFS operating in the visible (band R), is reported for a few systems. The maximum achievable Strehl ratio is usually above 0.5. Telescopes equipped with laser guide stars are indicated. The E-ELT value corresponds to the currently studied EPICS instrument

of operation, although not yet reaching operation at visible wavelengths (≈ 0.7–$0.5\ \mu$m).

The early ESO COME-ON system led to the ADONIS system (1992) on the La Silla 3.6-m, which was productive of new science and became the prototype for the VLT AO. The NAOS AO instrument was installed on the VLT in 2001. AO systems based on Shack-Hartman WFS were successively installed on the William Herschel telescope (NAOMI, 2002), the Keck (1998), the Hale 5-m at Palomar (2000), the Gemini North (2003) and South telescopes.

An alternate and efficient AO route was proposed by Roddier [26], namely the curvature sensor, which senses the second derivative (curvature) of the wavefront phase and uses a deformable curvature mirror. Along this second line, the 3.6-m Canada-France-Hawaii was equipped: first with the Hokeupa system (1994), to demonstrate the advantages in sensitivity of this new concept replacing photo-cathodes with avalanche diodes in the WFS, in parallel with the PUEO (1996) system. Both of these instruments soon became highly scientifically productive. The use of the same method followed on the SUBARU AO system (Hawaii) and MACAO (2003), which is currently installed on the VLT 8.2-m telescopes for interferometric use. At the end of the years 2000, the Large Binocular Telescope 8.6-m Magellan (Arizona) will be the first to be equipped with an adaptive secondary mirror having 672 actuators (Fig. 1), a decisive step paving the way to an optimization of AO systems in the future. For the VLT, an adaptive secondary with 1170 actuators, 1.1 m in diameter, is being prepared.

Since it provides nearly diffraction-limited images, AO offers new possibilities of exploiting these images. Coronography is being revived [27], especially for the direct detection of extra-solar planets near bright stars (Fig. 5); coupling AO with integral field spectroscopy of galaxies is achieved at the VLT with the SINFONI infrared spectrograph (2004), or with the planned FALCON spectrometer. The need of AO for optical interferometry with medium or large telescopes ($D \geq r_0$) has led to the MACAO system on the four 8.2-m VLT telescopes. For interferometric observations of stellar motions in the immediate vicinity (a few milli-arcsec) of the Galactic Center black hole, an extremely stable AO system is required at the interferometric recombination focus. This system will be fully integrated with the imaging camera (VLTI instrument GRAVITY), in order to reach a 10 μarcsec resolution.

Each of the three basic components of an AO system has progressed continuously since the 1980s: the wavefront sensor (WFS), the deformable mirror and the control loop. It is not possible to review the details here, but a few aspects can be pointed out. The demand for increasing values of N is high: to operate at shorter wavelengths; to obtain a very high value of S to reduce the residual image halo; to exploit larger telescopes (ELT) at their diffraction limit. Indeed N has constantly increased since the beginning of AO and values larger than 10^5 are aimed at for the ELTs. This may lead to considering a hierarchical correction, acting in several stages, separating tip-tilt (now done an all classical AO systems), higher-order aberrations and possibly even higher orders which require the correction of a much smaller phase excursion σ_ϕ.

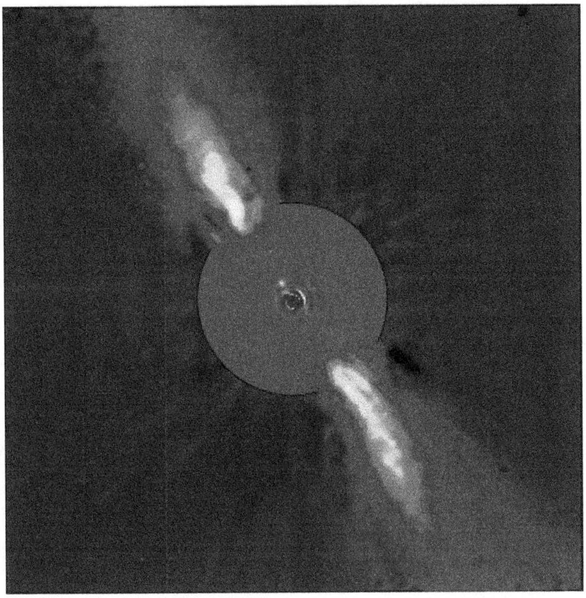

Fig. 5 Combining coronography and adaptive optics. Observation of a probable planet in the disc of the A star β Pictoris with NAOS on the VLT in the L' band (2008) [28]. Credit: Lagrange, A.M. & ESO

Beside the now classical piezo-electric deformable mirrors, the use of arrays of micro-mirrors or MEMS (Micro-Electric-Mechanical-System), in integrated optics design, makes it possible to consider feasible values of $N \gg 10^3$. Finally, the future control loops for advanced AO systems and large N values may require nearly 100 parallel processors and up to 10^{11} bytes of random access memory (RAM) when dealing with ELTs.

Progress in wavefront sensing is along several fronts. First, increasing the sensitivity of fast CCD cameras (e.g. L3CCD) to analyze the wavefront and to reach photon noise at the sensing wavelength (R or I bands): this leads to limiting magnitudes $m_R \approx 15 - 16$ for Shack-Hartman systems, and even higher values reaching 17.5 for curvature systems. Second, new concepts for wavefront sensing are proposed, such as the pyramid sensor [29] (2001) which is a modern version of the classical Foucault knife-edge test. Third, a better optical efficiency may be obtained by avoiding splitting the incoming light between the image camera and the WFS camera: a single camera could be considered for both functions (instrument SPHERE for the VLT).

The laser guide star (LGS) concept is well described in [4] and its astronomical applications began in 1995 with Lloyd-Hart et al. on the MultiMirror Telescope (Arizona) [30], using resonant retro-scattering on the 90-km high atmospheric sodium layer and correcting a star image (1996) for the first time. It was soon followed by other systems, especially the one (ALPHA) on the German 3.5-m telescope at Calar Alto. With the developed expertise, the next decade sees a number of systems coming to fruition on large telescopes: on the Keck (2004), the VLT (2005), Gemini North (2006) and South (2007), and Palomar (2008). If the aim of the correction is more modest, it becomes sufficient to correct for the seeing due to the low-altitude turbulence (ground layer adaptive optics or GLAO), in which case Rayleigh retro-scattering of the laser, obtained off tropospheric layers, is sufficient and much less demanding on laser power: this is the method demonstrated on the William Herschel Telescope (2004) and adopted on the Large Binocular Telescope for its LGS. In principle when using laser guide stars there would no longer be a limiting magnitude of available natural stars for closing the AO loop, a fact which opens the way to extragalactic observing programs. Yet the residual problem of tip-tilt aberration, which cannot be corrected for with a LGS, forces one to measure it on a natural star, of magnitude m_V or m_R as faint as 17–18. Several methods have been proposed to circumvent this problem. There is no doubt that LGS already does and will considerably expand the domain of astronomical AO.

Solar astronomy is not ignored by AO, but encounters severe challenges. Seeing during the day is usually worse than at night. It is necessary to use visible wavelengths to observe the solar photosphere, but then atmospheric parameters (r_o,τ_c, α_o) are much less favorable than in the near infrared. First observations began in the 1970s at Sacramento Peak (New Mexico), and have continuously progressed since: see the reviews in [1] and [4] and, for a recent one, Oskar von der Lühe in this Volume. The National Solar Observatory 4-m telescope (ATST, 2008) at Sacramento Peak is equipped with multi-conjugate adaptive optics [31].

Obtaining the best possible final images from the AO-corrected images, given a measurement of the Point Spread Function, has stimulated work on deconvolution and image processing using a priori information [32].

5 The future of astronomical adaptive optics

The future of astronomical AO is certainly bright. But the diversity of various observing goals which have to be addressed requires a diversity of solutions, depending on the expected degree of correction, corrected field of view, sky coverage and site qualities [34]. These aspects have been well reviewed in 2008 by Thierry Fusco [33]. In [35, 36], Norbert Hubin examines in detail the AO requirements for ELTs, since such requirements are very high and critical for a proper use of these giant telescopes.

Figure 6 is a summary of the various directions along which AO is developing, all aiming at covering the wavelength range 0.8 to 10 μm. They represent a complexification of SCAO (Single Control AO) which, as considered above, uses a natural reference star and a single deformable mirror (DM), leading to limited field of view and sky coverage.

The first direction is LTAO (Laser Tomographic AO), where several LGS are simultaneously used to increase the field-of-view, hence the sky coverage. Extreme AO (EXAO), aims at extremely high Strehl ratios (≥ 90 %), in order to reduce the incoherent speckle halo around a bright star and, associated with a coronograph, to be able to image faint sources (exoplanets, jets, accretion disks) around it, with dynamic ranges larger than 10^6. For example, this is

	Category #1	Category #2	Category #3
Corrected FoV (diameter)	Small (Isopl. Angle, i.e. 20" with 0.8 of max Strehl)	Medium (1')	Large (5')
Performance metric	70% Strehl (K) 33% Strehl (J)	50% Strehl (K) 11% Strehl (J)	16%(K), 5% (J) EE in 50mas (4 x seeing)
Obs. λ range	R–N	J–N	J–N
PSF uniformity	low	high	medium
Potential systems	SCAO (1DM, 75% Strehl*, NGS, low sky coverage)	MCAO (2 DMs, several LGS)	GLAO (1 DM, several LGS, special LTAO mode)
	LTAO (1DM, 60% Strehl, several LGSs, high 'sky coverage')	MCAO (3 DMs, several LGS), larger FoV	MOAO (1-2 DM, several LGS, restricted to 'islands', better Ensquared Energy)
	XAO (2DM, bright NGS)		

Fig. 6 An overview of possible AO categories for the future, especially for ELTs. This table, adapted from Hubin [36], explores the potential systems which could be realized, for spectral bands from the red (R) to 10 μm (N). All values are estimated for 0.5 arcsec natural seeing and a $D = 42$ m telescope. Abbreviations: SC = single correction; LT = laser tomography; X = extreme; MC = multi conjugate; GL = ground layer; MO = multiple objects; EE = encircled energy. Reproduction of this table courtesy of Norbert Hubin (ESO)

the goal of the SPHERE instrument to be installed on the VLT after 2010, and to prepare for the EPICS instrument under consideration for the planned 42-m E-ELT. To achieve such ultra-high performances, one must optimize the WFS analysis (avoid e-folding of high spatial frequencies, correct for telescope vibrations, detect any residual aberrations between the WFS and the image cameras), use a bright reference star and possibly two DMS.

While EXAO is needed to image the immediate environment of stars (within an isoplanatic field, i.e. 10–20 arcsec), the second direction searches for a wide corrected field (≈ 1 arcmin), especially needed for the observation of galaxies. A single correcting mirror cannot correct a large field of view, and the concept of MCAO (Multi Conjugate Adaptive Optics [24]) has been introduced to solve this problem with the use of several LGS and correcting mirrors (up to 3), conjugated with the different turbulent layers of the atmosphere. The Point Spread Function is uniform in the image, and the correction good ($S \approx 50\%$). The MCAO is currently being tested on the VLT with a MultiConjugate AO Demonstrator (MAD) and convincing results (Fig. 7).

Finally, one may also aim at an even larger field-of-view, and pay the price with a smaller correction ($S \approx$ 5–10%). This is achieved by GLAO (Ground Layer AO), which is a special tomographic mode (LTAO), using a single DM and several LGS. It exploits the larger isoplanatic field connected with the ground-layer turbulence. An interesting strategy is to select some areas of interest in

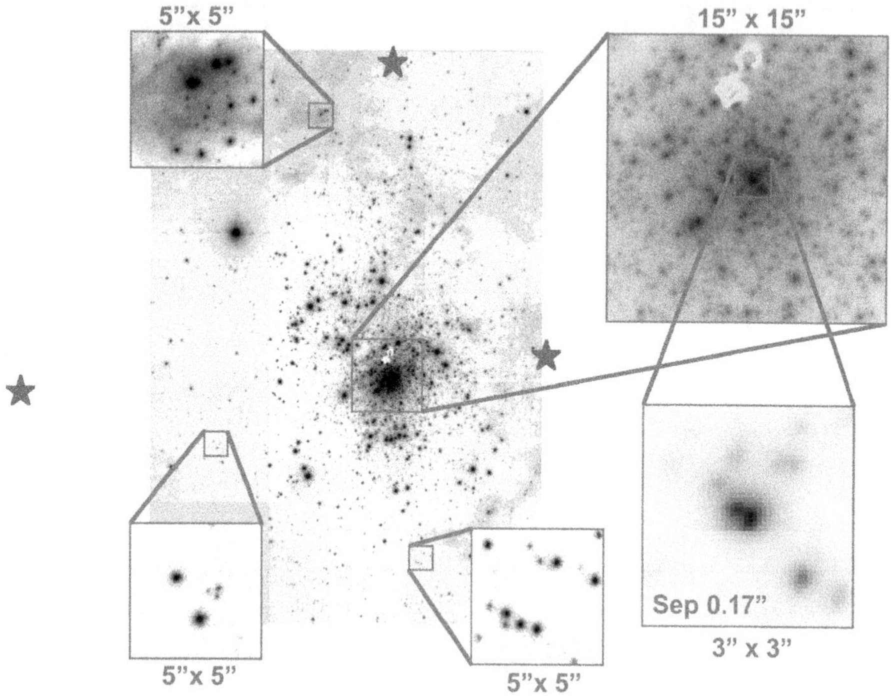

Fig. 7 Observation of the star-forming region R136 with the MCAO demonstrator on the VLT. *Left*: A field of 120 × 85 arcsec, mosaic of 3 images in the H band, and the position of the three artificial stars (⋆). *Right*: the two inserts show respectively 15"×15" and 3"×3" fields, with a double star (0.17" separation). Other inserts in the field of view show the quality of correction, with a FWHM varying between 90 to 140 mas. Reproduction of this figure courtesy of Norbert Hubin (ESO)

the field-of-view (e.g. the galaxies it contains), and to "address" the correction specifically for these areas or "islands of correction": this is the MOAO (Multiple Objects AO) strategy.

For all these AO modes, there appears to exist a large number of degrees of freedom, which may correspond to: the given position of the natural guide stars; the programmed position of the several LGS in the sky, both in the vicinity of the object of interest; the correction applied to the several DM; the position of the "islands of correction" in the image; and indeed the seeing parameters. Optimization procedures for the command of the single or multiple DM are clearly possible ([33] and references herein).

6 Conclusion

In less than two decades, adaptive optics has deeply transformed ground based optical astronomy: 10-m class telescopes are now achieving their diffraction resolution limit practically over the whole sky; optical interferometry with large telescope arrays has become possible with unprecedented sensitivity; the way is open for the next generation of optical telescopes in the 30–40 m range. Many technological challenges remain, but many solutions are also in sight. Sites with exceptional seeing conditions, such as Dome C in Antarctica [37], or possibly Dome A, will likely offer to adaptive optics new challenges and ultimate performances.

Acknowledgements For their contribution to this review, the author expresses deep thanks to his colleagues and friends: Jacques Beckers, Frank Eisenhauer, Eric Fossat, Thierry Fusco, Norbert Hubin, Marie Glanc, Laurent Mugnier, François Rigaut, Gérard Rousset at ESO, Observatoire de Paris, Observatoire de Nice, ONERA, Scottsdale. The editor kindly improved the style.

References

1. Beckers, J.: Adaptive optics: principles, performances and applications. Ann. Rev. Astron. Astrophys. **31**, 13–62 (1993)
2. Alloin, D.M., Mariotti, J.M.: Adaptive optics for astronomy. In: NATO ASI Series, vol. 423. Kluwer, Deventer (1994)
3. Hardy, J.W.: Adaptive Optics for Astronomical Telescopes. Oxford University Press, Oxford (1998)
4. Roddier, F. (ed.): Adaptive Optics in Astronomy. Cambridge University Press, Cambridge (1999)
5. Babcock, H.W.: The possibility of compensating atmospheric seeing. Publ. Astron. Soc. Pac. **65**, 229–236 (1953)
6. Linnik, V.P.: On the possibility of reducing atmospheric seeing on the image quality of stars. Opt. Spectrosc. **3**, 401–402 (1957) (in Russian)
7. Tatarski, V.I.: Wave Propagation in a Turbulent Medium. McGraw Hill, New York (1961) (originally published in Russian)
8. Fried, D.L.: Optical resolution through a randomly inhomogeneous medium for very long and very short exposures. J. Opt. Soc. Am. **56**, 1372–79 (1966)
9. Optical Telescopes of the Future: Proceedings of an ESO Conference, ESO, Garching, Geneva, 12–15 December 1977
10. Hardy, J.W., Lefebvre, J.E., Koliopoulos, C.L.: Real time atmospheric compensation. J. Opt. Soc. Am. **67**, 360–369 (1977)
11. Benedict, R. Jr., Breckinridge, J.B., Fried, D.: Atmospheric compensation technology. Introduction to the special issue of J. Opt. Soc. Am. **11**, 257–260 (1994)
12. Buffington, A., Crawford, F.S., Muller, R.A., Orth, C.D.: First observatory results with an image sharpening telescope. J. Opt. Soc. Am. **67**, 304–305 (1977)
13. Roddier, F.: The effects of atmospheric turbulence in optical astronomy. Prog. Opt. **19**, 281–376 (1981)
14. Ulrich, M.H., Kjaer, K. (eds.): Scientific importance of high angular resolution at infrared and optical wavelengths. In: Proceedings of an ESO Conference, ESO, Garching, March 1981
15. Sibille, F., Chelli, A., Léna P.: Infrared speckle interferometry. Astron. Astrophys. **79**, 315–328 (1979)
16. Woolf, N.: High resolution imaging from the ground. Ann. Rev. Astron. Astrophys. **20**, 367–398 (1982)
17. Angel, J.R.P.: Very large ground-based telescopes for optical and IR astronomy. Nature **295**, 651–657 (1982)
18. Roddier, F., Léna, P.: Long-baseline Michelson interferometry with large ground-based telescopes operating at optical wavelengths. II. Interferometry at infrared wavelengths. J. Optique (Paris) **15**, 363–374 (1984)
19. Working Group on Interferometry: Interferometric imaging with the Very Large Telescope. Final Report of the ESO/VLT Working Group on Interferometry, VLT Report, no. 49. ESO, Garching (1986)

20. Goad, J.W., Merkle, F. (eds.): Interferometric imaging in astronomy. In: Joint Workshop ESO-KPNO, Oracle, 12–15 January 1987
21. McCall, S.L., Passner, A.: Adaptive optics in astronomy. In: Adaptive Optics and Short Wavelengths Sources, vol. 6. Physics of Quantum Electronics. Addison-Wesley, Reading (1978)
22. Foy, R., Labeyrie, A.: Feasability of adaptive telescope with laser probe. Astron. Astrophys. **152**, 129–131 (1985)
23. Dicke, R.H.: Phase contrast detection of telescope seeing errors and their correction. Astrophys. J. **198**, 605–615 (1975)
24. Beckers, J.: In ESO conference on very large telescopes and their implementation. In: Ulrich, M.H. (ed.) ESO, Garching (1988)
25. Rousset, G., Fontanella, J.-C., Kern, P., Gigan, P., Rigaut, F., Léna, P., Boyer, C., Jagourel, P., Gaffard, J.-P., Merkle, F.: First diffraction-limited astronomical images with adaptative optics. Astron. Astrophys. **230**, L29–32 (1989)
26. Roddier, F.: Curvature sensing and compensation: a new concept in adaptive optics. Appl. Opt. **27**, 1223–1225 (1988)
27. Rouan, D., Riaud, P., Boccaletti, A., Clénet, Y., Labeyrie, A.: The four-quadrant phase-mask coronagraph. I. Principle. Publ. Astron. Soc. Pac. **112**, 1479 (2000)
28. Lagrange, A.-M., et al.: A probable giant planet imaged in the β Pictoris disk. Astron. Astrophys. **493**, L21-25 (2009)
29. Ragazzoni, R.: Pupil plane wavefront sensing with an oscillating prism. J. Mod. Opt. **43**, 189 (1996)
30. Llyod-Hart, M., Angel, Jacobsen, J.R.P., Wittman, D., et al.: Adaptive optics experiments using sodium laser guide star. Astrophys. J. **493**, 950–954 (1995)
31. Rimmele, T.R., et al.: The unique scientific capabilities of the advanced technology solar telescope. Adv. Space Res. **42**, 78–85 (2008)
32. Mugnier, L.M., Le Besnerais, G., Meimon, S.: Inverse problems in optical imaging through atmospheric turbulence. In: Idier, J. (ed.) Bayesian Approach for Inverse Problems. ISTE, Wiley, London (2008)
33. Fusco, T.: Optique adaptative et traitement d'images pour l'astronomie : de nouveaux enjeux et de nouvelles solutions. Habilitation à diriger des recherches, Université Paris Diderot (2008)
34. Hubin, N.: Adaptive optics programs in Europe. In: ELT Funding Group Schiphol Meeting. ESO, Garching (2006)
35. Hubin, N., et al.: Adaptive Optics for Extremely Large Telescopes. In: Scientific Requirements for Extremely Large Telescopes, Whitelock, P.A., Dennefeld, M., Leibundgut, B. (eds.). Proc. IAU Symposium, vol. 232. Cambridge University Press, Cambridge (2006)
36. Hubin, N.: ESO adaptive optics roadmap toward the European ELT. In: Science with AO-fed Instruments on Large Telescopes. Dunk Island, Anglo-Australian Observatory Conference (2008)
37. Trinquet, H., Agabi, A., Vernin, J., Azaouit, M., Aristidi, E., Fossat, E.: Night time optical turbulence vertical structure above Dome C in Antarctica. Publ. Astron. Soc. Pac. **120**, 203–211 (2008)
38. Weigelt, G.P.: Speckle holography measurements of the stars Zeta Cancri and ADS 3358. Appl. Opt. **17**, 266 (1978)
39. Chelli, A., Léna, P., Sibille, F.: Angular dimensions of accreting young stars. Nature **278**, 143–146 (1978)

Exp Astron (2009) 26:49–63
DOI 10.1007/s10686-009-9166-5

REVIEW ARTICLE

The development of astronomical interferometry

Andreas Quirrenbach

Received: 31 March 2009 / Accepted: 1 April 2009 / Published online: 24 April 2009

Abstract Astronomical interferometry was pioneered by Fizeau and Michelson in the 19th century. In the 1920s, the first stellar diameters were measured. The development of radio interferometry began in the 1950s, and led to the construction of powerful synthesis arrays operating at cm, mm, and sub-mm wavelengths. Modern computer and control technology has enabled the interferometric combination of light from separate telescopes also in the visible and infrared regimes. Imaging with milliarcsecond resolution and astrometry with microarcsecond precision have thus become reality.

Keywords Instrumentation: interferometers · Techniques: interferometric · Telescopes · Stars: fundamental parameters · History of astronomy

1 Introduction

The development of astronomical interferometry is intimately linked to the quest for imaging with ever-higher angular resolution, and with the desire to measure the positions of celestial objects with ever-higher accuracy. Even under the best of circumstances—when atmospheric seeing is unimportant or correctable—the angular resolution α of any telescope is limited by diffraction to

$$\alpha \approx \lambda / D \ , \quad (1)$$

where D is the diameter of the telescope aperture, and λ the observing wavelength. The aperture does not necessarily have to be filled; it can consist

A. Quirrenbach (✉)
ZAH, Landessternwarte, Königstuhl 12, 69117 Heidelberg, Germany
e-mail: A.Quirrenbach@lsw.uni-heidelberg.de

of separate elements spread over an area with diameter D, but with individual sizes much smaller than D. For Eq. 1 to hold, an interference pattern has to be formed, which means that the light from all sub-apertures has to be combined coherently. The minimum requirements are thus that internal optical pathlengths have to be controlled with a precision $\lesssim \lambda^2/\Delta\lambda$, and have to be kept constant to within $\lesssim \lambda/2\pi$ during the integration time required for obtaining a single data point. Here $\Delta\lambda$ denotes the width of the observing band, or the width of individual channels if multiple spectral channels are used. The ability to perform such coherent light combination is the main technological driver behind the development of astronomical interferometers operating in different wavelength regimes.

Interferometry is now in very widespread use at radio and mm wavelengths, and is gaining in importance in the visible, infrared and sub-mm regimes. Introductions into astronomical interferometry can be found in the book by Thompson et al. [56], in the collection of classical papers by Lawson [23], and in the reviews by Quirrenbach [39] and Monnier [33].

2 Interferometric techniques

2.1 Aperture synthesis imaging

The fundamental observable in most astronomical interferometers is the complex visibility $\Gamma = Ve^{i\phi}$. V corresponds to the contrast of the fringes

$$V = \frac{I_{\max} - I_{\min}}{I_{\max} + I_{\min}} \ , \tag{2}$$

and ϕ to the position of the interference pattern. The complex visibility is related to the sky brightness distribution $B(\xi, \eta)$ through a Fourier transform (the van Cittert-Zernike Theorem, see e.g. Shao and Colavita [50], Thompson et al. [56]):

$$Ve^{i\phi} = \Gamma(u, v) = \iint B(\xi, \eta) e^{-2\pi i(u\xi + \eta)} d\xi \, d\eta \ , \tag{3}$$

where ξ and η are the coordinates in the tangent plane of the sky, and u and v the two components of the baseline vector (i.e., the vector connecting two interferometer elements) in the aperture plane, measured in units of the observing wavelength. It is thus possible to derive the sky brightness distribution, i.e., obtain an image of the sky area observed, by computing the inverse Fourier transform of the visibility data. This does of course require sufficient sampling of the (u,v) plane, according to the Nyquist theorem. Interferometric imaging of complex objects thus requires taking many individual visibility measurements.

Unfortunately, astronomical interferometers are frequently plagued by large phase errors, induced by the Earth's atmosphere or unknown pathlengths within the interferometric equipment. Because the phases are corrupted, it is usually not possible to obtain interferometric images by a straightforward inverse Fourier transform. However, not all phase information is lost. If we consider phase errors ψ_i due to the pathlength from array element i to the beam combiner, or due to atmospheric turbulence above element i, the observed phases ϕ'_{ij} for three elements 1, 2, and 3, are related to the true phases ϕ_{ij} by the relations

$$\phi'_{12} = \phi_{12} + \psi_1 - \psi_2 \tag{4a}$$

$$\phi'_{23} = \phi_{23} + \psi_2 - \psi_3 \tag{4b}$$

$$\phi'_{31} = \phi_{31} + \psi_3 - \psi_1 \ . \tag{4c}$$

Adding these three equations yields

$$\Phi_{123} \equiv \phi'_{12} + \phi'_{23} + \phi'_{31} = \phi_{12} + \phi_{23} + \phi_{31} \ , \tag{5}$$

from which one can see that the so-called *closure phases* Φ_{ijk} are immune to the most common types of phase errors. They can thus be used together with the visibility amplitudes V_{ij} in image reconstruction algorithms (see e.g. Perley et al. [37] for details). These algorithms use a priori knowledge on the image (e.g., the condition that all pixels must be non-negative, the assumption that the source is contained in a certain area on the sky, or assumptions on the smoothness of the brightness distribution) to derive the most plausible image that is consistent with the measured visibility amplitudes and closure phases.

2.2 Interferometric astrometry

If constructive interference is to be observed at all wavelengths, the internal pathlengths within the interferometer must exactly compensate external pathlength differences (see Fig. 1). Since the external delay d_{ext} is related to the baseline $\vec{B}$ through the relation

$$d_{\mathrm{ext}} = \vec{B} \cdot \hat{s} = B \cos\theta \ , \tag{6}$$

where $\hat{s}$ is a unit vector in the direction towards the star, the angle θ between $\vec{B}$ and $\hat{s}$ can be derived from d; this corresponds to a one-dimensional measurement of the source position, provided that the length and direction of the baseline are accurately known. The second coordinate can be measured with a separate baseline at a roughly orthogonal orientation. The astrometric precision σ is given by

$$\sigma = \frac{1}{\mathrm{SNR}} \cdot \frac{\lambda}{2\pi B} \ . \tag{7}$$

Since high signal-to-noise ratios can be obtained for bright objects, σ can be orders of magnitude smaller than the resolution λ/B of the interferometer. The power of interferometry for high-precision astrometry is thus due to two factors: the straightforward relation (Eq. 6) between celestial positions and delays, which can be measured with high precision, and the possibility to achieve a precision much better than the resolution limit (Eq. 7).

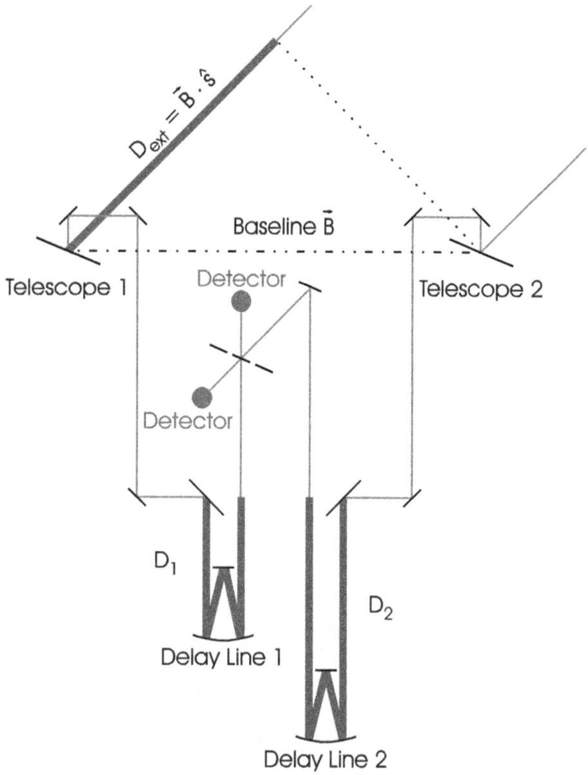

Fig. 1 Schematic drawing of the light path through a two-element interferometer. The external delay $D_{\rm ext} = \vec{B} \cdot \hat{s}$ is compensated by the two delay lines. The zero-order interference maximum occurs when the delay line positions are such that the internal delay $D_{\rm int} = D_1 - D_2$ is equal to $-D_{\rm ext}$. The beams from the two telescopes are combined with a plate that has 50% transmission and 50% reflectivity (or with a mixer, in the radio case)

At visible, near-infrared, and mm wavelengths, the precision of ground-based astrometry is limited by phase noise induced by the Earth's troposphere; in the cm and dm regime the ionosphere is the limiting factor. Two-color methods (also called dual-frequency techniques) take advantage of the dispersive nature of the atmospheric fluctuations to estimate their amplitude, thus improving the astrometric accuracy [9]. For some applications—such as astrometric searches for extrasolar planets—absolute parallaxes and proper motions are not needed; in these cases relative measurements between stars located close to each other on the sky are sufficient. Very high precision can be achieved in such "narrow-angle" measurements [49].

2.3 Nulling interferometry

For high-contrast applications such as the detection of radiation from extra-solar planets, it is necessary to reject the light from the much brighter central star very efficiently. In an infrared interferometer this starlight suppression can be achieved by introducing an achromatic π phase shift in one arm of the interferometer, so that there is destructive interference for light arriving on-axis; this concept was conceived by Bracewell [5]. The most important characteristic of such a nulling interferometer is the factor N by which the starlight is reduced compared to a constructive interferometer. To produce a deep null, the two arms of the interferometer must be made symmetric with very high precision (apart from the π phase shift, of course), and the wavefronts must be perfectly flat. Because even the best optical components do not deliver sufficiently high wavefront quality, one has to "clean" the wavefront with a modal filter, e.g., with a single-mode fiber. The fundamental mode transmitted by such a fiber corresponds to a perfectly flat wavefront; all other modes are reflected at the fiber entrance.

3 Early interferometric experiments

3.1 Astronomical interferometry in the 19th century

The first steps towards astronomical interferometry were taken in France by Fizeau and Stéphan, the director of the Marseille Observatory. In 1868, Fizeau discussed interference phenomena in the context of telescopic observations and remarked that "a remarkable and necessary relation [must exist] between the dimension of the fringes at that of the luminous source". He suggested that one could observe stars using a mask with two holes in front of a large telescope and that "it should be possible to obtain some new data on the angular diameters of these stars" [12]. Stéphan put these ideas to work at the largest reflecting telescope in existence at the time, the 80 cm reflector at the Observatoire de Marseille built by Foucault. He could not resolve any stars with a mask that had two holes separated by 65 cm and concluded that these observations "prove not only that the apparent diameter of the observed stars is smaller than 0.″158, but show also that the diameter is a very small fraction of the preceding number" [53].

Somewhat later, Michelson developed the mathematical theory of astronomical interferometry [29], and used it for measurements of Jupiter's planets [30]. Michelson thus seems to have invented astronomical interferometry independently, although he might have been influenced by Fizeau. This question has been discussed in more detail by Lawson [24].

3.2 The first stellar diameter measurements

Michelson and collaborators took the next step in technical sophistication. To overcome the resolution limitation imposed by the telescope diameter, they constructed an interferometer consisting of two mirrors placed in a periscope-like configuration on a 20-foot beam mounted across the top of the 100-inch telescope on Mt. Wilson. By varying the baseline length up to the maximum of nearly 6 m, and noting the baseline at which the fringes disappeared, they succeeded in making the first stellar diameter measurements [31]. The instrument was also used to determine a precise "visual" orbit of the bright binary star Capella [1]. Motivated by these successes, the construction of a 50-foot interferometer was begun on Mt. Wilson, but this instrument

Table 1 Selected radio interferometers

Name	Location	Year	Remarks
Cliff Interf.	Sydney, Aus	1945	Reflection at ocean surface
Ryle & Vonberg	Cambridge, UK	1946	First radio interferometer
Mills Cross	Fleurs, Aus	1954	Two arms of 457 m
One-Mile Interf.	Cambridge, UK	1962	Earth-rotation synthesis
WSRT	Westerbork, NL	1970	14×25 m, $B_{max} = 2.7$ km
VLA	Socorro, NM	1980	27×25 m, $B_{max} = 36$ km
MERLIN	UK	1980	Radio-linked array $B_{max} = 217$ km
EVN	Europe	1980	VLBI network
NMA	Nobeyama, Japan	1985	mm-array, 6×10 m
ATCA	Narrabri, Australia	1988	6×22 m
PdBI	Plateau de Bure, F	1988	mm-array, 6×15 m
VLBA	USA	1993	VLBI array, 10×25 m
GMRT	Pune, India	1995	m-waves, 30×45 m
HALCA	Earth orbit	1997	Space VLBI
DASI	South Pole	1999	CMB measurements
Sub-mm Array	Mauna Kea, HI	2003	Sub-mm-array, 8×6 m
CARMA	Owens Valley, CA	2006	mm-array, 6×10.4 m and 9×6.1 m
ATA	Hat Creek, CA	2007	42×6.5 m, $B_{max} = 300$ m
LOFAR	NL + neighbors	2007	30...240 MHz
ALMA	Chajnantor, Chile	2012	Sub-mm-array, 50×12 m
SKA	tbd	$\geq$2020	One square kilometer

See text for more details

never delivered any useful results, probably because the required mechanical stability could not be achieved at the time.

4 The development of radio interferometry

4.1 Radio interferometry in the 1950s

Following the early discoveries of cosmic radio emission by Jansky and Reber, and taking advantage of the enormous progress made on radio and radar technology during World War 2, Ryle and Vonberg [48] constructed a radio interferometer using dipole antenna arrays operating at 175 MHz. Around the same time, Pawsey, Bolton and Stanley in Australia placed a radio antenna on top of a cliff and measured the interference pattern between the direct beam from a rising source and the beam reflected by the ocean surface—this is essentially a Lloyd's mirror arrangement [36] (Table 1). This work was largely motivated by the desire to study the radio emission from the Sun, but it also resulted in the identification of the strong sources Tau A, Vir A, and Cen A with the Crab Nebula, M 87, and NGC 5128, respectively. The finding that other strong sources such as Cas A and Cyg A could not easily be identified with prominent visible-light objects was one of the drivers behind the development of increasingly powerful radio interferometers in the early

1950s, which were used for more extensive surveys, as well as more precise measurements of positions and angular sizes.

A technical breakthrough came with the invention of phase switching [45]: when the phase between the two interferometer elements is reversed periodically, the detector output alternates between $(V_1 + V_2)^2$ and $(V_1 - V_2)^2$. By measuring the difference $4V_1V_2$ between these two states one can thus obtain the correlation between the two individual antenna voltages V_1, V_2. This technique is much less susceptible to amplifier noise and gain fluctuations than a simple adding interferometer, and enables the detection of objects many orders of magnitude below the noise level in modern interferometer arrays.

Near Sydney, Mills et al. [32] constructed an interferometer consisting of two long, narrow antenna arrays arranged in the form of a cross; the outputs from the two "bars" of the cross were combined with each other in a phase-switching receiver. A survey of the Southern sky conducted with the Mills Cross yielded a list of 2200 sources; a successor instrument located at Molonglo performed a survey at 408 MHz with 2.8 arcmin resolution, which is still widely used today. In the Northern hemisphere, the Cambridge group produced the 3C and 4C surveys.

4.2 Modern radio telescope arrays

With a large number of radio sources of diverse types being cataloged, the task of determining their morphologies gained more and more prominence. Since mapping complicated structures requires measurements with many different baselines (see Section 2.1), this required a new approach: antennas moveable on tracks, and the systematic use of Earth rotation synthesis, which takes advantage of the variation of the projected baseline during the course of a day. The first instrument built according to this principle was the One-Mile radio telescope in Cambridge; it could achieve a very uniform coverage of the (u,v) plane by varying the distances between its three antennas from day to day, and observing for $\sim$ 12 hours in each of these configurations [46]. A more advanced facility was constructed nearby in 1971; with four moveable and four fixed telescopes on baselines up to 5 km and surface accuracy sufficient for observations at wavelengths as short as 2 cm it provided detailed images of Cyg A, showing for the first time the core and the hot spots in the lobes. A key technology necessary for the success of aperture synthesis techniques was the emergence of electronic computers that were powerful enough to perform the Fourier transforms. Ryle received the Nobel prize for his contributions to the development of radio interferometry and aperture synthesis; his Nobel lecture [47] provides a good overview of the work on these topics in Cambridge.

Inspired by the success of the One-Mile telescope, new facilities optimized for synthesis imaging were also built in other countries. The Westerbork Synthesis Radio Telescope in the Netherlands [2] followed the Cambridge design of a linear East-West array, which requires a full $\sim$ 12 hr synthesis observation for each image. In contrast, the 27 antennas of the Very Large Array (VLA) near Socorro in New Mexico [55] are arranged on the three arms of a Y, which provides an instantaneous two-dimensional coverage of the (u,v) plane and thus the capability to produce "snapshot" images of relatively bright sources within a few minutes, while still allowing for the possibility of full Earth rotation synthesis if higher image quality is desired.

From the 1980s on a number of mm-wave interferometers have been built that can image the continuum and line emission of objects as varied as nearby star-forming regions and distant galaxies in the young universe, with a spatial resolution of about $1''$. These instruments are now beginning to face some of the same difficulties that plague optical and infrared interferometry: fast phase fluctuations induced by atmospheric turbulence, the need to build antennae with smooth surfaces and small thermal and gravitational flexure for use at high frequencies, and the limited transparency of the atmosphere in the infrared / sub-mm / mm-wave region. Suitable conditions for these facilities can only be found at the highest and driest sites in the world.

4.3 Very long baseline interferometry

Whereas "local" radio interferometers with baselines up to a few hundred kilometers can be linked by cables, waveguides, or microwave transmission lines, a different approach is needed if even higher angular resolution is desired. In Very Long Baseline Interferometry (VLBI) arrays, telescopes separated by many hundreds or even thousands of kilometers are connected "virtually": the signal at each antenna is recorded on tape, the tapes from all antennas are shipped to a correlating center, and the "fringes" are formed long after the observation in a computer (e.g. Cohen et al. [8]). In this setup it is not possible to distribute the signal from a common local oscillator to all antennas; therefore each station has to be equipped with a very stable frequency standard such as a hydrogen maser.

Somewhat paradoxically and in apparent contradiction to Eq. 1, the highest angular resolution and best astrometric precision in all of astronomy can be achieved at radio wavelengths; this is of course due to the extremely long baselines of VLBI arrays. The discovery of apparent superluminal motion in compact extragalactic radio sources [59] was among the early successes of this technique. By turning Eq. 6 around one can use observations of compact sources to determine the antenna positions with high precision; VLBI is therefore used extensively for geodetic purposes, in particular to determine Earth orientation parameters. VLBI observations were initially conducted by ad hoc networks of independent radio telescopes. These networks have become more formalized over time, and an array of ten identical telescopes spread over the whole territory of the US (the VLBA) was constructed specifically for VLBI observations.

To overcome the limit on baseline length set by the diameter of the Earth, VLBI arrays have been formed that include one antenna in space. The first such experiment, conducted with the 4.9 m radio antenna of the TDRSS communications satellite, showed that a number of extragalactic radio sources are sufficiently compact at 2.3 and 15 GHz to be detected on baselines of up to 2 Earth diameters [28]. The Japanese HALCA mission [18] carrying an 8 m antenna conducted extensive observing programs at 1.6 and 5 GHz of extragalactic radio sources as well as pulsars and OH masers in conjunction with large ground-based arrays. Unfortunately, the 22 GHz channel of HALCA, which might have been scientifically most productive, failed; a successor mission is therefore planned that will extend the frequency coverage up to 43 GHz.

5 Intensity interferometry

Intensity interferometry differs from the usual amplitude interferometry in that it does not compare the amplitudes of the wavefront at different points, but rather the intensity measured with independent detectors. The mathematical foundations of intensity interferometry were laid by Hanbury Brown and Twiss in the early 1950s; they could in fact show that the correlation in the intensity of a celestial object observed with two telescopes is proportional to the square of the visibility in a Michelson interferometer with the same baseline [16]. The technique has the advantage that the tolerances on pathlength equalization between the interferometer arms are far less tight than in amplitude interferometry, but it suffers from a much poorer signal-to-noise ratio. Intensity interferometry was used to measure the diameters of a few radio sources, but did not have a large impact on radio astronomy because amplitude interferometry turned out to be not too difficult.

In the visible, however, the technical difficulties of stabilizing the optical pathlength are much more substantial, which made intensity interferometry much more attractive for this wavelength range. After a successful laboratory demonstration, Hanbury Brown and Twiss [17] therefore proceeded with the construction of an instrument consisting of two five-foot searchlight mirrors that could be separated by up to 9 m, and used it to measure the diameter of Sirius. A larger intensity interferometer with two 6.5 m reflectors that could be moved on railway tracks on a circle with 188 m diameter was later constructed at Narrabri Observatory [14]. The measurements of the diameters of a sample of hot stars carried out with this instrument [15] are still the best such data available today.

It is interesting to note that intensity interferometry has found applications far beyond astronomy. For example, correlations between pion pairs emitted from the interaction region in particle colliders can be analyzed with the formalism of Hanbury Brown-Twiss interferometry, and yield information on the collision geometry (e.g., Baym [4]). The method is also used in condensed matter physics, atomic physics, and quantum optics (e.g., Kleppner [21]).

Table 2 Selected optical/infrared interferometers

Name	Location	Year	Remarks
20-foot	Mt. Wilson, CA	1919	Mounted to 100-inch telescope
Intensity Interf.	Narrabri, Australia	1956	Diameters of hot stars
I2T	Obs. Cote d'Azur, France	1974	Fringes with separate telescopes
Mark I	Mt. Wilson, CA	1979	Fringe tracking
Mark III	Mt. Wilson, CA	1986	$B_{\max} = 31.5$ m, astrometry
ISI	Mt. Wilson, CA	1988	10 μm heterodyne
GI2T	Obs. Cote d'Azur, France	1985	Interf. spectroscopy
SUSI	Narrabri, Australia	1991	$B_{\max} = 640$ m
COAST	Cambridge, UK	1991	5×40 cm
IOTA	Mt. Hopkins, AZ	1993	3×45 cm, $B_{\max} = 38$ m
NPOI	Anderson Mesa, AZ	1994	Imaging and astrometry
PTI	Mt. Palomar, CA	1995	Dual-star astrometry
CHARA Array	Mt. Wilson, CA	2001	6×1 m, $B_{\max} = 330$ m
Keck Interf.	Mauna Kea, HI	2001	2×10 m, nulling
VLTI	Paranal, Chile	2001	4×8.2 m$+ 4 \times 1.8$ m
MRAO	Magdalena Ridge, NM	2010	6×1.4 m, $B_{\max} = 340$ m
LBT	Mt. Graham, AZ	2011	2×8.4 m in single mount

See text for more details

6 Modern optical/infrared interferometry

6.1 The first and second generation

The first direct interference fringes between separate telescopes were reported by Labeyrie [22] (Table 2). In addition to making measurements at visible wavelengths, his instrument at the Observatoire de la Côte d'Azur, called Interféromètre à Deux Télescopes (I2T), pioneered interferometry at near-infrared wavelengths [11]. The successor of the I2T, called GI2T (Grand Interféromètre à Deux Télescopes, Mourard et al. [34]) used two innovative "boule" telescopes with 1.5 m apertures, and was used to develop interferometric spectroscopy.

Active fringe tracking was first demonstrated by the Mark I interferometer on Mt. Wilson, CA [52], which evolved into the Mark II and Mark III instruments [51]. The Mk III was specifically designed to perform wide-angle astrometry, but a variable baseline that could be configured from 3 m to 31.5 m provided the flexibility needed for a variety of astronomical programs. Since full computer control of the siderostats and delay lines allowed almost autonomous acquisition of stars and data taking, the Mk III could observe up to 200 stars in a single night. This capability was an important factor for the calibration of instrumental effects, and for the scientific productivity of the instrument. The Mark III was scientifically very productive, yielding stellar diameter and limb darkening measurements [35, 43], binary orbits (e.g., Hummel et al. [19]), the geometry of Be star disks [41], and a demonstration of wide-angle astrometry [20].

Interferometry in the mid-infrared was pioneered by a group around Townes in Berkeley. Their Infrared Spatial Interferometer (ISI, Hale et al.

[13]), also located on Mt. Wilson uses heterodyne techniques similar to those familiar from radio facilities. It operates at wavelengths from 9 to 12 μm, using a CO_2 laser as local oscillator. The ISI has mostly been used to study dust shells around evolved late-type stars.

6.2 Present interferometric facilities

While important astronomical results could be obtained from the visibility measurements of the early single-baseline interferometers, an important objective of more recent facilities has been the development of imaging capabilities. The first closure-phase images from an optical synthesis array, showing the two components of the binary system Capella, were produced by the COAST instrument in Cambridge [3]. The power of synthesis imaging in the near-infrared was also demonstrated by Tuthill et al. [57], who were able to reconstruct the dust distribution in the Wolf-Rayet binary WR 104 from data obtained by masking the entrance aperture of the Keck telescope with a non-redundant arrangement of holes. The Navy Prototype Interferometer near Flagstaff, AZ, and the CHARA Array on Mt. Wilson have been designed with imaging capabilities in mind; both facilities support the simultaneous combination of light from up to six telescopes.

Some of the largest optical telescopes in existence can also be used in an interferometric mode. The two 10 m Keck telescopes on Mauna Kea, HI, form an 85 m baseline. The Keck Interferometer is equipped with a near-infrared beam combiner and with a 10 μm nulling instrument, which is used to characterize the zodiacal light surrounding nearby stars. Plans to expand the Keck Interferometer with 1.8 m "outriggers" complementing the two 10 m telescopes have unfortunately been abandoned.

The Very Large Telescope Interferometer (VLTI) operated by the European Southern Observatory on Cerro Paranal in Chile consists of four large 8.2 m "Unit Telescopes" as well as four smaller moveable 1.8 m "Auxiliary Telescopes". The first generation of VLTI instruments comprises a single-baseline mid-infrared instrument (MIDI, Leinert et al. [26]) and a near-infrared instrument with spectral resolution up to $R = 12\,000$ and capable of combining light from three telescopes (AMBER, Petrov et al. [38]). PRIMA, a facility that will enable precise narrow-angle astrometry and phase-referenced imaging with the VLTI [10, 42], is currently in the commissioning phase. A second generation of instruments [44] will further enhance the capabilities of the VLTI.

7 The future of astronomical interferometry

7.1 Ground-based interferometry

The rapid progress in computing technology is shifting the cost balance in radio interferometers towards arrays with more but cheaper individual antennas.

This trend is currently embodied by the Allen Telescope Array in California and by the Low-Frequency Array (LOFAR) with stations in the Netherlands and in neighboring countries. The long-term goal is the construction of an array with 1 km^2 collecting area operating at frequencies between 70 MHz and 10 GHz. Several precursor projects are currently under construction, and technology development programs are under way at a number of institutions around the world.

The next big leap in sub-mm and mm-wave interferometry will also be made through a truly global cooperation: Europe, Japan, North America, and Chile have entered a collaboration to construct the Atacama Large Millimeter Array (ALMA) at a site located at an elevation of 5,000 m above sea level in the Chilean Andes. At this elevation and in an extremely dry climate, the atmosphere is transparent in many frequency bands that are inaccessible from other sites. ALMA will consist of an interferometric array of fifty 12-m antennas distributed over an area 18 km in diameter, supplemented by twelve 7-m antennas forming a compact array and four 12-m "total power" antennas. ALMA will cover the wavelength range from 0.3 ... 10 mm. ALMA will be complementary to large optical telescopes by giving us a view of cold objects and those regions of the universe that are hidden from our view by large amounts of dust.

In the visible and near infrared, many completely new observations would become possible with an array consisting of large telescopes providing baselines up to a few kilometers [54]; it would for example be possible to obtain images of stellar surfaces with many pixels across their disks. Conceptual studies indicate that the construction of such a facility should become possible in the $\sim$ 2020 time frame (e.g., Quirrenbach [40]).

7.2 Space interferometry

Turbulence in the Earth's atmosphere makes interferometry in the visible wavelength region rather difficult; in the far-infrared, UV, and X-ray regimes ground-based observations are completely impossible because of atmospheric absorption. A number of interferometric space missions have therefore been proposed.

NASA's Space Interferometry Mission [58] will push the precision of astrometric measurements far beyond the capabilities of any other project currently in existence or under development. SIM is a single-baseline interferometer; after some design improvements it consists of two 50 cm telescopes on a 6 m baseline. In its "narrow-angle" mode (i.e., for measurements of a target with respect to reference stars within a field of $\sim 1°$), SIM will provide an accuracy of $\lesssim 1$ μas for each measurement. This makes it possible to carry out a high-precision survey of nearby stars, sensitive to the astrometric signature of Earth-like planets. In "wide-angle" mode SIM can measure parallaxes and proper motions with a precision of a few micro-arcseconds for objects down to $V \sim 20$.

Mid-infrared nulling interferometry has been studied by ESA and by NASA as one of the possibilities for a mission to characterize the atmospheres of terrestrial extrasolar planets. Such a mission must have an angular resolution of ~ 35 mas, corresponding to 0.7 AU at 20 pc, therefore an interferometer with baselines of at least 50 ... 100 m is required. The contrast between a terrestrial planet and its parent star at 10 μm is $\lesssim 10^{-6}$, thus an excellent depth and stability of the null are required. The spectral region from 6 to 20 μm contains a number of important spectral features that can be used to diagnose the presence of H_2O, CO_2, and O_3, which serves as a proxy for O_2 [25].

Concepts for space-based far-infrared ($\sim 25 \ldots 600$ μm) interferometers with cryo-cooled (~ 4 K) telescopes of modest size have been developed in the US and in Europe (e.g., Leisawitz et al. [27]). A structurally-connected interferometer could provide access to baselines up to tens of meters; a far-IR interferometer with components flying in formation could sample baselines up to 1 km.

A space-based UV-optical Fizeau interferometer dubbed "Stellar Imager" has been studied by NASA [6]. 20 to 30 mirror satellites forming baselines of 100 m to 1 km would send the light to a central beam combiner and thus provide 0.1 mas spectral imaging of the surfaces and—via spatially-resolved asteroseismology—interiors of solar-type stars.

Laboratory experiments have shown that X-ray fringes can be obtained with grazing-incidence interferometers in a geometry that would in principle be suited for astronomical applications [7]. A meter-class X-ray interferometer would provide a resolution of ~ 100 μas at 1 keV, sufficient to image the coronae of nearby stars, but the distance between the optics and the detector would have to be tens of kms. The resolution of an X-ray interferometer with baselines of a few 100 m would be ~ 100 nas, and could thus resolve the event horizon of the black holes at the center of our Galaxy, and in active galactic nuclei. In a few decades, astronomical interferometry could thus cover the whole electromagnetic spectrum from meter-waves all the way to X-rays.

References

1. Anderson, J.A.: Application of Michelson's interferometer method to the measurement of close double stars. Ap. J. **51**, 263–275 (1920)
2. Baars, J.W.M., van der Brugge, J.F., Casse, J.L., Hamaker, J.P., Sondaar, L.H., et al.: The synthesis radio telescope at Westerbork. Proc. IEEE **61**, 1258–1266 (1973)
3. Baldwin, J.E., Beckett, M.G., Boysen, R.C., Burns, D., Buscher, D.F., et al.: The first images from an optical aperture synthesis array: mapping of Capella with COAST at two epochs. A&A **306**, L13–L16 (1996)
4. Baym, G.: The physics of Hanbury Brown-Twiss intensity interferometry: from stars to nuclear collisions. Acta Phys. Pol. B **29**, 1839–1884 (1998)
5. Bracewell, R.N.: Detecting nonsolar planets by spinning infrared interferometer. Nature **274**, 780–781 (1978)
6. Carpenter, K.G., Schrijver, C.J., Karovska, M.: The Stellar Imager (SI) vision mission. In: Monnier, J.D., Schöller, M., Danchi, W.C. (eds.) Advances in Stellar Interferometry, vol. 6268, p. 626821. SPIE, Bellingham (2006)
7. Cash, W., Shipley, A., Osterman, S., Joy, M.: Laboratory detection of X-ray fringes with a grazing-incidence interferometer. Nature **407**, 160–162 (2000)

8. Cohen, M.H., Jauncey, D.L., Kellerman, K.I., Clark, B.G.: Radio inteferometry at one-thousandth second of arc. Science **162**, 88–94 (1968)
9. Colavita, M.M., Shao, M., Staelin, D.: Two-color method for optical astrometry: theory and preliminary measurements with the Mark III stellar interferometer. Appl. Opt. **26**, 4113–4122 (1987)
10. Delplancke, F.: The PRIMA facility phase-referenced imaging and micro-arcsecond astrometry. NewAR **52**, 199–207 (2008)
11. Di Benedetto, G.P., Rabbia, Y.: Accurate angular diameters and effective temperatures for eleven giants cooler than K0 by Michelson interferometry. A&A **188**, 114–124 (1987)
12. Fizeau, H.: Prix Bordin: rapport sur le concours de l'année 1867. C. R. Acad. Sci. **66**, 932–934 (1868)
13. Hale, D.D.S., Bester, M., Danchi, W.C., Fitelson, W., Hoss, S., et al.: The Berkeley infrared spatial interferometer: a heterodyne stellar interferometer for the mid-infrared. Ap. J. **537**, 998–1012 (2000)
14. Hanbury Brown, R., Davis, J., Allen, R.: The stellar interferometer at Narrabri Observatory I. A description of the instrument and the observational procedure. MNRAS **137**, 375–392 (1967)
15. Hanbury Brown, R., Davis, J., Allen, R.: The angular diameters of 32 stars. MNRAS **167**, 121–136 (1974)
16. Hanbury Brown, R., Twiss, R.Q.: A new type of interferometer for use in radio astronomy. Phil. Mag. Ser. 7, **45**, 663–682 (1954)
17. Hanbury Brown, R., Twiss, R.Q.: A test of a new type of stellar interferometer on Sirius. Nature **178**, 1046–1048 (1956)
18. Hirabayashi, H., Hirosawa, H., Kobayashi, H., et al.: The VLBI space observatory programme and the radio-astronomical satellite HALCA. PASJ **52**, 955–965 (2000)
19. Hummel, C.A., Armstrong, J.T., Quirrenbach, A., Buscher, D.F., Mozurkewich, D., et al.: Very high precision orbit of Capella by long baseline interferometry. Astron. J. **107**, 1859–1867 (1994a)
20. Hummel, C.A., Mozurkewich, D., Elias, N.M., Quirrenbach, A., Buscher, D.F., et al.: Four years of astrometric measurements with the Mark III optical interferometer. Astron. J. **108**, 326–336 (1994b)
21. Kleppner, D.: Hanbury Brown's steamroller. Phys. Today **61,8**, 8–9 (2008)
22. Labeyrie, A.: Interference fringes obtained on Vega with two optical telescopes. Ap. J. **196**, L71–75 (1975)
23. Lawson, P.R. (ed.): Selected Papers on Long Baseline Stellar Interferometry. SPIE Milestone Series, vol. MS 139. SPIE, Bellingham (1997)
24. Lawson, P.R.: Notes on the history of stellar interferometry. In: Lawson, P.R. (ed.) Principles of long baseline stellar interferometry. Course notes from the 1999 Michelson Summer School, pp. 325–332. JPL Publication 00-009. JPL, Pasadena (2000)
25. Léger, A., Mariotti, J.M., Mennesson, B., Ollivier, M., Puget, J.L., et al.: Could we search for primitive life on extrasolar planets in the near future? Icarus **123**, 249–255 (1996)
26. Leinert, C., Graser, U., Przygodda, F., Waters, L.B.F.M., Perrin, G., et al.: MIDI—the 10 μm instrument on the VLTI. Ap&SS **286**, 73–83 (2003)
27. Leisawitz, D., Baker, C., Barger, A., et al.: The Space Infrared Interferometric Telescope (SPIRIT): high-resolution imaging and spectroscopy in the far-infrared. Adv. Space Res. **40**, 689–703 (2007)
28. Levy, G.S., Linfield, R.P., Ulvestad, J.S., Edwards, C.D., Jordan, J.F., di Nardo, J., Christensen, C.S., Preston, R.A., Skjerve, L.J., Blaney, K.B.: Very long baseline interferometric observations made with an orbiting radio telescope. Science **234**, 187–189 (1986)
29. Michelson, A.A.: On the application of interference methods to astronomical instruments. Phila. Mag. **30**, 1–20 (1890)
30. Michelson, A.A.: Measurement of Jupiter's satellites by interference. Nature **45**, 160–161 (1891)
31. Michelson, A.A., Pease, F.G.: Measurement of the diameter of α Orionis. Ap. J. **53**, 249–259 (1921)
32. Mills, B.Y., Little, A.G., Sheridan, K.V., O.B.: A high-resolution radio telescope for use at 3.5 m. Proc. IRE **46**, 67–84 (1958)

33. Monnier, J.D.: Optical interferometry in astronomy. Rep. Prog. Phys. **66**, 789–857 (2003)
34. Mourard, D., Tallon-Bosc, I., Blazit, A., Bonneau, D., Merlin, G., et al.: The GI2T interferometer on Plateau de Calern. A&A **283**, 705–713 (1994)
35. Mozurkewich, D., Armstrong, J.T., Hindsley, R.B., Quirrenbach, A., Hummel, et al.: Angular diameters of stars from the Mark III optical interferometer. Astron. J. **126**, 2502–2520 (2003)
36. Pawsey, J.L., Payne-Soott, R., McCready, L.L.: Radio-frequency energy from the Sun. Nature **157**, 158–159 (1946)
37. Perley, R.A., Schwab, F.R., Bridle, A.H. (eds.): Synthesis imaging in radio astronomy. ASP Conf. Ser., vol. 6, 509 pp. Astronomical Society, San Francisco (1989)
38. Petrov, R.G., Malbet, F., Weigelt, G., Antonelli, P., Beckmann, U., et al.: AMBER, the near-infrared spectro-interferometric three-telescope VLTI instrument. A&A **464**, 1–12 (2007)
39. Quirrenbach, A.: Optical interferometry. ARAA **39**, 353–401 (2001)
40. Quirrenbach, A.: Design considerations for an extremely large synthesis array. In: Traub, W.A. (ed.) New Frontiers in Stellar Interferometry. SPIE, vol. 5491, pp. 1563–1573. SPIE, Bellingham (2004)
41. Quirrenbach, A., Bjorkman, K.S., Bjorkman, J.E., Hummel, C.A., Buscher, D.F., et al.: Constraints on the geometry of circumstellar envelopes: optical interferometric and spectropolarimetric observations of seven Be stars. Ap. J. **479**, 477–496 (1997)
42. Quirrenbach, A., Coudé du Foresto, V., Daigne, G., Hofmann, K.H., et al.: PRIMA—study for a dual-beam instrument for the VLT Interferometer. In: Reasenberg, R.D. (ed.) Astronomical Interferometry. SPIE, vol. 3350, pp. 807–817. SPIE, Bellingham (1998)
43. Quirrenbach, A., Mozurkewich, D., Buscher, D.F., Hummel, C.A., Armstrong, J.T.: Angular diameter and limb darkening of Arcturus. A&A **312**, 160–166 (1996)
44. Richichi, A., Delplancke, F., Paresce, F., Chelli, A. (eds.): The Power of Optical/IR Interferometry: Recent Scientific Results and 2nd Generation Instrumentation. Springer, New York (2008)
45. Ryle, M.: A new radio interferometer and its application to the observation of weak radio stars. Proc. R. Soc. A **211**, 351–375 (1952)
46. Ryle, M.: The new Cambridge radio telescope. Nature **194**, 517–518 (1962)
47. Ryle, M.: Radio telescopes of large resolving power. Nobel Lecture (1974)
48. Ryle, M., Vonberg, D.D.: Solar radiation at 175 Mc/s. Nature **158**, 339–340 (1946)
49. Shao, M., Colavita, M.M.: Potential of long-baseline infrared interferometry for narrow-angle astrometry. A&A **262**, 353–358 (1992a)
50. Shao, M., Colavita, M.M.: Long-baseline optical and infrared stellar interferometry. ARAA **30**, 457–498 (1992b)
51. Shao, M., Colavita, M.M., Hines, B., Staelin, D., Hutter, D.J., et al.: The Mark III stellar interferometer. A&A **193**, 357–371 (1988)
52. Shao, M., Staelin, D.H.: First fringe measurements with a phase-tracking stellar interferometer. Appl. Opt. **19**, 1519–1522 (1980)
53. Stéphan, E.: Sur l'extrême petitesse du diamètre apparent des étoiles fixes. C. R. Acad. Sci. **78**, 1008–1012 (1874)
54. Surdej, J., Caro, D., Detal, A. (eds.): Science cases for next generation optical/infrared interferometric facilities. Proceedings of the 37th Liège International Astrophysical Colloquium. Liège University (2005)
55. Thompson, A.R., Clark, B.G., Wade, C.M., Napier, P.J.: The very large array. Astron. J. **44**(Suppl), 151–167 (1980)
56. Thompson, A.R., Moran, J.M., Swenson, G.W.: Interferometry and Synthesis in Radio Astronomy, 534 pp. Wiley, New York (1986)
57. Tuthill, P.G., Monnier, J.D., Danchi, W.C.: A dusty pinwheel nebula around the massive star WR 104. Nature **398**, 487–489 (1999)
58. Unwin, S.C., Shao, M., Tanner, A.M., Allen, R.J., Beichman, C.A., et al.: Taking the measure of the universe: precision astrometry with SIM PlanetQuest. PASP **120**, 38–88 (2008)
59. Whitney, A.R., Shapiro, I.I., Rogers, A.E.E., Robertson, D.S., Knight, C.A., et al.: Quasars revisited: rapid time variations observed via very long baseline interferometry. Science **173**, 225–230 (1971)

Exp Astron (2009) 26:65–78
DOI 10.1007/s10686-009-9153-x

REVIEW ARTICLE

The submillimetre revolution

Ian Robson

Received: 12 December 2008 / Accepted: 10 February 2009 / Published online: 7 March 2009

Abstract Although this is a personal reflection of the rise of submillimetre and far-infrared astronomy I will attempt to cover all the major events and players. The review will be restricted to continuum techniques and will principally cover ground-based telescopes. Balloon-based platforms will also be discussed but airborne platforms will be covered by George Rieke. It is always important to remember that a telescope is only as good as the instrumentation it supports, and because there have been few true submillimetre telescopes constructed to date, this review will devote significant time to the instrumentation and technical developments that have been the underpinning of the submillimetre revolution.

Keywords Submillimetre astronomy · Telescopes · Instruments · Observing techniques

1 The problems of submillimetre astronomy

Submillimetre astronomy is fundamentally difficult, and development was initially slow due to a number of reasons: (1) the sources that shine predominantly in this regime are cold (<50 K) and so intrinsically weak emitters; (2) there were no purpose-built telescopes to collect the energy and so optical telescopes of relatively small collecting area were typically used; (3) the Earth's atmosphere is mostly opaque in the submillimetre, with only a few

I. Robson (✉)
UK Astronomy Technology Centre, Royal Observatory Edinburgh,
Scotland, EH9 3HJ, UK
e-mail: eir@roe.ac.uk
URL: www.roe.ac.uk

transmission ‘windows’ available, and even then, only from high, dry sites to counter the dominant absorption by water vapour molecules (see Fig. 1); (4) the early detectors were very insensitive. Advances in all of these areas, the growth of purpose-built telescopes at high, dry sites, ever-improving detector techniques, better control of filters to isolate the atmospheric windows and better sky-noise suppression have been the key factors in the long struggle for the submillimetre to emerge from the wilderness into mainstream astronomy.

2 In the beginning

In Europe, submillimetre astronomy began in London, at Queen Mary College in the late 1960’s, where John Bastin had assembled a group to work in this regime using a telescope situated on the roof of the Physics Building. This was an old World-War II searchlight dish, with anything but high efficiency, and at the great altitude of 110 ft above sea-level was restricted to observing the Sun from a few millimeters to about 0.8 mm. Progress was made using the 40-in. optical telescope at the Pic du Midi, on the French-Spanish border at an altitude of 10,000 ft. By this time the lessons of sky-subtraction from the infrared (see Rieke this volume) had been understood, and a twin-cone arrangement (like a pair of trousers) was used for the first time. The twin cones (the trouser legs) viewed adjacent areas of sky and the signals from both were chopped at around 1,000 Hz, giving a difference signal that was directly detected by absorption by the ‘Rollin’ detector. This was an Indium Antimonide (InSb) crystal, immersed (literally) in liquid helium in a glass dewar, which in turn was surrounded by a liquid nitrogen glass dewar. The detector was a broadband device (a bolometer) sensitive to a very wide range of wavelengths, and so filtering had to be understood before astronomical results could be interpreted. The output (AC) signals were fed into a phase-sensitive-detector (PSD), which, when fed with a reference signal from the ‘chopper’ resulted in only the difference signal between the ‘sky’ and

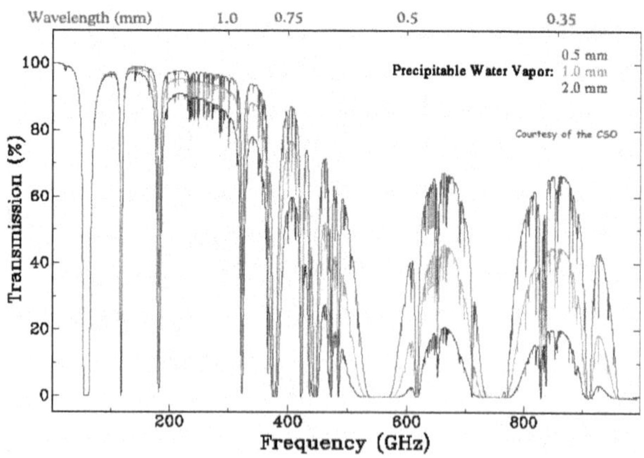

Fig. 1 Submillimetre atmospheric transmission spectrum (courtesy of the CSO)

'sky + source' being measured. The amplified output signal was then fed to some form of 'counter or integrator', a very early example being a 'Nuclear Enhancetron', a crude analogue signal-averaging device borrowed from the nuclear industry. These early experiments resulted in few detections apart from planets in the Solar System and the Orion Nebula.

2.1 Cosmic background studies

One of the big challenges at the time was the question of the '3K cosmic background' radiation discovered by [26]. While measurements in the cm-mm region had shown this to have a very good Rayleigh-Jeans slope the question as to whether the radiation 'turned-over' at around 1 mm to be the black-body expected from the big-bang fireball was paramount. While observations of interstellar molecules (see e.g. [29] for a review) had shown that eventually the radiation spectrum had to turn over, the shape and temperature of the spectrum was still mostly undetermined shortward of the expected peak. A number of groups were attacking this problem using balloon-borne experiments and polarising Michelson interferometers [24]. Although the signal from a 3K source is small, the ability to open up the solid angle meant that it should be detectable. Our first flight was in 1973 from Palestine, Texas but unfortunately the balloon burst on ascent at around 80,000 ft and the payload descended under essentially free-fall as the balloon had not been cut free and the parachute had not been deployed. Fortunately, the air-drag of the deflating balloon was sufficiently great and (combined with some good luck) the payload ended up crashing into a Texan forest, with the payload coming down equidistant between three trees, with the balloon and parachute being entangled in the tops of the trees, and the payload ended up oscillating up and down and never smashing into the ground. The payload was essentially intact and was successfully flown in Feb 1974 to produce a spectrum showing that the cosmic radiation did turnover and had a spectrum closely resembling 2.7 K [30]. Confirmation was soon achieved by the Berkeley group [40] 1 year later. While these showed the promise of the submillimetre regime, the drive was on to detect 'normal' astronomical sources associated with stars and galaxies. The search began for the best sites and facilities and in the early days there was quite a race between ground-based and balloon platforms for new discoveries.

2.2 The search for better sites, telescopes and detection techniques

In the 1970s a handful of groups were actively seeking out or developing new sites for submillimetre astronomy. Sites included: Mount Lemmon north of Tucson, Arizona at around 7,000 ft; Kitt Peak, west of Tucson at 7,000 ft; Izaña on the Island of Tenerife at about 7,000 ft; White Mountain at 12,700 ft in California; Mauna Kea in Hawaii at just under 14,000 ft and Mount Evans in Colorado at 14,260 ft. The telescopes ranged from an optical 24-in. (Mt Evans), an optical 88-in. (Mauna Kea); a 1.6 m submillimetre dish (White Mountain) to a high-frequency, 36 ft radio dish (Kitt Peak). Eventually, Mauna Kea would turn out to be the site of choice for submillimetre astronomy. During this time a number of advances were made in detection techniques. Cryostat design was improved, allowing portable, metal-based systems (Infrared Laboratories) that could also be 'pumped' to reduce the normal 4.2 K helium temperature to around 1.5 K and so increase detection sensitivity. These 'Low' cryostats became the industry standard, being reliable, simple to operate and allowing internal space for filters and rudimentary collecting optics. Filter technology was, however, still primitive, and shortwave 'blocking' filters were the norm for wavelength selection.

Detector technology was steadily improving and the InSb bolometer gave way to germanium devices, which had better sensitivity. However, the major breakthrough was the introduction of the germanium 'composite' bolometer, first suggested by Frank Low and described in Nishioka et al. [25]. In this case the absorbing element is a thin metal film (e.g. Bismuth), deposited on a dielectric substrate (e.g. Sapphire) for strength, to which the tiny, doped germanium crystal is glued. The bismuth is a much better absorber of the incoming radiation and in so doing its temperature is raised very slightly and the germanium acts as a thermometer, detecting this small rise through a change in resistance/voltage drop. The composite germanium bolometer (see Fig. 2) became the industry standard for well over a decade.

Photometer design was also improving, with better refocusing optics, chopping mirrors to remove the background sky, and more sophisticated integration and averaging techniques as the early portable computing systems became available. There were important experiments with telescope design, specifically the 'Infrared Flux Collector' at Izaña, Tenerife. This was Jim Ring's 60-in. prototype of a thin-mirror design for an infrared telescope. Instead of a

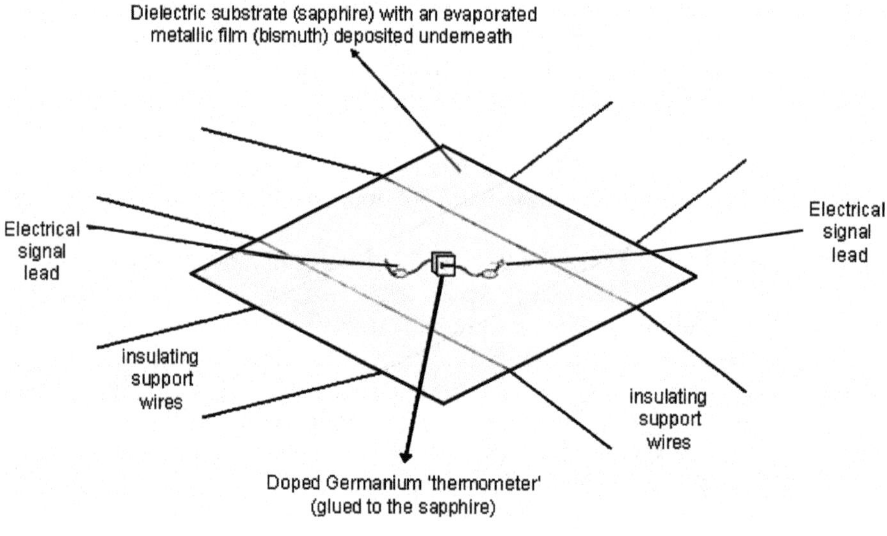

Fig. 2 Sketch of a composite bolometer

low aspect ratio (diameter/thickness) mirror of conventional telescopes, this concept was a very thin, light weighted mirror of much higher aspect ratio. This was much lighter and dramatically reduced the cost of the telescope, and led directly to the development of the United Kingdom Infrared Telescope (UKIRT). The Flux Collector provided a platform for experiments in photometer design but it was also abundantly clear that while the site was suitable for near-infrared astronomy, it was too low for submillimetre work.

Early progress was still slow and one trend was to use the 36-ft radio dish at Kitt Peak and work at very short millimetre wavelengths. This continued for about 5 years. In the beginning the detection threshold was around 20 Jy and planets and star-forming regions were the only detections, but by 1977 extragalactic sources were being detected at a few Jansky [33]. During this period the most successful submillimetre observations came principally from groups using White Mountain and Mauna Kea. Nevertheless, the success ratio was quite low and again, Galactic star formation regions were the order of the day with virtually all extragalactic sources apart from one or two being just beyond the threshold of detection (see e.g. [8, 12, 27, 28, 36, 37]. However, linked with mid-IR data, the presence of Starburst galaxies was clearly established (see Low et al. [23] for a review).

During the 1970s balloon-borne platforms were also a major source of far-infrared astronomical observations and surveys. A review of the technologies, platforms and discoveries can be found in Fazio [10].

3 The UKIRT years

The situation changed for the better with the introduction of UKIRT on Mauna Kea in 1987. This was a purpose-built, 3.8 m diameter infrared telescope on the world's best site at the time. UKIRT was designed for the infrared (like the IRTF) but submillimetre observations were allowed through peer review selection based on the science case. UKIRT had a chopping secondary, which made a huge improvement in detection sensitivity through sky-background suppression. In the early days we used our submillimetre photometer [31] at the Cassegrain focus, with a research student freezing in the dome, glued to the eyepiece of the photometer while balancing on the top of steps. Two major advances happened almost simultaneously, the introduction of TV guiding to UKIRT and the development of the 'Oregon photometer' [1]. The former meant that all observing could be done from the relatively shirt-sleeved environment of the control room and the photometer could be simplified and improved; the latter provided a breakthrough in detection sensitivity.

The Oregon photometer was developed by Ira Nolt and Jim Radostitz and used a small helium-three cryostat to cool the bolometer element to around 350 mK. This gave much-improved sensitivity, and along with the computing power of the UKIRT data reduction facility meant that the era of modern-day submillimetre astronomy had arrived. The detection sensitivity went from around 10 Jy to less than a Jansky and progress was rapid. Indeed, advances from this point onwards were essentially incremental until the advent of SCUBA in 1996. The Oregon helium-three photometer made a number of major discoveries, but it was far from being a common-user facility, being quite tricky to use and maintain, and this remained one of the main drawbacks to submillimetre astronomy moving into the mainstream. At this time the submillimetre scene was still restricted to a very small band of stalwarts who continued to pioneer the field, and were still beset by the vagaries of the weather, where even at Mauna Kea submillimetre conditions could never be guaranteed for a four- or five-night observing run. Nevertheless, major discoveries were made, especially in the extragalactic field (e.g. [4, 7, 9, 11, 32]).

The next breakthrough came with UKT14, the first truly common-user submillimetre photometer. UKT14 was designed at the Royal Observatory Edinburgh with all the latest knowledge of system improvements. The detection element was a composite germanium bolometer, cooled to helium-three temperatures, the detector being optically fed through carefully designed Gaussian beam optics to provide efficient illumination and control stray light. UKT14 came equipped with a ten-position filter wheel that allowed a wide selection of filters to be employed. The filters were also of the latest design of metal-mesh interference filters, tailored to match the atmospheric transmission windows at Mauna Kea and having relatively narrow pass-bands ($\lambda/\delta\lambda \sim 8$) that both minimised the sky background emission and gave much more accurate photometry for sources of unknown spectral index. Importantly, UKT14 also came with its own data reduction package and as such was a truly common-user instrument, no longer requiring the specialist observer to be at the tiller. This was the breakthrough that had been long awaited and UKT14 opened up the submillimetre to the general astronomical community almost at once (e.g. [39]).

4 Enter the JCMT and CSO

While this was going on, the James Clerk Maxwell Telescope (JCMT; Fig. 3) and the Caltech Submillimetre Observatory (CSO) were being constructed on Mauna Kea in what would become known as 'Millimetre Valley', although why this wasn't ever referred to as 'Submillimetre Valley' remains a mystery. The story of the JCMT is quite interesting as in the late 1970s we were site testing on Izaña for the UK Millimetre Telescope [14], which was planned as a spectroscopic telescope aimed at detecting interstellar molecules and probing interstellar chemistry. The submillimetre continuum group were not really part of this new project and had to fight tooth and nail to become accepted; the fact that we'd actually done some submillimetre observing and obtained astronomical data was a grudging point that couldn't be dismissed. Nevertheless, with funding tight, it looked as if Tenerife was going to be the preferred site, although we had made it clear that this would always limit the

Fig. 3 The James Clerk Maxwell Telescope with the membrane open

potential for the telescope and our interests would lie elsewhere (such as the KAO and high mountains). The UK were looking for another partner and the Netherlands were very interested but insisted that if they were to join the project, then it had to be a submillimetre facility and it had to go to a submillimetre site. The only possible site at the time was Mauna Kea, which we had tested some years earlier and was generally agreed to be the premier high-altitude developed site in the world. After much debate it was agreed that the 15 m diameter JCMT, as it was eventually renamed, would go to Mauna Kea. My role on the Steering Committee was then to ensure that the JCMT would be equipped with a chopping secondary that would allow continuum observations to take place rather than the slow, 1 Hz, nutating subreflector that was the ambition of the radio astronomers. The Netherlands provided a superb two-axis design that operated at around 8 Hz and enabled complex chopping patterns in two dimensions to be undertaken—hugely important in what was to follow. The original concept was that the JCMT would be protected by a membrane that would be opened every night for submillimetre observing, but this never looked realistic to me (and indeed never worked in practice) and so I ensured that the membrane material had very high transmission in the submillimetre so we could observe through it, as has been the norm ever since. Everything was then in place for the JCMT to be a success, which it undoubtedly has been. UKT14 was transferred to the JCMT in 1988 [6] and for the next 10 years it was the instrument of choice at the telescope and produced a host of new discoveries, including the first putative example of a class zero protostar [2], the breakthrough of many extragalactic sources including the confirmation of the IRAS, 'super-starburst' ULIRGs [34], measurements of radio-quiet quasars [17] and even rudimentary maps [18].

5 The SCUBA revolution

Clearly the name of the game, as at every other wavelength, is to move from single-pixel devices to imaging systems as soon as practical. The Submillimetre Common User Bolometer Array (SCUBA—[15] was the first to achieve this in a meaningful way and was by far the most successful of the first-generation submillimetre imaging systems. SCUBA was conceived in the mid-1980's following an application to produce an eight-element array (similar to what was being used at the NRAO 36 ft at Kitt Peak) as a prototype to understand better skynoise effects and skynoise cancellation techniques as well as undertake astronomy. This was turned down as not being ambitious enough and, with the JCMT under construction, a much more ambitious project was undertaken. This was to go for a 'camera' and simultaneously observe in two wavebands with a significant number of detectors at each wavelength. This initial concept was subsequently made more complex at the request of the users by the addition of a filter wheel for multiband photometry (at least seven wavebands) as well as long-wavelength single 'photometry' pixels. The main driver for these additions was for spectral energy distribution studies over as wide a range as possible (from 2.0 mm to 350 μm). SCUBA was approved by the JCMT Board in 1987.

The two SCUBA focal planes had 91 and 37 detectors observing at 350/450 and 850/750 μm respectively in a variety of combinations. The focal plane was composed of identical germanium composite bolometers, each hand-built, and positioned in a small integrating cavity at the end of its own feedhorn. The two arrays are shown in Fig. 4. The arrays were cooled to around 100 mK through a dilution refrigerator and the signals were taken out to room temperature via ribbon cables. These were a source of lengthy technical development to enable the best thermal and electrical properties to be used whilst at the same time ensuring that they could be attached (soldered) to the connectors—something

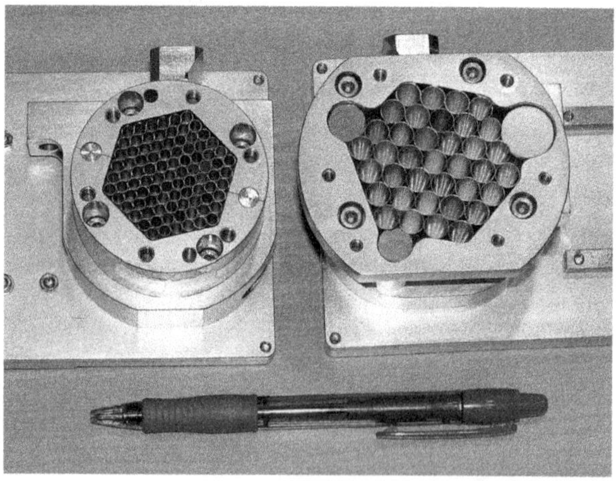

Fig. 4 The two SCUBA arrays showing the individual feedhorns

that was non-trivial. Sky-chopping was undertaken by the secondary mirror on the JCMT and special jiggle patterns were produced that enabled full sampling to be undertaken for either 'stare' modes, or for scanning in mapping mode. SCUBA mapped the sky some 5,000 times faster than UKT14, totally changing the name of the game. Because SCUBA was designed from the outset as a facility-class instrument for the 'common user' the data analysis software was developed along with the instrument. This enabled quick-look images to be obtained in real time, something that was invaluable during commissioning and a huge confidence builder in astronomical use. The provision of the high-quality, user-friendly, data analysis software was one of the key factors that made SCUBA such a huge success, and its importance and impact should never be underestimated.

As experience was gained with SCUBA, intensive work was undertaken on sky-noise subtraction and calibration. The former entailed looking at how the sky-noise varied across the array and then producing algorithms that could remove this noise, which was hugely important for stare observations of point-like or compact sources. It was also found that the skynoise on the long-wavelength array was closely correlated with that on the short-wavelength array, enabling short-wave skynoise to be better cancelled. The calibration formulation needed to determine two factors; the stability of the instrument and the highly variable attenuation due to water vapour along the line of sight. In the beginning it was difficult to separate these two factors, because submillimetre astronomical calibration sources are few and far between. The breakthrough came with the CSO sky radiometer, which provided an almost constant monitor of atmospheric attenuation at a wavelength of 1.3 mm, albeit at a single zenith angle. After much work enough data were eventually assembled to allow the extinction to be determined along the line of sight and hence a calibration correction to be available [3]. (Subsequently, a line-of-sight attenuation monitor was installed on the JCMT and inter-calibrated.) The solution of the calibration uncertainty was the final piece in the jigsaw that allowed an automatic, pipeline, data-reduction process to be achieved [21]; a crucial step for all future activities where survey-type observations are the norm and where data reduction by any other process would be impossible or unrealistic. With this step, submillimetre astronomy came into the mainstream and could be accessed by any user, and as we shall see, not even at the telescope.

The final step in the picture was to improve the efficiency of the facility through the introduction of queue-scheduled observing. This broke the mould of the standard pattern of astronomers having their fixed 'x-night' run in which to undertake their programme, where all too often the most highly ranked scientific proposals, requiring the driest of conditions, ended up having the entire run wiped out because it was too wet, only to see a much lower-ranked programme requiring only mediocre weather the following few nights having spectacular dry conditions. Clearly this was nonsense, but it took a considerable time before the community could be persuaded to agree to science-prioritised, weather-based queue-scheduling, and in the end this worked because of the collective enthusiasm of the UK and Canadian astronomers. In principle, queue-scheduling goes hand-in-hand with observations being undertaken by staff astronomers on behalf of the PI, but the JCMT was never staffed at this level and so community input and a degree of altruism had to be a modus operandi that everyone was comfortable with. However, the impact of completions of programmes requiring the driest of conditions improved hugely as a result of this operational change. Overall, for a host of reasons, the impact of SCUBA was enormous and this more than anything else changed the face of submillimetre astronomy. It is impossible to do justice to the impact but perhaps quoting just one paper is sufficient, the seminal paper from which 'SCUBA galaxies' evolved [19].

Although I have used the JCMT as the principal facility that moved submillimetre from the niche to the general purpose, we should not forget the other facilities and groups that were also ploughing the lonely furrow. The CSO and IRAM stand out in this regard; both have made major astronomical discoveries in their own right; indeed, where the JCMT has succeeded in continuum work, these two facilities have huge reputations in the spectroscopic field. Both facilities have also engendered the development of continuum array receivers, the CSO with SHARC, SHARC-II [5] and BOLOCAM [13] and from IRAM, MAMBO [22]. We now also have the APEX telescope operating on the Chajnantor Plateau at over 16,000 ft, where the 300-element LABOCA array [35] operating at 870 μm is now producing exciting science. We can look forward to the 50 m LMT [20] coming into operation. However, the next major step forward comes with the JCMT and SCUBA-2 (I will return to interferometry later).

6 The future

6.1 SCUBA-2

The success of SCUBA, especially in the discovery of the 'SCUBA Galaxies', showed that a true survey machine would be required—following in the footsteps of the optical and infrared. In late 1998, as Director of the JCMT, I issued a challenge to the UK ATC to come up with a design for a new instrument that would make use of the maximum field-of-view of the JCMT, would operate at helium-3 temperatures, would be sky-background limited, would have no moving parts, could operate at one wavelength and would cost less than $5M. After the initial concept was very strongly endorsed by an International Review chaired by Reinhard Genzel, this became SCUBA-2. However, it soon became apparent that many of the original guidelines would not be met – the field-of-view was the easiest but to achieve sensitivity under sky power-loading, 100 mK detector temperatures would be required and, following hugely enthusiastic endorsement by the JCMT user community, dual wavebands (like SCUBA) were endorsed. The key science drivers for SCUBA-2 were: large-scale extragalactic surveys; the study of the interstellar medium in local, grand-design spirals and other galaxies; a Galactic Plane survey for star formation; deep and unbiased surveys of molecular clouds. By 1999 the original proposal ended up as a $10M project at the Concept Review stage.

The key to SCUBA-2 would be a next generation detector; an array of pixels, essentially the world's first submillimetre 'CCD'. This came from promising experiments with Transition Edge Sensors, the latest and most sensitive detector element, and their construction being compatible with semi-conductor 'packaging', producing an 'array' of detectors on a 3-in. wafer by the group at the National Institute of Standards and Technology (NIST) in Boulder, Colorado. With a pixel size of 1 mm, this meant that the largest array that could be manufactured would be 40 × 40 columns, and technical reasons eventually led to the production devices being 40 × 32. Four of these arrays would be required at each wavelength to fill the 7 arc min diameter field-of-view. These arrays would be 'bare', with direct illumination just like a CCD or infrared array and so stray light would need to be extremely carefully controlled. On the other hand, the SHARC detectors had already shown that hornless devices would work. However, the actual detector is only half the issue, the other is the read-out of the data, and here multiplexing was necessary, requiring the development of a highly complex MUX wafer that would have to be bump-bonded to the detector array. The indium bumps would have to provide electrical as well as cooling paths for the detector, and as none of this existed or had been demonstrated, the entire array development path drove the schedule for SCUBA-2. The story of this development is far too complex to review here but see Holland et al. [16] for further details.

The natural project path to minimise risk was to develop the detector technology and then to proceed with the instrument, but this would mean a lengthy time before astronomy could be achieved with the telescope. The JCMT Board and funding agencies therefore took a very courageous decision, accepting the risk this involved, and authorised the construction of the instrument in parallel with the array development. As Harvey Butcher, the Chair of the Board commented at the time 'if you don't go for the big risks, you don't reap the big rewards'. Through this plan, astronomical results would be delivered 2 years earlier, in 2006. However, completing SCUBA-2 proved to be much more complex than anticipated, the array development always being on the critical path. SCUBA-2 was subsequently delivered to the JCMT in April 2008, but only with engineering-grade arrays, the science-grade arrays will follow in early 2009.

SCUBA-2 represents a huge step-increase in submillimetre capability, with over 10,000 pixels at the two wavelengths compared with 131 of SCUBA. The mapping speed is a factor of a million times faster than for a single pixel device such as UKT14, and over 2,000 times that of SCUBA. However, with this complexity of scale and size came the cost. SCUBA-2 will cost almost $26M by the time it is complete, but nevertheless, put in context this is a bargain. The costs per pixel of the three devices discussed above are: UKT14—$800k; SCUBA—$70k, SCUBA-2—$260. This dramatic cost reduction represents the

Fig. 5 SCUBA-2 under final commissioning in the lab at the UK ATC

advance of technology and ultimately astronomical value for money. SCUBA-2 and one 40 × 32 pixel engineering array are shown in Figs. 5 and 6.

6.2 Enter the interferometers and the need for surveys

One of the limitations of submillimetre astronomy when using a single dish telescope is the spatial resolution of the images. These are typically many arc seconds, at best around six and often greater than ten. This ultimately limits two factors: source identification and source structure. While high signal-to-noise observations can, in principle, overcome the former to obtain source positions to around an arc second in the most favourable cases, in reality life

Fig. 6 One 40 × 32 pixel SCUBA-2 850 μm sub-array on the 100 mK focal plane unit. The 'batwing' printed circuit board and four SQUID preamplifiers can also be seen

is never so accommodating, and confusion over source identification at other wavelengths for follow-up remains problematic. The second concern presents an intrinsic difficulty, and it is quite possible that many of the submillimetre 'point sources' are, in fact, double or multiple.

The solution, of course, is to use interferometers. At the short millimetre wavelengths, Plateau de Bure and BIMA stand out, while the Smithsonian Submillimetre Array (SMA) is currently unique. However, over the next few years another revolution will take place and that is ALMA. With ALMA there will be unprecedented sensitivity, exquisite spatial resolution and spectroscopic diagnostics. Of course we have to know where to point ALMA and this is where surveys come to the fore as preparatory tools working hand-in-hand with ALMA. Herschel will provide a large survey for follow-up work, but again, at the longer wavelengths of the submillimetre region, spatial resolution remains a problem for source identification. The BLAST [38] balloon-borne survey has already made important contributions to sky surveys, but with the same resolution problem. Therefore the realm of SCUBA-2 with its survey capability is the name of the game for the next big step. Beyond that, we are talking of the next generation of submillimetre-dedicated facility. The best game in town at the moment is the CCAT project; a 25 m diameter dish of very high surface accuracy (10 μm rms), designed to operate down to the shortest submillimetre wavelengths of 200 μm and to accomplish this by being sited at the best known location to date, the 18,400 ft location on Mount Chajnantor in Chile (overlooking the ALMA and APEX sites).

With the facilities about to come on-line on the ground and in space, and with future projects such as CCAT, the submillimetre revolution is here to stay and exciting new discoveries can be guaranteed in the years to come.

References

1. Ade, P.A.R., et al.: Infrared Phys. **24**, 403–415 (1984)
2. Andre, P., Ward-Thompson, D.W., Barsony, M.: Ap. J. **406**, 122 (1993)
3. Archibald, E.N., et al.: Mon. Not. R. Astron. Soc. **336**, 1–13 (2002)
4. Cunningham, C.T., et al.: Mon. Not. R. Astron. Soc. **211**, 543–547 (1984)
5. Dowell, et al.: In: Phillips, T.G., Zmuidzinas, J. (eds.) Millimeter and Submillimeter Detectors for Astronomy. Proc. SPIE, vol. 4855, p. 73 (2002)
6. Duncan, W.D., et al.: Mon. Not. R. Astron. Soc. **243**, 126–132 (1990)
7. Emerson, J.P., et al.: Nature **311**, 237 (1984)
8. Eve, W.D., Sollner, T.C.G., Robson, E.I.: Astron. Astrophys. **59**, 209–213 (1977)
9. Clegg, P.E., et al.: Ap. J. **273**, 58–63 (1983)
10. Fazio, G.G.: Infrared Phys. **19**, 341–351 (1979)
11. Gear, W.K., et al.: Mon. Not. R. Astron. Soc. **219**, 19–22 (1986)
12. Gezari, D.Y., et al.: Ap. J. **191**, L33 (1974)
13. Glenn, J., et al.: Proc. SPIE **3357**, 326 (1998)
14. Hills, R.E.: Infrared Phys. **18**, 819–825 (1978)
15. Holland, W.S., et al.: Mon. Not. R. Astron. Soc. **303**, 659–672 (1999)
16. Holland, W.S., et al.: In: Zmuidzinas, Holland, Withington, & Duncan (eds.) Millimeter and Submillimeter Detectors and Instrumentation for Astronomy III. SPIE, vol. 6275, p. 62751 (2006)
17. Hughes, D.H., et al.: Mon. Not. R. Astron. Soc. **263**, 607–618 (1993)

18. Hughes, D.H., Gear, W.K., Robson, E.I.: Mon. Not. R. Astron. Soc. **270**, 641–649 (1994)
19. Hughes, D.H., et al.: Nature **394**, 241–247 (1998)
20. Irvine, W.M., Carrasco, E., Aretxaga, I.: http://www.lmtgtm.org/book/lmtbook.html (2005)
21. Jenness, T., et al.: Mon. Not. R. Astron. Soc. **336**, 14–21 (2002)
22. Kreysa, E., et al.: Proc. SPIE **3357**, 319 (1998)
23. Low, F.J., Rieke, G.H., Gehrz, R.D.: Ann. Rev. Astron. Astophys. **45**, 43–75 (2007)
24. Martin, D.H., Puplett, E.: Infrared Phys. **10**, 105–109 (1969)
25. Nishioka, N.S., Richards, P.L., Woody, D.P.: Appl. Opt. **17**, 1562–1567 (1978)
26. Penzias, A.A., Wilson, R.W.: Ap. J. **142**, 419–421 (1965)
27. Rieke, G.H., et al.: Ap. J. **183**, L67–71 (1973)
28. Righini, G., Simon, M., Joyce, R.R.: Ap. J. **207**, 119–125 (1976)
29. Robson, E.I., Clegg, P.E.: In: Jauncey, D.L. (ed.) IAU Symposium 74 Radio Astronomy & Cosmology, p. 319. Reidel, Dordrecht (1977)
30. Robson, E.I., et al.: Nature **251**, 591–592 (1974)
31. Robson, E.I., et al.: Infrared Phys.**18**, 781–791 (1978)
32. Robson, E.I., et al.: Mon. Not. R. Astron. Soc. **213**, 355–364 (1985)
33. Rowan-Robinson, M., et al.: Astron. Astrophys. **62**, 249–254 (1978)
34. Rowan-Robinson, M., et al.: Mon. Not. R. Astron. Soc. **261**, 513–521 (1993)
35. Sirongo, G., et al.: The Messenger. **129**, 2–7 (2007)
36. Smith, J., et al.: Ap. J. **234**, 902–908 (1979)
37. Soifer, B.T., Hudson, H.S.: Ap. J. **191**, L83 (1974)
38. Truck, M.D.P., et al.: Ap. J. **681**, 415–427 (2008)
39. Ward-Thompson, D.W., et al.: Mon. Not. R. Astron. Soc. **241**, 119–133 (1989)
40. Woody, D.P., et al.: Phys. Rev. Lett. **34**, 1036 (1975)

Historical footnote to Ian Robson's talk at the Leiden observatory conference—400 Years of Astronomical Telescopes—entitled "The Rise of the Submillimetre" on Wednesday 1 October 2008

Harry van der Laan

That the location of the JCMT on Mauna Kea instead of on La Palma proved crucial for many developments in submm astronomy discussed in this talk, is emphatically evident in retrospect. It is, however, not well known how this came about. For the sake of the historical record this is summarized here.

In 1979, at the General Assembly of the IAU in Montreal, Harry van der Laan [HvdL], among other things chairman of the National Committee for Astronomy in Holland, approached Prof. F. Graham Smith, then director of the RGO to gauge the possibility of a strategic alliance between The Netherlands and the United Kingdom for the purpose of creating joint observing facilities in the Northern Hemisphere. After several hours of intense exchange in a brown Montreal pub, Graham and Harry, who knew each other from the early sixties in Cambridge, where Graham was a staff member and Harry a research student in Martin Ryle's group, agreed in outline on a UK/NL 80/20 collaboration.

Harry Atkinson [HA], then Head of the UK Science Research Council's Astronomy, Space and Radio Division, had also in 1979 been actively promoting within Europe the desirability of cooperation in ground-based astronomy. He had already played an important role in bringing the UK into the NL-US space project, IRAS. Towards the end of 1979 he approached Italian astronomers (including Giancarlo Setti and Franco Pacini) about possible cooperation. Setti said "great, but talk to Harry van der Laan"! HA also spoke to Henk van de Hulst, whom he knew well though the ESA Council – who also said, talk to HvdL. The result was a seminal meeting in

In consultation with Harry Atkinson.

H. van der Laan (✉)
Sterrewacht Leiden, The Netherlands

Geneva on 21 December 1979, including the two Harrys and Graham Smith. The fuse had been ignited.

Harry [HvdL] went home from Montreal and worked for many months persuading both his community and his research organization/the Dutch government to enter into a large commitment for nearly thirty years. His motive was to have access to telescopes in the visible/IR domains for combining with radio data from the WSRT. As soon as the Boards of SRC and ZWO had agreed and a Joint Steering Committee with Harry Atkinson (SRC) in the chair and Harry van der Laan as vice chair became active, HvdL came under pressure from Herman van de Stadt (Utrecht), Thijs de Graauw (SRON) and Harm Habing (Leiden). They had analyzed all available water vapour measurements on La Palma and on Mauna Kea and they argued vigorously that placing the submm telescope that was in the UK/NL plans on La Palma would be a terrible mismatch between telescope and submm site quality: very few submm observations could be expected on La Palma. The agitating trio pleaded for locating the telescope on Mauna Kea where the SRC was already operating UKIRT.

The agitation caused some unrest among UK astronomers and their spokesman Richard Hills (Cambridge) argued with HvdL not to jeopardize the whole thing by pushing for such a major change at this late stage. HvdL convinced HA that the Mauna Kea location was decisively superior. HA was very sympathetic, having already played a major role in placing there the UK Infra-red Telescope, UKIRT. The two Harry's then arranged for a meeting between the SRC and ZWO chief executives, Sir Geoffrey Allen and Professor Bob van Lieshout. This took place in Leiden and resulted in a mutual assurance that exploring the Mauna Kea option would not erode the commitment of either party to build the submm telescope. Soon thereafter the Boards agreed to the location shift, which would also shift the responsibility of the submm telescope from the RGO to the ROE (which was operating UKIRT), provided the capital costs for construction on Mauna Kea would not exceed the budgeted costs for construction on La Palma. The Joint Steering Committee accepted this condition and appointed a Tiger Team with Malcolm Longair (then director of the ROE) to seek savings in the submm design, especially that of the enclosure, to enable construction on Mauna Kea within the previous budget. The Tiger Team did manage to save several million pounds by a drastic simplification of the enclosure design (which therewith probably became superior to the original concept!) and the JCMT, as it came to be called, was built on Mauna Kea.

Altogether, a model for productive international collaboration and synergy!

Exp Astron (2009) 26:79–94
DOI 10.1007/s10686-009-9167-4

HISTORICAL REVIEW

Back to the future: science and technology directions for radio telescopes of the twenty-first century

James M. Cordes

Received: 16 December 2008 / Accepted: 3 April 2009 / Published online: 28 April 2009

Abstract The early days of radio astronomy showed incredibly diverse experimentation in ways to sample the electromagnetic spectrum at radio wavelengths. In addition to obtaining adequate sensitivity by building large collection areas, a primary goal also was to achieve sufficient angular resolution to localize radio sources for multi-wavelength identification. This led to many creative designs and the invention of aperture synthesis and VLBI. Some of the basic telescope types remain to the present day, now implemented across the entire radio spectrum from wavelengths of tens of meters to submillimeter wavelengths. In recent years, as always, there is still the drive for greater sensitivity but a primary goal is now to achieve very large fields of view to complement high resolution and frequency coverage, leading to a new phase of experimentation. This is the "back to the future" aspect of current research and development for next-generation radio telescopes. In this paper I summarize the scientific motivations for development of new technology and telescopes since about 1990 and going forward for the next decade and longer. Relevant elements include highly optimized telescope optics and feed antenna designs, innovative fabrication methods for large reflectors and dipole arrays, digital implementations, and hardware vs. software processing. The emphasis will be on meter and centimeter wavelength telescopes but I include a brief discussion of millimeter wavelengths to put the longer wavelength enterprises into perspective. I do not discuss submillimeter wavelengths because they are covered in other papers.

Keywords Astrobiology · Cosmology · Galaxies · General relativity · Radio telescopes · Digital processing

J. M. Cordes (✉)
Astronomy Department, Cornell University, Ithaca, NY 14853, USA
e-mail: cordes@astro.cornell.edu

1 Introduction

It has been 77 years since Karl Jansky first detected radio emission from the Milky Way, 70 years since Grote Reber's confirmation, and more than 60 years since the post-war migration of radar expertise to applications in radio astronomy and the invention of new infrastructure and techniques. Since then many discoveries have been made and Nobel prizes won as the new window on the universe opened. Included in them is the detection of the 21 cm hyperfine transition of atomic hydrogen by Ewen & Purcell (and Muller and Oort in the Netherlands) in 1951. Its deuterium counterpart at 92 cm took more than another half-century to detect [1]. Today, astronomy at radio wavelengths is fundamental to a multiwavelength and multi-messenger[1] program for astronomy and astrophysics that has evolved since the middle of the last century.

In this paper I discuss radio astronomy infrastructure starting around 1990. The history of radio telescopes prior to this time is covered in a companion article by W. Sullivan.

2 The quest for resolution and frequency coverage

By 1990 the radio spectrum was entirely observable from decameter wavelengths to submillimeter wavelengths, albeit with highly non-uniform sensitivity by any measure and incomplete coverage of the sky. Centimeter-wavelength arrays were providing deep radio images in all astronomical domains, from the solar system to the most distant active galactic nuclei. In addition, arrays dedicated to very long baseline interferometry (VLBI) had been established, including the European VLBI Network (EVN) and the U.S. Very Long Baseline Array (VLBA). At this time, the Green Bank Telescope (GBT) was being planned as a large clear aperture instrument that would provide stable spectral baselines for spectroscopy up to millimeter wavelengths. The Arecibo Telescope was upgraded during the 1990s by implementing a Gregorian system for correction of spherical aberration and other improvements that yielded the largest collecting area on the planet to wavelengths of 3 cm. This was the situation at the turn of the century.

3 The quest for sensitivity and field of view

So what else is needed? A great deal, as it turns out. Though the "phase space" of frequency and angular resolution are covered with current instruments, sensitivity and field of view (FoV) are lacking. At the highest angular resolutions

[1] I.e. photons, baryonic cosmic rays, neutrinos and gravitational waves. The future may include astronomy that uses dark matter particles as well.

from VLBI (0.1 to 1 milli-arcsec), fields of view are very small ($\sim$ 1 arc sec) and available collecting areas primarily allow detection of *non-thermal* sources. Much greater collecting area is needed to bring thermal sources, especially stars, into the realm of routine study. As discussed below, large-scale surveys of many types require an orders of magnitude increase in survey speed that can be provided only through innovations in radio cameras placed at the focal points of reflector systems and by phased array systems. These requirements drive the current activity in radio telescope development.

4 Science drivers

Current scientific forefronts that require observations at radio wavelengths include topics in astrobiology, astrophysics and cosmology, along with areas in fundamental physics. Some of these areas require usage of radio telescopes in concert with those at other wavelengths and also with non-photonic facilities, such as gravitational-wave observatories. Others have radio-only solutions. Next-generation radio telescopes are tools for all astronomers for discovery and understanding.

Stars and galaxies:

- *The first stars and first metals:* The first stars and black holes reionized the universe between redshifts of about 15 and 6. The process of reionization, extending back to pre-stellar epochs, can be identified through spectral-line mapping of the hyperfine 21-cm transition of hydrogen. The same first stars yielded the first metals. Highly redshifted carbon monoxide is a tracer for this process because it has a large dipole moment.
- *Formation and evolution of galaxies, including giant black holes:* The 21-cm HI line is also an excellent tracer of the history of star formation and of galaxy-galaxy interactions (tidal tails, mergers, harrassment). Continuum synchrotron emission is highly correlated with dust IR emission and is thus an excellent probe of star formation in galaxies. Accretion disks and jets from massive black holes in galaxies should evolve concomitantly.
- *Massive-black-hole formation and evolution:* The correlation between galaxies' halo and black-hole masses provides the incentive for tracing the time evolution of their co-evolution. As part of a multi-wavelength campaign, radio observations can identify and image active galactic nuclei and their jets to high redshifts. Jet energetics can be used to determine black-hole masses for a large sample.
- *Star formation:* The process of star formation and how the distribution of stellar masses is formed is still poorly understood. Panchromatic studies with high spatial resolution are needed to resolve individual regions that are in the process of collapsing, to relate these regions to larger-scale structures, and to study the disks and jets around protostars.

- *Thermal stars:* Detection of normal stars at radio wavelengths was the basis for original pessimism about a major role for radio telescopes in the study of the universe. Ironically, with telescopes now envisioned for the next decades, the detection of thermal emission will be possible, along with that of non-thermal emission from stellar activity. This opens up many new areas of study that will benefit from truly panchromatic observations.
- *Cosmic magnetism:* Magnetic fields are important in star formation and in the propagation of cosmic rays in the Milky Way and in other galaxies. Their origin is poorly understood. Zeeman splitting of radio spectral lines of H, OH and other molecules can be used as a powerful diagnostic for measuring field strengths seen by the gas. The Faraday effect—the rotation of the plane of polarization as waves propagate through ionized gas—provides a path-length-integrated measure of the magnetic field toward polarized sources. Through suitably large samples of objects, both the magnetic field of the Galaxy and magnetic fields on cosmological scales can be mapped and understood.

Cosmology, fundamental physics and extreme matter:

- *Precision cosmology, dark matter and dark energy:* The Hubble constant is now known to be within a few per cent of ~ 70 km s^{-1} Mpc^{-1} (e.g. [2]) through analysis of redshifts of galaxies and Type Ia supernovae along with analysis of the cosmic microwave background. One program for studying "dark energy," the apparent component of negative pressure that accounts for the apparent acceleration of the universe,[2] is to determine its equation of state through studies of baryon acoustic oscillations (BAOs). Surveys of large volumes containing large numbers of galaxies are needed, and such surveys are being done or are contemplated at optical, infrared and radio wavelengths. The ultimate precision of such enterprises hinges on knowing the Hubble constant more accurately than at present. High angular resolution analysis of water-vapor masers orbiting massive black holes detected at radio frequencies 22 GHz/$(1 + z)$ have the potential of providing the most accurate values of the Hubble constant [3]. These require detailed, multi-epoch spectroscopy using VLBI arrays. Dark matter in individual galaxies is quantified using rotational velocity curves determined by measuring Doppler shifts of spectral lines (e.g. the Hα line and the 21-cm line). Large surveys of galaxies and high-angular-resolution follow up will lead to progress in all these areas.
- *The nature of gravity and gravitational waves (GWs):* A powerful approach for testing General Relativity (GR) in the strong-field regime is to monitor radio pulsars orbiting other compact objects (white dwarfs, neutron stars and black holes), as demonstrated with the Hulse-Taylor binary pulsar and

[2] Note that alternatives to a cosmological constant include modification of the theory of gravity that underlies cosmological expansion or the presence of another constituent that alters the dynamics of the universe.

the more recently discovered double pulsar J0737-3039 [4]. Post-Keplerian analyses of the orbit allow determinations of the masses of neutron stars to better than 0.1% and provide tests on GR to similar precisions. Discovery of new binary pulsars with orbital periods shorter than the 2.4 h orbit of the double pulsar is a certainty because such binaries are continuously spiraling in owing to the emission of gravitational radiation. The more compact the binary, the better for testing GR. However, such binaries are more difficult to detect and require more sensitive telescopes that can probe more of the Milky Way.

Arrival time analyses of binary pulsars are modern-day versions of Römer's determination of the speed of light from monitoring of Jupiter's moons. The timing precision obtained in some cases is better than 1 μs, leading to the extraordinary constraints on GR and determinations of neutron star masses. The same precision provides the basis for using spin-stable pulsars as part of an array of objects to detect GWs with frequencies $\sim 10^{-9}$ Hz [5]. Backgrounds of such GWs may originate from massive black holes that merge as part of the galaxy assembly process. Current work indicates that plausble GW backgrounds are within reach of detection using the pulsar timing array approach [6]. The radio/gravity program complements GW observatories such as LIGO and VIRGO on the ground and LISA from space, which probe much higher frequencies than does the pulsar timing array.

Astrobiology:

- *Molecules:* To date, more than 140 molecules have been detected in the interstellar medium with as many as 18 atoms (naphthalene) including ten with deuterated forms. Molecular emission probes regions where star formation is taking place and more complex molecules may play a role in the jump-starting of life on newly formed planets. As with other wavelength bands, particularly the infrared, detailed study of molecules and searches for new molecular transitions in the radio band is a fruitful activity for the radio band.
- *Protoplanetary disks and exo-planets:* The disk paradigm for planet formation is well established but the complexity of planet formation in disks has been recognized only in the last 10 years or so. Observations in the optical and infrared bands probe dense disks and in some cases show evidence for gaps that signify an ongoing planet-formation process. However, dense disks that are optically thick at these and even at millimeter wavelengths will require study at centimeter wavelengths. With adequate sensitivity at wavelengths of a few centimeters along with angular resolutions of milli-arc seconds provided by VLBI telescopes, movies of the planet formation process can be made.
- *Search for extraterrestrial intelligence:* The radio band has been argued for many years to be the likely electromagnetic (EM) regime for detecting one class of techno-markers from other civilizations. The prototype signal

is put forward as an unmodulated or slowly modulated carrier signal with very narrow bandwidth. However, recent activity has also aimed to detect optical pulses that might be transmitted using pulsed lasers and would be easily detected amid the glare of the host star. Whether detectable emissions are intentional or simply leakage from technological activity we can only guess. However, large radio array telescopes now being considered provide spinoff capabilities for detecting techno-emissions of either kind out to significant distances in the Milky Way e.g. [7].

Exploration of the unknown: It is well appreciated that when new telescopes open up new regions of parameter space, such as wavelength coverage, angular resolution, and resolutions in time and frequency, new phenomena are discovered (e.g. [8]). Exploring current frontiers undoubtedly will combine complete censuses of phenomena, objects and processes that are known and will provide emergent knowledge through completeness. Merging objects, such as binary neutron stars, may be detectable in the radio band while also producing the well known class of gamma-ray bursts that have short durations, and events detectable with GW detectors. There are also unknown aspects of the universe that we can already envision which we have not yet discovered. Detection of black holes in mass ranges other than stellar and those found in galaxy centers comprise one such example. Another is the detection of other civilizations through techno-markers long discussed in the SETI community. There also remains the prospect of discovering entirely new classes of objects and phenomena, the unknown unknowns. Possible classes are EM signatures of events associated with dark-matter particles or as-yet unimagined kinds of technical activity by other civilizations. Other surprises may be associated with fundamental aspects of space time and the underlying number of dimensions.

On a practical basis, the exploration of the time domain, along with high-resolution coverage of other dimensions, may yield the greatest increase in phase-space volume [9]. The transient or Heraclitean universe ("you don't observe the same universe twice") is barely known across most of the EM spectrum except at high energies (X-and-γ rays). Some of the planned telescopes at optical and radio wavelengths will dramatically increase our knowledge of the time-variable universe.

5 Astronomical requirements

Wavelength/frequency coverage: High-redshift science demands that any given spectral transition be measured over a wide wavelength range. At radio wavelengths this implies that observations are necessarily made well outside the narrow protected bands that were defined long ago, such as the 21-cm atomic hydrogen (HI) band at 1.42 GHz, its deuterium counterpart at 0.327 GHz, and the 18-cm hydroxyl (OH) band at 1.66 GHz. Science goals for HI science now envision reaching redshifts of 15 or higher, requiring observations at frequencies below the FM band ($<$ 88 MHz). Frequencies below 1 GHz are heavily used for communications, so new facilities are planned mostly for very remote areas, such as desert areas in the southern hemisphere and also the back side of the Moon. Lines of water vapor with a rest frequency of 22 GHz and CO at 115 GHz and higher frequencies are also required. CO from the earliest galaxies ($z \sim 10$?) and water vapor to redshifts of about 1 imply observations at frequencies ~ 12 GHz. Continuum observations cover the entire radio band.

Increased sensitivity and survey speed: The broad range of science outlined above requires an increase of one to two orders of magnitude in *point-source sensitivity*. At the same, an even larger increase in *survey speed* is needed for large-scale surveys. The minimum detectable flux density ΔS is given by

$$(\Delta S)^{-1} \propto \sqrt{2BT}\left(\frac{A_{\text{effective}}}{T_{\text{system}}}\right), \tag{1}$$

where B is the bandwidth, T is the integration time, $A_{\text{effective}}$ is the effective collecting area and T_{system} is the effective temperature that describes the noise level of the combined sky + antenna + receiver system. The effective area is typically 60 to 80% of a reflector antenna's geometric area. Any and all factors in (1) are relevant to ongoing efforts to increase point-source sensitivity. These include

(1) designing antenna optics that minimize pickup of excess noise from the ground;
(2) building large numbers of antennas for arrays in cost-effective ways;
(3) using new devices that produce low intrinsic noise even at room temperature or using inexpensive cryogenic systems;
(4) increasing the available bandwidth through appropriate feed antennas and receivers and through digital backends that can process as much bandwidth as possible; and
(5) building stable systems and developing calibration methods that allow nearly arbitrarily long integration times.

The current state of the art with centimeter-wavelength arrays is to reach sensitivities $\Delta S \sim 1 - 10$ μJy after fairly heroic efforts. Future surveys with array telescopes now being contemplated aim to reach flux densities 10 to 100 times fainter.

Survey speed is the rate at which volume is surveyed or, equivalently, the rate at which solid angle is covered given observations that reach a specified distance:

$$SS \propto \frac{d\text{Volume}}{dt} \propto \Omega_i \left(\frac{A_{\text{effective}}}{T_{\text{system}}}\right)^2, \tag{2}$$

where Ω_i is the instantaneous solid angle covered by the instrument. For an antenna with a diffraction-limited beam solid angle Ω_1 and a feed array that provides multiple pixels, we would have $\Omega_i \approx N_{\text{pixels}}\Omega_1$ (ignoring whether the sky is Nyquist-sampled, etc.).[3] Obviously, survey speed improves dramatically if point-source sensitivity increases via the factor $(A_{\text{effective}}/T_{\text{system}})^2$ but the solid-angle factor introduces the possibility for a tradeoff between multiple-pixel systems and the number of antennas in an array. Development of radio cameras or their equivalent for increased survey speed is an area of very active research at present. The aim is to increase survey speed by several orders of magnitude.

Imaging and solid-angle coverage: Radio telescopes produce images of the sky in one of several ways (e.g. Fig. 1)

1. **Raster scanning:** Many telescopes are single-pixel systems. By scanning the telescope across the region of interest, a map is made with resolution $\sim \lambda/D$, where D is the aperture diameter.
2. **Radio cameras:** Multiple-pixel systems—analogous to detector array systems on optical and infrared telescopes—are of growing use with the advent of low-cost bolometer arrays at short radio wavelengths and MMIC receiver systems at wavelengths down to a few centimeters. At longer wavelengths, multiple feed horn systems are used but only with a few pixels owing to their large size and weight. Notable examples operating at 21 cm are the 13-beam system on the Parkes 64-m radio telescope and the 7-beam ALFA (Arecibo L-band Feed Array) system on the Arecibo 305-m telescope. Naturally, each pixel has a resolution $\sim \lambda/D$ equal to about 14 arcmin for Parkes and 3.5 arcmin for Arecibo.
3. **Phased-array feeds:** Phased arrays placed at the focus of an aperture can provide multiple pixels through electronic beam forming. No such systems have been used yet for astronomical observations but they are under vigorous development because they yield much greater and continuous instantaneous solid angle coverage than do feed horn arrays. The intention is to use phased-array feed systems on interferometric array telescopes, meaning that the advantages of the phased array for increasing the instantaneous sky coverage will be combined with the high resolution provided by aperture synthesis.
4. **Aperture synthesis:** By far the most used and most powerful imaging technique in radio astronomy is aperture synthesis using single-pixel feeds on individual antennas in an array of N concentrator antennas, providing a Field of View (FoV) $\Omega_i \sim \lambda/D$ sampled in pixels with angular resolution $\sim \lambda/b_{\text{max}}$ where b_{max} is the maximum antenna separation in the array.
5. **Aperture phased arrays:** Also under development are phased arrays that "look up" (rather than being at the focal point of a concentrating antenna). These are most feasible for rather long wavelengths ($\lesssim$ 50 cm) because individual antenna elements (dipoles) provide collecting area $\sim \lambda^2$ and

[3] This is equivalent to the solid angle covered by a CCD with the same number of pixels, each described by the point-spread function of the aperture.

many such elements are needed to provide the sensitivity, becoming expensive at small wavelengths. Such arrays have wide available FoVs owing to the broad angular response of a dipole ($\sim$ 1 rad). They can be used for aperture synthesis as above but anticipated applications include electronic beam forming for studies on individual compact sources. Applications may involve formation of multiple such beams in order to observe multiple targets or conduct several science programs at once.

High-dynamic-range imaging: Large-scale surveys of objects, particularly at wavelengths longer than about 10 cm, must address the source-confusion problem. With greatly increased sensitivity, the total range of flux densities that must be sampled exceeds 10^7, as may be seen from the log N-log S distribution of continuum flux densities of extragalactic sources [10]. Such imaging requires use of array telescopes with state-of-the-art calibration for instrumental, ionospheric and atmospheric fluctuations that perturb interferometric measurements. New approaches, including new algorithms and efficient use of large-scale computing resources, are needed for future telescopes at the longer wavelengths. The requirements are less stringent for shorter wavelengths (e.g. $\gtrsim$ 5 cm) because the sky is less filled with radio sources.

Sampling the time domain: Of increasing interest are time variations of astronomical sources across the electromagnetic spectrum and also using non-photonic tracers (gravitational waves, neutrinos and cosmic rays). Comprehensive surveys for burst sources have been made at gamma-and-X ray energies for decades, leading to the discovery of gamma-ray bursts, thermonuclear bursts, and magnetars, among other objects. From optical to radio wavelengths, however, no such surveys have been made. However, from targeted observations and serendipitous discoveries, we know that the sky is rich in transient sources at all bands, including gamma-ray burst afterglows. At the most commonly used radio wavelengths, to date, high-sensitivity measurements require large-aperture telescopes that yield small solid angle coverage because most single-aperture telescopes are deployed with single-pixel systems. Radio transients are known with time scales as short as 0.4 ns from giant pulses from the pulsar in the Crab Nebula [11] and from a variety of sources extending to days and longer. Undoubtedly, many individual sources have been missed and it is likely that new source classes remain to be discovered, such as *prompt* radio burst counterparts to gamma-ray bursts; coherent bursts from exoplanets; and perhaps bursts from evaporating black holes [12]. Along with a large range of time scales, radio transients cover a very wide range in amplitude and rate. Most likely, as yet undiscovered source classes involve bursts with very low rates or they have signatures in time and frequency that have not been adequately sampled.[4]

[4]An example of a frequency-time signature relevant to bursts with times $\lesssim$ 1 sec is the plasma-dispersion effect, which causes lower-frequency components to arrive later, a phenomenon well known from pulsar observations.

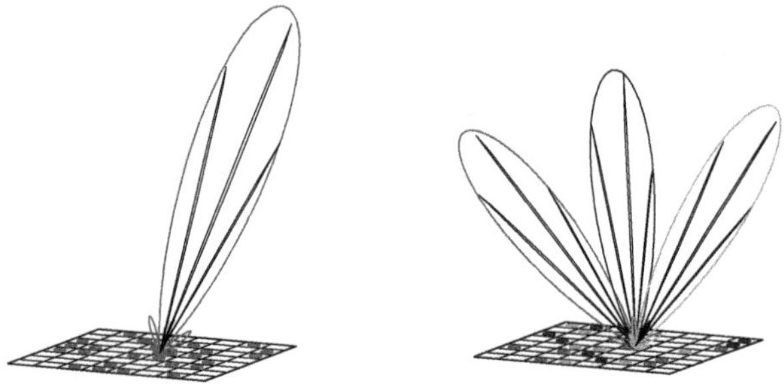

Fig. 1 Illustration of sampling of the sky with interferometric or beam-forming arrays, designated by the grid that is partially populated in each case with antenna elements, such as parabolic reflectors outfitted with a single-pixel or a multiple-pixel feed and receiver system. *Left*: The Airy-like pattern designates the diffraction-limited primary response of a single reflector with diameter D and angular width $\sim \lambda/D$. Internal, narrow beams represent the angular response from the array of reflectors in a beam-forming system or for aperture synthesis; these have angular widths $\sim \lambda/b_{\max}$, where $b_{\max}$ is the maximum separation of antennas in the array. With aperture synthesis, the three narrow beams shown are accompanied by all possible beams needed to cover the primary antenna response (or an even larger solid angle). *Right*: Similar except now three different primary beam patterns are shown, as would obtain from a multiple-pixel feed and receiver system used on each reflector (or if the array is divided into subarrays that point in different directions). Much of the current technology development for radio telescopes centers on building radio-frequency and digital components that allow such field-of-view expansion

Sampling the frequency domain: Interferometry, spectroscopy and time-domain science place different requirements on how radio frequencies are sampled. In all cases, narrow-band channelization is needed and provided by digital backend processors. Processors include correlation spectrometers and generalized Fourier-transform spectrometers (polyphase filter banks) that benefit from the increase in processor speeds and signal-path bandwidths associated with Moore's law. As a rule of thumb, science capabilities increase in accord with digital bandwidth increases, so state-of-the-art systems for radio astronomy push the envelope and thus represent a significant fraction of the total cost of a telescope. Moreover, these costs become, effectively, operating costs owing to the rapidity of changes in digital capability.

Managing radio frequency interference (RFI): The drive for redshift coverage implies that essentially complete wavelength coverage is needed across the EM spectrum. At radio wavelengths, this makes it mandatory to deal with terrestrial and satellite sources of interference. Relevant activities include working with regulatory agencies to minimize transmissions near existing radio observatories and to define radio-quiet zones for new instruments. Antenna designs can minimize scattering of terrestrial transmissions into the optics, like clear-aperture systems using off-axis parabolic sections, such as the 100-m Green Bank Telescope. Phased-array feeds may be able to null RFI from particular directions and receiver systems need to allow a large dynamic range in signal strength in order that RFI not prohibit detection of far weaker celestial sources; keeping receivers operating linearly with respect to input signal strength is essential. Finally, real time and post-processing in the frequency-time domain to edit RFI sources is an important element of implementing astronomically motivated analyses. In the context of arrays, formation of array beams or interferometric processing can aid RFI excision.

6 Enabling technologies and challenges

The general discussion above explains the extreme interest in technology development that will provide the astronomical requirements. Work is being done on essentially all elements of the radio astronomy system, from end to end, including

1. Design of highly optimized antenna optics that maximizes overall system gains while minimizing antenna sidelobes that pick up RFI and that limit imaging dynamic range. Such antennas need to accomodate multiple feed and receiver packages. Computer-aided EM design is an important part of this activity.
2. Currently, much work is being done to identify materials and methods of manufacture for constructing precise surfaces at low cost. Investigation of composite materials and metal hydroforming along with more traditional approaches is a very active area of development for large-N arrays. Closely related are telescope mounts, which must be able to handle large wind loads and provide repeatable pointing to better than 10% or even 1% of the primary beam angular diameter.
3. Broad-band feed antennas: traditional radio-astronomy feed and receiver systems have been narrowband (30% bandwidth or less) whereas science goals today require much more aggregate bandwidth. Vigorous work is being done on developing feeds and low-noise amplifiers that can cover 10:1 or larger instantaneous bandwidths. The challenge is to achieve low-noise performance over the whole band and provide good polarization performance. Broad-band feeds use log-periodic type structures to provide responses to waves of different wavelength.
4. Multiple-pixel feed arrays, phased-array feeds and aperture arrays: bolometer arrays have been developed with tens to thousands of elements and indeed serve as CCD equivalents, particularly in the submillimeter wavelength band. Phased-array feeds and aperture arrays, described above, are extremely promising but unproven technologies for radio astronomy at longer wavelengths. They will hopefully provide the instantaneous solid-angle coverage needed for large-survey-speed science. The challenge for both of these is the digital processing speeds that are driven by the desired bandwidth, the number of elements in the arrays, and the number and quality of phased-array beams that are needed.
5. Development of low-noise amplifiers for single-pixel and multiple pixel systems of all kinds. These are based, for example, on InP (Indium Phosphide) and GaAs (Gallium Arsenide) high electron mobility

transistors (HEMTs), SiGe (Silicon Germanium) BiCMOS transistors, and SIS (superconductor-insulator-superconductor) devices. Packaging in MMICs (monolithic microwave integrated circuits) is crucial to minimize contributions to T_{system} from interconnects between components and to minimize deployment costs on large-N systems.
6. Signal transport from antennas to a central processor using optical fiber provides the large bandwidths needed. Long-haul fiber networks are needed for VLBI arrays.
7. Digital processing capabilities will continue to follow Moore's-law growth in some respects, providing optimism for being able to process wide bandwidth signals from thousands of sensors. Tempering this optimism however is the need for smart algorithms that provide the required imaging dynamic range along with processing for non-imaging (spectroscopic and time-domain) science. Software for big-science projects does not follow Moore's law, thus providing a major challenge.
8. Large telescope projects are attractive because they involve new technology. On a social level, however, they provide the enormous challenge of how to build and maintain science-community support and involvement, and provide opportunities for both large-scale, survey science and for smaller-group, targeted science. With cost overruns and external pressures on science budgets, there is also the inevitable clash between maintaining existing telescopes or building new, smaller telescopes coevally with the mega-projects.

7 Existing telescopes and new instruments

Existing or soon-to-exist telescopes: The meter to millimeter portion of the EM spectrum is covered by existing, large single-aperture antennas and array telescopes. At centimeter wavelengths, the largest single apertures are the Arecibo telescope (305 m), the Green Bank Telescope (GBT; 100 m), the Effelsberg 100-m telescope, and the Lovell telescope at Jodrell Bank (76 m). These will be joined by the Chinese Five-hundred-meter Aperture Spherical Telescope (FAST) some time next decade. The GBT has an active surface as is also intended for FAST; the latter will also make use of a phased-array feed. At millimeter wavelengths, the Caltech Submillimeter Observatory (10.4 m, operating to 750 μm) and the IRAM (Instituto de Radioastronomia Milimetrica) Observatory (Pico Vileta) (30 m; 1 mm) will soon be accompanied by the Large Millimeter Telescope (50 m, operating to 1 mm).

Array telescopes for centimeter wavelengths with medium angular resolution are, in approximate order of decreasing wavelength, the Giant Meter Wavelength Telescope (GMRT) comprising 30 45-m reflectors in India, the Westerbork Synthesis Radio Telescope (WSRT) with 14 25-m antennas, MERLIN (Multi-Element Radio-Linked Interferometer Network) with 6 antennas of varying size that sometimes include the Lovell telescope, the Australia Telescope Compact Array (ATCA) with 6 22-m antennas, and the

Expanded Very Large Array (EVLA), comprising 27 25-m antennas. The shortest wavelength varies substantially between these arrays, being only 20 cm for the GMRT but 3 mm for the ATCA. Recent improvements include use of optical fiber for signal transport, allowing large increases in bandwidth. Also, receivers have been developed to allow continuous frequency coverage.

High-angular resolution arrays include the European VLBI Network (15 antennas), the Very Long Baseline Array in the U.S. (10 antennas), the Chinese VLBI Network (4 antennas), and the Australia Long Baseline Array (8 antennas). This list does not include VLBI arrays used for geodesy. Current innovations include migration from magnetic tape-based recording with off-line processing to hard-disk-based recording. In addition, real-time correlation using signal transport via optical fiber is beginning to make obsolete the distinction between "connected element" arrays and VLBI arrays. In Europe, a program of e-VLBI is making use of such connectivity with ever increasing bandwidths. VLBI with space antennas is now planned with the Japanese VSOP-2 mission (the second in the VLBI Space Observatory Program), which follows the VSOP mission that ran from 1997-2005. VSOP-2 consists of a 9-m off-axis paraboloid deployed in a 2.5×10^3 km orbit (apogee) that will provide 39 micro-arcsecond resolution when working with ground-based antennas at its shortest wavelength (0.7 cm).

Millimeter arrays include the IRAM 6-antenna array at Plateau de Bure, and the Combined Array for Research in Millimeter-wave Astronomy (CARMA) in California (6 10-m antennas + 9 6-m antennas and 8 3.5-m antennas).

Telescopes under construction and planned: There is tremendous activity taking place across the radio spectrum. FAST has already been mentioned above though it will appear on the same time scale as some of the telescopes mentioned here, and the EVLA and LMT will also be completed early next decade.

The meter-wavelength band is perceived to be an under-explored part of the spectrum as well as being the band of key interest for studies of the Epoch of Reionization using the redshifted 21-cm line of hydrogen. The Low-frequency Array (LOFAR) in Europe and the Murchison Widefield Array (MWA) in Western Australia target studies of the EoR, pulsars, the transient sky, and other science areas. The Precision Array to Probe Epoch of Reionization (PAPER) more explicitly targets the EoR. The Long Wavelength Array (LWA) in the U.S. (approximately co-located with the EVLA) addresses high-redshift radio galaxies and atmospheric science. By necessity, all of these low-frequency arrays must contend with the calibration problems posed by fluctuations in the Earth's ionosphere. These low-frequency instruments will inform the construction of a major low-frequency component of the Square Kilometer Array (SKA). They also may lead to construction of a low-frequency array on the back side of the Moon if radio-frequency interference prohibits detection of the EoR signal from terrestrial platforms.

At shorter wavelengths (60 cm to 3 cm), the Allen Telescope Array (ATA), the Australian Square Kilometer Array Pathfinder (ASKAP), and the Karoo Array Telescope (MeerKAT) are large-N arrays that are science instruments in their own right but also serve as prototypes and platforms for testing subsystems for the SKA. All are based on relatively small reflectors and a variety of innovations related to dish fabrication methods or feed designs. The ASKAP and MeerKAT instruments are being built on or near the short-listed SKA sites in Western Australia and South Africa, respectively, while the ATA is in Northern California.

The biggest projects to discuss are of course the Atacama Large Millimeter/Submillimeter Array (ALMA)[5] and the Square Kilometer Array.[6]

ALMA: ALMA is a large, international project that will revolutionize mm science with its high sensitivity and wavelength coverage. Science goals are in the areas of the first stars and galaxies, star and planet formation, and interstellar chemistry. It is now under construction on the Atacama plateau in Chile and will come on line in the first half of next decade. It will initially have 66 antennas of two sizes for a total collecting area of 6000 m^2 operating from 3.5 mm to 0.3 mm (84 GHz to 950 GHz). Later phases will probably extend the coverage to longer wavelengths (9.6 mm). ALMA uses very precise reflector antennas and mounts, superconducting mixers and, owing to the large processing bandwidth of 16 GHz for each antenna, requires more than 10^{16} operations per second to correlate the signals from the antennas. A related project now at a conceptual stage is CCAT (Cornell Caltech Atacama Telescope), a large single aperture for submillimeter wavelengths that would be used to conduct large-scale surveys whose results would be followed up using ALMA.

SKA: The Square Kilometer Array was conceived from the start as an international project that would provide sufficient sensitivity to detect Milky-Way-type galaxies in the 21-cm hydrogen line to redshifts of 1 to 2, requiring roughly a square kilometer of collecting area. Since the early 1990s, the science case has grown to include all the topics discussed above with an accordingly more ambitious list of technical requirements than initially discussed. Recent activity envisions that the full science case will be accomplished through phased deployment of antenna sensors of types now under research and development. The initial phase will cover the mid-range ($\sim$ 3 cm to 1 m wavelength) using reflector antennas and possibly aperture arrays, with parallel development of a long-wavelength array ($\gtrsim$ 1 m) contingent on current activity in detecting the EoR signal with the instruments described earlier in this section. The shortest wavelengths may not be deployed using reflector technology until well after the EVLA and ALMA have paved the way, science wise, during the next decade.

[5] ALMA web site: http://www.almaobservatory.org
[6] SKA web site: http://www.skatelescope.org

The two short-listed sites for the SKA are in Western Australia and in South Africa, with a final down select expected around 2012.

The SKA will provide the meter and centimeter complement to the large-scale optical and infrared telescopes also planned for the next decade, such as the European Extremely Large Telescope (E-ELT), the Thirty Meter Telescope (TMT), and the Giant Magellan Telescope (GMT). The SKA will also function as a synoptic survey telescope that will survey the sky comprehensively for steady and time-variable sources in the same manner as proposed for the Large (optical) Synoptic Survey Telescope (LSST) and now being conducted by the Panoramic Survey Telescope and Rapid Response System (Pan-STARRS).

Significant technology development for the SKA is ongoing in all areas, from reflectors, mounts, and dipole arrays to the digital signal processing that will occur at (now) staggering rates. Owing to the large scale of the project, it requires the global collective efforts of scientists and engineers with expertise in radio technology as well as engagement with industry at an early phase.

Acknowledgements I thank the organizers for putting together a stimulating program that both celebrated the invention of the telescope and its first astronomical use by Galileo and also looked toward the future. For useful conversations I thank Woody Sullivan and my colleagues in the National Astronomy and Ionosphere Center, the National Radio Astronomy Observatory, the U.S. SKA Consortium, and the SKA Program Development Office. This work was supported in part by a grant from the National Science Foundation to Cornell University.

References

1. Rogers, A.E.E., et al.: Deuterium abundance in the interstellar gas of the galactic anticenter from the 327 MHz line. Astrophys. J. Lett. **630**, L41 (2005)
2. Hinshaw, G., et al.: Five-year Wilkinson Microwave Anisotropy Probe (WMAP) observations: data processing, sky maps, and basic results. Astrophys. J. Suppl. **180**, 225–245 (2009)
3. Braatz, J.A., Reid, M.J., Greenhill, L.J., Condon, J.J., Lo, K.Y., Henkel, C., Gugliucci, N.E., Hao, L.: Investigating dark energy with observations of H_2O megamasers. In: Bridle, A.H, Condon, J.J., Hunt, G.C. (eds.) Frontiers of Astrophysics: A Celebration of NRAO's 50th Anniversary. Astronomy Society of the Pacific Conference Series, vol. 395. Astronomy Society of the Pacific, San Francisco (2008)
4. Kramer, M., et al.: Tests of general relativity from timing the double pulsar. Science **314**, 97–102 (2006)
5. Foster, R.S., Backer, D.C.: Constructing a pulsar timing array. Astrophys. J. **361**, 300 (1990)
6. Jenet, F.A., et al.: Upper bounds on the low-frequency stochastic gravitational wave background from pulsar timing observations: current limits and future prospects. Astrophys. J. **653**, 1571 (2006)
7. Loeb, A., Zaldarriaga, M.: Eavesdropping on radio broadcasts from galactic civilizations with upcoming observatories for redshifted 21 cm radiation. J. Cosmology Astropart. Phys. **1**, 20 (2007)
8. Harwit, M.A.: Cosmic Discovery: The Search, Scope and Heritage of Astronomy. Basic Books, New York (1981)
9. Cordes, J.M.: Axes of discovery: the time domain and the radio synoptic survey telescope. In: Bridle, A.H., Condon, J.J., Hunt, G.C. (eds.) Frontiers of Astrophysics: A Celebration of NRAO's 50th Anniversary. Astronomy Society of the Pacific Conference Series, vol. 395, p. 225. Astronomy Society of the Pacific, San Francisco (2008)

10. Condon, J.J.: Deep radio surveys. In: Afonso, J., Ferguson, H.C., Mobasher, B., Norris, R. (eds.) Deepest Astronomical Surveys. Astronomy Society of the Pacific Conference Series, vol. 380, p. 189. Astronomy Society of the Pacific, San Francisco (2007)
11. Hankins, T.H., Eilek, J.A.: Radio emission signatures in the crab pulsar. Astrophys. J. **670**, 693 (2007)
12. Rees, M.J.: A better way of searching for black-hole explosions. Nature **266**, 333 (1977)

Exp Astron (2009) 26:95–109
DOI 10.1007/s10686-009-9163-8

REVIEW ARTICLE

Realization of X-ray telescopes—from design to performance

Bernd Aschenbach

Received: 23 March 2009 / Accepted: 26 March 2009 / Published online: 22 April 2009

Abstract For more than 45 years the building of X-ray telescopes for solar and astronomical observations has been practised with significant performance improvement. The various techniques applied are reviewed emphazising the impact of proper mirror material choice, grinding and polishing improvements and the role of metrology.

Keywords Astronomy · X-ray telescopes · Imaging · Grazing incidence

1 Introduction

This symposium celebrates the 400th anniversary of the invention of the optical telescope. X-ray telescopes are much younger, not even 60 years of age. Riccardo Giacconi, the "father" of X-ray astronomy and the astronomically used X-ray telescope, has familiarized us with the early beginnings and the scientific success of both X-ray astronomy in general and X-ray telescopes in particular, emphasizing the history in the United States of America. In the late 1960's, early 1970's experimental studies on developing X-ray telescopes were initiated also in several countries in Europe, including the Netherlands, the United Kingdom, Germany and the former Tchechoslovakia, and Italy and Denmark later on. In my talk I will concentrate on the physical principles of X-ray telescopes, the technological and technical challenges and the

B. Aschenbach (✉)
Max-Planck-Institut für extraterrestrische Physik, Giessenbachstrasse,
85741 Garching, Germany
e-mail: bra@mpe.mpg.de

performance of the various approaches, which of up to today have resulted in more than ten space-borne missions with X-ray telescopes for astronomical observations.

2 Grazing-incidence telescopes

2.1 The physics of grazing-incidence reflection

One way to focus and image sources of light is by using reflecting curved surfaces. The interaction of light with matter can be described by the complex index of refraction which describes the change of the properties of the incident electromagnetic wave when crossing the boundary between the two materials involved. The index *n* reads:

$$n = 1 - \delta - i \cdot \beta \tag{1}$$

δ describes the phase change and β accounts for the absorption. The reflection coefficients for *p* and *s* polarization are given by the Fresnel equations:

$$r_p = \left(\frac{E_r}{E_i}\right)_p = \frac{n^2 \sin\alpha - \sqrt{(n^2 - \cos^2\alpha)}}{n^2 \sin\alpha + \sqrt{(n^2 - \cos^2\alpha)}} \tag{2}$$

$$r_s = \left(\frac{E_r}{E_i}\right)_s = \frac{\sin\alpha - \sqrt{(n^2 - \cos^2\alpha)}}{\sin\alpha + \sqrt{(n^2 - \cos^2\alpha)}} \tag{3}$$

E_r/E_i denotes the ratio of the amplitudes of the reflected and incident electric fields and α is the grazing angle of incidence as measured from the interface plane. For normal incidence, which is the standard in optical telescopes, $\alpha \approx 90°$. This approach is, generally speaking, correct as long as the assumptions for applying the Fresnel equations are fulfilled. The reflected intensity or reflectivity is then $R_p = r_p \times r_p^*$ and $R_s = r_s \times r_s^*$, where the asterisk denotes the conjugate complex value.

The components of the index of refraction for a vacuum matter transition are often called the optical constants of the material. In the optical wavelength range, for instance, the real part of the index of refraction is greater than one, but with decreasing wavelength its becomes less than one, which changes the interaction of light with matter dramatically. The reflectivity of the surface at normal incidence decreases rapidly and the mirrors lose efficiency starting in the UV wavelength band. However, if one applies Snell's law to the incident and refracted light, it turns out that the refraction angle measured from the surface normal is greater than 90° for $n_r = 1 - \delta < 1$, or that total external reflection occurs for grazing-incidence angles $\alpha \leq \alpha_t$:

$$\cos\alpha_t = 1 - \delta \tag{4}$$

or for $\delta \ll 1$:

$$\alpha_t = \sqrt{2\,\delta}. \tag{5}$$

For actual applications a trade-off is to be made in terms of the effective collecting area between the design of a normal-incidence telescope and that of a grazing-incidence telecope. The effective collecting area is the product of the wavelength-dependent reflectivity times the geometric area of the primary mirror projected on the front aperture. Depending on the number of reflecting optical elements involved, grazing-incidence telescopes tend to be more efficient for wavelengths shorter than about 30 nm. Furthermore the reflectivity at normal incidence drops so rapidly with decreasing wavelength that for observations at wavelengths shorter than about 15 nm grazing incidence is the only choice. This limit can be extended somewhat to even shorter wavelengths by the use of multi-layer coatings of the mirror but only over a fairly restricted wavelength band. Multi-layer coatings of several hundreds of bi-layers, each a couple of Ångstrôm thick, can also be applied to grazing-incidence mirrors, thereby extending the photon energy range to about 100 keV.

The index of refraction or the optical constants can be computed from anomalous dispersion theory. For wavelengths λ or photon energies sufficiently off-set from any electron binding energy a coarse estimate of δ can be made:

$$\delta = \frac{r_e}{2\pi}\,\frac{N_0\,\rho}{A}\,Z\,\lambda^2 \tag{6}$$

where N_0 is Avogadro's number, r_e is the classical electron radius, Z and A are the atomic number and weight, respectively, and ρ is the mass density. For heavy elements for which $Z/A \approx 0.5$, the incidence angle of total reflection for $\delta \ll 1$ can be estimated to:

$$\alpha_t = 5.6\,\lambda\,\sqrt{\rho} \tag{7}$$

with α_t in arcmin, λ in Å and ρ in g/cm^3. For X-rays, with λ of a few Å, α_t is about one degree. Equation (7) suggests the most dense materials as reflective coatings like gold, platinum or iridium, which all have been used for X-ray space telescope mirrors. However, these materials show a pronounced reduction of reflectivity at energies between 2 keV and 4 keV because of the presence of M-shell absorption, so that nickel, for instance, despite its lower density has sometimes been preferred, in particular, for observations below 4 keV.

The optical constants are related to the atomic scattering factors, the most up-to-date tables of which have been compiled by the Center for X-ray Optics (http://henke.lbl.gov/optical_constants/, [1]). These tables cover the energy range from 50 eV to 30 keV for the elements with Z = 1–92, and are a very useful data basis for designing grazing-incidence optics.

2.2 Grazing-incidence telescope configurations

At grazing incidence, imaging of an extended source or imaging over some extended field requires at least two reflections, i.e. two reflecting surfaces. Single mirrors like grazing-incidence parabolas suffer from strong coma, preventing true imaging. However, such a mirror can still focus, and parabolas have been used as 'light buckets'.

There are three different configurations of two-mirror systems, which are the Wolter type systems, the Kirkpatrick-Baez type systems, and the focusing collimator or 'lobster-eye' systems.

2.2.1 Wolter telescopes

In 1952 Hans Wolter suggested three different types of imaging telescopes for grazing incidence, which have become known as Wolter telescopes of type I, type II and type III [2]. The surfaces used encompass a paraboloid, a hyperboloid and an ellipsoid. Type I and type II make use of a paraboloid and a hyperboloid, type III combines a paraboloidal and an ellipsoidal mirror. In each case the two mirrors involved are arranged in a coaxial and confocal manner. The main difference between the three types is the ratio of focal length to total system length, i.e. the minimum physical length of the telescope.

The focal length of a type I system (Fig. 1) is practically given by the distance from the paraboloid/hyperboloid intersection plane (Knickfläche) to the system focus. Therefore the physical telescope length always exceeds the focal length by the length of the paraboloid. This system has been mostly used in space observations because of its compactness, simple configuration as far as the interface to the mounting structure is concerned, and because it provides free space to easily add further telescopes inside and outside. These telescopes with multiple components are called nested systems. They increase the collecting area substantially.

Single type I systems have been used for solar X-ray observations whereas for astronomical EUV and X-ray observations, for which collecting area

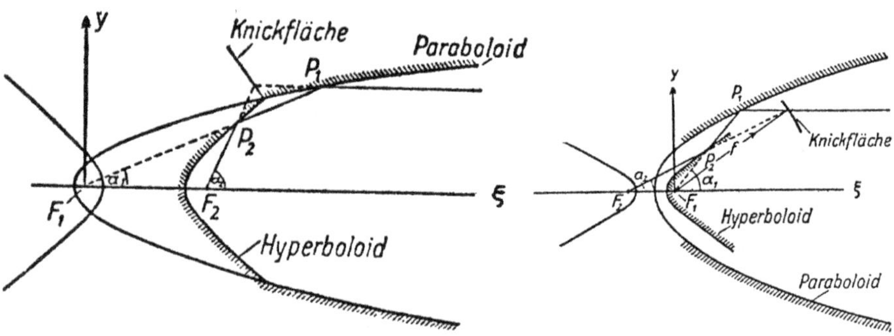

Fig. 1 Schematic of the Wolter telescope type I (*left*) and type II (*right*) [2]

is of utmost importance, nested systems have been used (the *EINSTEIN* observatory [3] and [4], *EXOSAT* [5], *ROSAT* [6], *ASCA* [7] & [8] and *Suzaku* [9], the *Chandra* [10] and [11] and *XMM-Newton* [12] observatories, as well as the *JET-X* telescopes of the *SWIFT* mission [13] and [14]). For instance, each of the three X-ray telescopes on board of *XMM-Newton* [12] accommodates 58 nested paraboloid-hyperboloid Wolter type I mirror shell pairs.

The Wolter type II system (Fig. 1) is a true telescopic system, for which the focal length can be much longer than the physical length of the telescope. These systems are useful for feeding spectrometers which require large dispersion.

The f-number is an important number for optical telescopes when imaging extended objects. The lower the f-number is the higher is the image brightness. Likewise f-numbers may also be defined for X-ray telescopes which can be computed using (1)–(7). It turns out that the f-number is inversely proportional to the angle of total reflection which in turn decreases linearly with increasing photon energy. Therefore telescopes optimized for the low-energy regime (<2 keV) are pretty fast and should make use of the Wolter type I design. The minimum effective f-number of the *ROSAT* telescope was 9. Telescopes for efficient observations of high-energy photons of up to 10 keV necessarily have much larger f-numbers (around 75 for *XMM-Newton* or 40 for *Chandra*), depending on how much emphasis is given to high energies. Type II's should be used if a very long focal length is required compared with the telescope length, because Wolter's "Knickfläche" (c.f. Fig. 1) can easily be positioned far in front of the entrance plane of the primary mirror. Even in the very soft X-ray domain f-numbers of less than 50 (e.g., the *CDS* telescope of the *SOHO* solar observatory) cannot be obtained.

Because of the intimate interdependence between f-number, grazing angle, telescope diameter and focal length, large diameter telescopes working at high energies can be constructed only with appropriate long focal distances, and because folding of the X-ray beam is unacceptable because of significant reflection losses, the distance between mirror module and focal plane becomes substantial. This created the idea of space formation flying of two spacecraft, one carrying the telescope and the other one far behind housing the focal plane instrumentation. If the separation is not too large an expandable optical bench might bridge the distance.

Wolter-type systems are free of spherical aberration, but still suffer from coma aberration, astigmatism and field curvature. In a second paper Wolter presented the equations for grazing-incidence telescopes which exactly obey the Abbe sine condition, eliminating coma completely. This is achieved by very small corrections (sub-μm to one μm) of the axial mirror profile from its nominal second-order shape. The exact surface shape has been derived by Wolter by extending the solutions to grazing incidence which Karl Schwarzschild had already obtained for normal incidence in 1905 [15]. Therefore these systems are named Wolter-Schwarzschild telescopes [16]. They surpass the Wolter systems in off-axis imaging performance if used at longer wavelengths, i.e. in the EUV and the soft X-ray band. Wolter-Schwarzschild type I telescopes were flown on the *EUV-Explorer* [17] and [18] and the *ROSAT-WFC* [19]. A Wolter-Schwarzschild type II system was feeding the spectroscopic telescope of the *EUV-Explorer*, and the *CDS* telescope on board of the solar *SOHO* mission [20] is of Wolter-Schwarzschild type II.

The maximum degree of nesting, and therefore the highest throughput relative to the entrance aperture area, is achieved with mirrors as thin as possible. Hundreds of thin foils or sheets representing the mirrors make up the telescopes used in the *ASCA* and *Suzaku* missions. The parabolic/hyperbolic shape of the Wolter type I mirrors is approximated by straight cones. The perfect image of an on-axis point source is lost but the imaging capability is preserved. Cone approximation of the Wolter type I configuration has also been used for the *BeppoSax* X-ray telescopes [21] & [22] (Figs. 2, 3, 4, 5, and 6).

2.2.2 Kirkpatrick-Baez telescopes

The first two-dimensional X-ray image ever obtained with grazing-incidence reflection was taken in the laboratory by Kirkpatrick and Baez [23]. The incident rays are focused to a line image by a parabolic mirror. On their path to the line focus the rays are reflected by a second parabolic mirror to the

Fig. 2 Bonding of one of the *ROSAT* hyperboloid mirrors to the central bulkhead of the telescope which eventually contains 8 separate, Zerodur-made paraboloid and hyperboloid mirrors

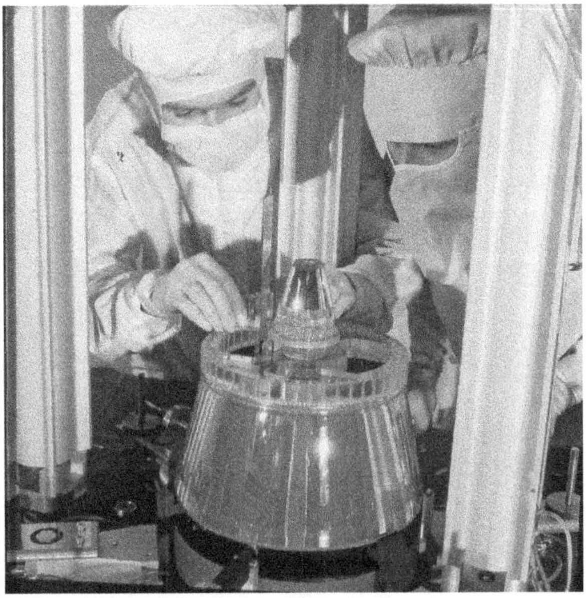

Fig. 3 Integration of the secondary mirror of the *CDS* telescope, which is of Wolter-Schwarzschild type II. Both the primary (*bottom mirror*) and the secondary (*top mirror*) as well as the carrying structure are made solely out of Zerodur. Grinding and polishing of the mirrors was particularly challenging because of the extreme asphericity of the surfaces. The telescope finally has an angular resolution of less than 2.5 arcsec HEW. This telescope has an aperture diameter of 275 mm, and despite the long focal length of 2578 mm the separation between front aperture and focal plane ist just 800 mm

point-like focus for rays parallel to the centre lines of the parabolas. The surface planes of the two mirrors are oriented at 90° to each other. In order to increase the collecting area (the frontal area) a stack of parabolas of translation can be constructed. However, in contrast to the single double-plate system, the image of a point-like source starts to become increasingly extended in size as the number of plates involved increases. Wolter type I telescopes bend the incident ray direction two times in the same plane, whereas the two bendings in Kirkpatrick-Baez systems occur in two orthogonal planes, which for the same incidence angle on the primary mirror requires a longer telescope.

A Kirkpatrick-Baez telescope has never been flown on a satellite mission, but a modification using flat plates instead of parabolas, still providing two-dimensional imaging, has successfully operated on sounding rocket flights delivering positive measurements of ordinary stars and clusters of galaxies [24].

2.2.3 Focusing collimator or 'lobster-eye' telescopes

The Wolter and the Kirkpatrick-Baez systems have in common a relatively narrow field of view which is practically limited to the grazing angle employed on the individual mirrors. Imaging systems of substantially larger field of view but at systematically reduced on-axis angular resolution have been proposed by Schmidt [25] and by Angel [26]. Such systems would be ideal for a wide field imaging monitor.

The principal layout of Schmidt's concept makes use of two stacks of plane mirrors, which are arranged in an upper and a lower stack and oriented orthogonally to each other. The mirrors within each stack are arranged in

Fig. 4 Riccardo Giacconi's "dream" mirror, the 1 m long and 1.2 m wide paraboloidal Zerodur mirror which makes up the biggest mirror of the 0.5 arcsec *Chandra* telescope. According to him this diameter of an X-ray mirror is of the size he has always been opting for since the beginning of telescopic X-ray astronomy. The biggest mirror of the earlier *Einstein* telescope had a diameter almost exactly half of it (Image credit: NASA/CXC/SAO)

such a way that their center lines inscribe a cylinder, where the two cylinders associated with the stack are at right angles to each other and the crossing of their center lines is at the origin of the coordinate system. A focus is formed half way between the mirrors and the origin of the coordinate system. Both sides of a mirror blade, i.e. the front and the back surface, are X-ray reflecting. The focusing is not perfect because of the finite height of the mirror blades. With such a device a full hemisphere of the sky could be observed simultaneously.

A variation of this design, which provides two-dimensional imaging, has been presented by Angel [26] (see also [27] and references therein). The device is composed of many small square-sided tubes with reflecting surfaces. The tubes are based on and distributed over the surface of a sphere. The axis of each tube follows a radius vector of the sphere. After a ray has been reflected twice within one tube but from adjacent walls a two-dimensional image is formed. The focal surface is a sphere with a radius which is half of that of the sphere carrying the tubes. This type of grazing-incidence optic is actually

Fig. 5 View to the rear of one of the three *XMM-Newton* Wolter I telecopes. There are 58 nested mirror shells produced in nickel via galvanic replication including gold plating. Parabola and hyperbola come in one piece with a thickness between 0.5 mm and 1.2 mm, spanning a diameter range from 35 cm to 70 cm with a length of 60 cm

realized in the reflective eyes of lobsters and shrimps, giving the name to this particular type of X-ray telescope. The optical principle is very similar to Schmidt's focusing collimator when shifting and merging the upper and lower mirror stacks of Schmidt's device into one section forming square-sided tubes.

Both in the Schmidt and the Angel design there are rays passing through the optics with only one reflection or none at all. They appear as a diffuse or line-shaped background of non-negligible brightness. In itself the imaging is not perfect, and, ultimately, the angular resolution of such a device is limited by the width of a single tube as seen by the detector, and at arcsecond resolution the diffraction imposed by the tube width has to be taking into account, favouring such a system for observing hard X-rays. Such a telescope would have great potential for continuous X-ray monitoring of large fields of the sky.

Fig. 6 One of the four X-ray telescope modules on board of the *Suzaku* satellite. The module houses a total of 175 shells or 1400 reflectors. The mirror shells are extremely thin (0.152 mm) aluminum foils, each of which is about 12 cm long [9]. This technique has been pioneered by Peter Serlemitsos in the late 1980's [7], and was also used for the fabrication of the *ASCA* telescopes [8]

Prototypes have been constructed and built in the Czech Republic by the group of Hudec et al. [28].

3 Design considerations

3.1 Parameters

The choice of a particular type of grazing-incidence telescope depends on the scientific objectives. First of all, the highest photon energy has to be considered, which according to (7) defines the optimum grazing angle. Such energies are found, for instance, in the astrophysically important Fe K-lines between 6.4 and 7 keV. The *XMM-Newton* telescope was designed to optimally cover this region and to provide an acceptable area at 2 keV.

Surveying telescopes should have a low f-number, which is equivalent to maximizing the product of the effective collecting area times the field of view. This requires a grazing angle as large as possible but compatible with the average photon energy within the band to be observed. The *ROSAT* telescope was designed along these lines.

Furthermore, such a survey telescope should have the best possible angular resolution over a field of view as wide as possible. Wolter type I telescopes show strong field curvature, which means that the angular resolution increases rapidly with growing field angle (off-axis angle) [29]. So-called polynomial telescopes tend to flatten the field [30] and [31]. They maintain the Wolter configuration but the second-order surface shape is replaced by higher-order polynomials at the expense of degrading the on-axis resolution. Alternatively, for nested systems, the primary mirrors of a Wolter telescope should be kept as short as possible without compromising the free entrance aperture.

3.2 Effective collecting area

These trade-offs are usually done by detailed ray-tracing taking into account the reflectivity of the coating, which usually changes substantially with energy. Figure 7 shows the effective collecting area of one of the three *XMM-Newton* telescopes. The widest of the 58 nested Wolter type I mirror pairs has a diameter of 70 cm. The geometric area of the entrance aperture is about 3100 cm^2 and the effective area at the lowest energies (c.f. Fig. 7) is about 1850 cm^2, equivalent to a throughput of less than 60%. This can be increased by the use of thinner mirrors like foils or thin sheets with which a throughput of more than 80% has been achieved. The on-axis angular resolution of such foil telescopes, however, is so far limited to about one arcminute.

3.3 Angular resolution, point spread function, encircled energy function

Even for a perfect telescope the light from a point-like object located on-axis is not concentrated in an infinitely small focal spot, but is distributed over

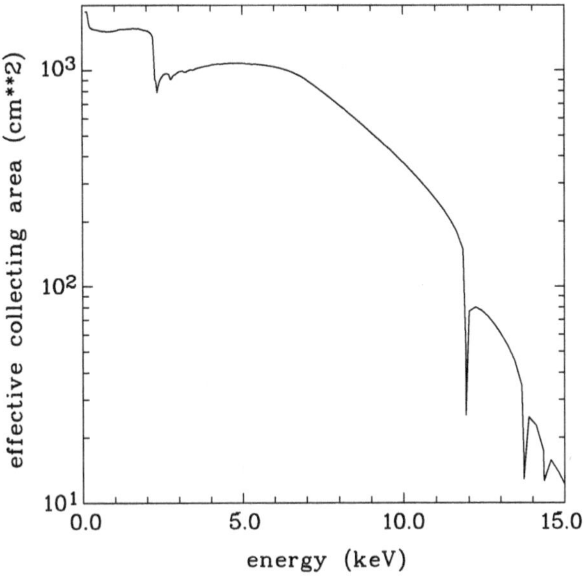

Fig. 7 On-axis effective collecting area of one of the three *XMM-Newton* X-ray telescopes, as designed. All 58 mirror pairs have been coated with gold, which causes the energy dependence of the area. The jumps and wiggles are due to the binding energies of the N-, M- and L-shell electrons

an extended image, the surface brightness of which is described by the point spread function (PSF). The encircled energy function (EEF) corresponds to the radial integration of the PSF, and describes the relative fraction of effective area within some radius.

At large field angles the image extent and structure are dominated by geometric aberrations inherent in the telescope design. In contrast, the on-axis PSF is determined by differences between the real and perfect shape of the reflecting surfaces, both in circumferential and in axial direction, as well as by alignment and mounting errors. For high-resolution telescopes, the most demanding factor is the control of the surface shape by appropriate metrology, through which a highly precise feedback to computer-controlled grinding and polishing of the relevant surfaces down to residuals of a few Ångstr̊om on either the mirrors or mandrels has become possible. Because of grazing incidence the tightest requirements have to be observed for surface errors along the mirror meridional or axial profiles. They need to be as low as possible over all spatial frequency scales which range from the full mirror length up to the micrometer range. The same holds for circumferential errors.

Low-frequency errors can be considered as geometric slope errors and they can be treated by geometric optics. At high frequencies the errors and their power spectral density distribution can be understood as scattering from randomly rough surfaces [32]. The fraction of scattered X-rays I_s/I_0 is given by:

$$I_s/I_0 = 1 - \exp\left[-(4\pi\ \sigma \sin\alpha\ /\ \lambda)^2\right] \tag{8}$$

The microroughness σ is the rms value of the surface height deviations. Microroughness reduces the image contrast, produces a loss of flux out of the image core and raises extended wings in the PSF. In the early days of grazing-incidence mirrors scattering was a major problem because of the low values for σ to be achieved for a reasonable high contrast image. For instance, for $I_s/I_0 = 0.1$ at X-ray wavelengths ($\lambda = 10$ Å) and grazing incidence ($\alpha = 1°$) a microroughness of $\sigma \leq 9$ Å is required. Values like 3 Å have become possible on highly aspherical grazing-incidence mirrors and have been realised on the *ROSAT* telescope mirrors.

In the early years of making X-ray mirrors scattering was not well understood, and only very extensive X-ray scattering measurements on flat samples showed that (8) was not only applicable to radio and optical wavelengths but also to the X-ray domain (c.f. Fig. 8, [33]). Scalar scattering theory and first order vector perturbation theory in the smooth-surface limit were shown to describe the observed scattering profiles when compared with metrology data. They were obtained with long-range, high-resolution, contactless profilometers, and later, in the late 1980's optical heterodyne interferometers with a vertical resolution capability down to the Ångstrôm level became available. These sophisticated metrology tools, including extremely precise roundness measurement devices, provided the necessary feedback to the manufacturers for improving grinding and polishing, which now has reached the level of sub-arcsecond X-ray imaging with mirror areas covering many square meters.

The shape of the PSF core is generated by geometric errors, the total of which accumulates to less than 0.1 arcsecond for the *Chandra* telescope. Figure 9 shows the on-axis EEFs for the *Chandra* telescope at various photon energies. Energy-dependent scattering, as expected from (8), appears to be present.

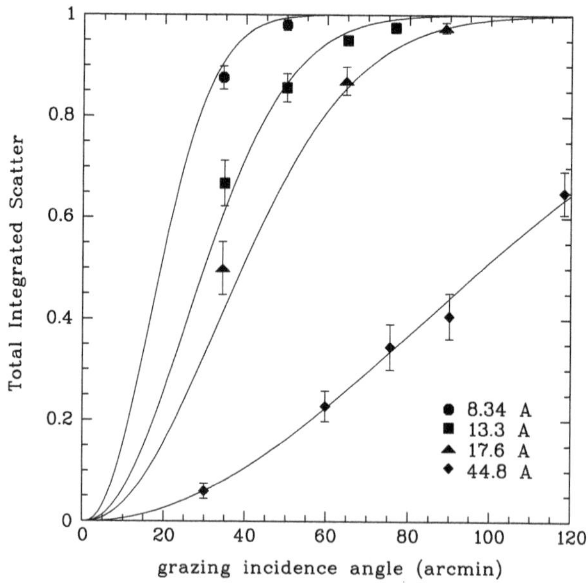

Fig. 8 Comparison of total integrated scatter (TIS) measurements with electromagnetic scattering theory (Equation 8, [33])

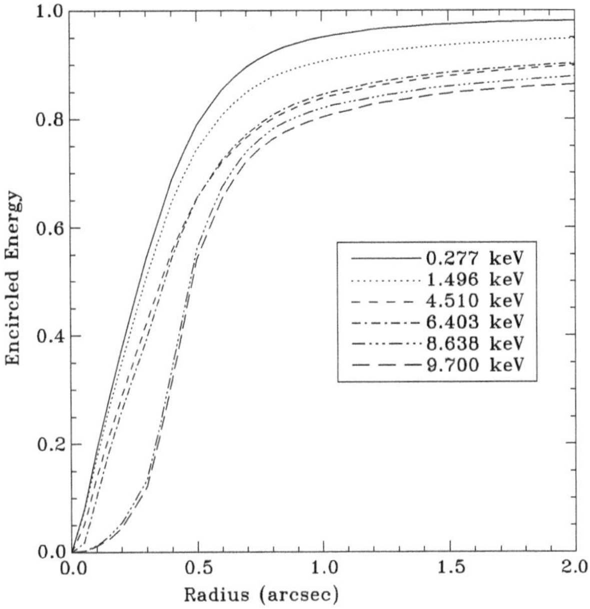

Fig. 9 Encircled energy function of the *Chandra* telescope for different photon energies [34]

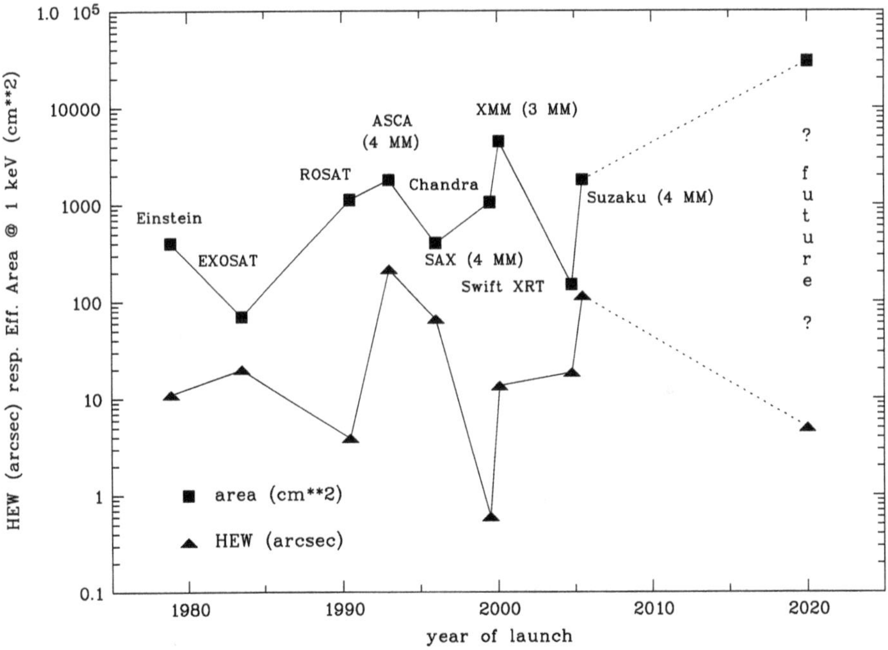

Fig. 10 Performance comparison of various X-ray telescopes, effective collecting area in cm^2 (upper data points) and angular resolution in terms of the half energy width (HEW) of the point spread function in arcsec at a photon energy of 1 keV (lower data points). MM stands for mirror modules. Dashed lines are pointing to a future super X-ray telescope with a collecting area of 3 m^2 and 5 arcsec resolution currently under study, which would explore the deep Universe. Needless to say that quite a number of other X-ray telescope space missions are being worked on

A summary of the angular-resolution and effective-collecting-area performance of the telescopes, which have successfully been used on astronomical satellites, is displayed in Fig. 10.

3.4 Baffling

Grazing-incidence telescopes have to be equipped with field stops to baffle the focal plane against stray light. Otherwise rays from outside the nominal field of view can pass through the telescope by a single reflection from either the primary or the secondary. There may even be portions of the sky which can be viewed directly from the focal plane detector. Such straylight increases the background light and produces ghost images. A complete suppression of straylight requires field stops in several different planes, which are best in front of the telescope aperture and down in the telescope's mirror section. For tightly nested systems with little if any available space between adjacent mirrors a single baffle system in front of the telescope can be used, which, however, reduces but does not completely eliminate stray light.

References

1. Henke, B.L., Gullikson, E.M., Davis, J.C.: X-ray nteractions: photoabsorption, scattering, transmission, and reflection at E = 50–30,000 cV, Z = 1–92. At. Data Nucl. Data Tables **54**(2), 181–342 (1993)
2. Wolter, H.: Spiegelsystem streifenden Einfalls als abbildende Optiken für Röntgenstrahlen. Ann. Phys. **10**, 94–114 (1952a)
3. Van Speybroeck, L.P.: Einstein observatory /HEAO-B/ mirror design and performance. SPIE Proc. **184**, 2–10 (1979)
4. Giacconi, R., Branduardi, G., Briel, U. et al.: The Einstein /HEAO 2/ x-ray observatory. ApJ **230**, 540–550 (1979)
5. de Korte, P.A.J., Giralt, R., Coste, J.N. et al.: EXOSAT x-ray imaging optics. Appl. Opt. **20**, 1080–1088 (1981)
6. Aschenbach, B.: Design, construction, and performance of the ROSAT high-resolution x-ray mirror assembly. Appl. Opt. **27**, 1404–1413 (1988)
7. Serlemitsos, P.J.: Conical foil x-ray mirrors: performance and projections. Appl. Opt. **27**, 1447–1452 (1988)
8. Serlemitsos, P.J., Jahoda, L., Soong, Y. et al.: The x-ray telescope on board ASCA. Pub. Astron. Soc. Jap. **47**, 105–114 (1995)
9. Serlemitsos, P.J., Soong, Y., Chan, K.-W. et al.: The x-ray telescope onboard Suzaku. Pub. Astron. Soc. Jap. **59**, 9–21 (2007)
10. Van Speybroeck, L.P.: Grazing incidence optics for the U.S. high-resolution x-ray astronomy program. Appl. Opt. **27**, 1398–1403 (1988)
11. Van Speybroeck, L.P., Jerius, D., Edgar, R.J. et al.: Performance expectation versus reality. SPIE Proc. **3113**, 89–104 (1997)
12. Aschenbach, B., Briel, U.G., Haberl, F. et al.: Imaging performance of the XMM-Newton x-ray telescope. SPIE Proc. **4012**, 731–739 (2000)
13. Citterio, O., Campano, S., Conconi, P. et al.: Characteristics of the flight model optics for the JET-X telescope onboard the Spectrum-X-Gamma satellite. SPIE Proc. **2805**, 56–65 (1996)

14. Burrows, D.N., Hill, J.E., Nousek, J.A. et al.: The Swift x-ray telescope. Space Sci. Rev. **120**, 165–195 (2005)
15. Schwarzschild, K.: Untersuchungen zur geometrischen Optik. Abh. Wiss. Göttingen Bd. IV Nr. 2 (1905)
16. Wolter, H.: Verallgemeinerte scawarzschildsche spiegelsysteme streifender reflextion als optiken für röntgenstrahlen. Ann. der Phys. **10**, 286–295 (1952b)
17. Bowyer, S., Green, J.: Design of the extreme ultraviolet explorer long-wavelength grazing incidence telescope optics. Appl. Opt. **27**, 1414–1422 (1988)
18. Bowyer, S., Malina, R.F.: The extreme ultraviolet explorer mission. Adv. Space Res. **11**, 205–215 (1991)
19. Willingale, R.: ROSAT wide field camera mirrors. Appl. Opt. **27**, 1423–1429 (1988)
20. Harrison, R.A., Sawyer, E.C., Carter, M.K. et al.: The coronal diagnostic spectrometer for the solar and heliospheric observatory. Sol. Phys. **162**, 233–290 (1995)
21. Citterio, O., Conti, G., Mattaini, E., Santambrogio, E., Sacco, B.: Optics for x-ray concentrators on board of the astronomy satellite SAX. SPIE Proc. **597**, 102–110 (1986)
22. Citterio, O., Bonelli, G., Conti, G. et al.: Optics for the x-ray imaging concentrators aboard the x-ray astronomy satellite SAX. Appl. Opt. **27**, 1470–1475 (1988)
23. Kirkpatrick, P., Baez, A.V.: Formation of optical images by x-rays. J. Opt. Soc. Am. **38**, 766–774 (1948)
24. Gorenstein, P., Gursky, H., Harnden, F.R., Jr., DeCaprio, A., Bjorkholm, P.: Large area soft x-ray imaging system for cosmic x-ray studies from rockets. IEEE Trans. Nucl. Sci. **NS-22**, 616–619 (1975)
25. Schmidt, W.K.H.: A proposed x-ray focusing device with wide field of view for use in x-ray astronomy. Nucl. Instrum. Methods **127**, 285–292 (1975)
26. Angel, J.R.P.: Lobster eyes as x-ray telescopes. ApJ **233**, 364–373 (1979)
27. Priedhorsky, W.C., Peele, A.G., Nugent, K.A.: An x-ray all-sky monitor with extraordinary sensitivity. Mon. Not. R. Astron. Soc. **279**, 733–750 (1996)
28. Hudec, R., Inneman, A., Pína, L., Hudcová, V.: Lobster eye: new approach to monitor GRBs in x-rays. AIP Conf. Proc. **662**, 494-496 (2003)
29. VanSpeybroeck, L.P., Chase, R.C.: Design parameters of paraboloid-hyperboloid telescopes for x-ray astronomy. Appl. Opt. **11**, 440–445 (1972)
30. Burrows, C.J.,Burg, R., Giacconi, R.: Optimal grazing incidence optics and its application to wide-field x-ray imaging. ApJ **392**, 760–765 (1992)
31. Conconi, P., Campana, S.: Optimization of grazing incidence mirrors and its application to surveying x-ray telescope. Astron. astrophys. **372**, 1088–1094 (2001)
32. Aschenbach, B.: Boundary between geometric and wave optical treatment of x-ray mirrors. SPIE Proc. **5900**, 59000D-1–59000D-7 (2005)
33. Lenzen, R.: Imaging properties of a Wolter-I type x-ray telescope with particular reference to contrast reduction through diffuse reflection. PhD thesis, Universität Tübingen, Germany, 123 pp. (1978)
34. Chandra Proposers' Observatory Guide: Rev.6.0, TD 403.00.006, pp. 36–38 (2003)

Exp Astron (2009) 26:111–122
DOI 10.1007/s10686-009-9164-7

REVIEW ARTICLE

Gamma-ray telescopes

400 Years of astronomical telescopes

Neil Gehrels · John K. Cannizzo

Received: 20 December 2008 / Accepted: 31 March 2009 / Published online: 23 April 2009

Abstract The last half-century has seen dramatic developments in $\gamma-$ray telescopes, from their initial conception and development through to their blossoming into full maturity as a potent research tool in astronomy. Gamma-ray telescopes are leading research in diverse areas such as $\gamma-$ray bursts, blazars, Galactic transients, and the Galactic distribution of ^{26}Al.

Keywords Gamma rays: General · Telescopes · Bursts · Blazars · Galactic transients

1 Introduction

The current panoply of high-energy astronomical observatories is impressive: *INTEGRAL*, *Swift*, *Fermi-GLAST*, *RHESSI*, *Suzaku*, and *AGILE*. These satellites have made fundamental contributions to their fields and in fact have had broad impact beyond their immediate purviews. These areas of study include $\gamma-$ray bursts (*Swift*, *INTEGRAL*, *Fermi-GLAST*), solar physics (*RHESSI*), supernovae (*Swift*), Galactic astronomy (*INTEGRAL*, *Fermi-GLAST*), Galactic transients (*Swift*, *Suzaku*), and AGN (*Suzaku*).

N. Gehrels (✉)
NASA/GSFC/ASD/Code 661,
Greenbelt, MD 20071, USA
e-mail: gehrels@milkyway.gsfc.nasa.gov

J. K. Cannizzo
CRESST/UMBC/NASA/GSFC/ASD/Code 661,
Greenbelt, MD 20071, USA

2 Background

A basic consideration of the theory of photon interaction as a function of energy divides naturally into three regimes: photoelectric interactions at $E \lesssim 1$ MeV, Compton scattering for $E \simeq 1$ MeV, and pair production for $E \gtrsim 1$ MeV (Fig. 1). This natural ordering has provided the impetus for three basic telescope designs, (i) coded aperture mask for low energies, in which the illumination of a non-redundant array grid casts a shadow on a detector, thereby giving a sky position (e.g., the BAT on *Swift*), (ii) a Compton telescope for medium energies (e.g., COMPTEL on *CGRO*, and (iii) pair telescope in which an incident high-energy γ−ray interacts with a medium and produces a $e^+ - e^-$ pair. The secondary phase consists of signal detection, which for the three cases is carried out by (i) scintillators (e.g., NaI, BGO, CsI, $LaBr_3$, and GSO), (ii) solid state detectors (e.g., Si, CdZnTe, Ge, HgI_2), and (iii) pair trackers (e.g., spark chamber, Si). Figure 2 shows the history of γ−ray technology, with examples of observatories based on these 3 strategies.

3 Telescope technologies

3.1 Low energy gamma ray telescopes

The earliest astronomical spacecraft (actually—unintentionally astronomical) to utilize collimated detectors were the *Vela* satellites [14]. Each carried six 10 cm^3 CsI scintillation counters distributed so as to achieve optimal nearly isotropic sensitivity. The detectors were sensitive in the energy range between

Fig. 1 A schematic showing of photon interactions indicating linear attenuation in an absorptive medium versus energy

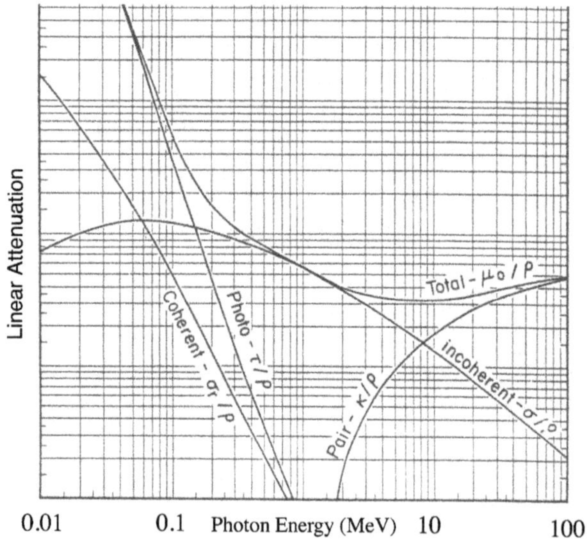

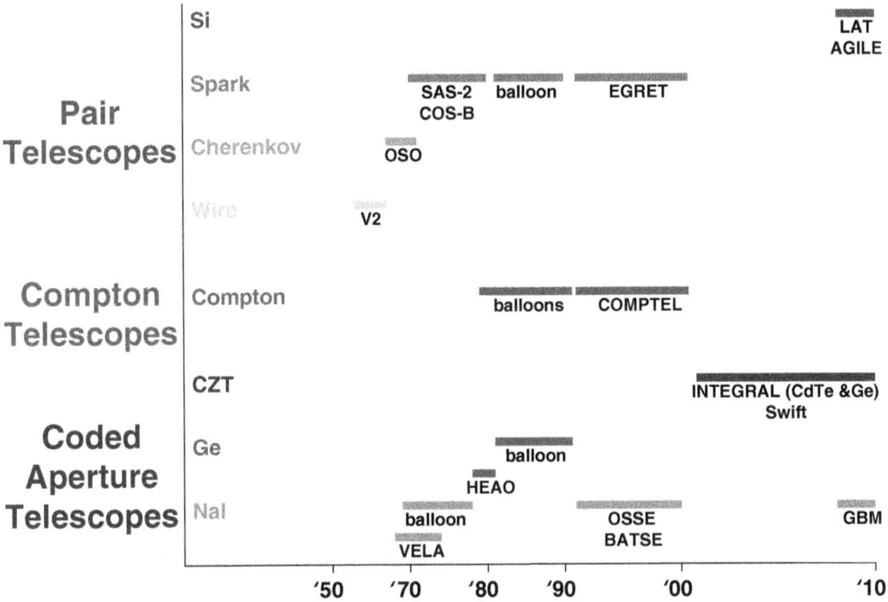

Fig. 2 A history of γ−ray telescope technology

0.2 and 1.0 MeV for *Vela 5* and 0.3–1.5 MeV for *Vela* 6, with a detector efficiency 0.17–0.50. The scintillators were shielded against direct penetration by electrons below ~ 0.75 MeV and protons below ~ 20 MeV. A high Z-shield (i.e., a shield to protect against nuclei with large atomic number) attenuated photons with energies below that of the counting threshold. No active anticoincidence[1] shielding was provided. The accumulated data included a background component due to cosmic particles and their secondary effects. The observed background rate was ~ 150 c s^{-1} for *Vela* 5 and ~ 20 c s^{-1} for *Vela* 6.

HEAO-I (1977-1979) had gas-proportional counters, with each counter possessing a multigrid collimator [27]. Each collimator was composed of a stack of etched Mo sheets interleaved with spacer frames. To maintain thermal stability a heat shield made of 2 μm Kimfol polycarbonate film and coated with 80 nm of Al was put in front of each collimator. Incident X-rays passed through the heat shield and Mylar film, with the net transmission of the two layers determining the response to soft X-rays.

Collimation of γ−rays and background reduction for the OSSE instrument [12] aboard *CGRO* (1991–2000) were achieved via active shielding and a passive W collimator. Each of the four OSSE detectors consisted of a NaI scintillation crystal shielded behind by an optically coupled 7.6 cm thick

[1] In anticoincidence counting, two counters are connected so that a pulse is recorded by one of them only if there is no simultaneous pulse in the other. This is useful in rejecting particles that do not originate from the source being studied.

CsI scintillation crystal in a "phoswich" (phosphor sandwich) configuration. Each phoswich was enclosed in an annular shield of NaI scintillation crystals providing anticoincidence for γ−ray interactions in the phoswich. The annular shield also enclosed a W slat collimator which sets the $3.8° \times 11.4°$ FWHM γ−ray FOV of the phoswich detector.

The BAT instrument on *Swift* (2004-) has a D-shaped coded mask[2] made of $\sim 54,000$ Pb tiles ($5 \times 5 \times 1$ mm) mounted on a 5 cm thick composite honeycomb panel which is mounted by composite fiber struts 1 m above the detector plane [8]. The BAT coded mask uses a random 50% open/50% closed pattern rather than the more common uniformly redundant array pattern. The mask area is 2.7 m^2, which yields a half-coded FOV of 1.4 sr.

3.2 Pair telescopes

OSO-3 (1967–1968) achieved the first extensive γ−ray observation in space, using a Cerenkov scintillator detector [2], and showed the Galactic disk to be a strong source of high-energy γ−rays. The detector consisted of 0.64 cm thick plastic scintillators and a 1 in thick lucite Cerenkov radiator over a 0.16 cm thick Pb sheet to convert γ−rays into $e^- - e^+$ pairs. The discriminator thresholds on all detectors were set using sea level muons. The energy threshold was 50 MeV (1%), rising to 13.5% at 200 MeV, and 17% at 1 GeV, where the percentages indicate the fraction of the total geometric area. The false alarm rate for singly charged particles was about 0.15%.

The γ−ray telescope flown on SAS-2 (1972–1973) consisted of an assembly with 16 spark chamber modules above a set of four plastic scintillators, with another 16 modules below the scintillators [7]. Thin W plates (which had a thickness of ~ 0.03 radiation lengths) were interleaved between the spark chamber modules, with an area ~ 640 cm^2. The large number of W plates and spark chambers provided material for an incoming γ−ray to be converted into a $e^- - e^+$ pair, which could then be identified and used to provide a direction for the incoming γ−ray, and a way of determining the energy of the electrons in the pair by measuring the Coulomb scattering. The sensitivity range of the instrument was about 30–200 MeV.

After *Vela*, COS-B (1975–1982) was the primary workhorse for studying γ−ray bursts [6]. COS-B consisted of a dome-shaped 1.1 $m^2 \times 10$ mm thick plastic scintillator viewed by nine photomultipliers for rejection of charged particles (the anti-coincidence counter, or ACO). The ACO maximum sensitivity was at 700 KeV.

The EGRET instrument [13] on *CGRO* consisted of an anticoincidence system, a spark chamber with interspersed conversion material to materialize the incident photons and determine the trajectory of the secondary $e^- - e^+$

[2]The concept of the coded mask was discovered by Aristotle. As he was walking under some trees, he noticed that patches of sunlight on the ground were in the shape of the Sun and not of the gap between the branches. The trees acted in the same way as a coded mask.

pair, a triggering device to detect the presence of charged particles with the correct direction, and an energy measuring device—a total-absorption spectrometer NaI crystal. The 2-cm-thick plastic scintillators were viewed by 24 photomultiplier tubes coupled to the lower perimeter so as to keep the EGRET FOV free of irregular mass distributions. The conversion of the incoming γ−rays into $e^- - e^+$ pairs occurred in an upper stack of 28 spark chambers interleaved with Ta foils with an average thickness of 90 μm (= 0.02 radiation lengths). The total depth for on-axis radiation was hence about 0.54 radiation lengths, leading to a conversion efficiency of about 30% at 100 MeV. The volume of the upper chamber was $\sim 80 \times 80 \times 45$ cm^3, so that an effective area ~ 2000 cm^2 resulted for the center of the FOV.

The LAT [1] on board *Fermi-GLAST* (2008–) is a pair conversion telescope with a precision converter-tracker and calorimeter, each consisting of a 4×4 array of 16 modules supported by a low-mass Al grid structure. A segmented anticoincidence detector covers the tracker array, and a programmable trigger and data acquisition system uses prompt signals from the tracker, calorimeter, and anticoincidence detector subsystems to form a trigger. The precision converter-tracker has 16 planes of high-Z material in which γ−rays incident on the LAT can convert to an $e^- - e^+$ pair. The converter planes are interleaved with position-sensitive detectors that record the passage of charged particles, thus measuring the tracks resulting from $e^- - e^+$ pair conversion. This information is used to reconstruct the directions of the incident γ−rays, Each tracker module has 18 x, y tracking planes, consisting of two layers (x and y) of single-sided Si strip detectors. The 16 planes at the top of the tracker are interleaved with high-Z convert material (W), i.e., material capable of generating interactions with nuclei of high atomic number.

3.3 Compton telescopes

The COMPTEL instrument [21] on *CGRO* combined a large FOV with imaging. It consisted of two detector arrays, and upper one of low-Z material and a lower one of high-Z material. In the upper detector an incoming γ−ray was Compton scattered, and the scattered photon subsequently made a second interaction in the lower detector. A time-of-flight measurement was utilized to confirm the causal relation. The locations and energy losses of both interactions were monitored. For events that were completely absorbed, the arrival direction of the γ−ray had to lie within a known angle of the direction of the scattered γ−ray, which allowed a determination of its original direction on the sky. Incompletely absorbed events yielded uncertainty cones that did not constrain the source position.

3.4 Focusing telescopes

Actual imaging of hard X-rays (5–80 keV) can be achieved using multilayer mirrors. The technology has been proven with the InFOCμS balloon instrument, which obtained a 2.6 arcmin resolution image of 4U 0115 [25]. The Pt/C reflector surface multilayer mirror had a diameter of 40 cm and a focal length of 8 m. The incident angle could lie in the range 0.105–0.356 deg. There were 255 nestings and 2040 reflectors. The overall telescope had an effective area of 49 cm^2 at 30 keV, an angular resolution of 2.4 arcmin (half-power diameter), and a field of view of 11 arcmin.

Currently under development is the *NuSTAR* instrument, a NASA SMEX mission slated for launch in $\sim$ 2011 which will contain an extensible boom for its multilayer mirrors, and CdZnTe detectors [10]. *HEFT*, a proof-of-concept balloon payload, contains three co-aligned conical-approximation Wolter I mirrors, each of which focuses hard X-rays/soft γ−rays onto a shielded solid-state CsZnTe focal plane. Each of the three optics has 70 shells, with radii between 4 and 12 cm. The reflectors are coated with W/Si multilayers, providing good reflectance up to the W K-edge at 69.5 keV. The total collecting area on-axis at 30 keV is 100 cm^2. *NuSTAR* consists of three co-aligned telescope modules with a 10-m focal length, and the optics and detectors are placed at either end of an extensible mast. The optics consist of 130 shells each, with radii between 5.5 and 16.9 cm. The shells are coated with a combination of W/SiC and Pt/SiC multilayers. Combined with low graze angles, the smooth response goes up to 80 keV. *NuSTAR* will achieve angular resolution of 40 arcsec (half-power diameter). The *NuSTAR* and *HEFT* focal plane detectors have the same dimensions, but the *NuSTAR* detectors are surrounded by an active CsI shield rather than graded-Z/plastic shield.

Looking further into the future, new methods are being discussed that could in principle take advantage of the very short wavelength of γ-rays to achieve ultra-high resolution imaging. The diffraction-limited angular resolution

$$\theta_d = 1.22\frac{\lambda}{d} = 0.12\left(\frac{500\ \mathrm{keV}}{E}\right)\left(\frac{5\ \mathrm{m}}{d}\right)\ \mu\mathrm{arcsec}, \qquad (1)$$

where d is the lens diameter. This ultra-high resolution could be applied to directly image the event horizons of black holes. One technique for accomplishing direct, high-resolution imaging is through Fresnel lenses [24]. Fresnel zone plates (FZPs) focus radiation by blocking with opaque annuli the photons contributing to spatial frequencies that would arrive at the image plane non-optimally. This technique is similar in spirit to "apodization" (literally—"reducing the feet"—referring to the diffraction pattern) used by optical astronomers in which a PSF is enhanced by minimizing the higher order maxima of the diffraction pattern, effectively by throwing more of the light into the central, on-axis maximum. Phase Zone Plates (PZPs) go one step better by replacing the opaque annuli with media possessing a thickness of refractive material which imposes a phase shift of π. The final step in improvement consists of Phase Fresnel Lenses (PFLs) in which the refractive material has a profile that is radial with respect to the central optical axis, so that all the radiation arrives at the focal point with exactly the correct phase. This is an exact analog of the optical astronomers' apodization. The use of PFLs for high-energy photons is achievable because all materials have a refractive index that differs from unity. For X-rays and $\gamma-$rays the refractive index $\mu = (1 - \delta)$, where δ is a small positive number, leading to a μ value slightly less than unity.

Another possibility for direct imaging is through the use of Laue concentrators [26]. This method takes advantage of the phase information carried within the high-energy photons by using Bragg diffraction, which refers to the interference between the periodic (wave) nature of light and the periodic structure of matter in a crystal lattice. In a Laue diffraction lens, a large number of crystals are oriented in an annular configuration so that incident photons are imaged onto a common focal spot. The background noise is kept low because of the resultant small detector volume. In a Laue-type crystal diffraction lens the crystals are positioned along concentric rings. In order to be diffracted the wavelength and incident angle of the incoming photon must obey the Bragg relation

$$2d \sin\theta_B = n\lambda, \tag{2}$$

where d is the crystal plane spacing, θ_B the incident angle of the photon, n the reflection order, and λ the photon wavelength. A crystal at a distance r_1 from the optical axis is oriented such that the angle between the incident beam and the crystalline planes is the Bragg angle θ_B. For a given focal distance f of the lens, the radius of the ith ring is given by

$$r_i = f \tan|2\theta_{B_i}| \approx f\frac{n\lambda}{d_i}, \tag{3}$$

where d_i is the crystalline plane spacing of the ith ring. The bandwidth

$$\Delta E \approx \frac{2E^2 d\Delta\theta}{nhc} \approx 40\,\mathrm{keV}\left(\frac{E}{511\,\mathrm{keV}}\right)^2\left(\frac{d}{d_{\mathrm{Ge}[111]}}\right)\left(\frac{\Delta\theta}{1'}\right), \tag{4}$$

where $\Delta\theta$ is the angular range over which the crystals reflect monochromatic radiation, and Ge[111] refers to the germanium $[hkl] = [111]$ crystalline plane.

4 History

The first $\gamma-$ray detector was launched on a V2 rocket as a suborbital flight. It was sensitive in the range 3.4–90 MeV and used wire ionization detectors. The astrophysical findings were upper limits [19, 20]. The first $\gamma-$ray maps were made by *Explorer-11* and *OSO-3*. The high-energy instruments on these satellites were of the cathode-ray type, and their design foreshadowed future directions in their use of converter, calorimeter, and plastic anticoincidence [15]. Figure 3 shows the first $\gamma-$ray map, a distribution characterized by a broad maximum toward the Galactic center, and an isotropic component with a softer energy spectrum. The spark-chamber era, spanning 20 MeV–10 GeV, was dominated by COS-B from 1975 to 1981, and by EGRET on *CGRO* from

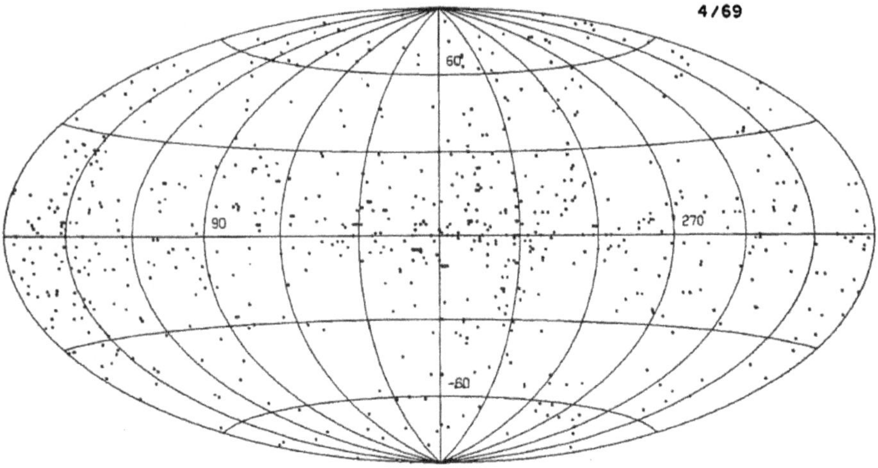

Fig. 3 The first γ−ray map, from Kraushaar et al. [15]

1991 to 2000. The current instrument continuing this thread is the LAT on *Fermi-GLAST*, with $\sim 10^6$ channels and 70 m^2 of Si surface. The Galactic map obtained by EGRET has yielded a plethora of Galactic sources and enriched our knowledge of their high-energy behavior. The third EGRET catalog (Fig. 4), which contains data from 1991 April 22 to 1995 October 3, has 271 sources ($E > 100$ MeV). The list includes a solar flare, the LMC, five pulsars, one probable radio galaxy detection (Cen A), 66 blazars, and 170 unidentified EGRET sources [11]. Model predictions indicate that *Fermi*-LAT should be able to detect 9900 sources at $> 5\sigma$ in a one-year all-sky survey. Early Compton telescopes (1–30 MeV) were flown on balloon programs at MPE in Germany, and UC Riverside and UNH in the US. This unexplored energy range is technically challenging due to scattering, and localizations on the sky are typically to $\sim 4°$. A major advance was the COMPTEL instrument on *CGRO* which operated from 1991 to 2000 [22]. A new window to astronomy was hereby opened: COMPTEL produced a reliable 1–30 MeV sky map, shown in Fig. 5 [23]. This was the first complete 0.75–30 MeV all-sky survey, which found 32 non-variable sources and 31 γ−ray bursters. Among the continuum sources were pulsars that had spun down, stellar black-hole candidates, supernova remnants, interstellar clouds, AGN, γ−ray bursters, and the Sun during solar flares. Line detections were also made: 1.809 MeV (^{26}Al), 1.157 MeV (^{44}Ti), 0.847 and 1.238 MeV (^{56}Co), 2.223 MeV (neutron capture).

The history of GRB research has been almost entirely written by high-energy, space-borne instruments. GRBs were discovered as a phenomenon in 1967 by the *Vela* satellites [14], which were initially launched +to verify the atmospheric nuclear test ban treaty. For many years GRB research was strongly impeded by the fact that the distances to the sources were completely

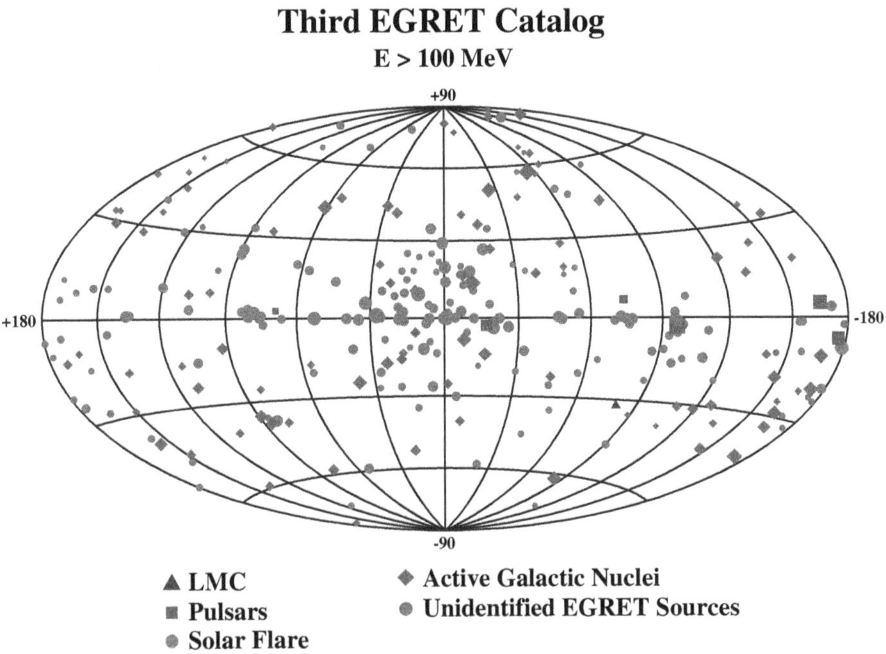

Fig. 4 The third EGRET catalog [11]

unknown. This began to change with the launch of *CGRO* in 1991 and the use of BATSE and OSSE [16, 17]. Within a few months, the uniform sky distribution of GRBs made it clear that there was no concentration of sources to the Galactic plane, and therefore GRBs had to lie at either much smaller

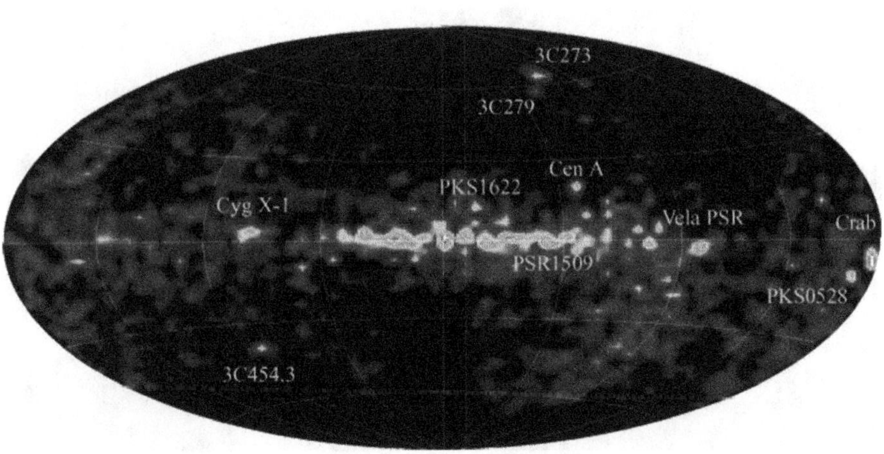

Fig. 5 The COMPTEL 1–30 MeV sky map

(~solar system), or much larger (~cosmological) distances. The tradition of outstanding GRB research begun by BATSE has been continued with the BAT on *Swift*, which is yielding $\sim 3 - 8$ arcmin localizations using a shadow mask design with CdZnTe detectors [3]. One of *Swift*'s major successes has been the first position determination of a short GRB [9].

The *INTEGRAL* sky as seen by the IBIS instrument has built upon the previous COMPTEL results. An all-sky map was produced which represented the first complete scan of the entire central equatorial zone of the Galactic plane at energies above 20 keV with a limiting sensitivity of $\sim$ 1 mCrab. in the central radian [5]. The resultant catalog also included a substantial coverage of extragalactic fields, and contained more than 200 high-energy sources. The mean position error for sources detected with significance above 10σ was $\sim$ 40 arcsec, enough to identify most of them with a known X-ray counterpart [5]. This is important given the high Galactic aborption inherent in most earlier surveys. Lebrun et al. [18] found that the 20–200 keV Galactic emission is resolved into compact sources. High resolution imaging ($\sim$ 10 arcmin) of the Galactic center at high energies is a very recent development in the field of high-energy astrophysics. Bélanger et al. [4] studied the persistent high-energy flux from the Galactic center using the the IBIS/ISGRI imager on *INTEGRAL* (Fig. 6). For the first time they detected a hard X-ray source, IGR J17456-2901, located within 1 arcmin of Sagittarius A* (Sgr A*) in the energy range $20 - 100$ keV. By combining the ISGRI spectrum with an X-ray spectrum corresponding to the same physical region around Sgr A* from *XMM-Newton* taken during part of the *INTEGRAL* observations, they inferred a 20–400 keV luminosity at 8 kpc of $L = (5.37 \pm 0.21) \times 10^{35}$ erg s^{-1}. They also set a 3σ upper limit on the flux at the electron-positron ($e^- - e^+$) annihilation energy of 511 keV from the direction of Sgr A* at 1.9×10^{-4} photons cm^{-2} s^{-1}.

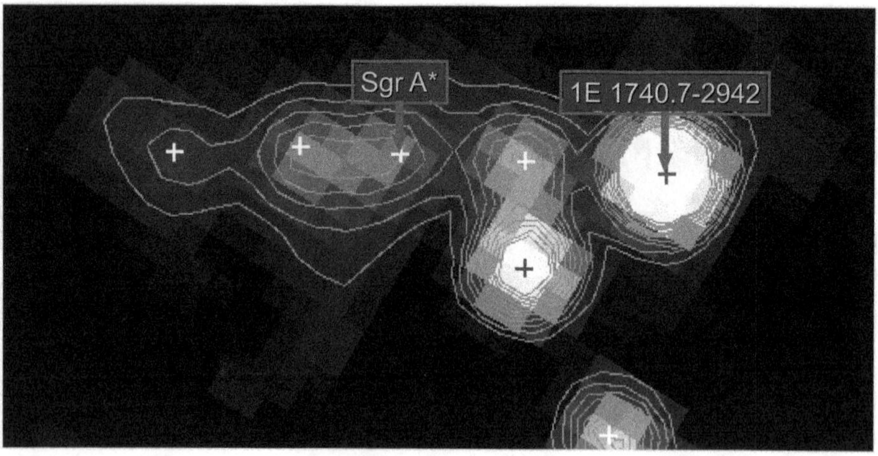

Fig. 6 View of the Galactic center by *INTEGRAL*-IBIS

5 Conclusion

Gamma-ray astronomy has taken enormous strides in the last few decades. Dramatic improvements in both spectral and imaging capabilities have thrust γ−ray telescopes into the forefront of astronomical research. For about a decade the prompt localizations so essential to GRB research were carried out largely by *CGRO*; now the torch has been passed to *Swift*. The Galactic center has now been explored in a new and interesting way, via γ−rays, and our knowledge of transient sources in the Galactic bulge and the diffuse Galactic background has deepened. Future missions that seek to take advantage of the ultra-high resolution potential of γ−ray observing at the diffraction limit offer the hope of direct imaging of nearby black hole event horizons - which would allow us to probe directly General Relativity in the strong field limit.

References

1. Atwood, W.B., et al.: The large area telescope on the fermi gamma-ray space telescope mission. Astro-ph/0902.1089v1. Astrophys. J. (2009, in press)
2. Badhwar, G.D., Kaplon, M.F., Valentine, D.A.: γ−ray observations from the OSO-3 satellite. Astrophys. Space Sci. **27**, 147 (1974)
3. Barthelmy, S.D., et al.: The Burst Alert Telescope (BAT) on the SWIFT midex mission. Space Sci. Rev. **120**, 143 (2005)
4. Bélanger, G., Goldwurm, A., Renaud, M., Terrier, R., Melia, F., Lund, N., Paul, J., Skinner, G., Yusef-Zadeh, F.: A persistent high-energy flux from the heart of the milky way: INTEGRAL's view of the galactic center. ApJ **636**, 275 (2006)
5. Bird, A.J., et al.: The second IBIS/ISGRI soft gamma-ray survey catalog. ApJ **636**, 765 (2006)
6. Boella, G., Gorisse, M., Paul, J., Taylor, B.G., Wills, R.D.: The ability of COS-B to measure gamma-ray bursts. Astrophys. Space Sci. **42**, 103 (1976)
7. Fichtel, C.E., Kniffen, D.A., Hartman, R.C.: Celestial diffuse gamma radiation above 30 MeV observed by SAS-2. ApJ **186**, L99 (1973)
8. Gehrels, N., et al.: The swift gamma-ray burst mission. ApJ **611**, 1005 (2004)
9. Gehrels, N., et al.: A short γ−ray burst apparently associated with an elliptical galaxy at redshift $z = 0.225$. Nature **437**, 851 (2005)
10. Harrison, F.A., et al.: Development of the HEFT and NuSTAR focusing telescopes. Exp. Astron. **20**, 131 (2005)
11. Hartman, R.C., et al.: The third EGRET catalog of high-energy gamma-ray sources. ApJS **123**, 79 (1999)
12. Johnson, W.N., Kinzer, R.L., Kurfess, J.D., Strickman, M.S., Purcell, W.R., Grabelsky, D.A., Ulmer, M. P., Hillis, D.A., Jung, G.V., Cameron, R.A.: The oriented scintillation spectrometer experiment - instrument description. ApJS **86**, 693 (1993)
13. Kanbach, G., et al.: The project EGRET (Energetic Gamma-Ray Experiment Telescope) on NASA's Gamma-Ray Observatory (GRO). Space Sci. Rev. **49**, 69 (1988)
14. Klebesadel, R.W., Strong, I.B., Olson, R.A.: Observations of gamma-ray bursts of cosmic origin. ApJL **182**, 85 (1973)
15. Kraushaar, W.L., Clark, G.W., Garmire, G.: Preliminary results of gamma-ray observations from OSO-3. Can. J. Phys. **46**, 414 (1968)
16. Kurfess, J.D., et al.: Oriented scintillation spectrometer experiment observations of Co-57 in SN 1987A. ApJL **399**, 137 (1992)
17. Kurfess, J.D., Bertsch, D.L., Fishman, G.J., Schönfelder, V.: Overview of the compton observatory instruments. AIPC. **410**, 509 (1997)
18. Lebrun, F., et al.: Compact sources as the origin of the soft γ−ray emission of the Milky Way. Nature **428**, 293 (2004)

19. Perlow, G.J., Kissinger, C.W.: A search for primary cosmic gamma-radiation. I. Phys. Rev. **81**, 552 (1951)
20. Perlow, G.J., Kissinger, C.W.: A search for primary cosmic gamma-radiation. II. Low energy radiation above and within the atmosphere. Phys. Rev. **84**, 572 (1951)
21. Schönfelder, V.: The imaging gamma-ray telescope COMPTEL aboard GRO. Adv. Space Res. **11**, 313 (1991)
22. Schönfelder, V., et al.: Instrument description and performance of the imaging gamma-ray telescope COMPTEL aboard the compton gamma-ray observatory. ApJS **86**, 657 (1993)
23. Schönfelder, V., et al.: The first COMPTEL source catalogue. A A **143**, 145 (2000)
24. Skinner, G.K., von Ballmoos, P., Gehrels, N., Krizmanic, J.: Fresnel lenses for x-ray and gamma-ray astronomy. SPIE J. **5168**, 459 (2004)
25. Tueller, J., et al.: InFOCμS hard x-ray imaging telescope. Exp. Astron. **20**, 121 (2005)
26. von Ballmoos, P., et al.: CLAIRE's first light. New Astron. Rev. **48**, 243 (2004)
27. Wood, K.S., Meekins, J.F., Yentis, D.J., Smathers, H.W., McNutt, D.P., Bleach, R.D., Byram, E.T., Chubb, T.A., Friedman, H.: The HEAO A-1 X-ray source catalog. ApJS **56**, 507 (1984)

Exp Astron (2009) 26:123–132
DOI 10.1007/s10686-009-9147-8

REVIEW ARTICLE

Past and future of ESO

L. Woltjer

Received: 20 October 2008 / Accepted: 20 January 2009 / Published online: 11 March 2009

Abstract In 1964, the European Southern Observatory (ESO) came into being as a response to the need of European astronomers for access to large telescopes in good observing conditions. The original plan called for a 3.6 m telescope in addition to a number of smaller instruments. The 3.6 m telescope took some 12 years to build, while ESO headquarters moved from Hamburg to CERN to Garching near Munich, and during the times of office of three Directors General, and it was built on La Silla in Chile, where superior sites had been found. After 1982, when Italy and Switzerland joined the original six member countries, ESO was able to expand by building the 3.5 m New Technology Telescope, and ultimately planning and constructing the VLT on Paranal. At present ESO is a stable organization that has been successful in its original aims, and in view of new plans can look forward in confidence.

Keywords ESO · Astronomers · European

1 The origins of the European Southern observatory

After the Second World War Europe had fallen far behind the US in particle physics and in astronomy. In both areas a pooling of resources through the creation of an intergovernmental organization provided the solution.

L. Woltjer (✉)
Observatoire de Haute-Provence, 04870 Saint-Michel l'Observatoire, France
e-mail: dem.wol@bluewin.ch

	Idea	Convention	Major instrument	Site
CERN	1948	1953	1959: 28 GeV synch.	1952 Genève
ESO	1953	1964	1976: 3.6-m telescope	1980 Garching 1964 La Silla

The six countries who supported the idea of ESO were France, Germany, the Netherlands, Belgium, Sweden and Denmark. Italy was still planning to build its own telescope and, at an early date, procured a 3.5-m blank. The UK participated in some early meetings, but subsequently withdrew to construct a telescope with Australia; in particular the Astronomer Royal, R.v.d.R. Woolley, was antagonistic to a collaboration with the Europeans, perhaps because of his South African origins and the Boer war. Fred Hoyle, after attending some meetings, stated that ESO would never become a functioning observatory. As a result, and because in France numerous governments succeeded one another, it took 11 years to arrive at the intergovernmental convention. Subsequently, it took another 12 years to complete the main instrument, a 3.6-m telescope.

2 The search for the best site

Before the convention had been approved, a site survey had been initiated. Since the southern Milky Way had not been studied in sufficient detail, a southern hemisphere site was preferred. The Dutch, who had many connections with South Africa, started there, but by the time the convention was ready, reports received from Chile indicated that sites there were superior. In fact, already a century earlier R. Stephenson, a British member of parliament, had commented on the exceptional clearness of the night sky above the Andes. He loaned his private yacht to C. Piazzi Smyth who in 1856 sailed to Tenerife to see if, as Newton had suggested, mountain tops were optimal locations for telescopes. In fact, he found that near the top of the 2,700 m high Pico de Teide stars could be seen four magnitudes fainter than at sea level.

In the thirties, C.G. Abbot of the Smithsonian Institution, who attempted to measure the solar constant, needed a site with a very low water vapor content through the atmosphere. He found Cerro Montezuma at practically the same latitude as Paranal to be the best place in the world for his purposes, far better than sites in Namibia. By 1960, G.P. Kuiper, one of the great astronomical entrepreneurs, had selected Cerro Tololo for the construction of the (inter)American observatory. Yale and Columbia at that time were looking for a site for their astrographic telescope. To avoid sharing a mountain with Kuiper they settled for El Leoncito in Argentina, east of the Andes, where conditions were less good owing to Atlantic disturbances drifting in.

Initially there followed much discussion of the possibility of ESO leasing part of a mountain on American territory. While the more Anglo-Saxon oriented Dutch favored this, O. Heckmann, the first ESO director, decided on a more independent solution. This led to the choice of La Silla some 100 km

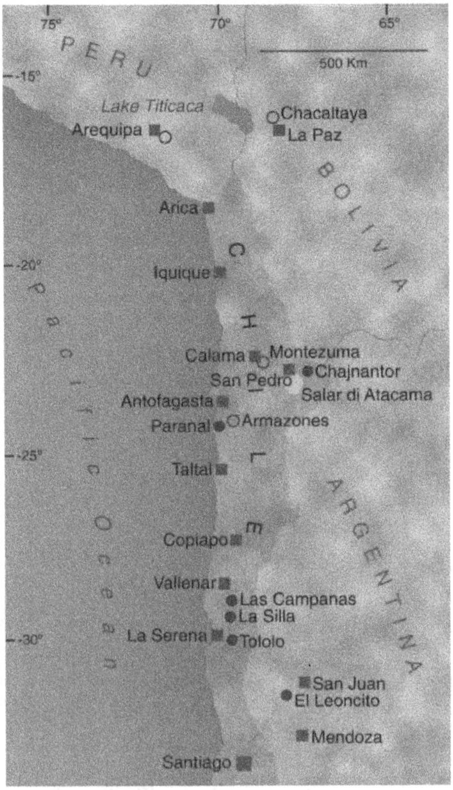

Fig. 1 Astronomical sites in Chile and surroundings. *Filled circles* are currently active observatories, *open circles* inactive ones; *squares* are geographical locations. (Courtesy ESO/E. Janssen)

further to the north and around 500 km from Santiago (Fig. 1). Heckmann was acquainted with padre Starischka, rector of the German high school through which many of the ministers in the Chilean government had passed. As a result of his assistance, in a few months a diplomatic convention at the highest level between Chile and ESO was concluded, subject to ratification by the Chilean parliament and the ESO Council.

3 The 3.6 m telescope and the move to Garching bei München

As mentioned above, it took another 12 years to develop the main instrument, the 3.6 m telescope. This was due to a gross underestimate of the technical effort. In fact, one engineer with some draftsmen were supposed to make the mechanical design. In 1969 A. Blaauw, the new Director General from 1970, realized that failure of the 3.6-m telescope project would endanger ESO's existence. As a result, a collaborative venture with CERN was organized which brought in the necessary engineering competence. The telescope was installed at La Silla in late 1976. Because of its history it was not innovative; it also had hardly any instrumentation (Fig. 2).

Fig. 2 Aerial view of La Silla. The 3.6 m is the biggest dome at the *top*, *left*; the NTT is located just *below*. The big building in the center is the hotel with restaurant and administration, surrounded by dormitories. Among the further telescopes are the SEST (*top right*), the 1-m photometric telescope (1966), the 1.5 m spectrographic telescope (1968), the 2.2-m MPG/ESO telescope (1984) and various smaller national telescopes. (Courtesy ESO/C. Madsen)

O. Heckmann, as first Director General, had placed ESO's European activities in Bergedorf where the Hamburg observatory, which he directed, was located. Even though most of ESO's very limited technical and scientific personnel had moved to Geneva, the administration had remained in Bergedorf. Since Germany was underprivileged in having no international scientific organizations on its territory, it was finally decided that ESO should move to Germany. Having the freedom to choose a site within Germany, I then chose Garching bei München, where ESO's headquarters were subsequently constructed. Much assistance was given by the Max-Planck-Gesellschaft and its President, Reimar Lüst. During the transfer nearly half of ESO's technical staff decided to look for alternative employment in Geneva and elsewhere, which led to further delays in instrumentation development.

4 Development of the NTT and the VLT

Completion of the 3.6-m signaled a decisive moment: was this telescope to be the end of ESO's ambitions in telescope technology or was it just the first in

Fig. 3 The four 8.2-m telescopes of the VLT and the 1.8-m mobile telescopes for interferometry which may be placed at various stations

a long-term program of technology development worthy of Europe's general potential? This then led me to propose the VLT, the Very Large Telescope, in the form of an array of four 8-m unit telescopes (Fig. 3). Council was concerned about the cost of such a project and initially kept its distance, while some astronomers thought that having a second 3.6-m telescope to increase the annual observing time would be more important. A careful management of the conservative European astronomical community was needed.

The situation changed in 1982 when Italy and Switzerland joined the original six member countries. This augmented the confidence in ESO as a European organization. It also improved the financial basis and allowed the technological work force to be enlarged and experimental projects to be initiated. This led to the 3.5-m **N**ew **T**echnology **T**elescope which demonstrated low-weight mirror technology, Ray Wilson's Active Optics, alt-azimuth mounts and new drive mechanisms. It was completed well below budget, for a third of the cost of the 3.6-m.

The choice of 3.5 m for the NTT mirror was dictated by the agreement with Italy for the slicing of the mirror which had been previously acquired. After that agreement had been reached, we realized that the Italian administrators thought that the slicing would leave two half circles! Some time later it was found that the interior of the blank had too many bubbles and a new Zerodur blank was obtained for the NTT.

Active Optics (see also the paper by Noethe in the present volume) had major benefits, both for the optical quality and for the cost. In the previous generation of telescopes the primary mirror had been thick and heavy to minimize deformations of the optical surface. The weight of such a mirror propagates through the whole telescope design from the tube to the bearings and the whole support structure. With Active Optics a much thinner mirror was supported by the pressure from computer-controlled actuators. These were ultimately controlled by the results from an Image Analyzer which determined the actual shape of the optical surface and the pressures to be applied to the mirror to improve it.

Also the introduction of the alt-azimuth mounting led to important simplifications and cost savings in the telescope. Equally important, it reduced the size of the enclosure. In fact, the diameter of the enclosure for one of the 8.2-m telescopes was no larger than that for the 3.6-m had been. Also an alt-azimuth mounting naturally provides two Nasmyth foci at which heavy instruments may be placed semipermanently. A switch between the two may be made by rotating the Nasmyth flat mirror, which takes no more than a minute. So as a function of atmospheric conditions one may select the most appropriate instrument leading to an optimal use of telescope time.

While the Adaptive Optics experiments in the laboratory had been successful, experience under actual observatory conditions was still lacking. So for the NTT the diameter-to-thickness ratio was chosen at 15, still a very conservative value. If Adaptive Optics would be fully successful, this was sufficient to give the mirror the necessary flexibility; if not, the mirror was sufficiently rigid that acceptable images could be obtained with simpler means. With the NTT producing spectacular images on the first night of operation at La Silla, no doubts remained about the validity of the Adaptive Optics, and for the VLT mirrors the diameter-to-thickness ratio was chosen at nearly 50.

One difficult problem remained: how to obtain the 8.2-m blanks. As a fallback option a mirror blank made of fused silica had been retained, while also various metals had been studied. A fused silica blank may be made in almost any size, since it involves the welding together of pieces of the material. However, a monolith of glass ceramic, if obtainable, seemed preferable for the very thin mirrors. For the NTT Schott in Mainz had produced a superb 3.5-m blank made of Zerodur, a nearly zero-thermal-expansion material with excellent polishability. Such a mirror would not change its focal length when the temperature of the telescope changes. The low thermal expansion results from mixing, on a microscopic scale, glass with a positive expansion coefficient with crystalline material with a negative one. To obtain the right characteristics it is necessary to let the material slowly crystallize at temperatures around 900°C. This process typically takes many months. Unfortunately, at the Schott factory no mirror blanks of more than 4 m in diameter could be made.

Long negotiations between ESO and Schott followed the funding of the VLT project, which resulted in a contract under which Schott would build a new factory with the necessary capacity. Permits had to be obtained and the new furnace could only be fired some 3 years after VLT approval. Subsequently, the first three blanks cracked, which was not surprising since no experience of dealing with such large, thin pieces of Zerodur existed. The first blank was delivered in mid-1993, some 2 1/2 years later than in the original project plan. Two years after that four excellent blanks had been delivered. Because of their extreme thinness special equipment was needed to lift or transport them and avoid breakage.

Subsequently, the blanks had to be polished; the contract had been awarded to REOSC near Paris. Again, a new factory with a new fully computer-controlled polishing machine had to be constructed in order to achieve an unprecedented accuracy of the shape of such large surfaces. The first polished mirror could leave REOSC in 1995 and begin its voyage to Chile. A very detailed analysis had to be made to ensure the transportability of the large mirrors.

Also the secondary mirrors posed difficulties. For observations in the infrared it is essential that the general background radiation due to the telescope and the atmosphere be subtracted precisely. If it is not, the weak signal from astrophysical sources would be overwhelmed. To obtain adequate subtraction one switches between two adjacent positions on the sky, one including the source and the other excluding it. The two different positions are obtained by slightly changing the orientation of the secondary mirror. Because the background is variable on subsecond timescales, the secondary mirror has to oscillate between the two orientations at around 10 Hz. This can only be achieved if the weight of the mirror is minimized. The solution adopted was to make the secondaries of beryllium metal, a lightweight, expensive material, which is very dangerous to handle. As a result, the cost of the secondary units exceeded the polishing costs of the 8.2-m mirrors.

The mirrors have to be mounted in a mechanical telescope structure, which in turn is placed in a rotating, fully climate-controlled enclosure (Fig. 4). Finally, all the motors of the mirror units, the telescopes, and the instruments have to be controlled by a highly complex software system.

Interferometry has become an important aspect of the VLT. Since with four fixed telescopes the number of baselines is rather small, it was envisaged to have also some smaller mobile telescopes. Their number varied with time as a function of ESO's budgetary situation, but finally, as pushed for by Pierre Léna, four movable 1.8-m telescopes have been installed. Also the construction of the central interferometric laboratory, where the beams from the various telescopes are combined, represented a major effort.

The confidence resulting from the accession of new ESO member countries and the quality of the VLT proposal (The Blue Book) allowed the project to be funded at the end of 1987 at just over 300 M€ in 2008 money. Upon completion actual expenditure on the VLT was some 15% above the planned budget in the Blue Book. This was largely due to additions to the project made after project approval, mainly the Cassegrain foci at all four telescopes and the beryllium mirrors. It is a tribute to the ESO technical staff and to their very detailed analyses that a project with so much innovation could be costed with such precision.

The high cost of the project made it worthwhile to look for sites superior to La Silla in the "heart of dryness" just south of Antofagasta where the 2,700 m

Fig. 4 One 8-m telescope. Just under the center piece the flat mirror is visible which sends the light through the elevation axis to one or the other of the Nasmyth foci; when the flat is removed, the light can reach the Cassegrain focus close to the human figure in the foreground. The thinness of the primary is clearly seen. At the *top* is the 1.2-m secondary mirror composed of beryllium. It may oscillate between two positions at some 10 Hz for observations in the infrared

high Cerro Paranal rises at no more than 10 km from the Pacific Ocean with its very stable atmosphere.

Earlier, in some countries instrumentation had been constructed for ESO by the national observatories, but the quality had been variable. However, many laboratories and observatories in the meantime had worked on space projects and became used to rigorous quality, budget and time control. The VLT proposal foresaw that much of the instrumentation would be made by (consortia of) national institutes with ESO paying only the hardware costs (see p. 321 of VLT proposal in March 1987, the so-called "Blue Book"). This not only relieved the ESO budget, but it also led to a more closely interconnected European astronomical community.

5 Further expansion of ESO and future prospects

The successful completion of the VLT and the anticipation of further telescope developments led the UK (which in 1987 had declined), Spain, Portugal, Finland, Czechia and Austria (the latter after 35 years of intermittent discussions) to join ESO for a total of 14 countries. So new possibilities opened up: for some years there was planning for a 100-m OWL telescope; in the meantime this has been downsized to 42 m. Is one single telescope, just slightly larger than the planned US 30-m, the obvious choice? Or would it have been preferable to go again for a multiple telescope like the VLT with its interferometric options? After all it is that way that ESO has made its unique reputation. ESO's success has been made possible by a reputation for building to budget. Something like the OWL saga seems hardly conducive to maintaining that reputation.

Figure 5 shows the numbers of astronomers (as measured imperfectly by the number of IAU members in the world) and the total telescope area available in the world respectively for Europe only. The cross indicates the effect of adding the 42-m E-ELT. The figure illustrates the rough proportionality between optical telescope area and numbers of astronomers. On average for each astronomer in the world there is around 0.1 m^2 of mirror in a fully steerable telescope. The figure also illustrates the catastrophic condition of European optical astronomy in the post-war years.

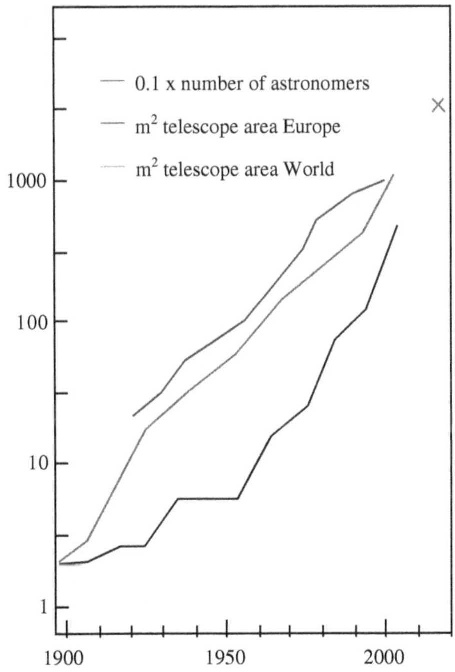

Fig. 5 Astronomers and telescopes, 0.1 × number of astronomers (*red*), m^2 telescope area Europe (*blue*), m^2 telescope area World (*green*)

In 1984 work was started on SEST, the 15-m Swedish-ESO Submillimeter Telescope, installed at La Silla at a modest cost of 10 M€. This brought ESO into the domain of millimeter astronomy. The follow up on this project is very much costlier: ALMA, the Atacama Large Millimeter Array, (50+ telescopes spread over 10×10 km) a collaboration of ESO, the US and Japan, was originally budgeted at 276 MUSD (year 2000) for both the US and ESO.

The European plans had originally been for an array of 15-m telescopes principally for cosmological purposes, while the US favored 8-m telescopes for sub-millimeter observations also for astrochemistry, which required a very dry and therefore high site; they had already explored the Chajnantor plateau at 5,000 m altitude. That site was chosen for ALMA, while the collaboration resulted in 12-m telescopes. ALMA is currently under construction. As a result, ESO is now spread over four sites, three of which are in Chile.

The extension of ESO's wavelength domain raises the question of its limits. In radio astronomy (few mm—tens of meters wavelength), astroparticle physics, and others, entirely satisfactory worldwide arrangements have been made, including for technology development. So it would be counterproductive for ESO to enter these fields.

One option in ESO's future could be to participate in astronomical ventures in Antarctica. How about joining China on Dome A, if it is as good as expected, in particular for interferometry? But the infrastructure would certainly be costly and a careful analysis of costs and benefits would be needed. Alternatively, the hills above Chajnantor might be worth further study.

In any case the 44 year experience has shown that the model of an intergovernmental organization chosen by the founding fathers has given ESO a stability that is far greater than a simple agreement between research councils could have done. With its nearly pan-European membership, it should have a very bright future.

Misc

A much more detailed description of ESO's development may be found in the following two books:

Blaauw, A.: ESO's Early History, ESO (1991)
Woltjer, L.: Europe's Quest for the Universe, EDP Sciences, Les Ulis, France (2006), which also covers developments outside ESO

The role of observatories in underdeveloped countries

Kevindran Govender

Abstract The Southern African Large Telescope (SALT) Collateral Benefits Programme (SCBP), housed within the South African Astronomical Observatory (SAAO), was established in conjunction with the decision to build SALT. The purpose of the programme is to ensure that tangible benefits are reaped for society from the investment in the telescope and from astronomy in general. This paper will explore the developmental role of the SAAO and SALT by looking at the activities of the SCBP, thus highlighting by example the role that observatories can play in underdeveloped countries. Due to the location of the author, the African perspective dominates this paper and a plan for astronomy in Africa will be discussed. This paper also briefly covers one of the global cornerstone projects of International Year of Astronomy 2009 (IYA2009) entitled "Developing Astronomy Globally", which is aimed at ensuring benefit from astronomy to developing countries.

Keywords Development · Observatories · SALT · Africa · International year of astronomy

1 Context

1.1 Astronomy in society

Astronomy is often regarded as an esoteric science with little visible impact on development. How does one justify the large investments in telescopes,

K. Govender (✉)
South African Astronomical Observatory, Observatory Road,
Observatory, Cape Town, South Africa, 7925
e-mail: kg@saao.ac.za

observatories and astronomical research while there are people, usually in close proximity to these observatories, living in poverty? Yet astronomy remains a subject of interest, in one way or another, to people of all backgrounds, cultures, nationalities and economic standing. Virtually every culture in the world has an established relationship with the stars, moon and sun dating back hundreds if not thousands of years ago. In fact, for as long as human beings have walked the earth we have looked up at the night sky and wondered about the things we saw. In Africa for example, people have used the stars for centuries, be it for navigation, agriculture, or even story telling. Today, astronomy as a field of study has developed into something that attempts to answer some of the biggest questions imaginable. Essentially, astronomy has come to form part of our humanity – our heritage as a thinking species.

Astronomy is also a spark that triggers the curiosity and wonder. Curiosity that is essential for the development of a person and thus the development of a people. Astronomical observatories are ambassadors of astronomy. When people wonder about the skies they usually assume that their questions are being investigated at an observatory – and they are usually correct, even though the things they wonder about have often been answered already. There is a problem of communication here: some concepts that are simply taken for granted by even a student of astronomy are grappled with by people who have not been exposed to astronomy education. Astronomical observatories, through their sheer presence and status within a society should try their hardest to communicate such concepts through education and outreach initiatives.

1.2 Astronomy in Southern Africa

The Southern African region has always played a significant role in astronomy on the African continent, with South Africa as the lead country both for number of publications and investment in astronomy. The three major observatories in Southern Africa are: the South African Astronomical Observatory (SAAO) which is home to the Southern African Large Telescope (SALT), the largest single aperture optical telescope in the Southern Hemisphere; the High Energy Stereoscopic System (HESS) in Namibia, the largest Gamma Ray detector in the world; the Karoo Array Telescope (MeerKAT) in South Africa, a demonstrator towards the bid to host the Square Kilometre Array – which will be the largest radio telescope array in the world.

Within the South African context the government's investment in astronomy has often been controversial though never significantly opposed. Being a country with a number of socio-economic and developmental challenges following a difficult history, South Africa has embarked on a visionary journey in recognition of the importance of astronomy for development. An extract from the White Paper on Science and Technology (1996) captures the vision behind the investment in "big science" projects:

> The well-being of scientific activity in South Africa is intimately linked to material factors. World-wide there is a clear trend for curiosity-driven

research to increase as a function of national per capita income. Nevertheless, there is a danger of adopting too economistic a viewpoint. Even at our current stage of development, there is a need to recognise the importance of the knowledge-generating function of research [...]. Human wonder and curiosity and the ability to recognise serendipitous discovery account for much of scientific progress. Basic enquiry [...] is absolutely essential, particularly at the universities and technikons, which deal with young minds. It is important that fundamental research activity not be regarded as impractical, because it is the preserver of standards without which, in the long term, the applied sciences will also die. Scientific endeavour is not purely utilitarian in its objectives, and has important associated cultural and social values. It is also important to maintain a basic competence in "flagship" sciences such as physics and astronomy for cultural reasons. Not to offer them would be to take a negative view of our future – the view that we are a second class nation, chained forever to the treadmill of feeding and clothing ourselves.

South Africa has also recently brought into law the Astronomy Geographic Advantage Act "*to provide for the preservation and protection of areas within the Republic that are uniquely suited for optical and radio astronomy; to provide for intergovernmental co-operation and public consultation on matters concerning nationally significant astronomy advantage areas; and to provide for matters connected therewith.*"

Such steps are significant for the sustainability of astronomical observatories and signify the important role that is seen to be fulfilled by having such facilities. The observatories in South Africa have in turn always played an active role in the development of education and the public understanding of science. The SCBP has developed into the main instrument for implementing this vision.

1.3 Astronomy for development: an African perspective

The SCBP has initially developed *four* areas: *Education*; *Development of Research*; *Public understanding of Science*; and *Development of Partnerships.* However, there has also been a recently growing fifth area: *astronomy for job creation* (specifically in the field of "astro-tourism").

1.3.1 Education

Education, which is probably the most sustainable form of development, remains one of the most important and crucial challenges facing the African continent. Astronomy, through its innate ability to stimulate curiosity, can be used as a rallying point for the development of a strong learning culture. In Africa, astronomy remains one of the most accessible of all sciences, with the biggest laboratory (a dark night sky) being most available to people in rural areas. Night-time satellite images of the earth clearly illustrate this. The widespread buy-in by the education and outreach sectors in South Africa has proven how powerful the impact of astronomy can be on education.

1.3.2 Development of research

South Africa aims to invest over 1% of its Gross Domestic Profit (GDP) in science and technology. In such an environment astronomy can *stimulate* research in many other fields. For example, the development of large telescopes has had an impact on technological capabilities, requiring high level optical, construction and data manipulation expertise. The mathematics and theoretical physics communities also tend to overlap or interact closely with the astronomy community. In fact the pre-requisite for following a career in astronomy is a solid mathematics or physics background. The development of astronomy thus leads inevitably to the development of those respective university research departments.

1.3.3 Public understanding of science

In many developing countries one finds that modern scientific knowledge amongst the public is not very strong: traditional knowledge or superstitions dominate. Astronomy plays a key role in addressing this often sensitive relationship between traditional and scientific knowledge systems since the sky is something that virtually every culture is already familiar with (in one way or another). Indigenous astronomy is usually a very interesting topic of conversation and an easy way to start a gradual process of introducing people to the modern understanding of the universe. For example, a conversation about beliefs surrounding an eclipse followed by a practical demonstration of the principle has been found to be very effective in stimulating scientific thinking. By bringing to the public the things we know about the universe and more importantly, how we know them (technologies employed, scientific method, etc), we stimulate a public understanding and appreciation of science and technology.

One must stress though that the conversation certainly goes both ways, and an acknowledgement of indigenous knowledge is important in building the relationship with and pride of a people. Indigenous astronomical knowledge in Africa (such as the constellation isiLimela, aka Pleiades, which indicated the start of the planting season) serves as proof of the advanced thinking and observations of our ancestors. In fact, archaeological finds in Timbuktu and other parts of Africa strongly suggest that Africans were doing astronomy hundreds if not thousands of years ago.

1.3.4 Developing partnerships

The 8th United Nations Millennium Development Goal is to "*Develop a Global Partnership for Development*". The need for partnerships are also highlighted in the sentiments and spirit of the African Union and regional co-operation bodies such as the Southern African Development Community (SADC), the East African Community (EAC), and the Economic Community for West African States (ECOWAS). By its very nature astronomy stimulates a need to form partnerships with other countries, not just for the large financial investments but also for the geographic distribution of observatories.

1.3.5 Job creation (astro-tourism)

The field of astronomy in South Africa is growing into a significant attraction for tourists. The number of tourists to the small town of Sutherland (home to the SAAO and SALT) has become so large that a number of small businesses have sprung up all around the town, ranging from accommodation to restaurants to souvenir shops. The demand for tours at SALT during certain times of the year far exceeds the capacity of the observatory which has resulted in the training and involvement of unemployed individuals from Sutherland, and thus to a small tourism business being formed by these people. As the lights of cities block out more of the night skies Africa has been holding on to one of its greatest untapped resources – a dark and clear night sky. A growing number of commercial African tourism companies are making the most of this valuable resource using "amateur" telescopes and attracting business from foreign visitors as well as corporate companies. In Africa there is great potential for "astro-tourism" to be developed just as much as "eco-tourism" already has been. The International Year of Astronomy, in bringing attention to the field, is the opportune time to explore this potential further.

1.4 The International Year of Astronomy 2009

In December 2007 the United Nations General Assembly declared that 2009 would be the International Year of Astronomy (IYA2009). The United Nations Educational, Scientific and Cultural Organisation (UNESCO), together with the International Astronomical Union (IAU), have embarked on leading the celebrations under the theme "*The universe, yours to discover*". IYA2009 marks the four hundredth anniversary of Galileo Galilei's first astronomical observation through a telescope. It will be a global celebration of astronomy and its contribution to society and culture, with a strong emphasis on education, public engagement and the involvement of young people. Events will take place at national, regional, and global levels throughout the whole of 2009.

The International Year of Astronomy 2009 is supported by eleven cornerstone projects. These are global programmes of activities centred on a specific theme and represent the means to achieve IYA2009's main goals.

For this event nearly all observatories and astronomical institutions around the world have begun on some level to prepare. In recognition of the role that this event has to play in developing nations, especially those that have neither observatories nor astronomy departments in their universities, the 11th cornerstone was established, entitled "Developing Astronomy Globally".

2 SALT Collateral Benefits Programme (SCBP)

2.1 Original SCBP

In 1998, the South African government approved the construction of SALT in Sutherland, Northern Cape Province. The government agreed to contribute 50% of the capital construction costs with the remaining 50% sourced from international partners. SALT would enable South Africa to remain internationally competitive in astronomy well into the 21st century.

The original SALT Collateral Benefits Programme was established during the *construction* of SALT and its objectives were directed at the benefits derived by society from *building* this large telescope. It was designed to maximise benefits from the investment of public funds through five main thrusts: industrial empowerment; educational empowerment; public outreach and direct educational benefits; science education visitor centres; and SALT as an African facility.

2.2 Current SCBP

In January 2006, with the SCBP division under new leadership and with SALT moving from the construction to operational phase, the role and purpose of the SCBP was re-evaluated. The new strategy was based on a number of documents including the business plans of the original SCBP, the National Research Foundation (NRF) and the South African Agency for Science and Technology Advancement (SAASTA). Recommendations made from external evaluations of the SCBP, together with input from the original champions of the programme, were also taken into account.

From the outset, the programme always aimed to be seen as being owned by the staff of the whole of the SAAO (and SALT) and managed by the SCBP Division. Thus one should not view the SCBP in isolation from the rest of the SAAO. An additional benefit of this programme is its usefulness to other scientific institutions. If proven to be successful, this strategy could serve as a model that could readily be implemented at any science-research-based institution.

Although the main focus of activities would be in the immediate vicinity of the SAAO sites (Karoo-Hoogland and Cape Town Metropolitan) the programme would seek to make an impact across the continent.

The new mandate was stated as follows:

> To utilize the knowledge, technology and other available resources within the SAAO and similar institutions in order to facilitate education enhancement, science communication and socio-economic development and thereby contribute to the improvement of the quality life of all people within reach, or, in the words of the parent organization, the NRF: *A prosperous South Africa and African continent steeped in a knowledge*

culture, free of widespread diseases and poverty, and proud contributors to the well-being of humanity.

The SCBP would focus on 3 areas: *education in mathematics, science, engineering and technology* to supply the country and the continent with well trained and motivated professionals in increasing numbers; *science communication and awareness* to engage with the public so as to disseminate knowledge in the fields of astronomy and space science; *socio-economic development* to contribute to a better quality of life for all people, especially the disadvantaged.

Various activities of this programme will be briefly described in the next 3 sub-sections under the 3 focus areas.

2.2.1 Education

The biggest impact that SALT and astronomy has had on the people in South Africa has probably been in education. SALT is an icon that is known by virtually every learner in the country – it is even a part of the school curriculum at various levels. To a young South African it means that there are great opportunities in their own country in the fields of science and technology. The SCBP uses this icon to inspire students towards careers in science and technology. Astronomy and SALT are also used as a tool for teaching concepts in mathematics, science and technology at school level. A brief description of a number of ongoing educational activities is given below. There is also a large number of specific events that are not mentioned here.

Ongoing Activities:

- *Educator and learner workshops:* The SCBP conducts workshops for both educators and learners on a regular basis. These vary from building telescopes and spectroscopes to a variety of hands-on activities that help one to understand astronomical concepts such as seasons, eclipses, etc.
- *Work shadow:* The work shadow programme links learners to technical staff and astronomers who they can "shadow" for a few days to learn about the respective career.
- *Science Clubs:* The SCBP coordinates the formation and networking of science clubs across the country. Although the focus is usually on astronomy all science clubs are part of the network.
- *Resource Development:* New educational resources are constantly developed or adapted from existing ones. These resources are shared with all collaborators nationally and internationally.
- *Outreach:* Telescopes are taken out to schools and science clubs who cannot travel to the observatory. Although usually for evening talks and stargazing, daytime activities are also possible, using telescopes with solar filters.

- *Universe Awareness (UNAWE):* UNAWE activities use astronomy as a tool for inspiration and are targeted at the very young children especially in rural areas.
- *Development of the Sutherland Schools:* The SCBP is closely involved in efforts to develop and enhance the schooling in the town of Sutherland. This comes in the form of various projects in partnership with the Department of Education as well as direct interventions such as extra lessons and teacher training.
- *Scholarships:* The SCBP coordinates school level and university level scholarships awarded to deserving students in the fields of astronomy and physics. The school level scholarship serves as encouragement to follow a science career. The SALT international partner universities also offer scholarships for South African students to pursue postgraduate studies overseas.

2.2.2 Science communication and awareness

Public understanding of science and technology, especially in the rural Northern Cape area where SALT is located, is generally very low. There has, in the past, been a feeling amongst the Northern Cape communities that the observatory and SALT were very far from their reality. In order to address this, the SCBP organizes events and activities to better the contact with the community. Star-gazing events in the town are accompanied by talks by astronomers/engineers. Open discussions are held about the community's belief systems as well as the science being done at the observatory. Free SALT tours are also offered for people in Sutherland. All this has greatly improved the public opinion about astronomy and also helps to address one of the major problems in rural South Africa – a lack of self confidence amongst the poor, especially regarding science and technology. Examples of some public outreach activities in Sutherland and across South Africa are listed below.

Ongoing Activities:

- *Tour programme at Sutherland:* Both day and night to the research telescopes and visitor centre in Sutherland. Two visitor telescopes (14 and 16 inch reflectors) are available for stargazing.
- *Public Visits to the Observatory in Cape Town.*
- *Development of resources for the public:* Materials such as information flyers and posters as well as astronomy-related memorabilia are developed for the public.
- *Distribution of resources to various expos and festivals:* SCBP coordinates with partners across the country to have astronomy represented at these numerous events and provides the necessary resources.
- *Media and Press releases:* Regular press releases. These include anything from recent research to astronomical events like eclipses.
- *Indigenous Astronomy:* Projects regarding the investigation and development of information concerning indigenous astronomy.

- *Visitor Centre Developments:* Exhibits at the visitor centre in Sutherland are constantly under development illustrating different concepts of astronomy in a user friendly manner.
- *Monthly "What's up" resources:* Online and print explanations of what can be seen in the sky that month.

2.2.3 Socio-economic development

This focus area, although probably the most important to people in poorer communities, is also the most difficult to address directly. The SCBP is constantly involved in discussions with the communities on activities and programmes that will help to raise the socio-economic status of the area. Close cooperation with the local municipality is needed. The biggest impact on the community since the inauguration of SALT has certainly been in the field of tourism. In the first year after inauguration the number of visitors to the small town of Sutherland jumped from a few hundred to over 13000 visitors per annum. This has resulted in a rapid increase in the number of guest houses, coffee shops and tourism-related businesses such as the Arts & Crafts project for the unemployed. SALT has been involved in a number of socio-economic development projects on various levels, from being a partner in funding proposals to actual implementation on the ground. Some of the highlights are listed below.

Ongoing Activities:

- *Close working relationships with various development stakeholders:* Partnerships have been established to enhance the development in the region. Partners include: Karoo-Hoogland Municipality; National Development Agency; Tourism Enterprise Programme; Succulent Karoo Ecosystem Programme; Military Service Corps.
- *Regular community meetings hosted by SCBP:* These are basically brainstorming sessions where the community has an opportunity to engage with the observatory and come up with ways in which the people of the area can benefit from the observatory being there.
- *Exploration of the various socio-economic roles of SALT in South African Society.*
- *Community Tour guides:* Unemployed individuals from the Sutherland community were selected and trained to conduct basic tours of the facilities. This opportunity increased the employability of the individuals and many found jobs soon after. More guides are brought in as some leave.
- *Tarring of roads:* There are ongoing negotiations which the observatory is involved in to improve the quality of the roads in the area. Better roads mean better accessibility of the observatory and thus more tourism for the town.
- *Sourcing funding for development:* The SCBP is involved in various partnerships, especially with the local municipality to source funding for development projects that would lead to job creation.

- *Arts and Crafts programme:* The tourism generated by SALT has led to initiatives such as the Arts and Crafts programme which enables individuals from the community to develop marketable products for tourists. The SCBP runs a small shop at the visitor centre which markets these products.

3 The role of observatories in Africa

The development of astronomy has to be addressed in two separate areas, one being the development of astronomy research and the other the development of the so called "education and public outreach". This paper mentions both areas but the focus will be on the latter. The case of South Africa is mentioned separately not only because of its strength in astronomy in Africa but also because it serves as an example of what could be achieved by other African countries, and thus other underdeveloped countries.

3.1 The case of South Africa

3.1.1 Astronomy research

The biggest programme in terms of the development of astronomy research in South Africa has been the National Astrophysics and Space Science Programme (NASSP) which is made up of a consortium of 11 universities and 4 research facilities training postgraduate students in Astrophysics and Space Science up to PhD level. The course is based at the University of Cape Town and lecturers come from universities and research facilities all over the country. NASSP trains students from all over Africa and many will return to their home countries to establish astronomy programmes there. For this reason the SCBP works closely with the NASSP students and has established many links with African countries through them. These astronomy research students can carry the programmes run by the SCBP to be used in their home countries.

3.1.2 Education and outreach

This section will cover a sequential roadmap of past activities and events in South Africa that have begun to shape astronomy education and outreach towards a brighter future. The entire astronomy community is involved in this; the SCBP has a coordinating role.

- *Astronomy outreach stakeholder meetings:* The Square Kilometre Array Project in South Africa (SKA SA) initially called together the astronomy community in the beginning of 2007 to discuss possible collaborations. This was followed by a number of stakeholder meetings hosted by SCBP and

SKA SA in an attempt to rally the community together both for IYA2009 and the consolidation of astronomy education and outreach efforts. Participants included observatories, planetariums, amateur astronomy groups, science centres and government departments.

- *Survey of astronomy education and outreach:* Soon after the initial stakeholder meetings an open survey was carried out to establish who was doing what across the country, and to seek input as to where astronomy education and outreach in South Africa should be heading.
- *AstroNet email list:* As the number of stakeholders increased the "AstroNet" email list was set up as a means of open communication amongst all concerned parties dealing with astronomy education and outreach. This email list has since grown dramatically and is now an invaluable resource for news and events in the field.
- *Website www.astronomy2009.org.za:* This website serves two purposes: consolidation of existing South African efforts and the coordination of activities for IYA2009. A section of this website was specifically set aside for the coordination of IYA2009 in Africa.
- *Consolidation of astronomical resources:* A database of available astronomy education materials in South Africa was created and published online.
- *AstroGuide:* As the number of people wanting to get involved in astronomy outreach increased, a resource was set up to provide guidance for any interested parties. The "AstroGuide" is a bulleted list of important points that one should cover when talking about astronomy.
- *AstroCD:* In recognition of the fact that many schools and organisations around Africa have access to computers but not necessarily the internet, a CD full of *free* electronic resources was put together. This CD contains local and international educational resources as well as free software used in the teaching of astronomy. The AstroCD became very popular among schools and astronomy stakeholders and hundreds of CDs to date have been distributed free of charge. The second version has recently been released.
- *"Essential Astronomy" Outreach Training Workshops:* In preparation for IYA2009 and to ensure greater reach of astronomy, a number of workshops were held across the country for education and outreach staff who dealt with educators, learners and the public on a regular basis. The idea was that the instructor for any general science education or outreach programme should be equipped with the tools and knowhow to use astronomy in their activities. This workshop proved incredibly popular and has since reached over 200 individuals from across South Africa and a number of other African countries including Kenya, Namibia, Nigeria, Uganda, Zimbabwe, Mozambique and Lesotho.

The aim of such activities is to enlarge and equip the pool of expertise in astronomy outreach and then to utilise this pool for such activities as "star parties" in public venues and the development of more curriculum-specific school resources.

3.2 A plan for Africa

A plan has been developed for "Astronomy Education and Outreach in Africa" which was built on input from a number of people across the continent (and beyond) who are involved in astronomy, space science or general science education in some way. This plan was developed for IYA2009 but it should last well beyond 2009. The need was influenced by the glaring lack of representation by African countries on the IYA2009 and astronomy scenes. In the build-up towards IYA2009 the developers of this plan wanted to ensure that Africa had a voice and reaped as much benefit as possible from this global event.

Since then a task group has been formed to take this plan forward. In building up the astronomy community in Africa, the task group aims to use the subject to spark an interest not only in science, engineering and mathematics disciplines, but also in education in general. Thus IYA2009 should be used as a launching pad for a network of African individuals and organizations who intend to work together into the future using astronomy to enhance education in Africa. With this in mind the theme that has been chosen for the whole plan is "Astronomy for Education".

4 Developing astronomy globally

This cornerstone project of the IYA2009 aims at developing astronomy on three levels through networking and partnerships with all those individuals and organisations around the world already active in the area. The three different levels or *focus areas* of development identified for this project are: *professional development* which will address tertiary education, research training and research infrastructure in order to build professional astronomy research capacity within the country; *development of the public understanding of astronomy* which will target the general public and take the form of activities and events that stimulate an interest in astronomy, as well as the development of an amateur astronomy community and the development of astronomy communicators; *school-level education development* which will address primary and secondary education. The following table (Table 1) maps out possible actions for each focus area depending on the respective levels of development in a particular country.

5 Conclusion

All activities mentioned in this paper have been driven by astronomical observatories in one way or the other. The role of observatories in underdeveloped countries is quite obviously tied in with the role of *astronomy* in those regions. The significance and relevance of astronomy for development has been demonstrated by example and forward thinking. It is clear that this

Table 1 Examples of actions to be taken depending on focus area and level of development

	Professional development	Development of public understanding of astronomy	School-level education development
Well established countries	- use as regional hub	- use as regional hub	- use as regional hub
Countries in need of support	- Link with IAU network - Student/researcher exchange programmes coordinated by regional hubs	- provide training for local "champions" - feed public info and stories to champions regularly - train facilitators to run public events - link with relevant cornerstones	- provide training for local "champions" - provide education resources and networks (link with Galileo Teacher Training and UNAWE Cornerstones for IYA2009)
Countries with non-existent programmes but with strong potential	- Provide guidance on student and research opportunities internationally (with clause to return to home country) - Combine with IAU Commission 46's programmes	- provide training for local "champions" - feed public info and stories to champions regularly - train facilitators to run public events - link with relevant cornerstones	- training workshop on "astronomy in the classroom" - link with activities of region - provide educational resources - introduction of astronomy into the school curriculum
Countries with non-existent programmes and with limited potential	- High-level discussions with government departments of science and education - Explore astronomy status in neighbouring countries and likelihood of collaborations		

field has the potential to make an impact on society, and as custodians of the field, this great potential falls into the hands of the observatories.

The only certain way of ensuring the sustainability of and investment in astronomy is to ensure that the developmental benefits of the field are realized. To realize these benefits observatories and astronomical institutes around the world should develop and support programmes similar to the SCBP. The SAAO has been exemplary in reaching towards their vision of ensuring tangible benefits to society by investing themselves in the SCBP. It is already evident that such programmes are paying off in the form of community support and thus governmental support for more astronomy.

However, besides what one can get out of such visionary actions, it is up to observatories, or their leaders, to take the initiative to fulfil one of the most fundamental missions of astronomy – satisfying human curiosity. Because in taking a step back it is clear that at the heart of astronomy is *human curiosity*. It has given rise to the field, it sustains the field, and it leads us to expand the field. Can we honestly say we live up to the goal of satisfying human curiosity if we do not impart what we learn to *all* humans on the planet? For this reason programmes like the SCBP need to be encouraged at all observatories, especially those in underdeveloped countries, so that we can impart what we discover to as much of the human population as possible.

Acknowledgements I wish to acknowledge the organizers of the conference for their dedicated and hard work and thank them for inviting me. I would also like to acknowledge Prof. George Miley for his incredible leadership and excellent discussions regarding the development of astronomy. Lastly, and perhaps most importantly, I wish to acknowledge Dr Peter Martinez for his guidance and mentorship regarding the development of astronomy in Africa. His willingness to share his many years of experience in the field has contributed significantly to the projects mentioned in this paper.

References

1. South African Department of Arts, Culture, Science and Technology: White paper on science and technology (1996)
2. Government of the Republic of South Africa: Astronomy geographic advantage act. Act No. 21 of 2007 (2007)
3. African Union: New partnership for Africa's development. http://www.nepad.org/
4. South African Department of Science and Technology: The ten-year plan for science and technology. http://www.dst.gov.za
5. South African Astronomical Observatory: African starlore. http://www.saao.ac.za/public-info/sun-moon-stars/african-starlore/
6. Holbrook, J.C., Urama, J.O., Medupe, R.T.: African cultural astronomy. In: The Timbuktu Astronomy Project: A Scientific Exploration of the Secrets of the Archives of Timbuktu (2008)
7. National Astrophysics and space science programme. http://www.star.ac.za/
8. The United Nations Millennium Project: http://www.unmillenniumproject.org/goals/index.htm
9. Martinez, P.: Building capacity for astronomy research and education in Africa. IAU Special Session no. 5 (2006)
10. The United Nations: Resolution adopted by the General Assembly 62/200, International Year of Astronomy (2009)

Exp Astron (2009) 26:133–147
DOI 10.1007/s10686-009-9158-5

REVIEW ARTICLE

Telescopes in the mirror of scientometrics

Virginia Trimble

Received: 24 February 2009 / Accepted: 25 February 2009 / Published online: 24 March 2009

Abstract Counting papers and citations is one way to estimate the significance of particular astronomical telescopes and other facilities in the long time gap between the verdict of history and the referee's report on your most recent proposal. This has been done for 2,184 observational astronomy papers published between 1960 and 1964 (with 14,237 citations in 1965–1969) and the numbers looked at in various ways. The extreme dominance of California in optical astronomy and of the UK and Australia in radio astronomy provides the background against which ESO, NOAO, NRAO, and A&A were founded, with equality of access to facilities having increased enormously in the intervening 40 years, but inequality of results having increased slightly. A number of other factoids about astronomical publications, the community, and their environments surfaced during the counting process, and a subset reported here, including a few pertaining to the more distant past.

Keywords Scientometrics · Citation analysis

1 Introduction: the astronomical world in 1960

The time frame for this study was determined by both personal and public issues. First, I started graduate school in 1964, marking (for me!) the cut between "history" and "current events" in astronomy. Second, the first five-year compilation of citation numbers from Science Citation Index covers

V. Trimble (✉)
Department of Physics and Astronomy, University of California,
Irvine, CA 92697 USA
e-mail: vtrimble@uci.edu

V. Trimble
Las Cumbres Observatory, Goleta, CA, USA

1965–1969. Third was the availability of journals—the "Russian" one in translation as Soviet AJ (though BAC had some papers in Russia (I read, but with difficulty), and the library stack contents at the Space Telescope Science Institute (STScI) and the University of California Irvine (VCI). Fourth was the rapid ramp-up of radio and space astronomy. And fifth the fading out of observatory publications as distinct from the journals we now read.

Between 1800 and about 1970, more than 200 observatories distributed their own publications—from Aarhus and Abastumai to Zo-se and Zurich (including some radio observatories)—though not all at once. At least 70 contained papers significant enough to be cited in the Kuiper compendium reviews. The L's alone included Lick (Trumpler on interstellar absorption), Lowell (Slipher's redshifts), Lund (Holmberg's binary galaxies), Liege, Lille, Leiden, Lemberg, Lembang, Lausanne, Leningrad, Lehigh, Le Hough (Peridier's private observatory, where de Vaucouleurs got his start), La Plata, Leander McCormick, Leyton, Lisbon, Lyon, London, and Louvain. An analysis of publications from the 1940s or earlier would therefore require access to one of the very few, very complete libraries of astronomy. And neither SCI nor ADS can provide lists of citations to these papers except from long after they were published. The last of these series to contain essential catalogues and such was probably that of the Dominion Astrophysical Observatory. Remarkably, quarterly issues from Skalnate Pleso in the Slovak Republic still arrive regularly.

The transition from observatory publications to broader-based journals was not entirely smooth. In 1939, Babcock was unhappy that his thesis (on the spectrum and rotation of M31) appeared as a Lick Observatory Bulletin rather than in ApJ (according to a letter he wrote when asked why he had moved so firmly away from extragalactic astronomy); and, in the same year, Popper was unhappy at his thesis being sent to ApJ when he would have preferred publication as an LOB (according to a remark he made on a conference tour bus when I mentioned the Babcock story).

The astronomical world in 1960 was, in many ways, very different from today's. Only the sun had been seen in ultraviolet, X-, and gamma rays; the AS&E (Giaconni et al.) and NRL (Bowyer et al.) rockets that saw Sco X-1 and Crab Nebula X-rays and the Arnold and Kraushaar gamma ray background belong to our period [3, 4, 7, 11]. In the radio regime, there had been very few optical identifications; log N – log S was still disputed (in unfriendly terms!); and no QSRS's (1963), nor CMB (1965), nor pulsars (1968) had yet been found. Rapid advances in 1960–1964 occurred in mapping and polarization of the Milky Way and emission from planets including spectra; sensitivity to polarization improved as well as angular resolution; OH was found in the ISM; and many more extragalactic sources received optical identifications.

Women were working in all the main wavelength bands, though not in enormous numbers. Among the space pioneers were Marcia Neugebauer (direct detection of the solar wind), Phyllis Freier (cosmic rays), and W.S. Cameron (solar system from Mercury spacecraft). There were women radio astronomers in most of the countries engaged in the field: Ruby Payne-Scott (Australia),

Patricia Leslie (UK), E.M. Berkhuijsen (Netherlands), L. Bottinelli (France), Z. Placova (Czechoslovakia), T.A. Lozinskaya (Russia), and N.H. Dieter and M.A. Kaftan-Kassim (USA). All appeared as authors, generally first authors, of papers in the current data base.

This applies also to the following women optical observers: Anne Underhill (NL & US), A.E Ringuelet-Kaswalder, Y. Andrillat, M-C. Lortet-Zuckerman (France), S. Torres-Peimbert (Mexico), M. Jaschek (Argentina), M. Hack (Italy), Choko Fujita (Japan), Wilhelmina Iwanowska (Poland), many Americans (Margaret Burbidge, Vera Rubin, Dorrit Hoffleit, and B.T. Lynds perhaps most widely known but also Barbara Middlehurst, K.C. Gordon, Sarah Lee Lippincott, Anne Pyne Cowley, and Edith Flather), and at least 18 first authors with surnames ending in ova, skaya, etc. in the Russian, East German, Polish and Czech journals, setting a lower limit to the number of women making significant contributions there. I mention by name only A. Szczepanowska (for reasons only G.B. Shaw fans will guess).

Some other things have not changed much. There were visa issues for conferences in the USA (Dirac having been denied admission); the Kefauver committee was worried about drug prices and drug advertising; the NASA budget and the fraction devoted to space science were bouncing around (though with a peak at something like 4% of GDP during the Apollo program that has never been equaled since); and fraudulent papers were sometimes withdrawn (e.g. Science 133, 941). The estimated cost of carrying out all the items in the Whitford report (astronomy for the 1960s) was, of course, a considerable underestimate. The initial figure, however, was only(?) $227 million. Keep in mind, though, that typical conference registration fees were $3–4 and hotel rooms cost $8–15 per night, even in New York.

2 (Relatively) deep time

"Deep time" can mean thousands of years in contrast to a four-year presidential term or millions of years in contrast to the 10^4 year duration of civilizations. In the present context it means items where I came across something interesting between 1608 and 1960.

The first modern astronomers, like sixteenth, seventeenth, and eighteenth century musicians, were churchmen, nobility, or royalty (remember Tycho's competitor for the parallax of SN 1572, the Landgrave of Hesse), or supported by them. They published in books and broadsheets (some shown in Van Helden's presentation) and wrote to colleagues and friends, sometimes in anagrammatical form. By the middle of the nineteenth century, there was a division between employees of government observatories and universities and wealthy amateurs (like Carrington the brewer and Draper the physician), and publication was more often in the form of articles in national academy and observatory publications. Now, of course, a large fraction of astronomers (and also of musicians) are university employees, who publish in a rather small number of weekly to monthly journals.

The first state-supported observatories were intended to assist navigation and published almanacs and ephemerides, for instance Paris 1679 (Connoissance des Temps 1679), London 1750 (Almanac 1767), San Fernando 1757 (publications 1773), Coimbra 1772 (publications 1904), and, well behind the pack, the US Naval Observatory, founded in 1830 (Almanac 1853). Meanwhile, private observatories and ones supported by educational institutions continued to proliferate (and in the niche of British University Observatories, I recommend the recently-published thesis of conference participant Hutchins [8]). The US had, perhaps, an advantage in its multi-tiered government structure that allowed dipping into pockets of the nation, the states, and the cities.

The first scientific societies and their publications date from the same period: The Royal Society of London, founded 1660 (and publishing Philosophical Transactions from 1665) and the Paris academy and its Comptes Rendues soon after. Purely astronomical societies and publications (sometimes coupled, sometimes not) come later. There is, I think, an interesting divergence between what happened in the United Kingdom, with a single Astronomical Society of London dating from 1820 (the Royal Astronomical Society after 1831) and its Memoires (1821) and in the divided areas called Germany and Italy, where there were very many local observatories, and the first journal (and now the oldest one in continuous publication, Astronomische Nachrichten, founded in 1821) was an extension of the letters-to-distant friends system, invented by Heinrich Christian Schumacher. He asked European astronomers to send him their letters in any European language; he then edited and assembled these and sent them back to everybody).

The RAS illustrates a very general principle that it takes about 100 people to establish a viable society with some sort of paper trail and regular meetings. Other organizations where you are like to find astronomer and for which this was true include the American Association for the Advancement of Science (1850, initially naturalists, geologists, and astronomers), the Italian society (1872, focused on spectroscopy and so arguably the first astrophysical society), Royal Astronomical Society of Canada (1868), Belgium (1880), France (1887), Astronomical Society of the Pacific (1895), American Physical Society, (1898), American Astronomical Society (1899), and the International Astronomical Union (1919).

National astronomical societies and some associated journals proliferated just before and just after World War I (Japan 1908, South Africa 1922, China 1922, New Zealand 1920, Poland 1925, Netherlands, 1921, USSR, 1924, and the Scandinavian journal Arkiv 1916). All of these survive in some form. Realignments after 1990 have added separate societies in the Czech and Slovak republics and parts of the former Yugoslavia and USSR, some with perhaps subcritical membership. Additional details of some of these "community" issues appear in [16].

Also stretched out over the entire four centuries is the gradual incorporation of women into astronomy. This occurred, I think, in three slightly overlapping stages. At first, the only choice was working with a father, husband, or brother. Caroline Herschel is the best known of these, but Sophia Brahe, Elizabeth

Hevelius, and Margaret Huggins are other examples. Next come formal, institutional helpers, the "computers" (also found in laboratories looking at cloud chamber tracks and in other projects with large quantities of data to extract or classify). Henrietta Swan Leavitt, Annie Jump Cannon, Williamina Paton Stevens Fleming, Antonia Caetana de Paiva Pereira Maury, and Donna Ebert (who worked with S. Chandrasekhar) on the theoretical side are astronomical examples of women whose work was important enough and independent enough that we remember them by name. The last of the traditional Mt. Wilson computers were still on duty when I started graduate school at Caltech in 1964. Finally comes the stage of full participation in research science, where the (female) astronomer asks a question, figures out how to answer it, and presents the work to her colleagues. Cecilia H. Payne (later Gaposchkin) was arguably the first of these with her 1925 thesis showing that virtually all stars have about the same chemical composition (and that H and He dominate). Bridgers across the phases include Maria Mitchell (who began working with her father, was a computer at USNO, and ended as professor at Vassar) and Ethel F. Bellamy of Oxford, who began working with an uncle, was a computer under Turner, and acquired solo responsibilities during World War II when most of the men were called to war service.

We end this section with Table 1, which shows total citation numbers in 1965–1969, first, to the works of four famous telescope pioneers; second, to five papers between 1912 and 1939 that I think were enormously important and did not appear in the journals now of high prestige (notice that John Herschel trumps William, and that only Trumpler's interstellar absorption left much of a footprint during this period); and third, to nine papers published in ApJ or AJ between 1953 and 1965 and singled out as highlights in the centenary

Table 1 Some giants' shoulders, citations 1965–1969

Galileo (all)	50
Huygens (all)	19
Wm Herschel	17
John Herschel	22
Slipher galaxy spectra (1912–1920) Lowell	4
Shapley Mt. W. Contr. 1918±	7
Payne 1925 Harvard Monograph	1
Hubble Proc. NAS 1929	3
Trumpler 1930 Lick $A_V = 1^m/_{kpc}$	30
Zwicky 1933 Helv. Ph. Acta DM	5
Babcock M31 rotn Lick 1939	5
Johnson & Morgan color system '53	184
Babcock solar mag. Cycle '53	29
HMS H = 180 1956	160
Wilson/Bappu effect 1957	46
Abell 1958 clusters of Gs	34
Eggen L-B S, MW history 1962	49
Neugebauer +'65, IR YSO cat.	55
Penzias & Wilson 1965	345
Gunn-Peterson '65	59
Hewish, et al., Bell pulsars 1968	99

volume edited by Abt [1]. Hewish et al. on pulsars is tucked at the end just to show that it is possible to accumulate many citations quickly. While two of the highly cited papers are cosmological (Penzias & Wilson finding the CMB and Humason, Mayall, & Sandage correcting the Hubble constant to a value of 180 km/s/Mpc) the third is Johnson & Morgan on color systems for stars.

3 The 1960–1964 papers and 1965–1969 citations

This section and the next are mostly about decisions: which journals to consult, which papers to count, which facilities to keep track of, and how to apportion papers and citations among facilities when more than one contributes to a paper. Which journals was entirely a matter of availability: ApJ (and Supplements), AJ, PASP; those that in 1969 folded into A&A (AN, Zs f Ap, Ann d'Ap, BAN, and BAC, with Bull. Astron., J. Obs., and Arkiv not available), MNRAS, Observatory, Astron. Zh (= Sov. AJ), Acta Astron., Nature, Phys. Rev. Letters, Science, PASJ, Icarus, and Australian Journal of Physics (the radio physics part). In the process of looking up citation numbers for these, I learned that some authors of papers in these journals had also published observational papers in physics and geophysics journals, in academy proceedings, and in observatory publications, but for none of them these most of the papers or (with a couple of very senior exceptions) their most cited.

As for which papers to count, I left out meeting abstracts (which appeared in AJ, PASP, Observatory, and Ann. d'Ap. regularly) but included meetings proceedings papers. In the days before Astro-ph and all, a few of these abstracts were cited a dozen or more times in 1965–1969, but, again, not many. Otherwise, every paper, even one-page ones, was included if it reported or analyzed observational data at any wavelength (including cosmic rays) and the telescope(s) used could be identified.

Table 2 shows the distribution by journals of numbers of papers, citations, and citations per paper, largest citation numbers, and numbers of papers with zero citations in each journal. This last addresses the claim "most papers aren't read by anybody." In fact, 22% went uncited (vs. 6% in recent data), but these are concentrated in five or six journals. Citation numbers were most difficult to determine for Sov. AJ. SCI recorded citations for both the Russian and translation versions, but the volume numbers, page numbers, and even years are not the same. I think, though, I caught them all. Self-citations are included for all papers, and one author (no, I won't tell you who) had only self-citations in the quinquennium.

What was an appropriate set of subtopics to look at? The choice was to use the same ones as in [17] and earlier papers in that series to facilitate comparison, though the result was one white dwarf paper (by Volker Weidemann), two on exoplanets (Peter van de Kamp reporting false alarms around Barnard's star), none on GRBs, and something of a pile-up under "stars." Topic assignments sometimes required a bit of *20–20* hindsight to decide, for instance, whether a set of radio angular diameters pertains to SNRs or AGNs.

Table 2 Distribution by journals of number of papers

Journal	Citations	Papers	*C*/*P*	No. 0's (%)	Largest C's
ApJS	377	21	17.95	2 (10)	52, 43
Aust J. Ph.	241	15	16.07	1 (7)	88, 45
PRL	958	66	14.52	12 (18)	85, 57
Nature	1,603	192	8.35	37 (20)	65, 64
Science	463	72	6.42	18 (25)	72, 60
ApJ	4,964	500	9.90	43 (9)	139, 109
AJ	1,341	222	6.04	46 (21)	50, 44
PASP	545	234	2.33	82 (35)	39, 30
Icarus	203	19	10.68	1 (5)	38, 30
MNRAS	1,426	147	9.70	14 (10)	137, 50
Observ.	176	94	1.87	32 (34)	12, 10
BAN	338	47	7.19	8 (17)	51, 25
Ann d'Ap	378	75	5.04	13 (17)	54, 29
Zs f Ap	197	67	2.94	18 (27)	30, 23
PASJ	270	67	4.03	15 (22)	14, 14
A Zh/Sov AJ	622	173	3.60	50 (29)	32, 31
Acta Astr.	47	45	1.04	29 (64)	9, 5
BAC	45	50	0.90	20 (40)	9, 8
AN	43	58	0.74	33 (57)	6, 6
Total	14,237	2,164	6.58	474 (22)	

The recent data excluded solar astronomy, which uses a nearly disjoint set of telescopes etc., but for 1960–1964 it turned out that radio and space astronomy had a very large component focused on the sun, heliosphere, magnetosphere, and sun-earth connections, and that these papers were the break-outs into Nature, Science, and Phys. Rev. Letters. They were thus included as separate topics: sun–earth radio and sun–earth space. "Service," incidentally, means catalogues, coordinate systems, and such.

Table 3 shows numbers of papers and citations by subfield, but notice that only the 1,248 papers (a bit more than half the total) with at least four citations in 1965–1969 are included. Those papers averaged 13.8 citations per paper

Table 3 Papers with ≥4 citations in 5 years

Category	*C*	*P*	*C*/*P*
Radio (solar–earth)	294	32	9.2
Radio other	4,083	305	13.4
Space (solar–earth)	669	44	15.2
Space other	1,172	53	22.1
Optical all	10,998	814	13.5
Optical service	374	21	17.8
Cosmology	513	24	21.4
Active galaxies	1,522	55	27.7
Other extragalactic	896	89	10.1
MW, ISM, YSO	1,636	126	13.0
NS/BH/SNe/SNR/CRs	898	48	18.7
Solar system	2,162	169	12.8
Exoplanets (vdK)	16	2	8.0
WDs (Weidemann)	9	1	9.0
Stars, star clusters, CBs, CVs, PNe	2,896	267	10.8

over the 5 years. In comparison with the average, cosmology, active galaxies (meaning both QSRS and radio galaxies), and neutron stars, supernovae, etc. and cosmic rays (for which the electron component was finally found in this period) count as hot topics; solar system and Milky Way are about average; and stars are already an endangered species. Coming down to 2001–2006, the solar system and stars have become less popular (in paper numbers and C/P), the Milky Way has roughly held its own, and extragalactic topics have increased both their fraction of the total astronomical literature and their relative citedness.

4 Facilities

How many telescopes (etc) contributed to the literature, and which ones did I keep track of individually? In 1960 there were, say Kuiper and Middlehurst [12], 100 sites with at least one optical telescope of 20″ or larger (a total of 155 such telescopes), and I kept track of each individually, though many have been summed in the following table. The KPNO 36″, Japanese 1.88 m, and others gathered their first photons during the period and were added to the ones tracked. There were also something like 200 smaller telescopes whose data appeared in at least one paper, and these were associated either with observatories having larger telescopes or grouped by geographical regions. Curiously, the total number of optical telescopes contributing to data published in 2001–2003 is not enormously larger.

The radio sites numbered about 75, only a subset appearing in [12] with perhaps 100 widgets. Cambridge most notably built and dismantled rapidly, though younger colleagues assure me that saying only ASA 800 film could be used to photograph them (hence the relatively poor images in both Sullivan's presentation and mine) will not be understood by their generation. Most of the radio telescopes in 1960 were in the UK and Australia; then the US roars in (as we seem to have been doing since at least 1917). There were European facilities at Nançay and Dwingeloo, Ondrejov and Bonn. And very many places engaged in only solar radio astronomy, e.g. Pisa, Cracow, Torun, Belgrade, Meudon, Berlin, Arcetri, Kiruna, Slough, Aldershot, Oslo, Nagoya, and Sagamore Hill. Again the total numbers 40 years later were only about 20% larger.

In space, very few things were used more than once. There were lots of balloons and aircraft; very many rockets (mostly Aerobees but also Skylarks, Journeymen, Atlas, and Veronique); and satellites with other primary purposes that turned out to be doing astronomy, including multiple Rangers, Luniks, Pioneers, Sputniks, Discovers, Vanguards, IMP, Injun, Mercury, Mariner, Alouette, Telstar, Echo, and Transit. This list has all the devices that contributed to the 1960–1964 papers and also some from Seaborn [13], which lists all of the space project launches (including failures) from 1957 to 1967 and is undoubtedly the most valuable free publication I ever picked up off an exhibitor's table! By the end of 1964, 51 Kosmos satellites and four Vela's had been launched, but we did not yet know that they were doing astronomy!

Over the next 40 years, the most conspicuous change was the rise of dedicated astronomical satellites. Solar (OSO) and ultraviolet (OAO) ones came first, but Uhuru in 1971 is the one you are perhaps most likely to remember or have heard of, right on down to Fermi (the republic formerly known as GLAST). For what it is worth, I thought the space X- and gamma-ray talks said too little about MIR/KVANT and other aspects of the Soviet program, but then so did I, for there were no "space" papers in Sov. AJ. MIR/KVANT of course belongs to a later period, as do even OSO-2 (Feb. 1965 launch and observation of solar UV, X, and gamma-rays) and OAO-2 (December 1968 launch for stellar UV observations).

5 Papers and citation numbers by telescope (etc...)

At this point we need (a) to do a bit more work, looking up citation numbers for 1965–1969 for all the 1960–1964 papers recovered from the journals listed in Section 3 and (b) to make another decision—how to apportion credit among the telescopes when more than one contributed to a given paper. Folks who care about a single telescope generally give it full credit for every paper that uses its data, but we adopt here the same algorithm described in [17]. Space, radio, and optical bands are considered separately, but within each band, all telescopes contributing receive equal partial credit for both the paper and the citations: Thus a paper with data from the Palomar 48″ Schmidt, the 200″, and the Mt. Wilson 100″ with 24 citations counts as one third of a paper and eight citations for each. If C/N is not an integer (25 citations and three telescopes, for instance), the extra one(s) go to the first telescope(s) mentioned by the authors, thus in the hypothetical example the Schmidt would get nine and the other two telescopes eight each.

Table 4 presents the optical numbers and Table 5 the radio ones. There is binning of multiple small telescopes on a single site and even by geographical area to keep the tables manageable. The "widget" class includes an early Ray Davis limit on solar neutrinos, Robert Dicke's solar oblateness measurements, and the Hanbury Brown interferometer at Narrabri. I separated high-altitude sites (Haleakala, White Mountain, Pic du Midi, Chalcultaya) to see whether the extra effort required to operate there paid off. The answer would seem to be yes, but there are only 18.5 papers involved.

There are no major surprises to be found in these tables. The Mt. Wilson 100″ yielded the most optical papers, but the Palomar 200″ the most-cited ones. Similarly at radio wavelengths, Jodrell and Australia produced more papers but Cambridge the most-cited ones. Overall, radio papers were cited almost twice as often as optical ones. This was not true for 2001–2003, which led to some awkwardness between authors and referees presumably from the radio community.

Table 4 Optical papers and citations, pro-rated

Facility	Citations	Papers	C/P	N_{max}	First light
Palomar					
200″	1,252	88.7	14.1	139, 100	1948
48″	459	79.9	5.7	69, 35	1948
Other	170	17.2	9.9	34, 22	
Mt. Wilson					
100″	1,127	164.9	10.8	89, 74	1917
60″	330	37.2	8.9	103, 25	1908
Other	10	9.8	1.0	6, 2	
McDonald					
82″	415	79.6	5.2	29, 27	1939
Other	178	20.3	8.8	25, 25	
Lick					
120″	261	49.6	5.3	34, 24	1959
Crossley	225	42.4	5.3	44, 20	1898
Other	90	28.0	3.2	15, 7	
Arizona	301	56.1	5.4	24, 21	
Other US	593	162.2	3.7	61, 40	
Canada	105	23.2	4.5	10, 10	
South Africa					
74″ Radcliffe	224	49.9	4.5	29, 16	1948
Other	248	64.1	3.9	21, 20	
Stromlo (etc.)					
74″	144	36.4	4.0	16, 11	1955
Other Aust.	169	44.8	3.8	18, 11	
USSR	346	108.1	3.2	47, 34	
E. Europe	109	109.7	1.0	10, 9	
W. Europe	422	118.9	3.6	35, 30	
Asia, So. Am. etc.	94	40.6	2.3	11, 10	
High altitude	293	18.5	15.8	43, 29	
Widgets	119	4.0	29.7	57, 51	
TOTAL	7,684	1,454	5.8		

Comparison of these numbers with 21st century ones has something to tell us about changes in equality of opportunity and achievement. In each time frame, the top seven or eight optical telescopes yield about 40% of the papers and 50% of the citations, but their geographical distribution is very different. All the early high achievers were in the USA, and California alone was responsible for half the citations and a third of the papers. Now the high-productivity telescopes are scattered among Chile, Hawaii, the Canary Islands, Australia, and outer space, and nearly all of the world's astronomers can apply for time on many of them, while Lick, Mt. Wilson, and Palomar had very restricted user communities. This casts some light on what Allan Sandage meant, in an oral history discussion of steady state cosmology [10]. When he said that he didn't think steady state was ever taken very seriously in California, he was referring not to theorists but to observers with access to the large telescopes.

The top 1% of citation-gatherers are listed in Table 6. They include space, radio, and optical studies, most from well-known facilities, and have garnered 12% of all the citations. The top 1% of the papers for 2001–2003 are again spread over all wavelengths, come mostly from well-known (and well

Table 5 Radio: All papers and citations pro-rated

Facility/location	Citations	Papers	C/P
Jodrell	446	41.97	10.6
Cambridge	592	35.34	16.8
Australia	623	41.94	14.9
Other European[1]	307	44.37	6.9
NRAO	272	33.39	8.2
OVRO	858	35.81	24.0
NRL	120	20.0	6.0
Harvard	134	17.0	7.9
Other US[2]	592	50.4	11.7
Canada, NZ, UK[3]	80	14.0	5.7
USSR	204	48.9	4.5
Total	4,228	383.1	11.0

(1) Dwingeloo, Nançay, Ondrejev, Bonn, (2) Goldstone, U. Mass, U. Mich, Ohio State, MIT, Florida State, Yale, Boulder, Clark Lake, others, (3) Ottawa, DRAO, Algonquin, Malvern, Christchurch. In addition, many places were doing solar radio: Pisa, Cracow, Torun, Belgrade, Meudon, Berlin, Arcetri, Kiruna, Slough, Aldershot, Oslo, Nagoya, Sagamore Hill. US papers are concentrated in later years. About 100 devices on 80 sites

supported) facilities; and are responsible for 15% of the citations. Thus while equality of access has improved over the 40 intervening years, achievement has become a bit more unequal.

Table 7 lists the most-cited paper from each of 23 lower-profile astronomical telescopes, perhaps illustrating the principle mentioned by other speakers that the right people with the right ideas are the most important astronomical

Table 6 The most cited papers (1%)

N	Journal	First author	Facility	Subject
139	ApJ	Greenstein	200″	Quasars
137	MNRAS	Conway	Cambridge, Jodrell, OVRO	Quasars
109	ApJ	Matthews	OVRO, 200, 60, POSS	Quasars
103	ApJ	Oke	Mt. W. 60	Calibrations
100	ApJ	Sandage	200″	H_0
96	ApJ	Kellermann	OVRO	Radio sources
89	ApJ	Kaplan	Mt. W. 100″	Mars atm.
88	Aust. JP	Bolton	Parkes, POSS, Stromlo 74″	Catalogue
85	PRL	Giacconi	Aerobee	Sco X-1
83	ApJ	Matthews	POSS < 200″, 120″, Stromlo 74″	AGN
72	Science	Bowyer	Aerobee	Crab
65	Nature	Schmidt	200″	3C273
64	Nature	Greenstein	200″	3C48
63	Nature	Hewish	Cambridge	Scintillation
61	ApJ	O'Dell	Pine bluff	PNe
60	Science	Neugebauer, M.	Mariner II	Solar wind
59	ApJ	Maltby	OVRA, POSS	Radio sources
58	ApJ	Heeschen	NRAO 300′	Radio galaxies
58	Nature	Bowyer	Aerobee	Crab, Sco X-1
57	PRL	Davis	C_2Cl_4 tank	Nu limit
54	Ann d'Ap	Lequeux	Nancay	Radio sources
54	PRL	Ness	IMP-1	IPM

1754 or 12% of all citations to top 1% papers vs 13–15% for 2001–2003

Table 7 Significantly cited papers from less obvious telescopes

Number (*N*)	Journal	First author	Facility	Subject
61	ApJ	O'Dell	Pine bluff	PNe
54	Ann d'Ap	Lequeux	Nancay	Radio sources
44	AJ	Kron	Crossley 36″	Star clusters
43	Nature	Metzger	Ranger 5, 3	Gamma bkgd.
38	AJ	Gehrels	Goethe Link 36″	Solar system
38	AJ	Eggen	Cape 24″, 18″	Stars
31	Nature	Smith, HJ	Harvard patrol	3C 273
30	Sov AJ	Moroz	Sternberg 1.25m	Solar system
29	MNRAS	Feast	Radcliffe 74″	Galaxies
29	MNRAS	Ingham	Chacultaya	Solar system
29	AJ	Eggen	Cape 24″ 18″	Milky way
29	ApJ	Murray, BC	White Mt. 19″	Solar system
28	MNRAS	Griffin, RF	Cambridge 36″	Stars
27	Ann d'Ap	Courtès	OHP 1.93	Supernovae
25	ApJ	Brandt	McD 10″	Galaxies
25	BAN	Van Woerden	Dwingeloo	Milky way
25	Ann d'Ap	Weinberg	Haleakala	Solar system
24	ApJ	Slettebak	Perkins 69″	PNe
23	ApJ	Collins	Washburn	PNe
24	Sov AJ	Basharinov	8 mm something	Solar system
24	AJ	Zabriskie	Slope Obs.	Solar system
23	Zs f Ap	Elsässer	Boyden 30 cm + Jungfraujoch	Milky way
22	MNRAS	Westerlund	Uppsala So. Sch.	Stars

facilities. For what it is worth, three of Tables 6 and 7 first authors were at the meeting '400 years of Astronomical Telescopes': Giacconi, Lequeux, and O'Dell. But the Gehrels of the Table is the father of participant Gehrels.

The binning of years and facilities has concealed many details. One of my favorites might be called "Hit the ground running?" Among the radio telescopes (etc.) represented, many first came on line during the quinquennium and so contributed nothing the first year or two and a great deal toward the end. Table 8 shows seven examples. The numbers are citations to the papers written during the individual target years. Parkes and Owens Valley are the most spectacular, but each of the seven seems to have got up and running very quickly. Optical telescopes today seem to take longer to shake down according to Table 9. The numbers are papers rather than citations and would have been zero for Gemini, HET, Magellan, and Subaru a year or two before, but not

Table 8 Hit the ground running? Total number of citations in 1965–1969 to papers making use of data from the following facilities, published in individual years

Facility	1960	1961	1962	1963	1964
Parkes	0	10	36	72	237
Jodrell 250′	9	39	120	179	120
NRAO 300′	0	0	0	22	105
OVRO Interf.	0	12	167	165	468
Goldstone	0	0	10	7	106
Arecibo	0	0	0	0	6
Ohio State Monst.	0	0	0	1	30

Table 9 Optical ramp-up comparison: Numbers of papers (pro-rated) attributable to data from new facilities by year

Facility	2001	2002	2003
Gemini	2.2	15.2	19.0
HET	1.5	4.8	8.2
Magellan	4.0	5.4	18.2
Subaru	18.7	40.5	19.9
VLT	69.8	106.7	169.1
Keck	104.5	129.7	131.4

for the VLT and Keck, which were in the process of increasing the effective numbers of mirrors available to the community.

6 Other changes and other topics one might explore

Any rational being who looks at science publishing over the 20th century will be most struck by "MORE!!!" More papers, more authors, more facilities, though fewer astronomical journals with the demise of observatory publications and the A&A merger. I haven't counted authors but there were 2,164 observational papers in 1960–1964 (433 per year, including solar-terrestrial radio and space ones) and 11,831 papers in 2001–2003 (3944 per year, excluding all solar work). More people and papers are not, of course, unique to astronomy. The founders of Physical Review Letters noted that in 1925, Zs. f. Physik, arguably the most prestigious of the time, had 285 papers by 367 authors, while the 1963 PRL had 1600 papers (and longer ones) by 2500 authors. The weekly PRLs of 2008 are on track to add up to more than 4,000 papers, and the Phys. Rev. total will be about 18,000 [14], with author numbers dominated by collaborations of 100 and more people, generally listed alphabetically.

Within astronomy, the fraction of single-author papers has declined from more than 50% to about 10% and the fraction of multi-national papers increased from 5% to 40%. Abt [2] shows that these trends are common to many sciences, though astronomy is uniquely multinational, perhaps because the sky is there 24/7 both north and south; and the unique structure of the IAU may have played a role (the other international unions have only nations and societies as members, vs. our more than 9,000 individual colleagues). Multi-wavelength papers have gone from rare (1.2% in 1960–1964) to common (26% in 2001–2003).

The average lengths of papers have climbed more or less monotonically by factors of two to five from 1900 into the 1980s in physics, math, chemistry and astronomy, and in the UK and Japan as well as in the US [15]. That trend continued in astronomy for another decade at least. Leveling off has been claimed, but I suspect proper allowance has not been made for the increasing range of tabular, graphical, and mathematical material that now appears on-line only. The average ApJ Letter is now longer than the average paper of the inter-war years; and no one can now claim to read "all" of anything.

Astronomy today is a sort of economic indicator, in the sense that the ratio of GDP of a country to its number of IAU members is bracketed within relatively narrow limits, which can be checked by anybody with an almanac (e.g. [5]) and an IAU membership directory (e.g. [6]). Let us take as our unit billions of $US in GDP per IAU member. Then the average (with very small dispersion) for 23 developed country members is 4.81; for 12 developing countries 6.86; for 15 members from Eastern Europe and the former Soviet Union 1.26; and for eight Moslem countries 16.1. For what it is worth, Israel at 1.86 resembles the places from which many of its immigrants came and South Korea and Taiwan at 10.5 and 10.2 resemble the Moslem group of rather prosperous countries with little tradition of astronomy. Notice that we have hereby discovered the obvious by a very difficult method!

Anyone with a set of old almanacs and past IAU membership directories (these go back to 1923 or so and appear in Transactions B) could find out whether the cost per astronomer has inflated faster or slower than the general consumer price index. If anyone decides to do this, I would very much like to know the answer. So, perhaps, would your Office of Management and Budget.

There have been, of course also enormous changes in the identities of the telescopes contributing to published data. Nothing that penetrated into space appears in both 1960–1964 and 2001–2003. At radio wavelengths, the Jodrell 250′, the Parkes 230′, the OVRO interferometer, Dwingeloo, Nançay, and Lebedev appear in both data sets (but all after varying degrees of modification); and a number of older sites, like Goldstone and MIT's Millstone Hill Radar installation, have almost complete new facilities. Both were originally designed to keep in touch with man-made stuff in space.

Optical longevity is also considerable. Although Mt. Wilson is gone from the journals, the Palomar 48″ Schmidt remains remarkably productive. The large mirrors at Palomar and Lick and the McDonald 82″ are still with us. Of the 72–74″ telescopes, the Radcliffe in South Africa remains its most productive research instrument; others at Mt. Stromlo, DDO, DAO, and in Japan carry on at some level. And a 2008 press release says that the Perkins 69″ (moved from Ohio to Arizona in 1961) is getting a new spectrophotometer and such. Particular telescopes at OHP, Pic du Midi, KPNO, Bosque Alegre, Asiago, Lowell, and in the Crimea are still yielding data for publication. And, of course, the telescope everyone loved to hate, the Great Melbourne, provided a few papers in 1960–1964 in its 1954 MSSO incarnation, and MACHO papers 40 years later, until fire swept the area in our own decade. There were also some telescopes larger than 20″ that did not contribute any papers during either period. Some were in places with awkwardly bright skies (Lisbon, Rome) and others in places difficult for other reasons. Indeed the only instrument I failed to trace at all is a 38.5″ multi-focus instrument that was built in what was then Elizabethville, Congo in 1960 [12].

Acknowledgements My hearty thanks go to many folks: George Herbig, Paul Hodge, Patricia Whitelock, Michael Feast, Patricia Henning, and John Mathews for help in chasing down some relatively obscure telescopes; Jet Katgert for showing all the images (some borrowed from references [9] and [12]) while I was speaking at the conference and for thoughtful, friendly, but firm discussion of the text; librarians Jeanine and Jill at UCI and STScI for help beyond the call of duty in accessing paper copies of all the 1960–1964 journals; and the SOC and LOC not only for the invitation but for agreeing to accept this contribution typed on paper!

References

1. Abt, H.A. (ed.): AAS centennial volume. Astrophys. J. **525**(lC), Part 3 (1999)
2. Abt, H.A.: Scientometrics **72**, 156 (2007). doi:10.1007/s11192-007-1686-z
3. Arnold, J.R., et al.: J. Geophys. Res. **67**, 4878 (1962). doi:10.1029/JZ067i012p04878
4. Bowyer, S., et al.: Nature **201**, 1307, Science **146**, 912 (1964)
5. Economist: Pocket World in Figures. Profile Books (2007)
6. Engvold, O. (ed.): Trans. IAU 25B. Astronomical Society of the Pacific (2007)
7. Giacconi, R., et al.: Phys. Rev. Lett. **9**, 439 (1962). doi:10.1103/PhysRevLett.9.439
8. Hutchins, R.: British University Observatories 1772–1939. Ashgate Publications (2008)
9. King, H.C.: The History of the Telescope. Charles Griffith & Co. (1955)
10. Kragh, H.: Cosmology and Controversy. Princeton University Press (1996)
11. Kraushaar, W.L., Clark, G.W.: Phys. Rev. Lett. **8**, 106 (1962). doi:10.1103/PhysRevLett.8.106
12. Kuiper, G.P., Middlehurst, B.M. (eds.): Telescopes. Univ. Chicago Press (1960)
13. Seaborn, H.T. (ed.): TRW Space Log **7**(4), (1967–68)
14. Sprouse, G.D.: Phys. Rev. Lett. **101**, 120001–1 (2008). doi:10.1103/PhysRevLett.101.120001
15. Trimble, V.: Publ. Astron. Soc. Pac. **96**, 1007 (1984). doi:10.1086/131467
16. Trimble, V.: BeamLine **32**(1), 42 (2002)
17. Trimble, V., Ceja, J.A.: Astron. Nachr. **329**, 632 (2008). doi:10.1002/asna.200810999

Exp Astron (2009) 26:149–161
DOI 10.1007/s10686-009-9152-y

HISTORICAL REVIEW

Early history space astronomy

Issues of patronage, management and control

Robert W. Smith

Received: 9 February 2009 / Accepted: 10 February 2009 / Published online: 12 March 2009

Abstract The history of space astronomy is usually written from the perspective of the remarkable scientific findings garnered by space astronomers and the ways these findings have enriched and guided new views of the universe. This paper examines some aspects of the history of space astronomy in the early years of NASA, but takes patronage, management and control as its key issues.

Keywords Space astronomy · Orbiting astronomical observatories · Large space telescope · Superconducting supercollider

1 Introduction

The history of observational astronomy can be divided into four main periods: the age of pre-telescopic astronomy (from the earliest times to the early seventeenth century); the first era of the telescope (from the early seventeenth century to the mid-nineteenth century); the combination of the telescope with spectroscopy and astronomical photography (from the mid-nineteenth century to the mid-twentieth century), and the modern era (from the mid-twentieth century on) characterized by the opening up of all wavelength ranges of the electromagnetic spectrum to astronomical investigations. In this paper, I will concentrate on aspects of the development of space astronomy in the fourth of these periods. The focus will not be on the cognitive changes in astronomy

R. W. Smith (✉)
Department of History and Classics, University of Alberta,
Edmonton, AL, Canada, T6G 2H4
e-mail: rwsmith@ualberta.ca

that were driven by scientific findings from space telescopes (remarkable as they have been), but on some of the central institutional and social issues that have arisen in the history of large-scale space astronomy projects due to the new forms of patronage. Central will be patronage and what might be described as the organizational context of space astronomy, that is, who has power, how do they get it and maintain it, and how has that context shaped what spaces telescopes have been built. Key too will be issues of management and control. Indeed, if we recast the relatively recent history of the telescope more generally so as to take these issues into account, then a somewhat different narrative emerges in place of the usual fairly straightforward story of ever bigger and more powerful instruments. At root this story contains a dose of technological triumphalism. It also embodies what in some accounts is presented as almost an organic and perfectly natural process of telescope evolution instead of a story of often highly contentious and contingent matters to do with intricate mixes of competition and collaboration.

In so doing I will also centre on the United States in the 1960s and 1970s, and in particular on optical and ultraviolet wavelengths. This means that unlike the situation for, say, X-ray astronomers who had to figure out ways to get their instruments into space or they could not study the X-ray universe, the astronomers I will be discussing usually had a choice about whether or not to pursue space astronomy. This, as we shall see, was an important factor in the early history of the Hubble Space Telescope. However, to throw what I have to say about patronage and management in the history of space astronomy into a sharper light I will begin by taking a few steps back into the nineteenth century.

2 Patronage

As other papers in this volume have reminded us, large, powerful speculum metal telescopes of the first half of the nineteenth century were generally the products of highly motivated individuals who sat outside the mainstream of positional astronomy. Makers such as the third Lord Rosse and William Lassell financed their initiatives with their own money and were the telescope managers when they were not the actual users (Fig. 1). Perhaps the most famous example of where that was not the case was the Melbourne telescope, completed in 1862. For this 48-in. speculum metal reflector the design flowed from a committee appointed by the British Royal Society. The telescope was constructed at the Grubb works in Dublin, Ireland and so the makers were very distant from the users. The telescope was a failure, at least in its early years [1].

When we examine the development of astronomy in the United States in the late nineteenth century and early twentieth century, we find a major remaking of the tools of American astronomy, a remaking powered by the patronage of wealthy foundations and individuals who did *not* use the telescopes they financed. Nor were they makers, rather, they wrote the checks. Remarkably, astrophysics, one of the least utilitarian of the sciences, became arguably the

Fig. 1 William Lassell's 48-in. reflector

most richly endowed of all the sciences in the United States [6]. When we turn to the key examples of the observatories of Yerkes, Lick, and Mount Wilson, we nevertheless see institutions where astronomers set the research agenda. These institutions were headed by astronomers who essentially also had control over the construction of new telescopes. The directors of these institutions had, in fact, far more authority than would be typical later in the era of large-scale government funding.

In many ways the leading astronomical observatory in the world in the first half of the twentieth century was the Mount Wilson Observatory, established in 1904. A 60-in. reflector was added to its armory within a few years, and this was followed in 1919 by the 100-in. Hooker telescope, the most powerful in the world. For many problems, astronomers elsewhere were simply outgunned by the users of the 100-in. reflector. One example of this is provided by the case of measuring the redshifts of spiral nebulae, what we would now term galaxies [13]. This enterprise had been pioneered by V.M. Slipher at the Lowell Observatory in the early 1910s, but by the mid-1920s his patient and long-running attack on the radial velocities of the spirals had run out of steam. For new measurements of such dim and small nebulae the combination of the Lowell 24-in. refractor and spectrograph was ineffective. The financial position of Lowell was such that a new state-of-the-art telescope to bring Slipher back to the cutting-edge of these researches was not on the cards. Instead, this line of research was effectively taken over by Milton Humason working with Edwin Hubble at Mount Wilson, who enjoyed big institutional advantages in conducting such researches. In what I have elsewhere called "the 'material economy' of American astronomy, Hubble had access to telescopes, instruments, colleagues, and calculators about which Slipher could only dream. Also, in the 'moral economy' of American astronomy in this period—that is, the set of usually tacit assumptions and traditions that regulate and structure a scientific community—there was some sharing of photographic plates by astronomers at different observatories. Younger astronomers might also secure a fellowship to be held at least in part at the Mount Wilson Observatory, but only very rarely did astronomers in other institutions have direct access to the Mount Wilson telescopes and then only when the research was of especial interest to the Mount Wilson astronomers" [13]. But Mount Wilson was financed by the Carnegie Institution of Washington, and as a private institution it directly controlled the time on its telescopes. Thus the best and biggest instruments were available to only a small group of astronomers. One of the well-springs of the national observatory movement in the United States after World War II was the very unequal distribution of time on state-of-the-art telescopes with high-quality instrumentation.

In the construction of the 100-in. reflector on Mount Wilson there had been conflict between George Ellery Hale and his chief optician George Ritchey [10]. This conflict had ultimately been resolved by Hale firing Ritchey. The leading astronomer, then, had had the power and authority to take such an action. With the construction of the 100-in. we also see the start of a committee system to work on particular problems and issues.

This sort of system was still more evident with the considerably more challenging 200-in. reflector that would eventually go into action on Palomar Mountain in 1949, though planning had begun in 1928. An effort of that magnitude called forth very different sorts of management techniques than those that had generally prevailed for even the biggest telescopes of the nineteenth century. The astronomer J.A. Anderson now took on the role of telescope developer and his career was devoted to that end. As King points out, "Dr. Anderson was, for fifteen years, chief optical expert at Mount Wilson. To him fell eventually the tasks of choosing the observatory site and the organization of the optical work [for the 200-inch] in the shops of the California Institute of Technology, Pasadena." But King also notes that the "Hale telescope and observatory resulted from the co-operative effort of a legion of experts" ([4], p. 402). Thus a different 'Big Science' scale and complexity of enterprise was fashioning new social roles for astronomers.

3 World War II and after

The history of American science and technology during and after World War II has drawn a huge literature. The events of the years of World War II and immediately after were momentous in terms of their effects on the scientific enterprise and the transformed relationship between science and the state. Certainly the effects were enormous, not just in terms of the vast outpouring of funds driven by the acceptance by national governments that science was critical to national security, but also in the numbers of scientists, the attitudes of these scientists towards state funding, the remaking of various disciplines, and new forms of scientific practice. After 1945, these changes were entrenched by the rapid onset of the Cold War rivalry between the United States and the Soviet Union. Many of the telescopes and instruments discussed in this volume were in effect a gift of the Cold War.

For astronomers, one of the most important changes that occurred after World War II was the breakdown of the older forms of relationship between patrons and leading astronomers, relationships usually founded on informal contacts and an easy exchange of information. The management relationships in the aerospace world that many astronomers encountered after 1958 and the founding of NASA (the National Aeronautics and Space Administration) were instead fashioned around 'Requests for Proposals,' contracts and deliverable "products." Working with NASA would present much more bureaucratic and hierarchical structures than the ones to which most astronomers were used, and many felt uneasy in this new world.

4 NASA and the orbiting astronomical observatories

From the time of its establishment NASA was determined to be in tight control of the projects it funded. A direct challenge to the fledgling agency, however, came quickly from the National Academy of Sciences, the leading group of elite scientists in the United States. In 1958 the National Academy created a "Space Science Board" that initially sought to lay out priorities for space science by asking for interested scientists to send proposals for space projects that would follow those undertaken for the International Geophysical Year. By the end of 1958, there were three U.S. government agencies involved in space science, ARPA (the Advanced Research Projects Agency), the Department of Defence, and NASA. Thus NASA has to carve out a role for itself with potential institutional rivals as well as deal with space scientists. Although there would be ups and downs in the relationship, NASA's early response to the Space Science Board was tepid, if not hostile. NASA would certainly seek advice from scientists outside the space agency and would devise its own advisory structures as well as listen carefully to the views expressed by the National Academy, but it was loath to hand control over to anyone [7]. Advice from the research arm of the National Academy in the form of the National Research Council would later become extremely significant. Perhaps the most important institutional form of this advice for astronomers would come from the so-called decadal surveys. One example of such a survey was that conducted in the years around 1990 and led by John Bahcall with what would become known as the Bahcall Committee. In time, such surveys would directly engage many hundreds of astronomers every decade or so in developing priorities for the American astronomical community, priorities that NASA as well as the patrons of ground-based astronomy would take very seriously. But in the space agency's early years, the first of these broad surveys, led by the astronomer Jesse Greenstein, was well in the future (its reports would be published in 1972) and astronomers outside the agency lacked this counter to NASA's own decision making.

One specific example of what NASA's desire to maintain control in its first decade or so meant for astronomers who jumped into the world of space astronomy after the founding of the space agency comes from the story of the 'Orbiting Astronomical Observatories.' NASA, it is important to remember, was established as one of the U.S. responses to the so-called Sputnik crisis that had been prompted by the launch of *Sputnik I* in October 1957. President Dwight Eisenhower might have earlier disliked space projects that were pursued largely for the sake of prestige, but the world was different after 1957 and the orbiting of *Sputnik I*. Even for a reluctant Eisenhower prestige became a policy driver for space efforts, particularly after *Sputnik I* was followed by *Sputnik II*. In fact, the race for prestige in the Cold War theatre offered by space would become a crucial factor for U.S. (as well as Soviet) space enterprises. When George Kistiakowsky, President Eisenhower's Science Advisor, spoke to the President about the "potential gains in national prestige if we establish the first astro-observatory on a satellite, the President was very much interested and said he would certainly be in favor of proceeding vigorously" ([5], p. 385).

For some astronomers the prospect of building such an astro-observatory presented a remarkable opportunity. It was one they could hardly have imagined even a few years before. Particularly attractive was that working space observatories promised vastly more data than offered by the very short-lived (5 min or so) flights of rockets above the obscuring layers of the atmosphere. But which missions and observatories to fly? Here the new space agency confronted a critical issue for government funded modern science. Given that there are always more worthwhile projects than money to support them, how should the space agency decide on the funding levels for the various areas of astronomy and how pick which instruments and space telescopes to build from a wide array of possible ones? In the 1960s, NASA strove to best the Soviet Union in the battle for prestige. While by far the most effort in terms of prestige went into the Apollo program, prestige also mattered for science projects. Homer Newell, the then head of NASA's space science program, claimed in 1966 that "for the initial exploration in virgin fields relatively simple instrumentation and limited-scope research programs often suffice to permit rapid exploration of new technologies and breakthroughs of understanding" ([8], p. 6) (Fig. 2). Plans for the sort of astro-observatory discussed by Eisenhower and Kistiakowsky swiftly coalesced around what would become the Orbiting Astronomical Observatory (OAO) program. The OAO's were being funded for more reasons than potential scientific results. For agency managers, 'simple' instrumentation was less attractive in terms of prestige than more challenging and difficult-to-build systems. With the OAOs focused principally, although not exclusively, on ultraviolet astronomy, the resulting spacecraft would need to build on long experience of observations in the optical region and so go well beyond the sort of relatively simple instrumentation Newell reckoned sufficient for the exploration of "virgin fields".

Fig. 2 Homer Newell (*left*) with William Pickering and Harris Schumeier at a JPL press conference in 1964

Ultraviolet astronomy, pursued by for the standards of the time advanced and sophisticated spacecraft, promised prizes in terms of prestige as well as significant new scientific results, or at least that is the argument that prevailed that was made by NASA managers ([12], p. 38).

NASA had in hand a draft project plan in December of 1958, that is within months of the formation of the agency (which was formally established on 1 October 1958), and by February 1959 had assembled a working group of astronomers to help plan the OAO program. From its deliberations as well as the work of NASA staffers and NASA contractors, a design began to evolve in which there would be a spacecraft common to the OAO series. The space agency issued 'requests for proposals' for the OAO spacecraft and associated instrumentation in 1960 and 1961. The newly established Goddard Space Flight Center, NASA decided, would provide overall project management.

Issues of control however soon loomed large. The OAOs would be, for the time, NASA's most complicated scientific spacecraft. Each of the spacecraft would weigh around 1,800 kg and involve some 50 km of wiring and 440,000 parts. The astronomers in fact had pressed for a series of smaller and less complex observatories. They wanted to build a base of knowledge and support with less demanding astronomical payloads before moving to more advanced designs. While the astronomers did persuade NASA to split solar from non-solar missions by using very different platforms so that the OAOs would centre on stellar astronomy, the space agency, driving hard for prestige as well as original scientific results, chose to forge ahead with the more challenging version of the project.

The principal investigators on the Orbiting Astronomical Observatory program were also taken aback at how the program was to be managed on a day-to-day basis. Even the prominent Princeton astronomer Lyman Spitzer Jr., who had participated in a variety of large-scale government-funded programs during World War II and later in his fusion research, was surprised (Fig. 3). In 1960, he wrote to NASA Headquarters. For the design of a complicated scientific system, Spitzer argued, the most efficient way to proceed was to give overall responsibility to a single organization and to preferably one person. For relatively simple payloads to be launched into space, Spitzer conceded that it had been feasible for the scientists to develop "little black boxes" while the engineers fashioned the spacecraft. The NASA project manager was then able to harmonize the integration of the boxes with the satellite. But, in Spitzer's opinion, this would not be an appropriate way of doing business for a project as demanding as the Orbiting Astronomical Observatories. Splitting the scientists from the engineers was problematic. He contended that it was preferable "to have the scientific experiment and the engineering design integrated under one individual (or institution). Then the scientific requirements can be altered to meet the detailed engineering difficulties as they arise, and the engineering plans can be guided to yield the maximum scientific usefulness... It is obvious, I believe, that the overall responsibility to which I refer must be given to someone who is intimately familiar with the objectives of the system, and with the possible uses of the scientific equipment. Hence this responsibility must be given not to an engineer but to an astronomer, preferably a senior astronomer who has had substantial experience in the coordination of large engineering programs" [14].

Fig. 3 Artist's impression of OAO I

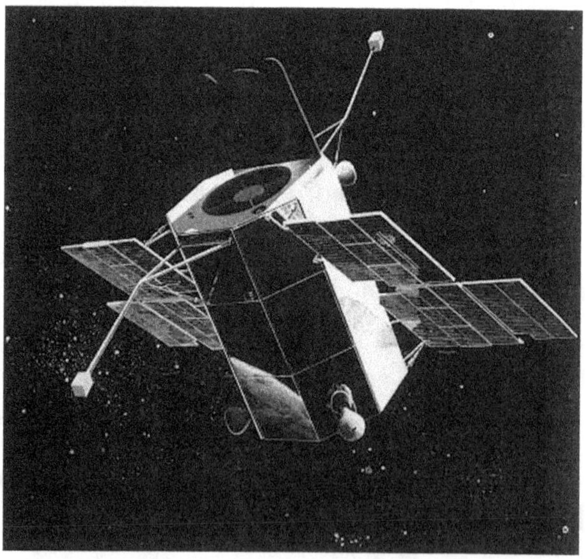

While NASA did respond to some of Spitzer's points and criticisms by others by offering consultations on the major technical problems of the spacecraft, the space agency was adamant in not granting official management responsibility for the development of the observatories to the astronomers. Had it done so, it would have meant, as far as NASA was concerned, that the agency had lost control over the Orbiting Astronomical Observatory program (Fig. 4). NASA was responsible to Congress for the funds that it spent, not the astronomers. Hence NASA's basic position, as one senior manager would later recall "was that operational responsibilities placed by law upon the agency could not be turned over to some other agency. Moreover, decisions concerning the space science program could not be made on purely scientific grounds. There were other factors to consider, such as funding, manpower, facilities, launch vehicles and the saleability of projects in the existing climate at the White House and on Capitol Hill—factors that only NASA could properly assess" ([9], p. 205).

What the leading astronomers outside of the space agency wanted was for NASA to run its scientific programs in similar ways to those they were familiar with from their interactions with the National Science Foundation in connection with ground-based observatories. One of these leading astronomers, Fred Whipple of the Smithsonian Astrophysical Observatory, had, for example, made his position clear in hearings before Congress in early 1958, that is, even before NASA had been established. In his view, "this national space agency should be the coordinating agency for the activities in space research. That does not mean that it should control this activity, but it should be in a position of coordinating." Space exploration, Whipple believed strongly, should not be under the control of the government or of the military. For Whipple, as DeVorkin has pointed out, "Co-ordination, not control, by an informed civilian government agency, letting contracts to universities and industry, and even to the military, was the ideal solution" ([3], p. 239).

Similarly, those U.S. government agencies responsible for high-energy accelerators had delegated to high-energy physicists important measures of control and oversight. At SLAC (the Stanford Linear Accelerator) engineers

Fig. 4 Lyman Spitzer photographed in the 1980s

reported to physicists, and when the design and construction of Fermilab began in the 1960s, physicists took important engineering roles and also assumed key management positions. Physicists designed, built and managed the laboratory, and again engineers reported to physicists. Hence in high-energy physics and ground based astronomy, the scientists had secured very important measures of overall management responsibility even within the context of patronage from the federal government. NASA, with its managers generally coming from very different sorts of cultures from those of high-energy physics and ground-based astronomy, would not follow suit.

5 Towards the Hubble Space Telescope

The issue of control would continue to be central for space astronomers after the early years of NASA. In post-war ground-based astronomy, astronomers had typically enjoyed large measures of autonomy. But to what extent was an individual space astronomer to be able to determine the direction of his or her research and to what extent were particular communities to set their own research directions and forms of scientific practice? These questions were certainly fundamental to those who were advocating what would end up as the Hubble Space Telescope.

In the 1960s, NASA's ways of doing business put off at least some potential space astronomers and bred a sense that advice coming from outside the agency was not really taken seriously. There was also concern at what many astronomers reckoned was a wasteful human spaceflight program combined with unease about increasing NASA involvement with universities. To address these matters, in 1965 NASA administrator James Webb appointed a committee to make recommendations on how NASA might engage more of the scientific community in its forthcoming big programs. These included what was known at the time as the Large Space Telescope, a telescope with a primary of 120-in. diameter that would work in the optical and the ultraviolet that would be sited either on the moon or on orbit that was already being actively discussed by some astronomers. In Webb's view, "We in NASA think it is essential that competent scientists at academic institutions participate fully in the next generation of space projects, and we believe that we will need new policies and procedures in order to enable them to participate" [15]. Norman Ramsey, a physicist from Harvard, was chosen to chair the committee.

After extensive deliberations, the Ramsey Committee backed the idea of a Large Space Telescope but decided to set it alongside other space observatories that would be designed to observe other wavelengths besides the optical and the ultraviolet. As might have been expected from a group of academic scientists, they argued that there should be strong scientific oversight of the orbiting telescopes project. What was the appropriate institutional mechanism to do this? The Committee proposed the formation of a consortium of universities. It would be called STAR, or Space Telescopes for Astronomical Research. Such a proposal was certainly in line with the emergence after World

War II of university consortia as the generally preferred means to manage costly new research facilities. By the time the Ramsey Committee reported, the use of university consortia had also become widely accepted as the tool to transform the concept of national laboratories or facilities (national in the sense of being open to all qualified scientists and not just members of the laboratory staff) into working institutions. But NASA balked at the Ramsey Committee's recommendations and the proposals for STAR got massively watered down until they ended up with the creation of a new advisory group, the Astronomy Missions Board, which, as events would demonstrate, carried little weight.

It is against this background then that we should see the efforts in the early 1970s by various influential astronomers concerned with a possible Large Space Telescope (what would in time be transformed into the Hubble Space Telescope, although for several years the Large Space Telescope became the Space Telescope before it was renamed the Hubble Space Telescope in 1983) to ensure that they, not NASA, control the telescope's scientific operations and, even better from the perspective of some of them, its development. NASA initially balked at the notion of giving an outside group responsibility for scientific operations let alone the telescope's development, and it was accepted by the space agency only with great difficulty. Here, though, the key point is that NASA did decide by the mid-1970s to create what became the Space Telescope Science Institute, overseen by a university consortium, to manage the telescope's scientific operations ([12], chapter 6).

There was, nevertheless, a continuing tension over what exactly the Institute's role was to be (a tension that would not disappear after the Institute was actually established) with some astronomers hoping for a decidedly wider role for an Institute. For them, the goal was a Space Astronomy Institute, not just a Space Telescope Science Institute. In 1975, for example, at a meeting of the Council of the American Astronomical Society attendees heard presentations on a "Space Astronomy Institute." The Space Astronomy Institute, in this plan, would play the same sort of role for space astronomy that the National Observatories played for ground-based astronomy.

One reason that NASA decided to form the Space Telescope Science Institute run by a university consortium was what elsewhere I have called coalition building [12]. In other words, in order to win approval from the Congress and White House for a Space Telescope, its advocates had to be able to assure their colleagues and its patrons that it would be technically feasible. But they also had to labour to make sure it would be politically feasible too, and for it to be politically feasible it needed the energetic support of a broad range of astronomers. The 'Space Telescope Science Institute' was thus in part the means to bring on board those ground-based astronomers who were uneasy at the prospect of getting involved in a big NASA project or who might be skeptical that outside advice would not be heeded by NASA. Such ground-based astronomers, it is important to note, had the option of continuing to pursue ground-based astronomy and not getting involved with space enterprises such as the Space Telescope.

Certainly in mid-1975 pressure from outside the space agency was building for an institute, and at that stage support for the Space Telescope was relatively flimsy. Thus for the telescope's supporters, drawing in more ground-based astronomers was an attractive proposition. Some form of Institute run by a university consortium that did not sit inside the fence of a NASA center seemed to the leading advocates among non-NASA astronomers to be one way to achieve that. As well as these political advantages, there were of course strong and obvious scientific and technical advantages to drawing in a wider body of interested astronomers.

More broadly, "the fundamental question of whether or not to approve the Space Telescope was constantly being reframed as the issues bearing on the decision were themselves reshuffled and repackaged, often because of the needs of coalition building. Moreover, the telescope's design and the associated programs to build it were central elements in these changes." Thus here was a very large scale project reconstituting "the institutional organization of the scientific enterprise and was itself reconstituted by its necessary financial, political and ideological ties to society at large" ([2], p. 15). Thus issues of management and control were at the very heart of the battle to win approval of the Space Telescope. Management, then, was not just something that was imposed on the perfect form of the telescope project in the manner of Platonic philosophy. Management and control issues were instead interwoven with scientific, technical, economic and political issues.

Space astronomy has certainly thrown up its share of scientific visionaries and charismatic leaders. But by the late twentieth century, the grand figures we can readily identify as the astronomer/managers and builders of the state-of-the-art reflecting telescopes of the nineteenth and early twentieth century had to a large degree been replaced by scientific communities and large and sometimes very large groups of scientists and engineers. The biggest sorts of space astronomy projects had become Very Big Science indeed.

6 Coda

As a coda to this paper, it is interesting to consider the case of the ill-fated U.S. very big science project, the Superconducting Super Collider (SSC), the design of which formally began in 1984. There the issue of the control of the workplace would prove to be hugely important. Writing on the history of high-energy physics in the U.S., Michael Riordan has argued that "Control of the scientific workplace became increasingly contentious during the 20th century as experimentation grew in scale from the laboratory workbench to particle accelerators and colliders that stretch for kilometers across the Earth's landscape. The need to find increasingly larger sources of funding to support their ever more costly instrumentation has required physicists to surrender a measure of control, at least temporarily, over their experimental activities" ([11], p. 125). In the earlier history of U.S. high energy physics, the physicists had very largely run the show, but with the SSC matters were quite different.

For assorted reasons, two antagonistic cultures developed in the management of the SSC, that of the physicists and that of managers and engineers more used to the projects pursued by the U.S. armed services. While the cancellation of the SSC in 1993 (by which time around $4 billion had been spent out of an estimate of well over $10 billion) occurred for various complex reasons, key factors in the SSC's demise were the battles between these two cultures. Control of the workplace *does* matter though it is rarely as crucial to a project's fate as in the example of the SSC.

References

1. Bennett, J.A.: Church, State and Astronomy in Ireland. 200 Years of Armagh Observatory. Armagh Observatory, Armagh (1990)
2. Capshew, J.H., Rader, K.A.: Big science: price to the present. Osiris **7**, 3–25 (1992). doi: 10.1086/368703
3. DeVorkin, D.: SAO during the whipple years. In: Orchiston, W. (ed.) The New Astronomy: Opening the Electromagnetic Window and Expanding Our View of Planet Earth. Springer, Dordrecht (2005)
4. King, H.C.: The History of the Telescope. Charles Griffin and Company, High Wycombe (1955)
5. Kistiakowsky, G.B.: A Scientist at the White House: The Private Diary of President Eisenhower's Special Assistant for Science and Technology. Harvard University Press, Cambridge (1976)
6. Miller, H.: Dollars for Research: Science and its Patrons in Nineteenth Century America. University of Washington Press, Seattle (1970)
7. Naugle, J.: First Among Equals. The Selection of NASA Space Science Experiments. NASA, Washington, DC (1991)
8. Newell, H.E.: Space astronomy program of the national aeronautics and space administration. In: Newell, H.E., Smith, H.J., Roman, N.G., Mueller, G.E. (eds.) Astronomy in Space, NASA SP-127. NASA, Washington, DC (1967)
9. Newell, H.E.: Beyond the Atmosphere: Early Yeas of Space Science. NASA, SP-4211, Washington, DC (1980)
10. Osterbrock, D.E.: Pauper and Prince: Ritchey, Hale and Big American Telescopes. University of Arizona Press, Tucson (1993)
11. Riordan, M.: A tale of two cultures: building the superconducting super collider, 1988–1993. Hist. Stud. Phys. Biol. Sci. **32**, 125–144 (2001). doi:10.1525/hsps.2001.32.1.125
12. Smith, R.W.: The Space Telescope. A Study of NASA, Science, Technology and Politics. Cambridge University Press, New York (1993) (with contributions by Hanle, Paul, Kargon, Robert, and Tatarewicz, Joseph; expanded paperback edition)
13. Smith, R.W.: Beyond the galaxy: the development of extragalactic astronomy 1885–1965, part 2. J. Hist. Astron. **40**, 71–107 (2009)
14. Spitzer, L.: Letter, Spitzer to N. Roman, March 18, 1960. Spitzer Papers, Princeton, file "OAO Correspondence 1960–1971"
15. Webb, J.: Letter, Webb to Norman F. Ramsey, January 14, 1966 in Appendix to "NASA Ad Hoc Science Advisory Committee Report to the Administrator 15 August 1966"

Exp Astron (2009) 26:163–177
DOI 10.1007/s10686-009-9141-1

REVIEW ARTICLE

Conceiving and marketing NASA's Great Observatories

Martin Harwit

Received: 30 October 2008 / Accepted: 6 January 2009 / Published online: 5 February 2009

Abstract In early 1984, the astronomical community's plans to launch a series of powerful telescopes able to study celestial sources at almost any wavelength were in trouble. The President of the United States had just declared his priority for a Space Station that was bound to be expensive, and Congress could not understand why yet another set of space observatories was needed when others were already being funded. To realize their aims, astronomers would have to advocate their needs much more effectively than in the past.

Keywords Astronomical observatories · Astronomy from space

1 Charlie Pellerin

Every 10 years, the US National Academy of Sciences conducts a survey to set astronomical priorities for the decade ahead. George Field, Director of the Harvard/Smithsonian Center for Astrophysics chaired this survey for the 1980s.

On March 8, 1984, Field wrote Senator Jake Garn, Chairman of the Senate Committee in charge of appropriations for NASA, pointing out that the 1980 decadal survey had "placed high priority on both the Advanced X-ray Astrophysics Facility (AXAF) and the Shuttle Infrared Telescope Facility (SIRTF)". NASA's plans for timely flights of both these observatories would

Unless otherwise designated, all documents cited below can be found in the Cornell University Kroch Library Archives, Division of Rare and Manuscript Collections: (14/7/2402 Martin Harwit papers, boxes 9, 10 and 11).

M. Harwit (✉)
Cornell University, Ithaca, NY, USA
e-mail: harwit@verizon.net

require that each underwent a Phase B study, for which the President's recently proposed FY85 (fiscal year 1985) budget was inadequate. Field wrote, "I strongly urge that your committee augment the NASA budget by approximately $10M to make these studies possible."[1]

The two missions also faced other threats. Nobody understood this better than Charles J. (Charlie) Pellerin Jr., Director of the Astrophysics Division of NASA's Office of Space Science and Applications (OSSA). There were two reasons:

In his January 25, 1984 State of the Union message, President Ronald Reagan had announced, "Tonight, I am directing NASA to develop a permanently manned space station and to do it within a decade. "A space station" Reagan assured his audience "will permit quantum leaps in our research in science, communications, in metals, and in lifesaving medicines which could be manufactured only in space...." Pellerin, like everyone within NASA knew that this directive would urgently refocus the agency's efforts. NASA had promised Reagan that the station would cost no more than $8 billion at 1984 rates. Every knowledgeable space engineer in the country knew that this was unrealistic.[2]

The plan envisioned the Space Station to be launched in 1991–92. The Station was to include two unmanned free-flying platforms, respectively dedicated to astronomy and Earth observations. An Orbital Maneuvering Vehicle, serving as an unmanned space tug for moving payloads around in the vicinity of Space Station, would permit astronauts from the Station to service these payloads. Pellerin now needed to revise his programs to fit in with these new Presidential priorities.

In addition, the astronomical community was split. Every so often an infrared astronomer would drop by Charlie's office to urge priority for SIRTF. Then an X-ray astronomer would call to argue for AXAF.

Charlie recalls, "The argument might have been worth having if one (of these) supplanted the need for the other. In fact, either mission increased the necessity for the other....I needed to create a fresh compelling strategy that would mobilize everybody behind both missions and the two already underway [the Hubble Space Telescope (HST) and the Gamma Ray Observatory, (GRO)]."[3] As the end of 1984 approached, Pellerin instructed his deputy, George Newton, to invite a number of astrophysicists to a one-day meeting at the start of the new year.

I had received George Newton's initial call and told him I'd gladly participate. He called again just before Christmas and asked whether I'd be willing to chair the meeting? I said "Yes." But after we hung up, I was puzzled. Normally,

[1]Letter from George Field to The Honorable Jake Garn, March 8, 1984.

[2]Google: 1984 State of the Union Message.

[3]C. J. Pellerin, Jr., "How NASA Builds Teams," a book in progress, copyright 4-D Systems, 2008, Chapter 16, Page 204.

a NASA division director appoints someone he knows well and can trust to chair his advisory committee without embarrassing him.

This was different. Pellerin and I had never met.

To those who had agreed to attend, a formal letter of invitation went out on December 21. It specified that the meeting should "stimulate discussion of the contributions of (each of NASA's proposed astrophysics missions) in context with those of others, and help illustrate the strength of the total program and its scientific justification....Output from the meeting is expected to be (a) unified understanding and its description in a set of graphic materials suitable for briefing informed lay people."[4]

In late afternoon of January 2nd, I arrived in Charlie's office at NASA headquarters. We spent the evening planning the next day's meeting, and found ourselves getting along well.

I had prepared a number of sketches in advance to meet Charlie's request for "graphic materials." They were simple doodles of the kind that we all draw on a blackboard to explain our thoughts to colleagues: no equations, no formulae—just line drawings with a few phrases. Charlie agreed that these might make for a good start.

2 The January 3, 1985 meeting of the Astrophysics Council

As the meeting got underway, the next day, I looked around the group Charlie had managed to assemble. In addition to five members of his own staff and three subcontractors handling logistics, Charlie had been able to call on Bob Brown from Marshall, Carl Fichtel and George Pieper from Goddard, George Field, Josh Grindlay, Robert Noyes, Irwin Shapiro and Harvey Tananbaum all from the Harvard-Smithsonian Center for Astrophysics, Riccardo Giacconi (Space Telescope Science Institute), Bill Hoffmann and George Rieke from Arizona, Ken Kellermann (NRAO), Jerry Ostriker (Princeton), Ed Salpeter from Cornell, and Ray Weiss (MIT). These were some of the scientists shaping astrophysics and the future of gamma-ray, X-ray, optical, infrared, and radio astronomy in the USA.[5] The day began with Charlie describing what he saw as a major issue, the presentation of a cohesive program that had a sound scientific rationale and could be explained to policy makers and to a wider community in simple and appealing terms. With GRO and HST well along in construction, it was time to focus on AXAF and SIRTF, particularly in the context of ongoing solar work and ground-based radio-astronomical efforts.

Shortly before lunch, I brought up the primary task Charlie had assigned us—the drafting of visual materials to help him to explain the astrophysics program.

I showed the straw-man viewgraphs, the doodles that I had discussed with Charlie the previous evening. The group then split into smaller teams and got

[4]Letter from Pellerin to Harwit, December 21, 1984.

[5]Minutes of the Astrophysics Program Coordination Group, January 3, 1985, page 1.

to work on constructing similar viewgraph sketches to illustrate the anticipated use of the four observatories to solve key astrophysical problems: the origin of the universe and structure of the cosmos; formation of galaxies and clusters; large scale motion and dark matter; quasars and active galactic nuclei; black holes; gas dynamics; star formation; and planetary systems.[6] By day's end, these viewgraphs had been projected and critiqued by the whole team. The sketches were given to Valerie Neal of the Essex corporation of Huntsville Alabama, who was there to support our work and whose doctoral studies fortunately had included the history of science. The group agreed to meet again in two months. In the meanwhile Valerie and I were to come up with a first draft of the booklet we had been asked to produce.[7]

3 A more permanent group

Although the Astrophysics Council had not been expected to convene again until March 26, Charlie called another meeting for February 22. By then, Valerie and I, assisted by an illustrator, Brien O'Brien, also of Huntsville, had assembled a first draft of drawings. These were circulated and discussed for further revision.[8] Charlie's original purpose had been to assemble a group that could solve his immediate problem of explaining the new thrust of the Astrophysics Division to his superiors at NASA, to Congressional staffers, to the astronomical community, and to the public; but the energy flowing out of the January 3 meeting had indicated that a more permanent council could also prove useful to him in other ways.[9]

Keeping the Astrophysics Council going, however, required broadening its membership to make it more representative of the astronomical community—and official consent. Congress is specific about who may or may not advise government agencies. Thus NASA was permitted to set up "Management Operations Working Groups," but could not arbitrarily institute its own advisory panels. A ruling on this issue was eventually handed down a few months later, in May 1985, by NASA's Logistics Management and Information Programs Division, which determined "that the Astrophysics Management Operations Working Group is not an advisory committee within the meaning of the Federal Advisory Committee Act (P.L.92-463) and hence not subject to the provisions of that statute." With this double negative, the Astrophysics Council was ruled legitimate but under a more official-sounding name, as the "Astrophysics Management Operations Working Group (AMOWG)." By then, however, habit proved hard to break and most of us continued to call it the "Astrophysics Council," a practice I'll continue to follow here.[10]

[6]Ibid. page 5.
[7]Ibid. page 6.
[8]Minutes of the Astrophysics Council Meeting, February 22, 1985, page 4.
[9]Ibid. pages 2–3.
[10]Letter from Richard L. Daniels to EZ/Director, Astrophysics Division, May 9, 1985.

4 The Great Observatories comic book

That winter of 1985, George Field happened to drop by Charlie's office. Charlie recalls, "I said, 'George, my only dilemma is what to call this program of observatories, it's so great.'" George replied, "Why don't you call it 'the Great Observatories.'" And we had a name![11] By late June, our brochure, in colorful comic-book style was ready and Charlie ordered a 15,000 print run. It proved immediately effective. In the brochure we tried to show the timelessness of astronomy, and astronomy's great role in public education. We alluded to some of the technical advances astronomy had contributed; but there were many fundamental questions we were unable to answer about how the universe had started, how stars and galaxies formed, and did black holes exist? We showed that the processes we needed to understand spanned huge temperature ranges, requiring telescopes covering comparably large spectral regimes. We explained that such telescopes could only work outside the Earth's atmosphere, opaque to the radiation the telescopes needed to detect. And we tried to impress that our inability to cover all these wavelength ranges could lead to our missing important discoveries.[12]

5 Marketing the Great Observatories

Meanwhile, Riccardo Giacconi, Director of the Space Telescope Science Institute at the time, but of course also an X-ray astronomer, had been invited to a small luncheon gathering at the White House. Among those meeting with President Reagan were Edward Teller; Jay Keyworth, the President's Science Advisor; Tom Paine, who chaired the National Commission on Space; Donald Regan, the President's Chief of Staff; and a small handful of others—five scientists and five administrators all told. Riccardo talked briefly about the astrophysics program. Edward Teller backed Riccardo's enthusiasm for the program by telling the President that astrophysics may be one of the most important scientific endeavors of our time. Subsequently, NASA Administrator James M. Beggs received a White House request for further "information on astrophysics." The meeting obviously had left a favorable impact.[13,14]

By law, NASA employees are prohibited from lobbying Congress for NASA funding. Charlie and his staff could merely "inform" the public about the promise of the Great Observatories. Members of the Astrophysics Council were under no such restriction. Those who could, walked the halls of Congress

[11] C. J. Pellerin, Jr., "How NASA Builds Teams," Ibid. page 206.

[12] The Great Observatories for Space Astrophysics, prepared under the auspices of the NASA Astrophysics Division, Dr. Charles H. Pellerin Jr. by Martin Harwit, Cornell University and Valerie Neal, Essex Corporation in consultation with the Astrophysics Management Operations Working Group; Graphic design and illustration by Brien O'Brien.

[13] Letter from Riccardo Giacconi to The Honorable Ronald Reagan, June 12, 1985.

[14] Minutes of the Meeting of the AMOWG, July 22, 1985, page 3.

to bring the Great Observatories to the attention of Congressional staffers and to speak out in favor of them. Nobody was more active than Harvey Tananbaum, who at the time was the team leader for the AXAF mission study and mirror development efforts.

In the fall of 1985, David Gilman in Charlie's Astrophysics Division and Diana Hoyt, on the staff of Congressman Lujan, ranking Republican on the House Committee on Science and Technology, arranged a series of lectures for Congressional staff. The format was a series of six one-hour talks on "Space Science for the Non-Scientist." The talks were presented in the late afternoon. On October 30, I gave the first of these on "Space Astronomy—An Historical Perspective," and this was followed, roughly at weekly intervals, by Claude Canizares (of MIT) speaking on "The Violent Universe—High Energy Astrophysics;" Blair Savage (from Wisconsin) on "The Visible and Ultraviolet Universe—Planets to the Edge of the Universe;" Michael Hauser (Goddard) on "The Infrared Universe—Cold Gas and Dust;" Eugene Parker (Chicago) on "The Sun—The Rosetta Stone;" and Irwin Shapiro (Harvard) on "The Future of Space Astronomy."[15]

6 The emphasis on Space Station

On March 26, 2 months after the Council's first meeting, and even before our brochure had been completed, we invited Dr. Burton Edelson, Associate Administrator for the Office of Space Science and Applications (OSSA) and Charlie Pellerin's immediate boss, to meet with us. We hoped he would accept our recommendation for the family of four observatories. George Field spoke first, on the importance of the Great Observatories concept and the accomplishments we could expect from AXAF and SIRTF. He urged a FY87 start for AXAF and a phase B study for SIRTF to commence at the same time. Edelson was clearly interested in the scientific aspects of the missions and asked many questions showing a personal curiosity about cosmology and life elsewhere in the universe as well as other matters.[16]

George Newton then described the role that Space Station and its co-orbiting platforms could play in the program. Space Station astronauts would readily reach these platforms, service the telescopes we had mounted on them, and return to their Space Station habitat.

Throughout the Astrophysics Council's discussions that year, Space Station was a given. At a meeting of the American Astronomical Society in January 1985, Charlie had spoken at length about the major missions the Astrophysics Division was hoping to launch, by the mid-1990s. "What will it take to make

[15]Speakers, Congressional Lecture Series, Space Science for the Non-Scientist, announcements of the individual talks are filed under October 30, November 6, 14, and 21, December 2, and December 5, 1985.

[16]Minutes of the Meeting of the Astrophysics Council, March 25 and 26, 1985, page 3.

this happen?" he asked, and then answered, "First we need the Space Shuttle and the Space Station, which together will allow us to launch large, complex, scientific payloads, and to keep them operating for extended periods, 15–20 years in the case of HST, AXAF and SIRTF. When the Space Station is operational, in about 1992, we will have a servicing crew available in space to fix things on-orbit which before could not be fixed; to replace our instruments, as new technologies give us better resolution or higher sensitivity detectors; to replace batteries and tape recorders in our spacecraft if they fail; to deploy solar arrays or antennae if they stick; and, possibly, even to fix the science instruments or to clean contaminated mirror surfaces, within the pressurized work areas of the Space Station. To use these new capabilities effectively we only need to make our systems simple to service, and to minimize our spares and support equipment by moving as far and as fast as possible towards common systems and common spares."[17]

In the era of 1985, and with President Reagan's personal dedication to Space Station, Charlie's vision seemed right on target. All the more remarkably, in our meeting with Burt Edelson, he advised caution. Burt listed the advantages but also disadvantages of a dependence on Space Station, and asked that we reconsider all these factors and write to inform him of our recommendation. I transmitted this in a letter 10 days later. We recommended "a promptly initiated cooperative venture with Space Station as the most promising step for the Astrophysics Division to undertake."[18,19]

But as 1985 progressed, a series of unsettling developments arose. At a September 20 meeting of the Council with William Raney, Director of the Division of Utilization and Performance Requirements in the Office of Space Station, Raney said he wanted us to know that the cost of launches to service observatories would be high. Thus far OSSA never had to take these into account; but in future someone would have to pay for orbital operations and repairs, and these were likely to be charged to OSSA. We expressed surprise at this, since one of the advertised benefits of Space Station for science projects had always been the potential for repairs, replacement and refurbishing.[20]

George Field later wrote me, "...the general tenor of the discussion of science on the Space Station was disquieting...the notion that OSSA would have to pay for servicing and repairs. What is Space Station for, if not this? The evidence suggests that as on the Shuttle, science comes after engineering—and this was supposed to be different, as I recall according to written documents

[17]Space astrophysics in the 1990's: The Decade of Achievement, Filed under February 12, 1985 because of the designation A. F. P. 2/12/85 at the upper right. The initials probably refer to Adrienne Pedersen of the BDM Corporation which supported the Astrophysics Division at the time.

[18]Minutes of the Meeting of the Astrophysics Council, March 25 and 26, 1985, page 3.

[19]Letter from Martin Harwit to Burton Edelson, April 5, 1985.

[20]Minutes of the Meeting of the AMOWG (Astrophysics Council) September 20, 1985, pages 3–4.

promulgated by Beggs and others....The Council should pursue this until satisfactory answers are obtained from Edelson's office or higher."[21]

7 An organization on hold

But NASA's leadership was suddenly thrown into turmoil by the abrupt resignation of NASA Administrator Jim Beggs, on December 4, 1985, when he was accused of criminal intent to defraud the Government, a charge from which he was fully exonerated but only after years under indictment. In place of Beggs, William R. Graham, who had recently been appointed NASA Deputy Administrator, was named Acting Administrator the same day.[22]

In the afternoon of Friday, January 3, 1986, a number of us met with Graham, but not before a series of preparatory meetings starting early that day had taken place. We first had to settle on some agreement with Burt Edelson on what the relation between Space Station and the Great Observatories would be; this had been dragging on, unresolved, ever since our first meeting with him the previous March.

Just before Christmas 1985, I contacted Herbert Friedman, one of the pioneers of X-ray astronomy, who now was Chair of the Assembly of Mathematical and Physical Sciences at the National Academy of Sciences. I had known him for 20 years, ever since he had encouraged me and helped me get started on infrared astronomy with cryogenically cooled telescopes in space, and had come to respect his ever-thoughtful advice. Herb also was a good personal friend of Burt Edelson, and on January 3, 1986 the three of us, joined by Charlie Pellerin, George Newton, and Harvey Tananbaum sitting in for George Field met for lunch and an informal discussion at the Cosmos Club.[23,24,25]

We reached agreement that we would go with the co-orbiting platforms for AXAF and SIRTF. But Burt Edelson wanted a white paper to show how this would work. He gave us a list of steps we would need to complete: He wanted endorsements from the Astrophysics Council; the Space and Earth Science Advisory Committee; the AXAF and SIRTF Science Working Groups and their communities; the Space Science Board; its Committee on Space Astronomy and Astrophysics; the Task Force on Science Uses of Space Station, chaired by Peter Banks; as well as Marshall, Goddard, and Johnson Space Flight Centers.

---

[21] Letter from George Field to Martin Harwit, October 15, 1985.

[22] Minutes of the AMOWG meeting, December 6, 1985, page 1.

[23] Postscript on a typed Note dated December 24, 1985 and headed "For George Field".

[24] Hand written notes some dated "3/1/85", some overwritten as "3/1/1986" (but all of them meaning January 3, 1986).

[25] Letter from Martin Harwit to George Field, January 7, 1986.

I said I'd make myself responsible for soliciting endorsements from the Council, SESAC and CSAA. Harvey volunteered to handle the X-ray community, and George Newton was going to get in touch with Bill Hoffmann about the approval of the infrared community. Herb Friedman thought he could get the Space Science Board to provide a two-page letter, whenever we let him know that we needed it. We thought that Giovanni Fazio or Hugh Hudson could ask the Peter Banks Committee for their endorsement. That still left open how the NASA Centers were to be approached.[26,27] At 4:15 P.M., that afternoon, Charlie Pellerin, Burt Edelson, Harvey Tananbaum and I met with Bill Graham, to whom we presented the discussions that had taken place over lunch, with Edelson taking the lead. As I wrote George Field the next week, on January 7, I thought that Graham looked terribly tired.[28]

A week thereafter, in great contrast, George Field, Charlie Pellerin and I met with Bill Graham, to make a more formal presentation. This time Burt Edelson was unable to join us and Graham was full of energy, and encouraging. He pressed us to assure wide support for the Great Observatories at all levels of society and added that he thought President Reagan might enjoy a briefing on the program." Two days later, Charlie wrote Graham, "If you are interested, we would be delighted to prepare (a) briefing package."[29,30,31]

8 The Challenger disaster and its wake

However, within less than two weeks all this fell moot. The Challenger Disaster of January 28, 1986 changed everything. Two months after the Disaster, at the April 17 meeting of the Council, Charlie discussed the impact on OSSA missions. We could expect a delay of 2 to 3 years for Shuttle payload space, particularly since military and commercial payloads would have priority.[32]

By our June Council meeting, there were questions about Shuttle launches from Vandenberg Air Force Base. A Vandenberg launch was necessary to take payloads into polar orbit. The Cosmic Background Explorer, COBE, would require such a launch. The Air Force was looking into multistage rockets capable of launching payloads comparable in size and weight to those designated for Shuttle launch. Other problems NASA was now facing were monthly costs in the $6–7M range just to keep HST ready for launch. The payload could not simply be stored; the gyros needed to be exercised to keep them in good condition.[33]

---

[26]Handwritten notes dated 3/1/86.
[27]Letter from Harwit to George Field, January 7, 1986.
[28]Notes attached to letter from Harwit to Field, dated January 7, 1986.
[29]Letter from Martin Harwit to William R. Graham, January 20, 1986.
[30]Letter from Martin Harwit to Burton I. Edelson, January 20, 1986.
[31]Letter from Charles J. Pellerin, Jr. to A/Dr. Graham, Jan 22, 1986.
[32]Minutes of the Astrophysics Management Operations Working Group April 17, 1986, page 2.
[33]Minutes of AMOWG meeting held on June 3, 1986, pages 2–3.

Charlie began pointing out the great value of Explorer Class missions: An Explorer Announcement of Opportunity had been sent out, and there was discussion of doubling the Explorer budget now, during the hiatus in Shuttle launches. However, Jeffrey Rosendhal on Dr. Edelson's staff warned that AXAF could feel the competition for funds if the Explorer program was to receive a significant infusion of money.[34]

By the August 1986 Council meeting, Charlie talked about options open to the Astrophysics Program in view of budgetary developments and a likely decline in the number of Shuttle flights per year. Delta and Titan launches might have a major role to play, and few of the major payloads could be launched by Deltas. If the launch cost then had to be paid with Astrophysics Division funds, they would require approximately $60M per Delta and $250M per Titan launch.

COBE by then was at an advanced stage of development and, with modifications of a Delta support system, could be taken into a polar orbit from Vandenberg Air Force Base. Payment for the launch vehicle, however, remained to be worked out. The Extreme Ultraviolet Explorer EUVE and the X-ray Timing Explorer (XTE) would be next in line.[35]

The final end to any association with Space Station surfaced on April 4, 1987, when The New York Times reported, "President Reagan, who has enthusiastically supported a proposed space station, today scaled back the plans... to cut the rapidly rising estimated costs of the space station and thus to reduce political opposition in Congress."[36]

9 Harvey Tananbaum's political action committee

Over the summer of 1986, Harvey Tananbaum had begun setting up an Ad Hoc Planning Group for the Great Observatories, a broadly representative resource group that would write to members of Congress, potentially give talks on the Great Observatories, and particularly AXAF and SIRTF, to publicize the effort among both colleagues and other supporters. A meeting held at the Center for Astrophysics in Cambridge, on October 3 that fall, discussed progress on AXAF and SIRTF, set objectives for the planning group, analyzed the activities of different Congressional committees, plans to involve fellow scientists, and other ways to go forward.[37,38]

As an aid to those wishing to make presentations, Harvey, helped by Kathy Lestition on the SAO staff, made available an immaculate Great

[34] Ibid page 4.

[35] Minutes of the August 15, 1986 AMOWG meeting, pages 2–3.

[36] Google: excerpted from the New York Times, Saturday, April 4, 1987, "President Scales Back Plans for Space Station over Cost".

[37] Minutes of the August 15, 1986 Meeting of the AMOWG, page 6.

[38] Letter from Harvey Tananbaum to Martin Harwit, 22 September 1986.

Observatories slide set, comprising 175 35-mm slides dealing not just with the space missions, but with astronomy and astrophysics as a whole, and how the Great Observatories fit into this grander scheme. These slides were useful to any member of the Ad Hoc Planning Group seeking to give a polished presentation. To accompany this slide set Christine Jones, Carolyn Stern, and Bill Forman of SAO wrote a book "New Windows on the Universe, The NASA Great Observatories", published in 1987. It devoted nearly a page of explanatory background text to each slide, whether it dealt with one of the Great Observatories and how it functioned, or the significance of gamma ray bursts, eta Carinae, Sunspots, dark matter, or radio or X-ray jets.[39]

10 Liaison with the solar systems exploration division

Meanwhile Charlie was seeking closer ties with, and support from the Solar System Exploration Division within OSSA, headed by Geoff Briggs. He asked me to work with Jim Head of Brown University, my counterpart in the planetary community working with Briggs, to come up with a recommendation regarding common interests for both SIRTF and the planned SOFIA airborne observatory. On July 24, 1987 Jim and I co-chaired a small meeting at Brown University in Providence, RI, that included Bob Brown and Warren Moos, both members of the Astrophysics Council, and David Morrison, Bradford Smith and Jurgen Rahe, representing planetary sciences and Geoff Briggs's Management Council. By the end of the day, we formulated a letter to Charlie and Geoff, in which we cited a number of areas of potential mutual interest to astrophysics and planetary science and recommended the establishment of a small group that could regularly meet to seek closer ties.[40]

11 International cooperation

With the difficulties the Challenger Disaster posed, international cooperation to advance the astrophysics program was also coming up in discussions. At the August 1986 meeting of the Council, I was asked to write Dr. Roger Bonnet, Scientific Director at ESA, whether he might be willing to attend an upcoming Council meeting. Bonnet agreed and, with Edelson, attended the February 1987 meeting of the Council. He impressed us with the strong impact that

[39]"New Windows on the Universe: The NASA Great Observatories," slide set and text prepared by C. Jones, C. Stern and W. Forman, Harvard-Smithsonian Center for Astrophysics.

[40]Letter from James W. Head III and Martin Harwit to Charles Pellerin and Geoffrey Briggs, July 24, 1987.

cancellations and uncertainties at NASA were having on ESA programs. It was a somber but constructive exchange.[41,42,43,44]

At a December 12, 1986 meeting, Council members had raised the question of finding a launch for the German-led ROSAT mission. Unable to launch AXAF soon, the US X-ray community was eager to participate in this German-led venture. By mid-1987, Charlie informed the Council that negotiations were underway to obtain either the earliest Delta or earliest Atlas/Centaur launch for ROSAT. ROSAT was eventually placed into orbit on a Delta-II Heavy rocket on June 1, 1990, and successfully operated until February 12, 1999. At this same meeting, Charlie also revisited the possibility of combining SIRTF with the ESA Infrared Space Observatory.[45] But this did not lead anywhere. By then ESA was already far advanced in its plans, and the US infrared astronomy community preferred to work on its own mission design.

12 The President's Science Advisor

During Bill Graham's brief tenure as NASA Acting Administrator, he had exhibited a keen interested in the Astrophysics program. When Graham was named the President's Science Advisor in late 1986, Harvey Tananbaum and I arranged to visit him on December 23. The main thrust of that meeting was to promote interest in AXAF. But Graham said he would also enjoy a discussion of the current problems in astrophysics and cosmology. This was an opportunity to have George Field participate, and on February 27, 1987, George, Harvey, and I again met with Graham and Richard G. Johnson on his staff. George gave a great presentation and we all had an enjoyable time. As I wrote George, a few days later, "That we got asked back was an excellent boost. That Dick Johnson cared enough to ask us to lunch and gave us good advice may be even more significant."[46,47,48]

13 A changing atmosphere at NASA

By April 1987 the atmosphere at NASA was also changing. In May 1986, James C. Fletcher, a former NASA Administrator had been brought back to permanently replace Jim Beggs. And on April 6, 1987, Lennard Fisk succeeded Burt Edelson as Associate Administrator of OSSA. His style could not have

[41] Minutes of the August 15, 1986 Meeting of the AMOWG, page 6.
[42] Letters from Martin Harwit to Roger Bonnet, August 22, 1986.
[43] Ibid., November 26, 1986.
[44] Letter from R. M. Bonnet to Martin Harwit, November 11, 1986.
[45] Minutes of the June 23, 1987 AMOWG meeting, page 3.
[46] Letter from George Field to Dr. William Graham, 31 December, 1986.
[47] Minutes of the AMOWG Meeting of February 24, 1987, page 2.
[48] Handwritten letter from Martin to George, dated March 1, 1987, final paragraph.

been more different from that of Edelson. Where Edelson had attempted to reach decisions through consensus that rarely could be found within NASA or the scientific community served by NASA, Fisk was analytical and decisive. He willingly considered advice, but was guided by a structured long-term vision for the tasks OSSA should be pursuing.

Helpful to Pellerin also were the good relations he had long enjoyed with both Fletcher and Fisk. When the Astrophysics Council first became a permanent fixture in the Astrophysics Division and its membership needed to be broadened, one of the first letters of invitation Charlie had sent out went to Jim Fletcher, whose advice as former NASA Administrator would have been invaluable.[49] Fletcher never did accept, but the invitation was symptomatic of the relation between the two that had sprung up during Fletcher's earlier term as NASA Administrator. Fisk also was a long-term friend. In the 1970s, when Len and Charlie had both been conducting energetic particle studies at the Goddard Space Flight Center, Fisk had often been a helpful mentor to the younger Pellerin then working toward a PhD while a full time researcher at NASA.[50]

This changed atmosphere at NASA led to a more favorable view of the Great Observatories and resulted, over the summer of 1987, in NASA's advocating AXAF as its prime science mission for a new start in FY89, to be proposed in the President's budget submission in early 1988.[51]

14 Stepping down from the council

By early summer 1987, I had been offered the Directorship of the National Air and Space Museum and told Charlie I felt I needed to leave the Council to take on this new job.[52]

When I departed Cornell for Washington, that August, hundreds of pages of correspondence, minutes of Council meetings, lists of memberships of the many national, NASA, and National Academy committees, with which the Council had interacted, were left behind at Cornell, where a young archivist volunteered to archive these and everything else I had filed in an ever-growing number of filing cabinets during 25 years at the university. I came across these papers on a first visit to those archives last winter and they have formed the basis for the current paper.

Half a year after my starting work at the museum, during the lunch hour on February 1, 1988, the phone rang. It was Bill Graham. "I thought you would

[49]Letter from Charles J. Pellerin, Jr. to James C. Fletcher, March 14, 1985.

[50]Conversation of Charles J. Pellerin with Martin Harwit, at Pellerin's home, Boulder, CO, May 26, 2008.

[51]Letter from James C. Fletcher to the Hon. Edward P. Boland, Chair Subcommittee on HUD-Independent Agencies, Committee on Appropriations, House of Representatives, October 14, 1987 (NASA Headquarters Archives, AXAF-1 File 5604).

[52]Minutes of the AMOWG Meeting of June 23, 1987, page 4.

like to know," he began, "that the President's FY89 budget went public today, and AXAF is in it slated for a new start." That was fantastic! And Bill obviously was as delighted as I was. I later learned that the new start for AXAF had been in considerable doubt at the last minute.[53] Perhaps Bill Graham's personal enthusiasm for the mission had carried the day.

At any rate, with AXAF now officially started, and with interest for SIRTF established in both the Astrophysics and Solar System Science Divisions, it looked as though the Great Observatories concept was being accepted throughout the relevant bureaucracies. There would be years of hard work ahead, but at least the past 3 years had established acceptance of a new way of planning space missions.

15 Looking back

Looking back two decades later, several questions may be asked:

1. Did the Great Observatories deliver what our comic-book presentation had promised?
2. Could we have done better in our efforts to promote the missions involved?

I believe that the response to the first question is positive, especially where HST, AXAF (now renamed Chandra), and SIRTF (now renamed Spitzer) are concerned. These three missions have now been loosely coordinated for many overlapping years:

We had said we needed to look further back in time to see how galaxies formed and evolved. Hubble and Spitzer have now shown how galaxies form through violent mergers of initially small fragments, some of which already contained heavy chemical elements when the Universe was only a few hundred million years old. With Hubble we study the visible light reaching us from un-obscured stars and clusters in these mergers. With Spitzer we observe the mergers' deeply dust-shrouded regions, including their active galactic nuclei whose X-ray emission Chandra registers.

We had guessed that planets exist around other stars. By now Spitzer has even obtained a coarse spectrum of a planet.

We had wondered about black holes and quasars and explosive outflows. These still are under intense scrutiny but, for example, even features like the jet emanating from the quasar 3C273 have by now been observed by Chandra, HST and Spitzer.

The coordinated program Pellerin envisaged has yielded striking insights. Of course there is still much to be done. We still do not understand dark matter or dark energy; but we have begun to obtain at least some insight into the

[53] "Revealing the Universe, The Making of the Chandra X-ray Observatory, Wallace Tucker and Karen Tucker, Harvard University Press, Cambridge, MA, 2001, pp 102–104.

nature of dark matter by observing the collisions of galaxies with Chandra, HST and ground-based optical observatories.

Regarding whether we could have done better: That is more difficult to decide. We certainly misjudged situations, made mistakes, and suffered under misconceptions. We felt constrained by President Reagan's political directives to work with Space Station. I find it amazing how seriously we tried to tie the fates of the Great Observatories to Space Station and the co-orbiting platforms that were to accompany it so that astronauts could refurbish our planned observatories to make them permanently active and productive in space. What saved us from ourselves, in the end, were two events and one miscalculation:

First, when budgetary realities sank in, the co-orbiting platforms were quickly abandoned by the Space Station project. Second, with the Challenger disaster and the many-months-long hiatus of launches that followed, it became clear that, where possible, it might be better to launch science missions on unmanned rockets, rather than Shuttle. While HST and Chandra were still launched on Shuttle, and HST has been serviced by astronauts, none of the Great Observatories have required Space Station operations. In the end, all this worked out well for the scientific community, but it goes to show how political directives can lead to engineering solutions that in hindsight are bizarrely complex when far simpler technology will do.

Finally, our miscalculation was that the Great Observatories should be permanent, or at least long-lived. It never occurred to us that the cost of mission operations, and support for the large scientific community required to handle the high data rates produced by these extremely powerful missions, would be unaffordable if the life expectancy of the observatories was too long. The NASA Senior Reviews conducted every other year have shown that we need to phase out some observatories to pay for others that either complement them or exceed their capabilities. The question of when a given observatory has peaked in its scientific yield and should be phased out is complex and needs careful scrutiny. Some observatories are worth continuing for a long time, whereas others appear ready sooner for replacement by missions capable of providing information totally inaccessible to date. These decisions are always wrenching. But, back in 1985, I think we failed—even as we loudly advocated the need for them—to see how immensely powerful the Great Observatories would turn out to be and how quickly they would fulfill their initial promise.

Few of us also took into account that there will always be a limit on how much society can afford to expend on science. Feeding our children remains more immediate than flying observatories in space and, one way or another, society tends to keep this in mind as it funds astronomical efforts.

Acknowledgements Work related to this paper was started at Durham University's Institute of Advanced Study, where I held a Distinguished Fellowship in the fall of 2007. I thank the directors and staff of the IAS for their wonderful hospitality. I would not have been able to write this paper without the dedication with which Ms. Elaine Engst archived my papers at Cornell University, two decades ago. The staffs of Cornell University's Kroch Library archives and the NASA History Division archives helped me greatly with their guidance. I thank Harvey Tananbaum for his critical reading of an earlier draft of this paper.

Exp Astron (2009) 26:179–199
DOI 10.1007/s10686-009-9149-6

REVIEW ARTICLE

Future technologies for optical and infrared telescopes and instruments

Colin Cunningham

Received: 20 January 2009 / Accepted: 21 January 2009 / Published online: 4 March 2009

Abstract The theme of this conference is the evolution of telescopes over the last 400 years. I present my view on what the major leaps of technology have been, and attempt to predict what new technologies could come along in the next 50 years to change the way we do astronomy and help us make new discoveries. Are we approaching a peak of innovation and discovery, and will this be followed by a slow decline? Or are there prospects for even further technology leaps and consequent new discoveries? Will global resource and financial crises bring an end to our great ambitions, or will we continue with bigger telescopes and more ambitious space observatories?

Keywords Telescopes · Infrared · Optical · Technology

1 The seeds of time

The process of innovation in Optical and Infrared (IR) Astronomy has been punctuated by the introduction of disruptive technologies which have changed the way astronomy is done on the ground and in space. This paper shows that over the 400-year history of the astronomical telescope the tension between science drive and technology push has resulted in significant steps forward, although the gestation periods for the development of new technologies and their practical introduction into telescopes and instruments have often been prolonged. Through a survey of current ideas for new technology from telescopes through adaptive optics to instruments and detectors, I attempt to

C. Cunningham (✉)
UK Astronomy Technology Centre, Royal Observatory Edinburgh,
Blackford Hill, Edinburgh, EH9 3HJ, UK
e-mail: Colin.Cunningham@stfc.ac.uk

define the critical technologies which should be developed. I conclude with some speculation on where we might go in the next 50 years, which of course will lay me open to future ridicule. As Banquo said in Macbeth: 'If you can look into the seeds of time, and say which grain will grow and which will not, speak then unto me'.

2 Technology development in astronomy

There are several ways that technology development can have an impact on the capability to do new astronomy:

- It can open new parameter space, which leads to new science. Examples are space telescopes providing access to new wavebands such as X-ray and gamma-ray.
- It can reduce cost, and so make new capabilities and consequent new science affordable. Good examples of this are the introduction of segmented primary mirrors, which has enabled the 8 m barrier to be surmounted at reduced cost per square metre, and the introduction of array detectors in the IR and submillimetre (submm) bands which enabled imaging and spectroscopy to become feasible at these wavelengths.
- It can reduce the time needed to reach a science goal—for instance the introduction of multi-object fibre-based spectrometers enabled huge redshift surveys which would not have been possible without such a multiplex gain.
- It can reduce the risks to a project; that is the likelihood of not delivering the performance needed to make science gains, or not meeting cost budgets or time constraints. Examples of such technologies are harder to find—most new technologies are likely to increase risk. However, the introduction of computer-aided design and modelling is an example of new technology which reduces risk by making it easier to predict performance before a system is built, at least in principle.

2.1 Disruptive technology

From the beginning of astronomy as a science, discoveries have been made following the introduction of new technology. It is instructive to use the concept of disruptive technology as applied to commercial markets [1]—where a new technology can completely change the market place. Disruptive technology can be defined as new technology that has a serious impact on the status quo and changes an entire way of doing things. For example, in the commercial world gramophone records were replaced by compact discs, without making fundamental changes to the market, but the next new technology, personal digital devices such as the I-Pod, resulted in major changes to the process of distributing music, with internet downloads almost replacing physical retail shops. Similar events can be seen in the history of astronomy:

1609 The capability of making precision glass lenses for eyeglasses, and critical innovations in Holland such as the appreciation of performance enhancement by stopping down the aperture led to Galileo's telescope and discovery of the moons of Jupiter, with profound implications in support of the Copernican model.

1663–1672 James Gregory and Isaac Newton's new concepts for reflective telescopes, combined with techniques for grinding and polishing metal mirrors in speculum led to William Herschel's telescopes and the consequent discoveries of Uranus and a whole new 'zoo' of nebulae.

1864 Léon Foucault invented the silvered-glass reflecting telescope, implemented in the 80 cm telescope in Marseille [2]. Together with his knife-edge test which made mirror manufacturing a much more controlled process, this gradually became the preferred technology for large telescopes, particularly after the limits of large refractors were reached.

1880–1888 Charles Piazzi Smyth took up Newton's suggestion that astronomical observations would be improved in the "serene and quiet air, such as may perhaps be found on the tops of the highest mountains above the grosser clouds" [3] by carrying out observations on Mount Teide in Tenerife, and indeed did find significantly improved observing conditions. The first major observatory on a high mountain site was the Lick Observatory in 1888, which was used to discover four new moons of Jupiter. Of course, modern astronomy from the ground has since been revolutionised by high mountain sites such as Mauna Kea, especially at infrared wavelengths. Thus mountain astronomy can be seen as a disruptive technology in that it changed how we do things, although without any major technological innovation.

1908–1948 The next major developments which could be regarded as disruptive were the series of large reflecting telescopes built by George Ritchey and George Ellery Hale, including the Hooker 100 in., which culminated in the 200 in. telescope, only completed 10 years after Hale's death. The 200 in. telescope was based on the technologies of low-expansion Pyrex glass, ship engineering and passive flexure compensation. These large reflectors were very productive, with the most famous work being Edwin Hubble's 1929 confirmation of the expansion of the universe, based on Vesto Slipher's original red-shift measurements of 1915.

1975–1990 The next really disruptive technology was the introduction of electronic array detectors to replace photographic plates in the visible and single detectors in the IR. Because of the increased sensitivity and linearity of Charged Coupled Devices (CCDs) and

IR Arrays, it was possible to improve the sensitivity of the then current 2–4 m telescopes by more than an order of magnitude, so that there was little need to build bigger aperture telescopes during this period.

1975 The introduction of computer control, pioneered at the Anglo Australian Observatory on an equatorial mount telescope, made altitude-azimuth telescopes and ultimately active telescopes practical, and paved the way for the 8–10 m telescope era. The BTA-6 (Bolshoi Teleskop Azimutalnyi) Russian 6 m telescope was one of the first to rely completely on computer-controlled technologies, although many other problems contributed to its lack of scientific impact.

1990 After a long gestation period, beginning with Hermann Oberth's proposal in 1923, via Lyman Spitzer's seminal paper in 1946, space telescopes such as the Hubble Telescope again changed the paradigm of doing astronomy, and resulted in a broad range of discoveries and beautiful images which sparked the public imagination. Perhaps the most important was the discovery of the anomaly of Supernovae distances, and the hypothesis of Dark Energy accelerating the expansion of the universe.

1990–1992 In the 1980s it began to be realized that there was little scope for further improvements in sensitivity for ground-based telescopes as detectors (at least at visible wavelengths) were almost perfect. Therefore, the only way to improve performance was to start looking at new technologies to enable larger aperture telescopes to be built. Jerry Nelson began to research segmented mirrors. Nelson's ideas culminated in the phenomenally successful Keck telescopes, which led to many discoveries, from galaxies at the edge of the universe using gravitational lenses to the recent images of giant planets. Roger Angel took an alternative approach (again based on a suggestion from Newton!) using spun cast honeycomb monolithic mirrors in borosilicate glass, also very successfully, leading to the Large Binocular Telescope.

1997 An alternative to these approaches was developed by the European Southern Observatory (ESO) and in particular by Ray Wilson. He realised that it was possible to automate the actuators which had been routinely used to take out static errors in primary mirrors, and if the primary mirror could be made thin enough, and correction for position and tilt of the secondary automated as well, then slow wavefront errors induced by gravitational flexure could be sensed and corrected. This active-optics technique was pioneered on the ESO New Technology Telescope (NTT), and subsequently adopted by the four 8 m telescopes making up the Very Large Telescope (VLT), the two Gemini telescopes and Subaru as described by Lothar Noethe in this volume. You could argue that the most disruptive technology of the twentieth century was the introduction of fast computer control which allowed alt-az telescopes, active optics and segmented telescopes to be practical.

1989–2008 Newton pointed out in 1704 "...there would be certain bounds beyond which Telescopes could not perform. For the Air through which we look upon the stars, is in perpetual tremor" [3]. Any technology which could overcome these effects from the ground is by any definition disruptive. In 1953 Horace Babcock [4] proposed a scheme base on wavefront sensing with a rotating knife-edge and a type of liquid deformable mirror, but practical technology was not available until a French team in 1989 built the COME-ON system [5], and the US Navy built their system on Maui, which would remain classified until 1991. Adaptive optics has now matured on the VLT, Gemini, and Keck telescopes, to the stage when revolutionary high-resolution images and astrometry are possible, enabling star dynamics near the centre of our galaxy to reveal evidence of a massive black hole [6], and recent direct imaging of gas giant planetary systems by the Gemini and Keck telescopes.

2.2 Technology planning: roadmap development

What new technology that is emerging now could revolutionise how optical and IR astronomy is carried out? It is noteworthy from the last section that the gestation period of novel technology from concept to widespread adoption and scientific productivity is usually measured in decades, so it is likely that the ideas presented here would take a considerable time to make an impact. How can we assess the readiness of these technologies to ensure a sensible balance of risk and performance improvement? One well-known method used by NASA and many aerospace companies is the use of Technology Readiness Levels (TRLs) [7]. NASA expects to reach TRL 6 or 7 by 5 years from launch in order to control risk. We have seen that it often takes many decades to climb the TRL scale in optical and IR astronomy on the ground.

It is worth considering the important factors involved in inserting this technology into a successful project. There is a degree of conservatism involved in large astronomy projects, particularly space ones. It is commonly accepted that successful projects need to balance three aspects: time, cost and performance; and that the task of a project team is to maintain constructive tension between them to ensure timely and effective attainment of science goals within resource constraints. How can we encourage innovation and manage risk in this project environment?

One way is through the application of strategic roadmapping techniques, as pioneered by Motorola. In an industrial situation, this is normally done by a top-down approach, first considering the market a company wishes to penetrate, then the product which could do this, and finally the technology and resources needed to develop a product which can be sold at a competitive price in a realistic timescale. The analogues in a scientific environment are

Fig. 1 Technology Roadmap

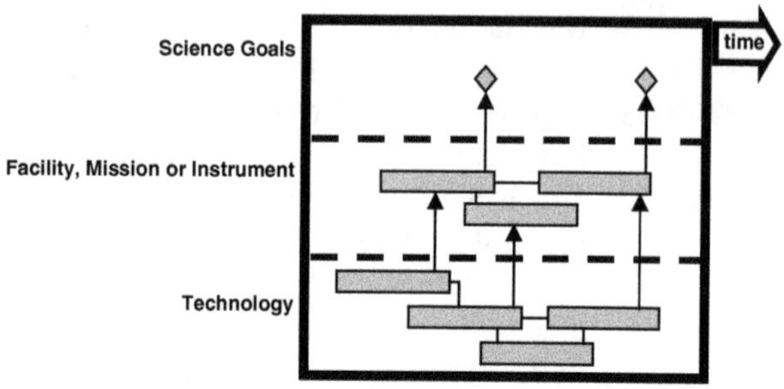

science goals standing for markets, and facility or instrument for product (Fig. 1). Techniques are available to develop roadmaps for these situations [8]. However, this top-down approach is rarely followed completely in practice even in an industrial context. Novel technology often comes from another research sector or market and can have a revolutionary effect on science capability.

Can we define science needs, then facility requirements, and follow this up by deciding what technology should be developed and when? This is the aim of the decadal surveys in the USA [9], at least at the top two levels. However, we have seen that many advances in astronomy have been brought about by the introduction of technology developed for other purposes—good examples being computer systems in the 1960s and 1970s resulting in automatically guided and active telescopes, CCDs developed originally for computer memory applications and IR detectors developed originally for military applications and even the telescope itself, as we see elsewhere in this volume.

Technology adoption in astronomy is a much less linear process, with iterative feedback between technology capability and science needs. Some technologies push us into new parameter space, as when new wavebands are opened up by the availability of new detectors.

We must also remember that there are political, sociological and economic imperatives which contribute to what we can do in astronomy. This could be regarded as the top level driver of the roadmap chart. If we do not take these issues into account, new facilities will not be built and new discoveries will not happen. In the era of big-budget astronomy from the James Webb Space Telescope (JWST) to the Atacama Large Millimetre Array (ALMA), political drivers influence the motivation for governments to spend money on astronomy. One motivation is to gain or maintain world standing in science, which is particularly important for the new economies of China and India. It is often said that the purpose of the big space agencies is not to do science, but to feed industry with contracts and motivation to develop technology with stretching requirements, ultimately providing economic benefit through innovation. A further motivation is to inspire young people into careers in science and technology, again for economic goals. The strive to reach fainter and more distant objects drives us to bigger and more expensive facilities, and it is not possible to succeed in these ventures without taking care of issues outside the purely scientific objectives.

2.3 Science drivers

Leaving aside these considerations, let us take a look at the current astronomy science roadmaps, and see if they can form a basis for technology planning. Three recent exercises in astronomy planning come up with clear consensus on the top-level goals for future missions and facilities:

NASA 2006 Strategic Plan

Discover the origin, structure, evolution, and destiny of the universe, and search for Earth-like planets [10]

ESA 2008 Cosmic Vision

What are the conditions for life and planetary formation? How did the Universe begin and what is it made of? [11]

ASTRONET 2008 European Science Vision

The Extremes of the Universe

- What is Dark Matter & Dark Energy?

Galaxy Formation & Evolution

- How were galaxies assembled?
- How did our Galaxy form?
- How do galaxies form and evolve?

Origins and evolution of stars and planets

- Where are most of the metals throughout cosmic time?
- How do we and the solar system fit in? [12]

The ASTRONET exercise has taken these top level astronomy objectives, and used them to develop a prioritised roadmap for future facilities based on wide consultation in Europe [13]. The EU-funded OPTICON Key Technologies Network is using this science and facilities roadmap to produce a plan for the technology developments Europe needs for telescopes and instrumentation over the next 15 years. It is important to remember that the message that 'the needs of science drive technology requirements' while true to some extent, ignores the real-world process where innovation in astronomy often comes from the introduction of technology developed for completely different purposes, usually with much more generous funding than astronomy Research and Development.

3 Predicting the next 50 years

Producing a science-facilities-technology roadmap for the next 10 to 20 years is challenging, but we can see that it is possible and can be a useful exercise to guide decision making processes on where to invest. Looking further into the future is fraught with perils, but I will try anyway. Remember what Niels Bohr (or was it Yogi Berra?) said: "Prediction is very difficult, especially if it's about the future".

Let us take a look at how accurate previous predictions have been. A hundred years ago, the respected senior figure in the US astronomical establishment, Edward Pickering [14] said "Have we at length reached the limit in size?..... It is more than doubtful, however, whether a further increase in size is a great advantage. Much more depends on other conditions, especially those of climate, the kind of work to be done and, more than all, the man behind the gun"—this in the era when his great rival George Ellery Hale was planning the bigger and bigger telescopes which would culminate in the 200 in. at Mt. Palomar. Some of Pickering's other predictions were nearer the mark: he suggested that we might build two large observatories at +30° and −30° in latitude, operated by service observing, with a central data reduction centre. However, he also proposed that the big scientific challenge was in determining stellar distances, and that methods which might be adopted included measuring differential absorption in the interstellar medium where he thought there was some evidence of variations in speed of light with wavelength. This idea along with this statement "If we could determine our motion with reference to the ether, we should have a fixed line of reference to which all other motions could be referred" were to very rapidly be seen to be relics of the previous century!

So, it is easy to look foolish when making predictions—and no doubt the present author is about to fall into that trap. However, other predictions have been more prescient. Just over 50 years ago, Henry King in his classic book on the history of the telescope [15] said: "Thus we return to the old problem of escaping from the damaging optical effects of the earth's atmosphere—and there is no escape as yet....in the future, when interplanetary travel becomes possible, observing stations and, ultimately, observatories may be established in space". In this he not only foresees the series of space telescopes we are so familiar with, but hints at the need for adaptive optics on the ground.

In order to put some perspective on the predictions I am about to make, let us look at some fundamental questions. Firstly, what can we measure about our universe? Martin Harwit's seminal book on cosmic discovery addresses this in depth [16]. In summary, there are only four ways we can obtain information from the universe outside the earth. Firstly, we can measure the properties of electromagnetic radiation arriving at our telescopes: primarily intensity in particular wavebands; and wavelength and hence energy of photons. At longer wavelengths, we can directly measure amplitude and phase, and finally we can measure polarisation, linear and circular and orbital angular momentum, although, as Harwit has pointed out [17], the latter has yet to be detected from non-artificial sources and we do not know if it could encode any useful astrophysical information.

At first sight then, the task of the telescope and its instruments seems straightforward: measure these properties and their variation in space and time, and from this information deduce the physics of astronomical phenomena, by observing: photon emission, thermal and energetic; photon absorption;

eclipses; gravitational lensing behavior; angles between objects (astrometry); and finally Doppler shift, and hence dynamics of objects.

You might ask why this is all so difficult—and expensive—but of course the answer is that the intensity of photons carrying information about most astrophysical phenomena is very low and highly contaminated by sources of background radiation.

Other probes of the universe do not measure electromagnetic radiation, although authors occasionally stretch the definition of the telescope rather a long way to include measuring some of these phenomena. We can measure particles or phenomena arriving at the earth or at a space facility, such as neutrinos, gravitational waves, cosmic 'rays' and magnetic fields. We can also collect matter arriving at the earth or by sample-return missions: meteorites and particles, or we can send a probe to another solar-system body and carry out experiments.

The latter two options have obviously very limited reach, and astronomy has been dominated for the last 400 years by measurement of electromagnetic radiation. Maybe in 50 years time that will no longer be the case, and neutrino and gravitational wave observatories will become dominant, but it would seem unlikely that such a radical change will occur. I will concentrate on measuring electromagnetic radiation, and for the purposes of keeping the paper reasonably focused, restrict myself to optical and IR wavelengths. I will follow the photons as they are collected by the telescope, are processed by the optics and instruments, and then are detected and processed by the software needed to convert measurements into useful data.

3.1 Ground-based telescopes

Novel technology has had a major influence on the development of telescopes, as we have seen. The biggest recent impact has been due to the introduction of segmented primary mirrors with the Keck Telescopes, which have enabled the concepts of 30–60 m telescopes to become feasible. It is worth noting that the first segmented optical telescope was built by Guido Horn-d'Arturo at the Bologna Observatory between 1935 and 1952. His telescope had a non-steerable primary mirror of 1.8 m diameter consisting of 61 segments: an example of the long gestation period we commonly see in innovation for astronomy—although Jerry Nelson was not aware of this work when he came up with his segmented telescope concept.

The rival technology of spun-cast honeycomb mirrors from Roger Angel's team has also been extending to form the equivalent of a 24 m primary mirror by combining seven 8 m monoliths as the basis of the Giant Magellan Telescope.

Building the currently envisaged Extremely Large Telescopes (ELTs) provides sufficient challenges to make it hard to envisage what might follow. In terms of single primary mirror concepts, it would seem that the limitations on size of the segmented primary mirror concept are primarily economic: the ESO Overwhelmingly Large Telescope (OWL) was dropped mainly because it was not seen to be affordable. The lessons learned in constructing the Thirty Metre Telescope (TMT) and the European-ELT may enable an even bigger ground-based telescope to become affordable in the more distant future—perhaps even on a cold dry site like in the Antarctic, where the balance between thermal IR performance and spatial resolution could beat what is possible in space.

The alternative to building bigger single-aperture telescopes in order to gain spatial resolution is to build interferometers. Again, this has a long history, starting with Michelson's interferometer added to the Hooker telescope in 1919. Progress in interferometry from this date until relatively recently was severely limited by lack of suitable high-precision beam-combiners, path-length compensators and fringe-tracking detectors. The availability of such technologies has now brought optical/IR interferometry into the mainstream—for example ESO's VLTI and the Keck Interferometer. Major technological advances were made in aperture synthesis imaging interferometry (as pioneered in radio astronomy) by the COAST interferometer [18], and are now being taken forward by the Magdelana Ridge Observatory Interferometer [19].

Critical technologies which are needed to take interferometry to the next levels of performance are: integrated optical devices to enable beam combining and even path length compensation to be made in monolithic devices which require no adjustment and can be more compact than discrete optics, lower-noise fringe-tracking detectors, nanometric metrology and low-dispersion optical fibres. Many of these technologies are being adapted from devices developed for communications applications by the photonics industry. Considerable work has been done on 2-D beam combiners fabricated in silicon [20], and it is also now possible to make 3-D structures using femtosecond lasers to modulate the refractive index of glasses and other materials [21].

Looking further into the future, are there prospects for disruptive technologies which could change the whole paradigm of optical and infrared telescopes? This is where we get into dangerous speculative areas, and the prospects for foolish conjecture become more likely. In the 400 years since Lipperhey, the only new basic telescope concepts have been those of Gregory and Michelson. One area where we can expect continued rapid advance in technology is in electronics and computing. What would be possible if extremely fast analog-to-digital (a–d) converters and data processors became available—perhaps in the form of optical rather than electronic computers? Would we then be able to build a phased-array telescope in the infrared or even visible bands—in other words the equivalent of a Square Kilometer Array (SKA) in the visible? If so, we could envisage a telescope which could image the whole sky by electronic beam steering rather than pointing an optical telescope at a small field. It could also be a high-resolution spectrometer, with no dispersion losses.

The key to this would be to devise an array of tiny antennas not very much bigger than the wavelength of the light, and then measure the phase and amplitude of the incoming photons with our hypothetical extremely fast optical amplifiers and a-d converters. Of course, even at radio or submillimeter wavelengths, such direct detection is not possible, and the incoming waves must be mixed with a longer wavelength local oscillator to form an 'intermediate frequency' (IF) signal which is of low enough frequency to be analysed. This is also possible at shorter wavelengths, and has been done at 10 μm in the Berkeley Infrared Spatial Interferometer [22]. However, if we attempt this technique at even shorter wavelengths, there are two physical restrictions which provide a more fundamental block to this idea than the unlikelihood of the required fast amplifiers, a–d converters and correlating computers being available.

One problem is the very narrow bandwidth of a heterodyne receiver. With known mixing and detection techniques at THz frequencies (e.g. ALMA), the maximum bandwidth is unlikely to be more than 20 GHz. Even with huge advances in IF amplifier design, it is hard to see bandwidths of more than 50 GHz, and this translates to less than 0.2 nm at 1 μm wavelength. This means that some complicated observing scheme would be needed to measure spectral lines to a reasonable signal/noise ratio. The more fundamental limitation to sensitivity of detectors which measure phase and amplitude, as opposed to integrating detectors like photodiodes, CCDs or bolometers is the quantum noise limit. Measuring phase and intensity at the same time means that the uncertainty principle has to be taken into account, and this rapidly dominates as frequency increases, whereas the ultimate noise limit for direct detectors is only dependent on the number of photons detected. Hale et al. [22] show that the ratio of signal/noise for direct detection against heterodyne detection is:

$$\sqrt{\frac{e^{h\nu/kT}-1}{1-\varepsilon}}$$

where T is the temperature of the optics and atmosphere, h is Planck's constant, ν is the frequency of the detected radiation, k is Boltzmann's constant, and ε is the quantum efficiency, including transmission through the atmosphere and optics. At a wavelength of 10 μm, and at room temperature with a QE of 0.9, this ratio is 37 times in favour of direct detection, although if atmospheric phase errors are included, that could tip the balance more in favour of heterodyne detection. However, at 1 μm the ratio is more than 10^{11}, so a NIR phased-array heterodyne telescope will never be competitive.

Recently, another concept has been suggested for the radio frequency SKA. The Fast Fourier Transform (FFT) Telescope recently proposed by Max Tegmark and Matias Zaldarriaga again relies on huge improvements in computing power to provide a powerful all-sky interferometer. They suggest that this could eventually be used at all wavelengths to provide the ultimate telescope. Considerations mentioned above should also apply to this FFT telescope, and it would not be competitive with 'analog' telescopes at visible and IR wavelengths even if future computing power made it possible. However, radical ideas for new telescope concepts may eventually prove practical—remember that the Astronomer Royal Richard Woolley said in 1956 that manned space flight would never be funded.: "It's utter bilge. I don't think anybody will ever put up enough money to do such a thing.... What good would it do us? If we spent the same amount of money on preparing first-class astronomical equipment we would learn much more about the universe".

Another radical telescope concept which new technology may now make practical at visible wavelengths is the intensity interferometer invented by Robert Hanbury Brown and Richard Twiss [23]. It is based on measuring the variation in correlation between photon streams measured by two telescopes looking at the same source as the telescope spacing is changed. New fast detectors may make measurement of quantum fluctuations possible, using similar systems to that of Hanbury Brown and Twiss, opening up new discovery space, as suggested by Dravins [24].

A radical concept which might at first sight be deemed to be rather impractical has been proposed by Thomas Ditto, a rather maverick figure somewhat outside the scientific establishment—but it is worth remembering that truly disruptive ideas nearly always come from outside the establishment. His idea [25], is based on Newton's experiment where he first understood the nature of light by splitting it into its constituent wavelengths using a prism, and showed that it could not be further split. Ditto proposes a telescope with no primary mirror, just a giant grating replacing Newton's prism. With this he suggests spectroscopy of millions of objects could be obtained simultaneously. His idea has engendered some serious debate, but it is not yet clear whether it is at all practical.

It seems most likely that in 50 years time we will still be building telescopes which are fundamentally no different from Newton's and Michelson's. Perhaps we will find new ways of building bigger and more affordable telescopes and arrays, and place them in new environments such as the Antarctic or even on the moon, but they would still be recognizable to Newton or Gregory.

3.2 Space telescopes

As others have pointed out at this conference, space telescopes were first seriously proposed by Hermann Oberth in 1923, and by Lyman Spitzer in 1945. However, it was the launch of Sputnik by Russia in 1957 which changed everything, and led to the birth of NASA, the race to the moon, and consequent availability of technology to allow construction of the NASA Great Observatories, not forgetting the ESA telescopes such as ISO and XMM-Newton. We are about to go into a second wave of exciting missions with the launch of the Herschel Space Observatory in 2009 and the James Webb Space Telescope a few years later. But what could come after these missions?

One parallel with the past is that new technology developed for Moon (and now Mars) missions could make a big change in what sort of space telescopes are practical. The ARES V heavy-lift vehicle (Fig. 2) could launch up to 71,000 kg within a fairing of cross section up to 10 m, making a fully monolithic 8 m UV/Optical telescope possible, replacing and considerably enhancing the UV capabilities of the Hubble Space Telescope which will not be covered by

Fig. 2 ARES V and 8 m telescope (Harley Thronson/NASA, JF&A)

JWST [26]. It would also enable a deployable Optical/IR telescope similar to JWST but of up to 30 m in diameter, given sufficient funds.

Another enabling technology which might result from the Moon/Mars programmes is robotic assembly of giant telescopes and interferometers in space, which could bring the real possibility of direct imaging and atmospheric spectroscopy of earth-like planets, with the obvious potential for discovery of signs of extraterrestrial life.

Of course, another opportunity which could be opened up by new missions to the Moon is the prospect of lunar telescopes. As well as the idea of radio arrays on the far side, there have been suggestions for a liquid mirror telescope based on a metal-coated organic liquid, spinning within an inflatable radiation shield [27], and Peter Chen's idea of fabricating spun-cast mirrors in situ from a composite material based on a mixture of resins transported from the earth and 'moon dust'.

3.3 Adaptive optics

Going back to earth (in several senses!), it is clear that the technology with the most impact on ground-based astronomy over the last few years is adaptive optics (AO), with recent highlights being the Galactic-centre and planet-imaging work mentioned previously. The major disadvantages of AO, limited sky coverage and limited field of view, are being addressed by the introduction of laser guide stars and multi-conjugate and tomographic techniques [28].

The roadmap towards AO in the era of ELTs appears to be relatively clear, and has been well described by Hubin and Ellerbrook [29]. A major technology requirement in the short term is for high-power sodium laser guide stars: gated to deal with spot elongation, of up to 50 W power, with high reliability and low service cost. Another limitation on AO system performance is wavefront sensor detectors, where we would like faster, larger format, and low-noise CCDs, as well as IR options for Mid IR and obscured regions like the galactic centre. The third key technology is in deformable mirrors, where a range of requirements for the large adaptive mirrors of up to 2.5 m to be built into ELTs and for the smaller deformable mirrors to be built into post-focal systems or instruments. For the large mirrors we need high stroke of up to 100 μm within tight thermal constraints, and possibly new materials for the active surface, and for the smaller mirrors, very high density with up to 10,000 actuators and potentially operation at low temperatures.

The fourth technology is high power real-time controllers, where it can be expected that the normal rate of progress of commercial systems will fulfill our requirements, but this is not to be taken for granted. It may be necessary to move to new architectures which may need to be adapted from other fields, possibly including graphical processors developed for gaming systems.

The inventive AO community will probably continue to come up with novel concepts, but note that of the current concepts (SCAO, MCAO, LTAO, GLAO, MOAO and EXAO[1]) only the first is in routine operation, with MCAO only available in demonstrator mode on the VLT with natural guide stars. It is also possible that for some applications where spatial resolution and low cost is more important than sensitivity that Lucky Imaging will have a role to play [30]. It is clear that AO systems are central to the success of future telescopes on the ground, and will take up a growing fraction of the funding budgets. As AO becomes integrated into the instruments, it will become more important to think of the whole telescope, AO and instrument chain as a system, in order to maximize performance and understand error budget tradeoffs, by integrated modeling techniques. It may be that the limiting factor will be the availability of experienced AO practitioners and systems modellers.

The major thrust for AO over the coming years will be towards higher-order correction to get towards feasible systems in the visible, consolidation of efforts towards wide field correction in the near infrared and cryogenic operation in the mid IR to reduce background. One long term potential might be towards a really low-background adaptive telescope on a high cold site, with a deformable secondary mirror and minimum number of warm optical surfaces, or even an adaptive primary mirror which might be feasible if actuator, sensor and controller costs could be reduced significantly [31].

3.4 Instruments

The telescopes and adaptive optics systems we have considered so far have what might seem to be the simple task of providing optimum image quality over a specified field of view. In order to do science with this image we need instruments. As we have already seen, the fundamental measurements that we carry out in optical/IR astronomy are straightforward in principle: we measure the intensity and energy of photons—and sometimes the amplitude and phase—and observe how this varies in time and space. To do this, we

[1]Single Conjugate AO, Multi Conjugate AO, Laser Tomography AO, Ground Layer AO, Multi Object AO and Extreme AO.

may divide the spectrum up in space or time by dispersion, filtering (including Fabry-Perot etalons), or by a Fourier Transform process, then spatially sample an appropriate field of view (FoV). Why is this often so difficult?

Of course, the answer is that there are many conflicting pressures on the design of an instrument, such as pixel size versus noise, image quality versus field of view, and detector and background noise and hence operating temperature versus mass and space limits. It is the task of the instrument scientists and design engineers to optimize these conflicting parameters, and it is often new technology which enables us to break through these barriers.

One area of recent rapid progress which is likely to continue to provide great benefits in instrument design and scientific performance well into the ELT era is what the EU-supported OPTICON programme has termed 'Smart Focal Planes'. We have seen 4 m telescopes make the best use of their available wide FoV by multiobject and integral field spectroscopy in order to conduct red-shift surveys and investigate galaxy dynamics by 3-D spectroscopy. MUSE and KMOS, now being built for the VLT, take these concepts to the next level, and we will see that AO-equipped ELTs will have proportionately wide fields of view and higher spatial resolution, putting pressure on instrument designers to respond. We now have many alternative concepts to fibre systems for cryogenic infrared multi-object spectroscopy (MOS) instruments on ELTs, such as EAGLE for the E-ELT [32] and IRMOS for the TMT [33]. The technologies which will be developed to select subfields, and analyse them spectrally, range from microrobots to MOEMS mirror arrays and precision-machined or replicated reflective image slicers [34].

Size and mass will be critical limiting factors for future instruments on ELTs. In principle, a diffraction-limited instrument for a bigger telescope should be no larger than for a smaller, as long as the field of view is scaled down in the inverse ratio of the aperture [35]. However, many science cases call for a relatively wide field, and it probably will never be possible to fully correct the atmosphere over wide fields, especially at visible wavelengths. This will result in pressures towards big and heavy instruments—for example the WFOS proposal from TMT, which at 8 m diameter and 10 m height is close to the size of current telescopes. Some technology in terms of light-weighting, new materials, novel volume phase holographic (VPH) and immersion gratings, highly aspheric reflective optics and active flexure compensation will assist [36], but to really break out of this constraint a new concept of building spectrometers has been proposed based on photonic devices. As with previous disruptive technologies, the origins are outside astronomy, in this case mostly from the telecommunications industry.

3.4.1 Photonic devices

We have already seen that waveguide beam combiners are being adapted for astronomical interferometry. Some of the other devices developed for communications and industrial instrumentation that are being investigated for astronomy include Bragg fibre gratings for OH-line suppression [37], photonic crystal fibres with tuned properties to provide wider monomode bandwidths and easier coupling to the telescope focal plane [38], and 3-D waveguides for mode transitions.

We can imagine these photonic technologies being combined with an integrated detector to make compact integrated spectrometers, possibly with built-in OH suppression. There are some promising concepts under development: array waveguide spectrometers [39] and integrated Lippmann interferometric spectrometers [40], with a joint research programme about to start with OPTICON FP7 funding [41]. Eventually, it may be possible to use such spectrometers to devise a wide-field multi-object spectrometer with up to a thousand spectrometers with integrated detectors, perhaps mounted on miniature self-propelled robots, patrolling a field and communicating with wireless links, possibly with inductively coupled power.

3.5 Detectors

The introduction of electronic detectors, CCDs and hybrid IR arrays, meets the criteria for disruptive technology. The huge increases in quantum efficiency, linearity and dynamic range which they brought made it possible to make rapid gains in overall sensitivity for long periods of the late twentieth century without the need to increase telescope aperture, but once these detectors started to approach 'perfect' performance, the impetus towards 8–10 m telescopes became clear. At visible wavelengths, there is little scope for further improvement, apart from extending the pixel count of electron multiplying CCDs to enable a factor of three improvement in signal to noise over conventional CCDs in spectroscopic applications [42] and avalanche photodiode arrays for high-time-resolution astronomy with resolution at the millisecond level [43]—or perhaps even faster as proposed for Quanteye [44].

In the near and mid-IR range there is still considerable scope for improvement of detectors—summed up as bigger, cheaper, better. The instruments for the European ELT will need 60–100 2k × 2k IR arrays. At current prices this would swamp the instrumentation budgets, so reducing the cost is very important. Alongside that, improvements in noise performance would be good, as would a larger format of at least 4k × 4k pixels, without reduction of the current generic pixel size of 18 μm—or even some increase to reduce the need for very fast camera designs.

These are all incremental developments, which can be envisaged now, given adequate investment. A more radical approach which would drastically change the way instruments are designed would be if the cryogenic detectors currently used at the wavelength extremes of X-ray and submm could be applied in the visible and IR. These devices are energy-sensitive, and hence can measure spectra directly if in a photon-counting regime. If a detector with a spectral resolving power of a few hundred could be developed, then we could envisage a truly solid-state 3-D spectrometer with no dispersive elements—and see how this could become part of a compact multi-object spectrometer. Even a low-resolution energy sensor might be used to simplify spectrometer design by reducing the need for order sorting. The current best performance in energy-resolving detectors using Superconducting Tunnel Junctions (STJs) is the European Space Agency's S-CAM, which has a resolving power of 8–11 at 500 μm and is limited to 10 × 12 pixels due to a lack of multiplexing options for STJs [45].

The theoretical limit for energy resolution of Transition Edge Superconducting devices (TES) in the short wavelength end of the spectrum (10 eV or 124 nm) is about 0.1 eV, which corresponds to a resolving power of 100 [46]. Therefore, it seems that the most likely use for energy-sensitive detectors would be where there is a need for a combination of low-resolution spectroscopy and high time resolution, for instance in the study of white dwarfs or other high-energy astrophysical phenomena.

Many other novel detectors are being developed which could provide advantages in specialised areas of astronomy: quantum dots, quantum well interference photodetectors (QWIPs), quantum well intrasub-band photodetectors (QWISPs), carbon nano-tubes, kinetic inductance devices (KIDs) and single electron transistors. However, it seems likely that more conventional CCDs and hybrid IR detector arrays will continue to dominate optical and infrared astronomy for the next 50 years, although there is a possibility that CMOS imager arrays will be improved sufficiently to replace CCDs.

3.6 Data handing

We have already seen that a relatively unsung enabling technology for astronomy in the last century was the rapid development of computers. It is clear that this will need to continue into the era of ELTs, where there are particular challenges for real-time control in both the active and adaptive optics systems, which will need Moore's law to keep on its trajectory. Data handling is also a major challenge, in particular for the wide-field survey telescopes, where we are already planning for 28 Tbytes of calibrated images per night on Large Synoptic Survey Telescope. This is one area where there is no chance that astronomy can be a driving market—we will need to ride on the coat-tails of the IT and communications industries, as we have done so successfully over the last few decades.

4 Key technologies for optical/IR astronomy

The table below shows my view of the critical technologies that need to be developed or adopted for optical and infrared astronomy in the relatively short-term future, say the next 5–15 years. Of these, I would place the first three in the disruptive category, as they would make the biggest impact on how we build instruments and telescopes.

Space Technologies	*Formation flying, heavy launch vehicles and robotic assembly could enable multi-craft interferometers and bigger telescopes*
Photonics	*Integrated optics and photonic fibres could make radical changes to interferometry and spectroscopy*
Energy sensitive detectors	*Most current instruments depend on dispersive elements to measure spectra. Superconducting detectors which measure photon energy could enable more efficient and more compact instruments*
Novel Materials	New materials emerging from the automotive and aerospace industries can help us develop ultra-lightweight and extreme-shape mirrors which could enable higher-performance ground and space telescopes and instruments
Novel Optical Components	New types of filters and gratings could help us to make more compact or higher throughput instruments.
Nanotechnology	New engineering materials with molecular level structures for high-performance optical coatings, metamaterials to enable novel optical devices, and carbon nanotubes for new types of detectors
New Manufacturing methods	New methods of grinding and polishing for hard and light materials, with parallel developments in precision metrology, for large refractive and reflective components
Fast Photon counting detectors	High Time Resolution Astronomy is in its infancy, but new fast detector arrays combined with the large collection area of the next generation of telescopes could open up whole new areas of science

5 The future of astronomy and technology

Bearing in mind that these technological promises are of no use unless they enable new science, it is worth asking where the new parameter space will be, when we have JWST, ELTs and perhaps nulling interferometers in space, like Darwin.

Will it be more sensitivity, with bigger, colder telescopes? Or higher spatial resolution through longer-baseline interferometry or larger single-aperture telescopes with better adaptive optics? Or will the emphasis be on high-contrast imaging and spectroscopy for exoplanet research? Or will the unexplored region be high time resolution made possible by bigger apertures and faster detectors? Or perhaps the current rich resource of wide-field surveys will prove more productive than the push for higher spatial resolution. Where will the balance be between space and ground, as improvements in adaptive optics result in better spatial resolution on the ground, balanced against the sensitivity gains in space with no losses from atmospheric lines and much lower thermal background?

But perhaps the idea of a continuous rate of improvement in discoveries in astronomy is wrong. It could be that the rapid rate of discovery in astronomy that we have experienced over the last century is reaching a peak, and over the next 50 years will go into a decline. Martin Harwit made an interesting prediction in 1981 [17], based on a study of discovery rates of new phenomena, and in particular the rate at which phenomena are discovered for a second time with completely different methods. He made the analogy with collecting baseball or football cards—as soon as you have a duplicate you know the total population is not infinite. He defined new phenomena in astronomy as those which opened up new fields with their own community of interest, as demonstrated through conference series and perhaps specialist journals. The analogy with discovery of new species of butterflies in a finite jungle is that this rate will reach a peak and then start to decline as it becomes harder to find the last few species. His prediction puts the peak at 2020, but of course it was made before the introduction of the great space observatories and the 8–10 m telescopes on the ground. It would be interesting to update this prediction, but I am not aware that anyone has tried this rather dangerous exercise.

Where will Astronomy be on the 450th anniversary of Lipperhey? Will we have run out of astronomical discoveries to be made at affordable cost? Perhaps the advent of the Square Kilometer Array will have turned us all into radio astronomers, or maybe the really unique discoveries will come from gravitational wave or neutrino observatories. Or new technologies like nano-engineered membrane telescopes in space and photonic instruments will have helped us to understand current mysteries like dark matter and dark energy.

Or maybe we are living in the Golden Age of Astronomy now, and governments, the public and benefactors will run out of patience with investment in bigger and bigger facilities—due to resource crises, global warming, and financial meltdown. It could be that the leading nations in astronomy 100 years after Sputnik will be the emerging economies like China and India, where the combination of rapid economic growth, huge numbers of well-educated scientists and engineers, and government ambitions may enable them to take dominant positions.

Whatever the source, it is nice to imagine that the headline news in 2058 could be the first images of a blue earth-like planet, and measurement of ozone and methane in its atmosphere. And perhaps we will also understand dark matter and dark energy—or even have discovered that gravity varies with time!

Acknowledgements I would like to acknowledge the work of the Opticon Key Technologies Network, funded by research funding from the European Community's Sixth Framework Programme under contract number RII3-CT-001566. I would also like to thank Dainis Dravins of Lund University for stimulating discussions and Harley Thronson of NASA for information on the potential of Ares V.

References

1. Christensen, C.M.: The Innovator's Dilemma. Business Press, Harvard (1997)
2. Tobin, W.: The Life and Science of Léon Foucault. University Press, Cambridge (2003)
3. Newton, I.: Optics. Royal Society, London, (1704)
4. Babcock, H.W.: The possibility of compensating astronomical seeing. Publ. Astron. Soc. Pac. **65**, 229–236 (1953) doi:10.1086/126606
5. Rousset, G., et al.: First diffraction-limited astronomical images with adaptive optics'. Astron. Astrophys. **230**, L29–L32 (1991)
6. Gillessen, S., et al.: Monitoring stellar orbits around the Massive Black Hole in the Galactic Center. Astrophys. J. (in press) arXiv:0810.4674v1
7. Mankins, J.C.: Technology Readiness Levels: A White Paper, NASA, Office of Space Access and Technology, Advanced Concepts Office. NASA, Washington, D.C. (1995)
8. Phaal, R., Farrukh, C.J.P., Probert, D.R.: Developing a technology roadmapping system. In: Anderson, T.R. (ed.) Technology Management: a Unifying Discipline for Melting the Boundaries, pp. 99–111. PICMET, Portland (2005)
9. Bahcall, J.: The Decade of Discovery in Astronomy and Astrophysics. National Academies Press, Washington, D.C. (1991)
10. NASA: NASA Strategic Plan. NP-2006–02–423-HQ. NASA, Washington, D.C. (2006)
11. Bignami, G., et al. (eds.): ESA: Cosmic Vision: Space Science for Europe 2015–2025, ESA BR-247 (2005)
12. de Zeeuw, P.T., Molster, F.J. (eds.): A Science Vision for European Astronomy, ASTRONET (2007)
13. Bode, M.F., Cruz, M.J., Molster, F.J. (eds.): The ASTRONET Infrastructure Roadmap: A Strategic Plan for European Astronomy. ASTRONET (2008)
14. Pickering, E.: The Future of Astronomy. Popular Science Monthly (1909)
15. King, H.C.: The History of the Telescope. Griffin, London (1955)
16. Harwit, M.: Cosmic Discovery. Harvester, Brighton (1981)
17. Harwit, M.: Photon orbital angular momentum in astrophysics. Astrophys. J. **597**, 1266–1270 (2003) doi:10.1086/378623
18. Baldwin, J.E., et al.: The first images from an optical aperture synthesis array: mapping of Capella with COAST in two epochs. Astron. Astrophys. **306**, L13–L16 (1996)
19. Creech-Eakman, M.J., et al.: Magdalena Ridge Observatory interferometer: progress towards first light. Proc. SPIE **7013**, 70130X (2008) doi:10.1117/12.789859
20. Malbet, F., Kern, P., et al.: Integrated optics for astronomical interferometry - I. Concept and astronomical applications. Astron. Astrophys. Suppl. **138**, 135–145 (1999) doi:10.1051/aas:1999496
21. Thomson, R.R., et al.: Optical wave-guide fabrication in z-cut lithium niobate (LiNbO3) using femtosecond pulses in the low repetition rate regime. Appl. Phys. Lett. **88**, 111109 (2006) doi:10.1063/1.2186389
22. Hale, D.D.S., et al.: The Berkeley Infrared Spatial Interferometer:) A Heterodyne Stellar Interferometer for the Mid-Infrared. Astrophys. J. **537**, 998–1012 (2000) doi:10.1086/309049
23. Hanbury Brown, R., Twiss, R.: A test of a new type of stellar interferometer on Sirius. Nature **178**, 1046–1048 (1956) doi:10.1038/1781046a0
24. Dravins, D.: Photonic Astronomy and Quantum Optics. In Phelan, D., et al. (eds.) High Time Resolution Astrophysics, pp. 95–132. Springer, New York (2008)
25. Ditto, T.D., Ritter, J.M.: Million object spectrograph. Proc. SPIE. **7018**, 70182R (2008) doi:10.1117/12.790244

26. Thronson, H.A., Postman, M., Stahl, H.P., Traub, W.: Astronomy Enabled by Ares V Heavy Lift," NASA Future In-Space Operations White Paper. NASA, Washington, D.C. (2007)
27. Angel, R., et al.: A Lunar Liquid Mirror Telescope (LLMT) for deep-field infrared observations near the lunar pole. Proc. SPIE **6265**, 62651U (2006) doi:10.1117/12.669994
28. Marchetti, E., et al.: MAD Star Oriented: laboratory results for Ground Layer and Multi-Conjugate Adaptive Optics. Proc. SPIE **7015**, 70150F (2008) doi:10.1117/12.787240
29. Hubin, N., Ellerbrook, B.L. et al.: Adaptive Optics for Extremely Large Telescopes. In: Whitelock, P., et al. (eds.) Scientific Requirements for Extremely Large Telescopes. Proc IAU Symposium, vol. 232, pp. 60–85. CUP, Cambridge (2005)
30. Mackay, C.D., et al.: High-resolution imaging in the visible from the ground without adaptive optics: new techniques and results. Proc. SPIE **5492**, 128 (2004) doi:10.1117/12.550443
31. Ulmer, M.P., et al.: Replication as an alternative approach for large segmented telescopes. Proc. SPIE **4837**, 786 (2003) doi:10.1117/12.456663
32. Cuby, J.-G., et al.: EAGLE: an MOAO fed multi-IFU in the NIR on the E-ELT. Proc. SPIE **7014**, 70141K (2008) doi:10.1117/12.787682
33. Eikenberry, S.: IRMOS: The near-infrared multi-object spectrograph for the TMT. Proc. SPIE **6269**, 62695W (2006) doi:10.1117/12.672266
34. Cunningham, C.R., et al.: Progress on smart focal plane technologies for extremely large telescopes. Proc. SPIE **5904**, 59040T (2005) doi:10.1117/12.617217
35. Cunningham, C.R., Evans, C.J., Monnet, G., Le Louarn, M.: ELT instrumentation for seeing-limited and AO-corrected observations: a comparison. Proc. SPIE **6986**, 69860K (2008) doi:10.1117/12.801279
36. Cunningham, C.R. et al.: Smart instrument technologies to meet extreme instrument stability requirements. Proc. SPIE **7018**, 70181U (2008)
37. Bland-Hawthorn, J., Englund, M., Edvell, G.: New approach to atmospheric OH suppression using an aperiodic fibre Bragg grating. Opt. Express **12**(24), 5902–5909 (2004) doi:10.1364/OPEX.12.005902
38. Corbett, J., Allington-Smith, J.: Coupling starlight into single-mode photonic crystal fiber using a field lens. Opt. Express **13**(17), 6527–6540 (2005) doi:10.1364/OPEX.13.006527
39. Bland-Hawthorn, J., Horton, A.: Instruments without optics: an integrated photonic spectrograph. Proc. SPIE **6269**, 62690N (2006) doi:10.1117/12.670931
40. Le Coarer, E., et al.: Wavelength-scale stationary-wave integrated Fourier-transform spectrometry. Nat. Photon **1**, 473–478 (2007) doi:10.1038/nphoton.2007.138
41. Allington-Smith, J.: The AstroPhotonica Europa partnership. Proc. SPIE **7018**, 70182 (2008) doi:10.1117/12.789840
42. Ives, D.J., et al.: ULTRASPEC—An Electron Multiplication CCD camera for very low light level high speed astronomical spectrometry. Proc. SPIE **7021**, 70210B (2008) doi:10.1117/12.790376
43. Phelan, D., Morrison, A.P.: Geiger-mode Avalanche Photodiodes for High Time Resolution Astrophysics. In: Phelan, D. et al. (eds.) High Time Resolution Astrophysics, pp. 291–310. Springer, New York (2008)
44. Barbieri, C., et al.: QuantEYE, the quantum optics instrument for OWL. In Whitelock, P. et al. (eds.) Scientific Requirements for Extremely Large Telescopes. Proc IAU Symposium vol. 232, pp. 506–507. CUP, Cambridge (2005)
45. Verhoeve, P., et al.: S-Cam 3, optical astronomy with a STJ-based imaging spectro-photometer. Nucl. Instrum. Methods Phys. Res. Sect. A **559**(2), 598–601 (2006) doi:10.1016/j.nima.2005.12.198
46. Romani, R.W., et al.: Transition Edge Cameras for fast Optical Spectrophotometry. In: Phelan, D., et al. (eds.) High Time Resolution Astrophysics, pp. 311–325. Springer, New York (2008)

Exp Astron (2009) 26:201–210
DOI 10.1007/s10686-009-9171-8

REVIEW ARTICLE

Perspectives and challenges for future telescopes

Tim de Zeeuw

Received: 12 April 2009 / Accepted: 6 May 2009 / Published online: 4 June 2009

Abstract At the celebration of 400 years of telescopic observations, it is appropriate to consider future ground-based facilities enabled by continuing technological developments, and driven by astronomical questions which are amongst the most fundamental in science and are of enormous interest to the general public.

Keywords Astronomy · Telescopes

1 Introduction

Astronomy is the study of everything beyond the Earth. The objects are far away, and hence they are generally small and faint so that only limited information on their nature can be obtained. This leads naturally to searches and surveys with small telescopes, followed up by measurements with the increased sensitivity and resolution provided by larger telescopes. Combining different types of observations is crucial. Today, astronomers take images, spectra, and time series covering the entire electromagnetic spectrum, from gamma-rays to the longest radio waves. The atmosphere allows ground-based observations through the optical-infrared and radio windows. Space telescopes are required for all other wavelengths. The rapid advances in technology in the past decades have led to a continued increase in angular resolution and sensitivity across all wavelength regimes (Fig. 1), now enabling observations of objects over 95% of the age of the Universe and the detection and characterization of planets orbiting other stars. It has also become possible to explore Solar system

T. de Zeeuw (✉)
ESO, Karl Schwarzschild Strasse 2, 85748, Garching bei Műnchen, Germany
e-mail: tdezeeuw@eso.org

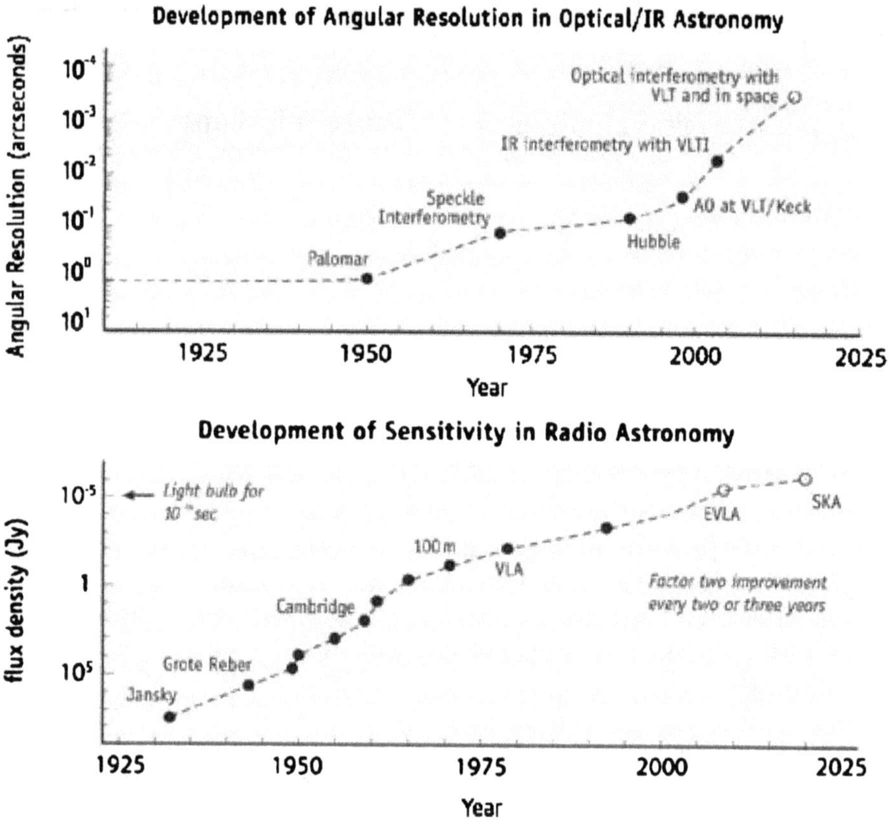

Fig. 1 The increase in angular resolution at optical/IR wavelengths and in sensitivity at radio wavelengths

objects in situ through dedicated space missions, and to use particles to study specific objects in the Universe. The expectation is that direct observation of gravitational waves from celestial objects will be achieved soon.

These developments promise tremendous progress towards answering key questions such as the nature of dark matter and dark energy; physics under extreme conditions including black holes, supernovae and gamma-ray bursts; the formation and evolution of galaxies from first light to the present; and the formation of stars and planets, including the origin of life, and even the question how we and the Solar system fit in. These are fundamental scientific questions and the answers are of interest not only to the broad science community but also to society. Answering them requires observations with existing and future telescopes on the ground and in space, as well as numerical simulations, interpretative and theoretical effort, and laboratory experiments.

Realization of the many proposed new facilities will require substantial investments by national and international funding agencies and, in some cases, private donors, with significant additional long-term commitments for operations. The United States has a long tradition of prioritization of such plans, through the series of Decadal Surveys, initiated in the 1960s. Similar strategic exercises are carried out regularly in many other countries. Recently, a number of national funding agencies, together with ESA, ESO and the Science Directorate of the European Commission, created ASTRONET in order to develop long-term strategic plans for European astronomy. This resulted in an integrated Science Vision, published in 2007, which identifies the most promising techniques and facilities to make substantial progress on the key science questions. It was followed by an Infrastructure Roadmap, published in 2008, which serves as the basis for an Implementation Plan. In recognition of the timescale needed to build large infrastructures, the strategy has a 20-year horizon. The priorities for future space observatories match those in ESA's Cosmic Vision, and the ground-based plans are in line with the priorities set by the ESO Council.

In what follows, the perspectives and challenges for future ground-based facilities are summarized, starting with the plans for new radio observatories and followed by an outline of the next generation of telescopes for the optical-infrared part of the spectrum, i.e., the direct descendants of the invention of Lipperhey and Galileo four centuries ago.

2 Radio observations

The opening of the radio window on the Universe in the middle of the twentieth century, and the subsequent discovery of the 21 cm emission of atomic hydrogen predicted by Henk van de Hulst, was a major milestone. Today there are many radio observatories all over the world, including giant single-dish antennas as well as powerful interferometers. They are linked into VLBI networks to achieve tremendous angular resolution.

In the past decades, a powerful drive occurred to go to shorter wavelengths, already foreseen by Fred Hoyle in *The Inferno*. This requires observations taken at high altitude to minimize the amount of water vapour above the antennas which blocks the (sub)millimeter signals from distant objects. Nobeyama, IRAM and CARMA (which combines the predecessor arrays in Owens Valley and at Hat Creek at a higher location) are the major current millimeter facilities. Mauna Kea hosts the CSO, JCMT and (e)SMA submillimeter observatories at 4,200 m. The quest for observations from even higher altitudes led to the establishment of the scientific reserve on Pampa La Bola and Llano Chajnantor (Fig. 2) in Northern Chile, with ASTE, NANTEN, APEX and various cosmic microwave background experiments at altitudes between 4,500 and 5,000 m. Plans for even higher facilities include the 25 m submillimeter CCAT to be placed at 5,600 m on Cerro Chajnantor.

Llano Chajnantor is also the site of the Atacama Large Millimeter/submillimeter Array (Fig. 3). ALMA evolved from separate regional plans to a global partnership between Europe, North America (USA and Canada), East Asia (Japan and Taiwan) and the host state Chile. The ALMA science

Fig. 2 Llano Chajnantor at 5,000 m

requirements are to detect CO and CII in a normal Milky Way galaxy at a redshift of three in less than 24 h, and to measure the spatially-resolved properties of the dust emission and gas kinematics in the nearest protoplanetary disks, all with a resolution to match or exceed that of the Hubble

Fig. 3 Artist's impression of ALMA as it will look in 2012

Space Telescope, the future James Webb Space Telescope and the diffraction limit of 8–10 m optical telescopes equipped with adaptive optics (Fig. 4). The top-level requirements lead to an array of 54 antennas (12 m) plus a compact array of 12 antennas (7 m). The maximum baseline of 14 km will provide a resolution of better than 10 milli-arcseconds at the highest frequencies. The wavelength range from 0.3 to 10 mm will be covered by receivers in up to ten bands. This range contains many diagnostic lines to characterize the molecular gas and also covers the peak of the dust emission. It will be a transformational facility for astronomy, and will be perfect for ground-based follow-up at high resolution and high sensitivity of observations carried out with the Herschel Space Observatory.

Looking beyond ALMA, there is much activity also at the longer radio wavelengths, in a drive to further increase sensitivity, resolution and field of view. The Square Kilometer Array has been a dream for decades and is steadily evolving towards a realistic long-term program. Many radio observatories are constructing so-called 'SKA Pathfinders', including ASKAP, ATA, EVLA, LOFAR, LWA, MeerKAT, MWA, and others. These will provide scientific advances in many areas, and also test key technology needed for the next step. There is general agreement that the SKA should be constructed in the Southern Hemisphere, in Australia and/or South Africa. Phase 1 (0.1 km^2) will build on experience with the Pathfinders, so that detailed design is foreseen to start in 2011 and construction could commence in 2015, if funding is available

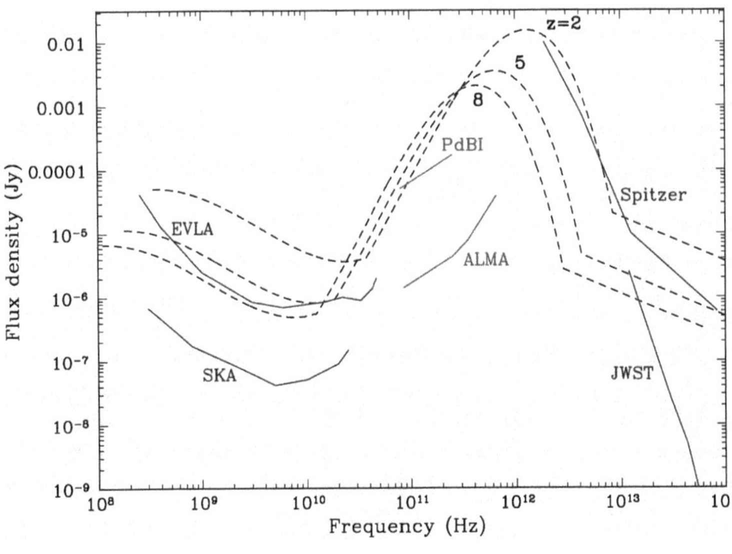

Fig. 4 To show the reach of the different facilities, the spectral energy distribution of the ultra-luminous infrared galaxy Arp220 is plotted for redshifts of two, five and eight. The sensitivity limits are indicated for the six future facilities SKA, EVLA, ALMA, PdbI, JWST and Spitzer. Since the peak of the emission gets redshifted into the sub-millimeter, ALMA will be able to detect those objects even at very early times of the Universe

and the governance has been defined. Phase 2, the extension to the full square kilometer, could then follow in 2020$^+$.

3 Optical-infrared observations

Telescopes of various sizes grace mountaintops around the world, some in very remote places. Giants of the past, including the Perfect Machine on Mount Palomar, are now considered medium-sized telescopes, and efforts are ongoing to rationalize the smaller ones, in parallel with the development of more powerful facilities. The ReStar initiative in the United States considers telescopes of up to 6.5 m, and aims to treat them as a system, avoiding duplication of instruments on undersubscribed telescopes. Europe has had an access program with similar goals for many years, funded by the European Union. These efforts naturally focus individual telescopes on unique science or techniques. Examples include the Sloan Digital Sky Survey, high-angular-resolution observations enabled by adaptive optics in the visible (WHT/SOAR), exo-planet searches (e.g., HARPS on the ESO 3.6 m at La Silla), and spectroscopy for the Galactic objects to be observed by ESA's astrometric mission Gaia.

Solar telescopes comprise a special class of medium-sized telescopes. A number of smaller facilities managed by the NSO in the United States are planned to be closed after the new flagship, the 4 m Advanced Technology Solar Telescope is completed on Haleakala. The design is ready and the project awaits NSF funding approval. European solar astronomers have initiated a design study for a similar 4 m class telescope in the Canary Islands, the EST.

A number of new survey telescopes are coming on line. Pan-Starrs1 has started taking data on Mauna Kea, and may be expanded into Pan-Starrs4. VISTA and the VST on Paranal will soon start 5-year programs of large public surveys, working together with data centres in Europe, coordinated by ESO. This builds a European survey capability which will deliver eminent science, provides crucial ground-based data in support of future space missions, and prepares the way for the next step. This could well be the proposed US-led Large Synoptic Survey Telescope, with an 8.4 m primary mirror and a large field of view which would provide deep images of the entire sky above Cerro Pachon in Chile every few nights. There are also strong scientific arguments for an 8 m-class wide-field spectroscopy facility for observations of the deep Universe, follow-up of faint objects in the Galaxy identified by Gaia, and support of the proposed space missions JDEM and EUCLID. Implementation will depend upon the required field of view for these studies. If this is less than about 30 arc minutes, one of several 8 m-class telescopes could be used, perhaps in campaign mode with an existing instrument. If the optimal field of view is much larger, then this can possibly be done on Subaru but may require a dedicated new facility.

Few astronomers would have predicted 25 years ago that today there would be over a dozen telescopes with primary mirrors larger than 8 m in diameter. These telescopes have transformed the field, and tremendous discoveries have been made with them, including key follow-up spectroscopy of discoveries by HST and other flagship space missions. Mauna Kea hosts the two mighty Keck telescopes as well as Subaru and Gemini-North. Initial steps are being taken to develop this into a more integrated system, to avoid duplication of effort and exploit the natural strengths of each of the telescopes. Gemini-South, optimized for observations in the near-infrared, as well as ESO's flagship, the Very Large Telescope and Interferometer on Paranal, are sited in Northern Chile (Fig. 5). Other facilities include the Large Binocular Telescope on Mount Graham with an interferometric option, the Gran Telescopio Canarias on La Palma, the Hobby-Eberly Telescope in Texas and the Southern African Large Telescope.

The VLT was designed from the start as an integrated system of four 8.2 m telescopes, including the possibility to combine the light from individual telescopes for interferometry, enabling very high angular resolution. The entire first-generation instrument suite was completed in 2007. The spectrometers provide full coverage in resolution, up to 10^5, over the accessible wavelength range. The Paranal arsenal also includes turn-key adaptive optics systems and a rapid-response mode to react to fast transient events detected by, e.g., SWIFT. Recently, the near-infrared imager HAWK-I was added as a 'generation-1.5' instrument.

Fig. 5 Aerial view of the VLT on Paranal

The VLT started operations 5 years after the Keck telescopes, but the decision to take the time to build a fully-integrated system consisting of four 8.2 m telescopes providing a dozen foci for a carefully-thought-out complement of instruments together with a series of 1.8 m Auxiliary Telescopes for the interferometer and two survey telescopes was the right one. The combination of a long-term adequately funded instrument and technology development plan with an approach where most of the instruments are built in collaboration with institutions in the ESO member states, with in-kind contributions in labor compensated by guaranteed observing time, has arguably created the most advanced ground-based optical observatory in the world. The operations model distinguishes visitor and service mode, and provides world-leading observing efficiency on a site where nearly 90% of the nights are clear.

Paranal will continue to increase in power over the next decade. The survey telescopes will come on line in 2009. On the VLT, X-Shooter will commence operations in 2009, with KMOS, SPHERE and MUSE to follow between 2010 and 2012, together with a multiple laser guide star system and an adaptive secondary mirror on UT4, and one or more additional instruments, including an ultra-stable high resolution spectrograph at the combined focus of the four 8.2 m telescopes (as foreseen in the original VLT design). The VLTI will be equipped with the second-generation instruments GRAVITY and MATISSE, to be followed by VSI, the latter perhaps with two additional Auxiliary Telescopes, if external funding can be found.

The success of the 8–10 m class telescopes has naturally led to renewed interest in even larger telescopes. These giants will focus on imaging and spectroscopy of exoplanets (with the goal of reaching Earth-like objects), and on the study of the first stars and galaxies, the evolution of cosmic parameters, and the nature of dark matter and dark energy.

Three distinct projects are in the design phase. The Thirty Meter Telescope Corporation established by Caltech, the University of California and a consortium of Canadian institutions, aims to build a telescope with a 30 m primary mirror consisting of about 500 1.45 m segments. The site would be either on Mauna Kea or in Northern Chile. The Giant Magellan Telescope would consist of seven 8.4 m segments, to be put on Las Campanas in Chile. The collaboration includes the Carnegie Institution of Washington, a number of US universities, and partners in Australia and Korea. Both projects have some of the construction funding pledged by private donors, and hope to obtain the bulk of the operations funding from the NSF after the ongoing Decadal Survey concludes in 2010. If construction can be funded in a timely manner, the TMT could be ready in 2018, with the GMT following soon after (Fig. 6).

The third project is the European Extremely Large Telescope, with a 42 m primary mirror consisting of nearly 1,000 segments of 1.45 m. Adaptive optics capabilities will be incorporated in the telescope in an innovative way. The design study has strong industrial engagement, and will be completed in the course of 2010. The project builds on the entire expertise of ESO and the institutions in its member states. Construction could commence in early 2011, and be completed by 2018. This would provide scientific synergy

Fig. 6 **a–c** Artist's impressions of the TMT (*top left*), the GMT (*bottom left*) and the E-ELT (*right*)

with ALMA, JWST and the SKA. When equipped with a 2 cm/s-accuracy spectrometer, the E-ELT will be able to carry out a direct measurement of the expansion of the Universe. The E-ELT is the top-priority new program for ESO, it is the highest priority ground-based project in the near-term for astronomy in Europe (ASTRONET) and it is one of the two astronomy projects on the ESFRI list of high-priority research infrastructures for Europe (the other being the SKA, to come later). Operations costs for the E-ELT can be covered by the ESO annual income, but part of the construction cost still needs to be raised. Additional funding is under active discussion in the ESO member states with encouraging indications. This would also enable ESO involvement in world-class ground-based facilities beyond the E-ELT.

The E-ELT would be the world's biggest eye on the sky, and provide a step beyond the VLT which is as large as the leap made by Galileo going from the naked eye to the first telescope.

4 Challenges

Construction of the flagship astronomical facilities of the future will be a technological challenge. This provides industry with opportunities to show-case their capabilities, but also requires strong and effective management.

The United States has a world-leading space science program through NASA, and much of radio astronomy is nationally led by the NRAO. The optical-infrared effort can be characterized as ‘free-market’, with a lot of competition between powerful private institutions and federally funded organizations catering to the wider community of astronomers. The Solar community has organized itself differently, and has decided on a national strategy leading to the ATST. In Europe, there is a high-quality space science program with relatively modest funding, through ESA and national agencies. Radio astronomy is supported by a network of essentially national institutions. The ground-based optical-infrared program in the South is world-leading through ESO, which is now expanding into the radio regime with APEX and ALMA. Some of the larger European countries have 8–10 m telescopes in the North, but there is general agreement that an extremely large telescope has to be organized through ESO. In Japan, the NAOJ has a key role in providing access to Subaru and ALMA, and also plans to join an extremely large telescope project.

ALMA bridges the traditional scientific divide between the optical/infrared and radio regimes, and is a global partnership involving ESO, NAOJ and NRAO, strong institutions from three continents with very different funding systems. Developing this into an effective partnership operating a transformational single facility in a remote location is challenging, but success will bode well for future large facilities.

The cost of future facilities such as TMT, GMT and SKA are such that they are likely to require large partnerships, and attention to effective governance will be important. Many of the 8–10 m class telescopes were built by partnerships, organized in various ways, and the experience with ALMA provides further useful guidance. Drawing on these experiences suggests that a tightly drawn collaboration with a single managing agency is potentially an extremely powerful approach. In the case of the E-ELT project, ESO already offers such a governance structure for its member states, making the management more straightforward.

5 Conclusions

There are tremendous technical and scientific opportunities to answer fundamental questions about the Universe. This requires the best use of existing facilities including rationalization or specialization of medium-size telescopes, as well as the full science harvest of the facilities now under construction, including ALMA, and the construction of new world-class ground-based telescopes (ELT’s and later the SKA), which also provide synergy with powerful space observatories. The development of a long-range European plan for astronomy, and similar efforts elsewhere including the US Decadal Survey, needs to be followed up by actions from the funding agencies to make these facilities a reality.

Introduction to Poster Presentations

Remko Stuik

400 years of exciting developments of Astronomical Telescopes is a long period of time to summarize in only three and a half days conference. The Poster Presentations have succeeded in both extending topics already addressed in several the invited talks as well as in addressing new topics.

Even though the invention of the telescope has been attributed to Hans Lipperhey, already 6 centuries before, the Book of Optics by Alhazen was published, showing first principles of optics. Several pages of the translation of this book are shown in the Poster by Høg. Once the telescope had become public knowledge, its existence spread very rapidly over the world, as shown in the Poster by Zoomers. In paintings of J. Brueghel the Elder, as presented by Selvelli, telescopes play a prominent role.

Observations from space have only been possible since World War II. Already quickly after the war, efforts were made to use space for observations not possible from the ground, as demonstrated in the Poster by Morton on the precursors of the Space Telescope. During the last several decades, the impact of space-based observations on the understanding of the universe has been overwhelming. Several poster presentations address a number of these space telescopes, like ISO (presented by Salama), AKARI (presented by Onaka) and XMM (presented by Gonzales-Riestra) being addressed, as well as to see presentations on space missions still to be launched, like GAIA (presented in several Posters by Blasco).

With the opening of new wavelength areas, the development of new types of optics became necessary. Hudec presents in two posters the development of X-ray optics in the Czech republic. Von Ballmoos focuses on optics for Gamma-Rays, where imaging is generally considered even harder.

The invited talks focused mainly on the history and large facilities, but throughout the centuries, smaller observatories have played an important role. The posters by Godunova and Annuk discussed two of these observatories; the Terskol and Tartu observatories respectively. The latest trend in observatories is developing remotely operated telescopes, allowing the users to observe from home, reducing the need to travel to remote locations. Two of these remote observatories are being discussed: The amateur astronomy club in Alkmaar, the Netherlands have developed their own robotic telescope, as presented by Nieuwenhout, while Nekola presented small robotic telescopes for follow-ups of High-Energy events in his Poster.

As organizer of this conference I would like to thank all Poster Presenters in delivering their excellent contributions and all the participants in their enthusiastic discussions during the poster session.

Remko Stuik
Leiden Observatory, Leiden University
e-mail: stuik@strw.leidenuniv.nl

650 Years of Optics:
From Alhazen to Fermat and Rømer

Erik Høg[1]

ABSTRACT: Under house arrest in Cairo from 1010 to 1021, Alhazen wrote his Book of Optics in seven volumes. (The kaliph al-Hakim had condemned him for madness.) Some parts of the book came to Europe about 1200, were translated into Latin, and had great impact on the development of European science in the following centuries. Alhazen's book was considered the most important book on optics until Johannes Kepler's "Astronomiae Pars Optica" from 1604. Alhazen's idea about a finite speed of light led to "Fermat's principle" in 1657, the foundation of geometrical optics.

Keywords: History of optics; history of astronomy; history of astrometry

Ibn al-Haytham

The opinions in Antiquity about light and how we see the objects around us followed Plato who understood light as rays emitted from the eye towards the surroundings. Since we see remote objects immediately upon opening our eyes the rays must propagate with infinite speed. Euclid said a hundred years later, around 280 BC, that light moves along a straight line, and he formulated the laws about reflection in a mirror. Hero from Alexandria proposed about 60 AD the general hypothesis that light takes the shortest path between two points, and on this basis he was able to reach the same results as Euclid.

In the Middle Ages the centre of natural sciences moved to the Arabic world where Alhazen formulated views about light which made great impact in Europe. Therefore his work and the effect on science in Europe should be mentioned at the present celebration of 400 years of telescopes. Alhazen's Book of Optics in a printed version from 1572 is found in the Leiden University Library in the Latin translation from the Kitab al-Manazir (Book of Optics). The ideas of light of Alhazen and the European "perspectivists" are explained and some pages from the book in Leiden are shown in Figures 1 and 2.

Alhazen realized that light has its origin outside the observer, that the rays on their way hit the objects, and that we see an object when the rays from the object enter our eyes, an idea already proposed by Aristotle. Alhazen described the eye and its functioning, and he made mathematical descriptions of the properties of light. He proposed that light moves with finite speed, and that it moves more slowly in dense media. His astronomy was a theoretical attempt to fit the spheres of the celestial bodies into each other, a task he criticized his predecessors, especially Ptolemy, for not having solved.

Alhazen , whose full Arabic name is Abu 'Ali al-Hasan ibn al-Hasan ibn al-Haytham, or just Ibn al-Haytham, was born in Basra about 965 and travelled to Egypt and Spain. He worked in Cairo and died there in the year 1040. According to Steffens (2007) and

[1] Erik Høg
Niels Bohr Institute, Juliane Maries Vej 30, 2100 Copenhagen Ø, Denmark
Erik.hoeg@get2net.dk

Wikipedia (2008), he conducted research in optics, mathematics, physics, medicine, and development of scientific methods. His main work, The Book of Optics, was written while under house arrest in Cairo during eleven years (1010-1021). According to Wikipedia, in his over-confidence about the practical application of his mathematical knowledge, he had assumed that he could regulate the floods caused by the overflow of the Nile. Ordered by the sixth Fatimid caliph, al-Hakim, to carry out this operation, he quickly perceived the insanity of what he was attempting to do, and retired in disgrace. Fearing for his life, he feigned madness and was placed under house arrest until al-Hakim died. During the arrest and after he devoted himself to his scientific work until his death.

According to medieval biographers, Ibn al-Haytham wrote more than 200 works on a wide range of subjects; at least 96 of his scientific works are known of. Most of his works are now lost, but more than 50 of them have survived to some extent. Nearly half of his surviving works are on mathematics, 23 of them are on astronomy, and 14 of them are on optics, with a few on other subjects. Not all of his surviving works have yet been studied.

OPTICAE
THESAVRVS.
ALHAZENI
ARABIS
libri ſeptem, nunc primùm
editi.
EIVSDEM liber DE CREPVSCVLIS
& Nubium aſcenſionibus.
ITEM
VITELLONIS
THVRINGOPOLONI
LIBRI X.
Omnes inſtaurati, figuris illuſtrati & aucti, adiectis etiam in
Alhazenum commentarijs,
A'
FEDERICO RISNERO.

BASILEAE,
PER EPISCOPIOS. M D LXXII.

Ex Bibliotheca Viri Illuſt. Iſaaci Voſſii.

Fig. 1 The title page of Kitab al-Manazir (Book of Optics) in Latin translation, a version printed in 1572 and found in the Leiden University Library 6767 A 13

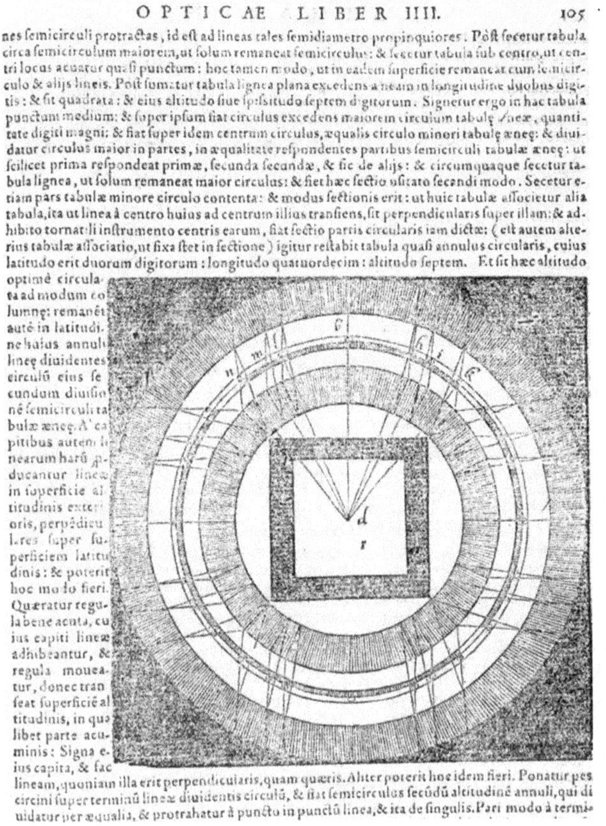

Fig. 2 A page from the Book of Optics, selected mainly for its nice figure, the meaning of which, however, we do not understand. According to an expert in Latin, Dr. Christian Marinus Taisbak, the text speaks in great detail about a bronze plate and a wooden plate in which certain lines shall be drawn with great care, but for what purpose we could not deduce

Alhazen in Europe

Some parts of the Book of Optics came to Europe about 1200, were translated into Latin, and had great impact on the development of European science in the following centuries. Alhazen's book was considered the most important book on optics until Johannes Kepler's "Astronomiae Pars Optica" from 1604. Surprisingly, Alhazens's book was almost unknown in the Islamic world until the 1320s, according to Denery (2005) from whom I am quoting in the following.

The Latin translation of the Book of Optics exerted a great influence, for example, on the work of Roger Bacon, who cites him by name, and on Kepler and Fermat. It brought about great progress in experimental methods. His research in catoptrics centred on spherical and parabolic mirrors and spherical aberration. He made the important observation that the ratio between the angle of incidence and refraction does not remain constant, and investigated the magnifying power of a lens. His work on catoptrics also contains the important problem known as Alhazen's problem. Alhazen has sometimes been called the "father of optics" and "the first scientist".

The scientists discussing optics in Europe at those times are called "perspectivists" after Roger Bacon's book "Perspectiva" from about 1270, but the word perspective has here a very different meaning from that in the Renaissance art of painting. Perspective meant the science itself about seeing, and the perspectivists thought that optics gave a deep insight into how we get to know anything about the world. It begins with the emission of light and colours from the objects through air to the eyes and then to the brain. This view is based on Alhazen's book.

Roger Bacon, a Franciscan theologian, describes emission of light and colours as a “multiplication of species”, which is in fact the title of one of his books, and an idea going back to Aristotle. The word species is explained as force or likeness, and species are considered as the source of all natural action and causation. The species is the real source to our sensing and intellectual understanding of the world. Through the species the surrounding medium is assimilated to the object. For example, a flame creates species in the surrounding air. These species heat the air and therefore assimilate the air to the nature of fire, but the air does not become fire, etc. etc., according to Denery (2005) pp.86-96. Bacon says that the visible object in its true nature and essence enters the eye and reproduces itself inside the eye.

A great hurdle for the understanding of how we see the objects was, with our words: the image formation in the eye. Light comes to our eyes from all directions, according to Alhazen, but how can the eye distinguish the directions to the various objects? Alhazen gave an answer which we know is wrong, but which Bacon adopted: the species multiply themselves in all directions, but only the species arriving perpendicularly to the surface of the eye are really sensed. The other species are refracted in the lens of the eye and neutralize each other. Clearly, the species must be understood in their remote historical context, and not as an “anticipation” of the modern photons.

Alhazen's ideas were not bettered until the time of Kepler and Snell; Willebrord Snell van Royen and René Descartes formulated the law of refraction mathematically in respectively 1621 and 1637. Pierre de Fermat, however, could not accept Descartes' justification or demonstration of Snell's law which was based on an analogy with mechanical phenomena. Based on Heron’s ideas about the shortest path and Alhazen’s assumptions about a finite speed of light, smaller in dense media, he formulated in 1657 “Fermat’s principle” which expresses that light follows the path which takes the shortest time. Fermat then derived the law about light moving in straight lines, as well as the laws of reflection and refraction, thus perfecting geometrical optics.

The legacy of Ibn al-Haytham in the Book of Optics about the different finite speeds of light spanned six centuries, culminating in “Fermat’s principle” in 1657 and, finally, the measurement of the finite speed in cosmic space by Ole Rømer in 1676.

Acknowledgement: I am grateful to John Heilbron for comments on previous versions of this paper. I gratefully acknowledge permission of the Leiden University Library to include the photographs from the Book of Optics.

References

Alhazen 1021, Book of Optics, Extracts from a version printed in 1572 found in the Leiden University Library in the Latin translation from the Kitab al-Manazir (Book of Optics).

Denery II Dallas G. 2005, Seeing and being seen in the later medieval world, Optics, Theology and Religious Life. 202 pp.

Steffens B. 2007, *Ibn al-Haytham: First Scientist* (Profiles in Science). Greensboro, NC: Morgan Reynolds Publishing, 2007.

Wikipedia 2008, Ibn al-Haytham In: Wikipedia, the free encyclopaedia.

The Great Melbourne Telescope: A Contextual History

Richard Gillespie[1]

Abstract. The Great Melbourne Telescope promised much when erected in 1870. Its design was overseen by leading British astronomers. Its maker, Thomas Grubb of Dublin, was a highly skilled instrument maker and designer. The Royal Society committee superintending the design and construction proclaimed that Grubb had created a telescope that would make great discoveries through studies of the southern hemisphere nebulae. Yet the telescope never lived up to expectations.

Historians have tended to blame one or more aspects of the telescope design: the speculum mirror, its focal length, or wind vibration. Yet none of these issues, taken separately or collectively, are sufficient to account for the telescope's modest performance. Instead the telescope needs to be placed into its technical, institutional, astronomical and cultural contexts, for it was the interplay between these many aspects that shaped the telescope's working life.

Keywords

Great Melbourne Telescope - Melbourne Observatory – Reflecting telescopes - Speculum mirrors

[1]Richard Gillespie
History & Technology Department, Museum Victoria, GPO Box 666, Melbourne, VIC 3001, Australia
E: rgillespie@museum.vic.gov.au
T: +613 8341 7380
F: +613 8341 7722

The Great Melbourne Telescope: A contextual history

Great hopes were held for the Great Melbourne Telescope (GMT) when it was erected in Melbourne, Australia in 1869. Its design and construction was overseen by leading British astronomers. They agreed that its maker, Thomas Grubb in Dublin, had excelled in creating a telescope that would make great discoveries through studies of the southern hemisphere nebulae. The primary aim was to detect if any changes had occurred in the southern nebulae since Herschel's observations in the 1820s; were they sufficiently changed to confirm the 'nebular hypothesis' or were they resolvable into stars?

Yet the telescope never lived up to early expectations. Historians have tended to blame one or more aspects of the telescope design: the speculum mirror, its focal length, or wind vibration (Gascoigne 1995, King 1955, Perdrix 1992, Ritchey 1904). Yet none of these issues, taken separately or collectively, are sufficient to account for the telescope's modest performance. Instead the telescope needs to be placed into its technical, institutional, astronomical and cultural contexts, for it was the interplay between these many aspects that shaped the telescope's working life.

The Southern Telescope

The idea of erecting a large telescope in the southern hemisphere first emerged in the late 1840s; it was proposed by British scientist Edward Sabine. He found an enthusiastic supporter in Romney Robinson, director of Armagh Observatory. Robinson was a close associate of Lord Rosse, whose 3-foot and 6-foot reflectors at Birr Castle in Ireland suggested the possibility of resolving all nebulae into stars. In the 1830s, John Herschel had spent several years at the Cape of Good Hope, cataloguing southern stars and nebulae. But Robinson argued that a much larger telescope was needed in the southern hemisphere to test the resolvability of nebulae and record their changes.

Following a report by Robinson and Rosse on their nebulae observations at Birr Castle, the British Association for the Advancement of Science at its meeting in Belfast in 1852 established a Southern Telescope Committee with the Royal Society of London. The Committee endorsed a design by Thomas Grubb of Dublin, who had previously made a 15 inch reflector for Robinson at Armagh. Grubb proposed a 4-foot reflecting telescope, mounted equatorially; like in Rosse's telescopes the mirror would be speculum metal. One eminent member of the Committee, Astronomer Royal George Airy, opposed the design, arguing that the difficulties of mounting such a large mirror equatorially were far beyond Grubb's understanding, and that Robinson had turned the whole matter into a 'Grubb-job' (Airy to Sabine, 25 Oct 1853, Airy Papers, RGO6, Cambridge University Library). In the event, no government funds were forthcoming and the telescope project was mothballed.

The Great Melbourne Telescope

The Impetus to build the telescope now shifted to Melbourne, Australia. William Wilson had been appointed the foundation professor of mathematics at the new University of Melbourne in 1855. He had previously taught in Belfast and founded an observatory there. The committee that selected Wilson for Melbourne included Romney Robinson and was chaired by John Herschel. Once in Melbourne, Wilson threw himself into finding support to build the Southern Telescope at a new observatory in Melbourne, seeking funds from the gold-rich Victorian Government, while corresponding with Sabine, now president of the Royal Society in London (Wilson 1855-6).

Sabine coordinated the British scientists' discussions over the final telescope design. Robinson, Herschel and others opted for the ten-year-old Grubb design (Royal Society of London 1871). Although Jean Foucault had recently built a 33-inch silver-on-glass primary mirror, the British scientists felt the technology was unproven and it would be a 'most hazardous experiment' (Herschel to Sabine, 22 Feb 1863, Herschel Papers, M1938, Royal Society of London). The Anglo-Irish technology was preferred to the French.

Construction of the telescope was undertaken by Thomas Grubb in new telescope works at Dublin, supervised by his 21 year-old son Howard Grubb. The southern telescope was to be the project that firmly established him internationally as a major telescope maker. Making the speculum mirrors was the most complex and dangerous part of the job. Each speculum required 1370 kg of alloy to be melted in a furnace, poured from the crucible into the mould, then drawn into an annealing oven for 24 days. Two mirrors were then ground and polished to the correct shape by a machine designed by Grubb (Robinson & Grubb 1869).

Figure 1. Erecting the Great Melbourne Telescope at Melbourne Observatory, 1869. Source: Museum Victoria

The Royal Society's telescope committee inspected the completed telescope in early 1868, and concluded that it was a 'masterpiece of engineering' (Rosse, Robinson & De La Rue 1868). Nests of triangular levers at the back of the speculum distributed weight in all positions of the telescope, preventing flexure and distortion of the image. Although the telescope weighed over 8 tonnes, one person could raise the telescope from horizontal to vertical in only 20 seconds. Most importantly, the optical powers of both specula were excellent; for example Nebula 46M 'was revealed as a ring, bright even on the dazzling background of surrounding stars' (Rosse, Robinson & De La Rue 1868). Robinson in Armagh was lyrical about the telescope: 'What strange nebular forms it may reveal, what new configurations of stars, what new cosmical laws it may unfold, what new elements it may disclose, we shall soon know…' (Robinson & Grubb 1869, 156).

Trouble in Melbourne

The Great Melbourne Telescope, as it was now formally called, arrived in Melbourne in November 1868, in the care of its trained observer. The Royal Society committee had toyed with the idea of sending either the young Howard Grubb or John Herschel's son as observer, but in the event selected a young Cambridge mathematician, Albert Le Sueur. He was present in Dublin for all the critical stages in the construction and testing of the telescope.

By mid 1869 the telescope was ready for use, and Le Sueur was soon writing enthusiastically to Robinson and Herschel, reporting great changes in the nebula in Argo since Herschel had observed it in the 1830s. Herschel replied that this suggested that the Nebula was indeed a vast cosmical cloud interposed between Earth and the stars (Herschel to Robinson, 21 Aug 1869, Herschel Papers).

But while Robinson and Herschel were enthusing, a letter from the Government Astronomer at Melbourne Observatory, Robert Ellery, was on its way to George Airy at Greenwich Observatory. Ellery reported to Airy: 'I am not satisfied with the performance of this Great Instrument, and I do not think the mirrors are at all perfect in figure. I have not yet seen anything like good definition of a planet with the lowest power. Of course we get plenty of light even with the Nebulae, but no definition with bright objects' (Ellery to Airy, 13 Aug 1869, Airy Papers). Worse still for Grubb, reports of difficulties with the telescope appeared in British newspapers; suddenly Grubb's reputation as a telescope maker was in the balance.

The British scientists leapt to the defence of the telescope, pointing out that it had been in excellent condition when tested in Dublin by the Royal Society committee. Privately, though, the British scientists began to question Le Sueur's abilities and actions: he had clearly used the wrong substance to clean the mirrors, and his reports from Melbourne betrayed a lack of understanding of some of the technical decisions made during construction, even though he had been present. Thomas Grubb confided to Ellery that he had always found Le Sueur a 'very rough experimenter' (Grubb to Ellery, 27 Oct 1870, VPRS 780, Box 3, Public Record Office of Victoria).

Initially Robert Ellery had left Le Sueur to manage all aspects of the telescope, and to correspond with the English scientists. But as he was drawn into the controversy, he became more actively involved in the operation of the telescope. Le Sueur resented what he thought was interference; he seemed to believe that he had been given sole charge of the instrument by the Royal Society. In May 1870 Le Seuer resigned, less than 12 months after the telescope had been erected; he returned to his native Jersey, not making any contact with the English scientists, and disappeared from view.

With Le Sueur gone, Thomas Grubb and Ellery now formed a rapprochement, based on the desire to get the telescope working effectively. Letters flowed constantly between the two, as they gradually sorted out difficulties with the complex instrument. In retrospect they should have always anticipated that it would take some time for the telescope to be adjusted, and for the observers to become acquainted with its peculiarities. By July 1871 Grubb could write to Ellery that 'it gives me much pleasure to learn that the Great Melbourne Telescope is now being worked satisfactorily' (Grubb to Ellery, 27Jul 1871, VPRS 780, Box 3).

Grubb still felt the need to protect his reputation, however. Word of the difficulties with the telescope had reached European and American astronomers. Grubb prepared a pamphlet, carefully outlining the mistakes that had been made by Le Sueur and refuting other criticisms of the design (Grubb 1871). None other than George Airy was deeply impressed by Grubb's evident mathematical as well as technical skill, conceding that 'Grubb has risen greatly in my estimation' (Airy to De La Rue, 10 Jan 1871, Sabine Papers, M1940, Royal Society of London). Coming from Airy that was praise indeed. The Royal Society printed copies of Grubb's paper and the other correspondence, and circulated it privately to astronomers around the world (Royal Society of London 1871). Grubb's reputation remained intact, and was even enhanced by the controversy; Thomas and son Howard would soon win the contract to construct a 27-inch refractor for the Vienna Observatory.

Putting the telescope to work

Robert Ellery now threw himself into working the telescope, corresponding with the Grubbs and Robinson. He purchased new eyestops to replace those altered inappropriately by Le Sueur, and began experimenting with polishing smaller mirrors, preparatory to re-polishing the large specula. Le Sueur was replaced as observer by Joseph Turner, a local photographer and amateur astronomer. Ellery admitted to Robinson that Turner 'would not set the Houses on fire'; he was an excellent photographer and good sketcher, but lacked mathematics and 'his eyes are not strong enough I am afraid' (Ellery to Robinson, 15 Jul 1873, VPRS 776, Public Record Office of Victoria).

Figure 2. The Great Melbourne Telescope, 1870s. Source: Museum Victoria

Despite his limitations, Turner would prove to be the only solid observer that the telescope ever had as its companion, and for the next 10 years he produced excellent drawings and observations of Herschel's nebulae, and spectroscopic analysis of their chemical makeup (Melbourne Observatory 1885). It was also used for other work, including study of lunar eclipses, comets, and observations of Jupiter and its moons.

Although more sensitive photographic plates were being developed in the 1870s, attempts to photograph nebulae were unsuccessful, owing to the long exposure times required and the difficulty in keeping the telescope tracking precisely on the faint objects. As the telescope was not protected by a dome, wind could also be a problem, although this is only occasionally cited as a problem in the observation diaries.

The results on whether nebulae had changed since Herschel observed them in the 1830s were necessarily equivocal. Turner used the position and magnitude of stars in the vicinity of the nebulae as constants; if they seemed the same as in Herschel's drawings, yet the nebulae itself seemed to have changed, this suggested that real changes had occurred. Yet the differences in optics, observer and atmospheric conditions could plausibly account for most of these apparent changes.

While the telescope was being built in Dublin, the English astronomer William Huggins had used the newly invented spectroscope to examine nebulae; the spectra clearly showed that some nebulae were clouds of gas, not faint stars. While a spectroscope was used on the Great Melbourne Telescope, Turner continued to rely primarily on his sketches. By the time of Turner's death, new sensitive photographic plates were being used at other observatories in association with smaller silvered-glass reflectors, and producing photographs of nebulae that started to eliminate the idiosyncrasies of the observer.

Repolishing the mirrors

By the late 1870s the main speculum was clearly tarnishing and the resultant images of nebulae fading (although for a while Turner tried to explain this in terms of Herschel's nebulae becoming fainter). Ellery had avoided attempting to repolish the main mirrors, as his earlier experiments with polishing smaller test mirrors had convinced him of the difficulty of the task. Ellery asked Howard Grubb what it would cost to repolish them in Dublin. Grubb clearly did not want to take on the task, and proposed that a new glass or metal mirror be made.

Repolishing was a thankless task. After months of work, sometimes involving four staff full-time, the mirrors were no better; if anything they had lost their figure, and had imperceptibly small ridges, which meant that light would not be focused into hard points in the eyepiece. Ellery and Grubb discussed placing a new 24-inch refracting telescope on the GMT mounting; Grubb recommended that it be placed inside the tube, an indication that he no longer considered the original telescope worth saving.

Two events meant that the idea was never pursued. Melbourne Observatory joined the Carte du Ciel project to photograph and chart the heavens, and was allocated a substantial zone of the southern skies. Observatory resources were increasingly channelled into this project, and a new telescope was purchased from Grubb for the purpose. At the same time, the 1890s depression cut into government expenditure, and several staff were forced into early retirement. The Great Melbourne Telescope was left idle.

Accounting for failure

Historians of astronomy have tended to argue that the Great Melbourne Telescope was a flawed telescope from the start; a hybrid design that, while pointing to the future with Grubb's advanced equatorial mounting and drives, carried a large metal mirror from the past. Certainly with hindsight we can see that research astronomy in the second half of the 19th century increasingly used the new techniques of photography and spectroscopy. While a spectroscope and camera were mounted on the telescope, it was not well suited to their use.

But the history of the telescope can also be understood by considering the institutional context of the Melbourne Observatory. The primary astronomical function of the colonial government observatory was positional astronomy – observing transits of known stars in order to maintain an accurate time service, establishing star catalogues for the geodetic survey, establishing longitude with increasing accuracy. In addition Ellery participated in many international projects, including eclipse expeditions, transits of Venus, observing new comets, searching for new satellites of planets, photographing sunspots, and measuring daily and hourly variations in the earth's magnetism. In collaboration with the astronomers in Adelaide and Sydney, Ellery established meteorology in Australia, and provided a daily weather service.

The Great Melbourne Telescope had to fit into this busy round of daily work, yet in the 1870s the Melbourne Observatory had a scientific staff of only five, a messenger and a workman. When Ellery became more involved in the operation of the telescope following Le Sueur's departure, he complained to a correspondent that 'The Great Reflector is as much trouble as all the other work together', and in a letter to Robinson in Ireland sardonically called it 'the Monster' (Ellery to James Simms, 28 Dec 1872, Ellery to [Robinson] 19 May 1874 VPRS 776).

Such monsters were more the province of wealthy gentleman astronomers, such as Lord Rosse at Birr Castle, or of astronomers such as the Herschels who had found royal patronage to support their work, and did not have a busy government observatory to run. Rosse in particular had the wealth and estate workers to experiment endlessly, making improvements in his great reflectors. Astronomy, at least in Britain, could be broadly divided into two institutional and cultural camps: the professionals engaged in government observatories and focusing on positional astronomy; and the gentlemen in privately funded observatories, undertaking more specialist research programs on nebulae, comets or double stars. Placing the Great Melbourne Telescope at Melbourne Observatory was creating an institutional hybrid.

The Telescope reborn

The telescope lay unused for decades until, following the closure of Melbourne Observatory in 1945, Mount Stromlo Observatory purchased it for the price of scrap. At Mount Stromlo the telescope was given a new 50-inch glass mirror, and became an integral part of Mt Stromlo's work from 1959 into the 1970s. In the 1990s the telescope was rebuilt with two detector mosaics for the MACHO project, to search for evidence of dark matter. Then in January 2003 a bushfire swept across Mt Stromlo, its firestorm destroying the majority of the telescopes and buildings.

Since the early 1980s Museum Victoria had progressively been retrieving discarded parts of the Great Melbourne Telescope as Mt Stromlo rebuilt the instrument. This included one of the primary mirrors, the telescope tube, eyepieces, right ascension axis circles and clamps, bearings, declination axis parts, and the polishing machine. The remaining parts of the telescope at Mt Stromlo will be dismantled in November 2008 and returned to Melbourne, including the cube, polar axis, primary mirror support system, declination axis spacer and north polar axis bearing.

Museum Victoria, Astronomical Society of Victoria and Royal Botanic Gardens, Melbourne have agreed to undertake a feasibility study to restore the Great Melbourne Telescope and return it to its original building at the former Melbourne Observatory site, which is now part of the Botanic Gardens. The Astronomical Society and Botanic Gardens already run public viewing nights using a Troughton & Simms 8 inch refractor and an 8 inch reflector. A detailed conservation plan has yet to be completed, but it is proposed to restore the mechanical integrity of the Great Melbourne Telescope to its 1869 condition, although it is likely that a modern optical system will be put in to enable its effective use as a publicly accessible telescope. 'The Monster' will be reborn.

References

Gascoigne, S.C.B.: The Great Melbourne Telescope and other 19th century reflectors. Hist. Rec. Aust. Sci. 10, 223-45 (1995)

Grubb, T.: The Great Melbourne Telescope: an examination of and reply to the official reports from Melbourne respecting the instrument, its erection at Melbourne, &c, &c., privately printed (1870)

King, H.C.: The History of the Telescope. London (1955)

Melbourne Observatory: Observations of the Southern Nebulae made with the Great Melbourne Telescope from 1869 to 1885, Part 1. Government Printer, Melbourne (1885)

Perdrix, J.L.: The last great speculum: the 48 inch Great Melbourne Telescope. Aust. J. Astron. 4, 149-163 (1992)

Ritchey, G.W.: On the modern reflecting telescope and the making and testing of optical mirrors. Smithson. Contrib. Knowl. 34, no 3 (1904)

Robinson, T.R., Grubb, T.: Description of the Great Melbourne Telescope. Philos. Trans. R. Soc. Lond. 159, 127-161 (1869)

Rosse, Lord, Robinson, T.R., De La Rue, W.: Report of the Committee on the Melbourne Telescope. Proc. R. Soc. Lond. 16, 313-316 (1868)

Royal Society of London: Correspondence concerning the Great Melbourne Telescope. In three parts: 1852-1870. Royal Society, London (1871)

Wilson, W.: Report on the steps taken in England to provide a telescope for observing the nebulae of the Southern Hemisphere. Trans. Philos. Inst. Vic. 1, 138-152 (1855-6)

On the telescopes in the paintings of J. Brueghel the Elder

Pierluigi Selvelli[1] · **Paolo Molaro**[1]

Abstract We have investigated the nature and the origin of the telescopes depicted in three paintings of J. Brueghel the Elder completed between 1609 and 1618. The "tube" that appears in the painting dated 1608-1612 represents a very early dutch spyglass, tentatively attributable to Sacharias Janssen or Lipperhey, prior to those made by Galileo, while the two instruments made of several drawtubes which appear in the two paintings of 1617 and 1618 are quite sophisticated and may represent early examples of Keplerian telescopes.

Keywords history of astronomy · ancient instruments

1 Introduction

The presence of spyglasses and other astronomical instruments characterizes three paintings of J. Brueghel the Elder (1568-1625) composed between 1609 and 1618. The paintings were drawn while the artist was appointed as a painter at the court of the Archduke Albert VII of Habsburg (1559-1621), spanish Governor of the catholic part of the Netherlands, and of his spouse, the Archduchess Isabella Clara Eugenia, "Infanta" of Spain and daughter of Felipe II.

The three paintings can be considered as high quality, realistic photographies and are rich in naturalistic details. This allows a detailed examination of the instruments therein depicted and an investigation of their nature.

2 The painting with the early spyglass

The painting "Extensive Landscape with View of the Castle of Mariemont" is conserved at the Virginia Museum of Fine Arts (VMFA) in Richmond, VA, USA. The Mariemont Castle, located near Bruxelles, was under reconstruction for several years, and a study

[1] INAF-Osservatorio Astronomico di Trieste, Via Tiepolo 11, I-34143, Trieste, Italy
Tel.: 39-040-3199218
Fax: 39-040-309418
E-mail: selvelli@oats.inaf.it, molaro@oats.inaf.it

Fig. 1 A detail of the painting "Extensive Landscape with View of the Castle of Mariemont" by J. Brueghel the Elder, ca. 1609-1612. Virginia Museum of Fine Arts, Richmond. The Adolph D. and Wilkins C. Williams Fund. Photo: Ron Jennings. The actual length of the spyglass is estimated to be about 40 cm.

of the development of the building, by comparison with paintings of ascertained dates, has made it possible to estimate that the date of the painting's completion was in the years 1608 - 1611 . In the central part of the painting one can see the important detail of the archduke Albert VII looking through a spyglass. The instrument has a cylindrical shape and appears metallic, probably made of tin, with two gilded rings at either end. (Fig. 1). By comparison with other figures and objects one can estimate that its length is about 40-45 cm and its diameter about 5 cm.

To our knowledge, this painting represents the most ancient reproduction of a spyglass, and we guess that the depicted instrument actually represents one of the first spyglasses ever built. We know from Daniello Antonini, who was serving with the Archduke in Brussels, that Albert VII owned copies of early spyglasses. Antonini, on September 2, 1611, wrote a letter to his former master Galileo (Bibl. Naz. Fir., Mss. Gal., P. VI, T. VIII, car.37) telling him that the Archduke had obtained some spyglasses from the "first inventor", although of lower quality than those owned by Galileo. Thus, it is likely that the "tube" held by the Archduke represents one of these spyglasses mentioned by Antonini. Concerning the manufacturer, it could be either Sacharias Janssen or Hans Lipperhey, both considered "fathers" of the telescope. According to Pierre Borel (Van Helden, 1977), Sacharias Janssen made some instruments measuring about 40 cm in length, and offered the best ones to Prince Maurice of Nassau and the Archduke Albert. Alternatively, according to Antonius Maria Schyrley de Rheita, the Marquess Ambrogio Spinola, Commander of the Spanish Army in the Flanders, bought a spyglass in The Hague near the end of 1608, probably made by Lipperhey, and offered it to Archduke Albert (Van Helden, 1977). Spinola was in The Hague in that period as a representative of the Spanish Governor, for peace negotiations with Maurice of Nassau, "Stadtholder" of the seven provinces .

The presence of the spyglass in the painting can be used to exclude the year 1608 in the range of the dates for its completion, since the first spyglasses appeared in autumn 1608, while the landscape with the trees and other details in the painting indicates (late) summer as the season.

3 The telescope(s) with drawtubes

The Allegory of Sight is one of the series of paintings of Jan Brueghel known as *Allegory of the senses* or *The five senses*, made in collaboration with Peter Paul Rubens, which is conserved at the Museum of El Prado, in Madrid.

The painting (oil on wood) is drawn with great accuracy and depicts a hall in the ancient royal Palace of Brussels, on the hill of Coudemberg, residence of the Archdukes, where their collection of paintings, precious items, and scientific instruments, most of them related to astronomy, were kept. The painting had been completed by 1617 as testified by the date that appears on a roll of several papers lying over a book entitled "Cosmographie", in the lower part of the painting, next to the author's signature. One can note various astronomical instruments such as a large astrolabe, an armillary sphere, a pedestal globe, a proportional compass, map dividers and sundials, which testify the interest of the Archdukes for science, and astronomy in particular. Each instrument has been meticulously characterized with true Flemish skill so that even the minutest details are accurately reproduced.

The telescope that appears between Venus and Cupid, (Fig. 2) consists of a main tube and seven drawtubes, all of which appear to be made of metal (probably silver). Each of the intermediary drawtubes terminates in an enlarged collar that appears to be made of the same metal. The lenses are housed in large rounded terminals. The instrument is fixed into a curved metal sleeve support attached to a brass joint which can be adjusted for angle. The pedestal consists of a turned column terminating in a simple saucer-shaped round base.

A comparison with other objects depicted in the painting indicates a maximum and minimum width for the drawtubes of about 7.5 and 2.5 cm, respectively, and an estimated total length, if the tubes were all drawn, close to 170 cm.

It should be be noted that a similar telescope is reproduced in a larger painting, also conserved at the museum "El Prado", named "The Allegory of the Sight and the Sense of Smell" (oil on canvas). This painting, completed around 1618-1620, was commissioned by the City of Antwerp from J. Brueghel and several other painters (following the "kunstkammer" style that was fashionable at that time), to celebrate the visit of the Archdukes to the city. The painting, which actually is a copy of the original that was lost in the fire of the Castle of Coudemberg near Bruxelles in 1731, includes several instruments that are very similar or identical to those reproduced in the previous painting.

The main difference between the two telescopes resides in the number of draw tubes (eight instead of seven) and in the color of the rings (black instead of silvery), see Fig.2 and Fig. 3. Also the pedestals are different in the two cases. Thus, despite

Fig. 2 A detail of the painting "The Allegory of Sight" by J. Brueghel and P.P. Rubens, 1617. Madrid, Museo Nacional del Prado. The actual size of the telescope is estimated to be about 70 cm.

the overall similarity in form, which clearly indicates the same origin (maker), the two telescopes are, apparently, two separate instruments. We note that Bedini (1971) described this telescope as constructed with cardboard tubes covered with white or a light-colored vellum. This seems hardly compatible with the metallic aspect of the tubes, as suggested by their color and reflectivity.

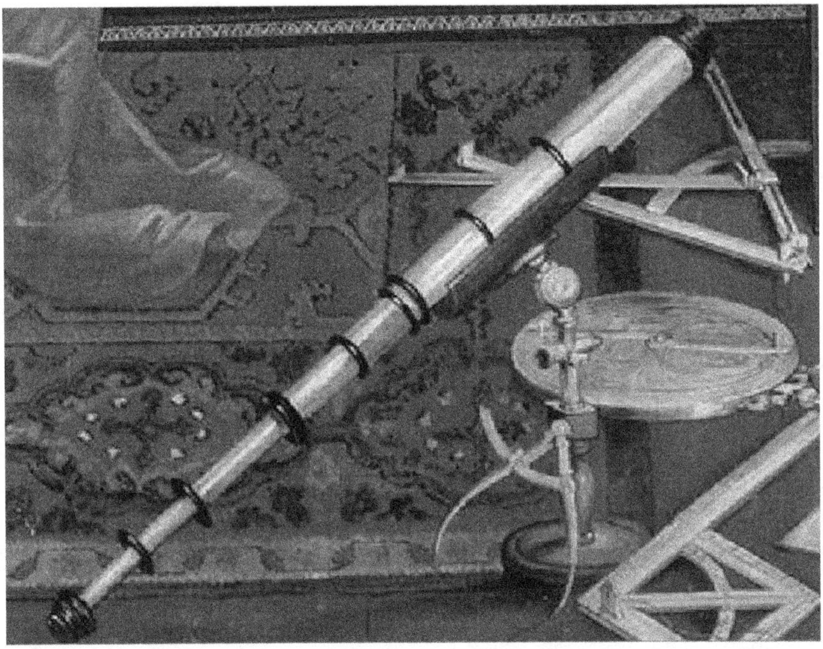

Fig. 3 A detail of the painting "The Sight and the Sense of Smell" by J. Brueghel et al., ca. 1618. Madrid, Museo Nacional del Prado. The actual size of the telescope in the painting is estimated to be of about 75 cm.

4 Dutch or Keplerian?

The origin and development of the "astronomical" telescope, the one consisting of two convex lenses, is uncertain and open to question, like the Galilean one. It was theoretically described by Kepler in his "Dioptrice" of 1611 but it is not clear when the first "astronomical" telescope was manufactured. "Rosa Ursina", completed by Scheiner in 1630, is the first book containing a reference to an astronomical telescope. In 1646, Fontana, in his "Novae Coelestium Terrestriumque Observationes" claimed to have manufactured an astronomical telescope in 1608, offering also a declaration of father Zupo stating that Fontana had shown a telescope made by two convex leness to him and to father Staserio already in 1614.

A definite upper limit for the manufacturing of the instrument depicted in Figs. 2 and 3 is set by the paintings' completion in 1617 and 1618 respectively. What is quite surprising is that their technology appears quite sophisticated for the epoch and there is no record of similar instruments until about three decades later (cf. the "Catalogue of Early Telescopes" by Van Helden, 1999). An examination of illustrations of early telescopes yields that the closest resemblance in shape and length is with the illustrations reported in C. Scheiner's works (*Disquisitiones Mathematicae* 1614, and *Rosa Ursina* of 1631).

We argue that these instrument may represent early examples of a Keplerian telescope, which makes their presence even more striking. Three circumstantial considerations seem to support this view:

– The presence of a quite large terminal (eyepiece) seems to indicate a compound eyepiece, quite incompatible with a Galilean (Dutch) mounting. In this case, the negative lens needs the eye to be brought as close as possible so that the eye's pupil becomes the aperture stop and the exit pupil. For the same reason, if the large terminal were just a lens-screen, this configuration would be barely compatible with a Dutch mounting.
– The overall length of the telescope is estimated to be about 180 cm. Even with low powers this would imply a very small field of view.
– The first records of Keplerian telescopes are related to the Habsburgs. Cristoph Scheiner in his "Rosa Ursina" claimed that he made a Keplerian instrument in 1617 and showed it to the Archduke Maximilian III, brother of Albert VII. Scheiner, in a letter on January 4, 1615 wrote of a "newly invented instrument" which actually could be the astronomical one, and the date is consistent with the former period. According to a different source, Maximilian III, about 1615, received a telescope with two convex lenses and Scheiner added a third one, thus manufacturing a terrestrial Keplerian telescope (we note, incidentally, that Scheiner actually used a Dutch telescope for his observations of sunspots in 1610). Likely, Albert VII obtained it from Maximilian or heard of it and obtained a similar one for his collection.

5 The monkey's tube

In the painting of Fig. 2, on the floor, just behind Cupid, and held by a monkey, one can note a tube that "prima facie" looks like an early spy-glass. Belloni (1964) described it as an "optical tube", while Bedini (1971) interpreted it as a microscope, of the Janssen type.

We recall that Willem Boreel, born in Middelburg and Dutch ambassador in France, when mentioning that a microscope made by Janssen was presented to the Archduke Albert VII, reported a description made by Cornelis Drebbel according to whom *the microscope had a tube made of gilded brass, resting on three dolphins made of brass which were supported by the disc of the base made of ebony* (Van Helden, 1977).

A close inspection on a high-quality reproduction of the painting has instead clearly shown that the tube appears to be made of metal (tin or silver) without any evidence of the above-mentioned supporting structures. The object is clearly a spyglass and its aspect and size clearly indicate that, most likely, it corresponds to the spyglass depicted in Fig. 1.

This confirms that the "tube" represents an early Dutch instrument, as suggested by Selvelli (1997), on the grounds of the painting of Fig. 2 only.

Acknowledgements We gratefully thank Inge Keil and Franz Daxecker for helpful information and valuable discussion.

References

1. Bedini, S.A. "Physis" Vol. 13, 149, *1971*
2. Belloni, L. Rendiconti Istituto Lombardo B98, 238, *1964*
3. Selvelli, P. L'Astronomia, 175, 36 *1997*
4. Van Helden Transact. of the American Philosophical Society 67, no. 4, *1977*
5. Van Helden Catalogue of early telescopes, Giunti, Firenze, *1999*

List of Participants

#	*Name*	*Affiliation*
1	Abrahams, Peter	Independent, United States
2	Angel, Roger	University of Arizona, United States
3	Annuk, Kalju	Tartu Observatory, Estonia
4	Aschenbach, Bernd	MPE Garching, Germany
5	Baars, Jacob	MPI for Radio Astronomie, Germany
6	Baes, Maarten	Universiteit Gent, Belgium
7	Baneke, David	Leiden Observatory, Netherlands
8	Banshidhar Sharma, Sandy	TNO Space and Science, Netherlands
9	Bennett, Jim	University of Oxford, United Kingdom
10	Blaauw, Adriaan	Leiden Observatory/Kapteyn Institute Groningen, Netherlands
11	Blasco, Carmen	ESA ESTEC, Netherlands
12	Boland, Wilfried	NOVA, Netherlands
13	Bolt, Marvin	Adler Planetarium, United States
14	Boudin, Nathalie	ESA ESTEC/Cosine, Netherlands
15	Boyle, Alison	Science Museum, London, United Kingdom
16	Braam, Ben	TNO Science & Industry, Netherlands
17	Brandl, Bernhard	Leiden Observatory, Netherlands
18	Broeders, Jan M.	Optische Fenomenen, Germany
19	Bruns, Don	Stellar Products, United States
20	Cordes, James	Cornell University, United States
21	Cornet, Gerard	Netherlands Institute for Space Research, Netherlands
22	Cunningham, Colin	UK Astronomy Technology Centre, United Kingdom
23	David, DeVorkin	Smithsonian Institution, United States
24	de Zeeuw, Tim	ESO, Germany
25	Drost, Jeanne	Leiden Observatory, Netherlands
26	Erickson, Edwin	NASA Ames Research Center, United States
27	Franx, Marijn	Leiden Observatory, Netherlands
28	Gehrels, Neil	NASA/GSFC, United States
29	Genzel, Reinhard	Max Planck Institute for Extraterrestrial Physics, Germany
30	Giacconi, Riccardo	Johns Hopkins University, United States
31	Gillespie, Richard	History & Technology Museum Victoria, Australia
32	Godunova, Vira	ICAMER Observatory, Ukraine
33	Govender, Kevindran	South African Astronomical Observatory, South Africa
34	Groothuis, Charlotte	Netherlands
35	Hagoort, Klaas	Airborne Composites BV, Netherlands
36	Harwit, Martin	Cornell University, United States
37	Hasinger, Günther	Max-Planck-Institut für extraterrestrische Physik, Germany
38	Heck, André	Strasbourg Astronomical Observatory, France
39	Henrichs, Huib	University of Amsterdam, Netherlands
40	Høg, Erik	Copenhagen University, Denmark
41	Hogenhuis, Harm	TNO Science & Industry, Netherlands
42	Holwerda, Benne Willem	Space telescope Science Institute, Netherlands
43	Holwerda, M.J.	Netherlands
44	Hudec, Rene	Astronomical Institute AS CR, Czech Republic
45	Hutchins, Roger	Independent, United Kingdom
46	Icke, Vincent	Leiden Observatory, Netherlands
47	Israel, Frank	Leiden Observatory, Netherlands
48	Jager, Rieks	NOVA / SRON, Netherlands
49	Jameux, David	ESA ESTEC, Netherlands
50	Jolissaint, Laurent	Leiden Observatory, Netherlands
51	Kamphues, Fred	Mill House, Netherlands
52	Kamphues-Chan, Penny	Mill House, Netherlands
53	Kanters, Edwin	Airborne Composites BV, Netherlands
54	Kaper, Lex	University of Amsterdam, Netherlands
55	Katgert, Jet	Leiden Observatory, Netherlands
56	Kendrew, Sarah	Leiden Observatory, Netherlands
57	Khanna, Ramon	Springer Verlag GmbH, Germany
58	Koschny, Detlef	ESA ESTEC, Netherlands
59	Kuijken, Konrad	Leiden Observatory, Netherlands
60	Le Poole, Rudolf	Leiden Observatory, Netherlands

#	Name	Affiliation
61	Lemke, Dietrich	Max-Planck-Institut für Astronomie, Germany
62	Lena, Pierre	Université Paris Diderot & Observatoire de Paris, France
63	Lequeux, James	Observatoire de Paris, France
64	Longair, Malcolm	Cavendish Laboratory, Cambridge, United Kingdom
65	Lub, Jan	Leiden Observatory, Netherlands
66	Maas, Ad	TNO Science & Industry, Netherlands
67	Mainieri, Vincenzo	ESO, Germany
68	Martinez Galarza, Juan Rafael	Leiden Observatory, Netherlands
69	McCray, Patrick	University of California, Santa Barbara, United States
70	McDonald, Don	Boulder Metric Inc., United States
71	Mekking, Jeroen	TNO Space and Science, Netherlands
72	Molster, Frank	Leiden Observatory, Netherlands
73	Monnet, Guy	ESO, Germany
74	Morton, Donald	National Research Council of Canada, Canada
75	Nekola, Martin	Astronomical Institute AS CR, Czech Republic
76	Nieuwenhout, Frans	AWSV Metius, Alkmaar, Netherlands
77	Noethe, Lothar	ESO, Germany
78	Nye, Derald	Kirmser-Wakabayashi Observatory, United States
79	O'Dell, C. R.	Vanderbilt University, United States
80	Ödman, Carolina	Leiden Observatory, Netherlands
81	Ojero-Pascual, Eduardo	Serco/ESA, Spain
82	Pinkau, Klaus	Max-Planck Institut, Germany
83	Quirrenbach, Andreas	Landessternwarte Heidelberg, Germany
84	Rahmati, Alireza	Leiden Observatory, Netherlands
85	Ravensbergen, Simon	Eindhoven University of Technology, Netherlands
86	Rieke, George	Steward Observatory, United States
87	Robson, Ian	UK ATC, United Kingdom
88	Rottgering, Huub	Leiden Observatory, Netherlands
89	Sabau-Graziati, Maria Dolores	INTA, Spain
90	Saenz, Adriana	ASD, Netherlands
91	Salama, Alberto	European Space Agency, Spain
92	Saucedo Morales, Julio	Universidad de Sonora, Mexico
93	Selvelli, Pierluigi	INAF, Italy
94	Serre, Denis	Leiden Observatory, Netherlands
95	Sinclaire, Peter	ESO, Chili
96	Smith, Robert	University of Alberta, Canada
97	Spyromilios, Jason	ESO, Germany
98	Stark, Ronald	NWO, Netherlands
99	Strano, Giorgio	Istituto e Museo di Storia della Scienza, Italy
100	Stuik, Remko	Leiden Observatory, Netherlands
101	Sullivan, Woody	University of Washington, Seattle, United States
102	Suzuki, Atsuto	KEK, High Energy Accelerator Research Organization, Japan
103	Talavera, Antonio	SERCO/ESAC, Spain
104	Thiele, Hans	Kayser-Threde GmbH, Germany
105	Trimble, Virginia	UC Irvine and LCOGT, United States
106	Trümper, Joachim	Max-Planck-Institut für Extraterrestrische Physik, Germany
107	Van de Stadt, Inge	AWSV Metius, Netherlands
108	van Delft, Dirk	Leiden University/Museum Boerhaave, Netherlands
109	van den Heuvel, Edward	University of Amsterdam, Netherlands
110	van der Hucht, Karel A.	International Astronomical Uniion, France
111	van der Kruit, Pieter C.	Kapteyn Institute, Groningen, Netherlands
112	van der Laan, Harry	Leiden Observatory, Netherlands
113	van der Veld, Liesbeth	Leiden Observatory, Netherlands
114	Van Helden, Al	Utrecht University, Netherlands
115	van Langevelde, Huib	Joint Institute for VLBI in Europe, Netherlands
116	Véron, Philippe	Observatoire de Haute Provence, France
117	Vertegaal, Louis	NWO, Netherlands
118	Voelk, Heinz	Max-Planck-Institut fuer Kernphysik, Heidelberg, Germany
119	von Ballmoos, Peter	Centre d'Etude Spatiale des Rayonnements, France
120	von der Lühe, Oskar	Kiepenheuer-Institut, Germany

#	*Name*	*Affiliation*
121	Waelkens, Christoffel	Institute for Astronomy K.U.Leuven, Belgium
122	Woltjer, Lodewijk	Observatoire de Haute-Provence, France
123	Zinnecker, Hans	Astrophysikalisches Institut Potsdam, Germany
124	Zoomers, Henk	Asian Historical Consultancy, Netherlands

AUTHOR INDEX

www.ingramcontent.com/pod-product-compliance
Lightning Source LLC
Chambersburg PA
CBHW071350220926
51613CB00008B/43
* 9 7 8 9 0 4 8 1 2 2 3 2 5 *